Physiology of Herbicide Action

Malcolm Devine
University of Saskatchewan
Stephen O. Duke
United States Department of Agriculture
Carl Fedtke
Bayer AG

P T R Prentice Hall
Englewood Cliffs, New Jersey 07632

Library of Congress Cataloging-in-Publication Data

Devine, Malcolm.
Physiology of herbicide action / Malcolm Devine, Stephen O. Duke, Carl Fedtke.
p. cm.
Includes bibliographical references and index.
ISBN 0-13-369067-9
1. Weeds—Physiology. 2. Plants, Effect of herbicides on. 3. Weeds—Control. I. Duke, Stephen O. II. Fedtke, Carl. III. Title.
SB611.D528 1993
632'954—dc20 92-104
CIP

Editorial/production supervision
and interior design: Laura A. Huber/Karen Bernhaut
Acquisitions editor: Betty Sun
Cover design: Wanda Lubelska Design
Manufacturing buyer: Susan Brunke
Prepress buyer: Mary Elizabeth McCartney

A Simon & Schuster Company
Englewood Cliffs, New Jersey 07632

The publisher offers discounts on this book when ordered in bulk quantities. For more information, write:

Special Sales/Professional Marketing
Prentice Hall
Professional & Technical Reference Division
Englewood Cliffs, New Jersey 07632

Printed in the United States of America

10 9 8 7 6 5 4 3 2 1

ISBN 0-13-369067-9

PRENTICE-HALL INTERNATIONAL (UK) LIMITED, *London*
PRENTICE-HALL OF AUSTRALIA PTY. LIMITED, *Sydney*
PRENTICE-HALL CANADA INC., *Toronto*
PRENTICE-HALL HISPANOAMERICANA, S.A., *Mexico*
PRENTICE-HALL OF INDIA PRIVATE LIMITED, *New Delhi*
PRENTICE-HALL OF JAPAN, INC., *Tokyo*
SIMON & SCHUSTER ASIA PTE. LTD., *Singapore*
EDITORA PRENTICE-HALL DO BRASIL, LTDA., *Rio de Janeiro*

Contents

Preface

Weeds constitute a serious and continuing limitation to crop production in all agricultural systems. In many parts of the world, herbicides have replaced hand labor as the primary method of weed control. Consequently, herbicides contribute significantly to the production of most of the major food and fiber crops. To date, almost 400 herbicides have been registered or are in the registration process, and these form the active ingredients of thousands of commercial products. However, there is still a need for new active ingredients. Changes in weed communities, the development of herbicide-resistant species, and changing toxicological and environmental fate requirements demand that more effective, selective, and environmentally benign herbicides be developed.

Since the discovery of the first synthetic organic herbicides in the 1940s, there has been great interest in understanding the mechanisms by which herbicides interfere with plant growth. Among the herbicides whose mechanism of action is currently understood, there are about 100 herbicides that inhibit photosystem II electron transport, 37 that inhibit branched chain amino acid synthesis, 32 that are active auxins, and 28 that interfere with microtubular synthesis or function. Several important sites of molecular action of herbicides, including acetolactate synthase, acetyl-CoA carboxylase and protoporphyrinogen oxidase, have been discovered within the last 10 years. In addition, our understanding of the mechanism of action of older herbicides, such as photosystem II inhibitors, has increased greatly during this time. Concurrently, there has been a great increase in our understanding of basic plant physiological and biochemical processes. In some instances, indeed, the availability of herbicides with very specific mechanisms of action has contributed substantially to our understanding of plant physiology and biochemistry.

In recent years, the volume of research published on the physiology and biochemistry of herbicide action has increased dramatically. This can be attributed to several factors: in general, there has been an increase in interest in the molecular action of herbicides within the scientific community; the use of modern experimental techniques, including protein crystallography, molecular modeling, specific and sensitive enzyme assays, and the myriad techniques associated with molecular biology has greatly expanded the ways in which new information can be obtained; and finally, although it has yet to lead to the discovery of a new herbicide, the use of "rational" approaches to herbicide discovery has prompted industries to become much more involved in the study of basic aspects of herbicide action.

Given the large and growing body of information available on herbicide action, it is becoming increasingly difficult to maintain broad familiarity with the subject. University students and teachers, government researchers and pesticide specialists, and scientists in the private sector require access to this information, yet in a unified and comprehensive manner. In this book we have attempted to meet this need, and to integrate the relevant information on herbicide action in plants into a single volume. However, no book on this subject can be complete, and new and exciting discoveries in the field of herbicide action are being made as this book goes to press. We hope that the reader can find this book a useful source of information on many aspects of herbicide action, as well as a starting point for more advanced reading and research on this topic.

In writing this book we have relied heavily on the advice and encouragement of many colleagues, including W. Oettmeier, K. Tietjen, W. H. Vanden Born, and K. C. Vaughn. To them and to many other unnamed colleagues who have assisted us, we offer our sincere thanks.

Malcolm D. Devine
Saskatoon, Canada

Stephen O. Duke
Stoneville, MS, USA

Carl Fedtke
Leverkusen, Germany

Chapter 1

An Introduction to Herbicide Action

1.1 INTRODUCTION

In simplest terms, herbicide action might be described as the physiological and biochemical interaction of a herbicide with a plant. Yet when we start to examine herbicide action, we see that it is not a question of a single interaction, but of multiple interactions at various levels within the plant. In many instances—but not all—we can identify a single "target site" or site of action within the plant. This is the site (usually an enzymatically active protein) to which the herbicide binds, or with which it interferes in some other manner, resulting, ultimately, in death of the plant.

It would be misleading, however, to consider herbicide action solely in terms of interaction at a target site. Overall, herbicide action may be thought of as comprising two phases; the first phase involves movement of the herbicide to the target site, while the second phase involves the metabolic consequences resulting from interaction at that site. Phase 1 starts with application to the plant, either directly to the foliage or via the soil to the roots. The herbicide then enters the plant at either (or both) of these sites. Entry into the plant is quickly followed by a series of steps that precede the arrival of the herbicide at its site of action. These include entry into cells, diffusion over relatively short distances (i.e., cell-to-cell movement), long-distance transport, metabolic conversion of the herbicide (either activation or deactivation), and entry into subcellular organelles. The interaction of the herbicide at the target site can be viewed as the first step in phase 2; this is followed by a series of toxic consequences that result in death of the plant (Scheme 1.1).

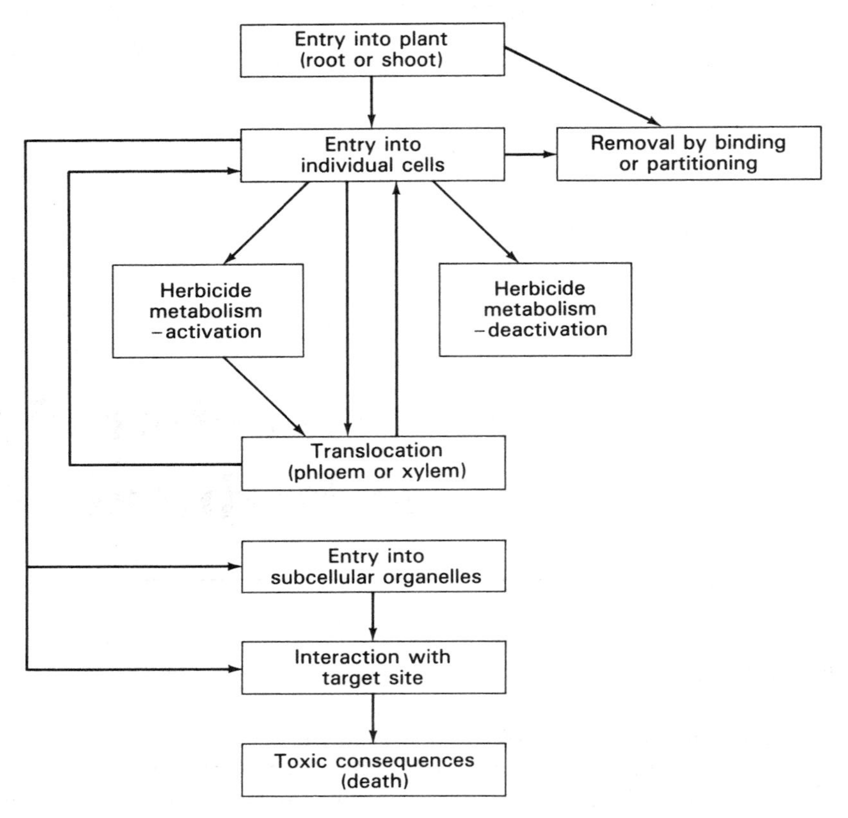

Scheme 1.1 Flow chart showing the sequence of events from herbicide entry into a plant to death of the plant.

The organization of this book reflects this chain of events. Starting with entry into the plant (Chapters 2–4), herbicide behavior is examined in terms of translocation in the phloem and xylem (Chapter 5), herbicide metabolism (Chapter 6), and herbicide interaction with various target sites in plants (Chapters 7, 8, 10–15). This interaction can be considered the mechanism of action of the herbicides in question. Other relevant aspects of herbicide action are considered in the concluding chapters.

The approach taken within the "mechanism of action" chapters is to consider herbicide action in terms of the different known target sites in plants. Thus, the focus is on target sites first, with the herbicides that interfere with those sites as a secondary consideration. For example, inhibition of photosynthetic electron transport is the focus of Chapter 7; within this chapter there is a listing of the many different structural groups of herbicides that interfere with that process, and how these groups may differ in their action. It is impossible to give equal treatment to all structural groups, because some herbicide groups, and some individual herbicides in particular, have been the subject of much more research than others. The emphasis here is not so much on the chemistry of the inhibitors but on the physiological and biochemical consequences of the inhibition caused by herbicides.

1.2 WHY DO PLANTS DIE?

As we consider the different sites of action within the plant, much of the discussion will be concerned with herbicide interaction with these sites. This can take various forms, from details of herbicide binding to receptor proteins to quantitative determination of the inhibition of biosynthetic processes. This knowledge of herbicide interaction with particular target sites does not, however, tell us why plants die. In some instances we can answer this question with reasonable confidence. For example, the biochemical and physiological events associated with oxygen toxicity (Chapter 9) are very well characterized. Consequently, we can say that we know how herbicides that inhibit photosynthesis—and photosystem I electron acceptors, in particular—kill plants. In one sense it may be inadequate to describe the mechanism of action of these herbicides simply as inhibition of photosynthesis.

With many other herbicide target sites, however, the ultimate cause of plant death is not known. Inhibitors of amino acid biosynthesis cause depletions of levels of particular amino acids, but it is difficult to relate this to plant death. Amino acids are continuously recycled through protein degradation, and depletion may not be obvious for some time after treatment. In addition, depletion may be significant in some tissues or organelles, but not detectable on a whole-plant basis. Presumably, inhibition of amino acid biosynthesis eventually leads to depletion of particular proteins, and the lack of some essential protein functions causes the plant to die. It is extremely difficult, however, to determine which protein losses are responsible for death of the plant. It may not be *one* particular protein that is involved, but the alteration of a network of interdependent proteins and/or syntheses.

Another possible cause of plant death is the accumulation of toxic precursors or intermediates as a consequence of the inhibition of biosynthetic pathways. In general, accumulation of a particular product (or products) may serve as evidence of inhibition of a later step in the pathway in which that product is an intermediate (Scheme 1.2, reaction 1). If this is the case, the biological activity of the precursors can be examined to determine their contribution to phytotoxicity. Identification of precursors requires detailed knowledge of the biochemical pathway involved and its regulation; inhibition of one step in a pathway may result in accumulation of the immediate precursor, or of other intermediates in the same pathway. For example, inhibition of acetolactate synthase (Chapter 13) leads to accumulation of α-ketobutyrate in treated tissues; this product is toxic at high concentrations. In some instances the reaction precursor may be diverted into other products (Scheme 1.2, reaction 2). Alternatively, subsequent to inhibition of a biosynthetic pathway, end-product depletion may lead to deregulation of that pathway, with wide-ranging

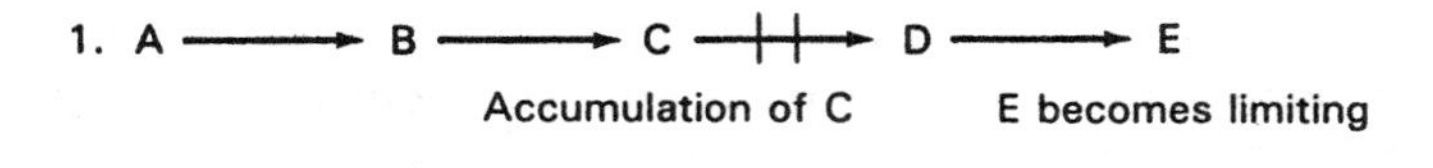

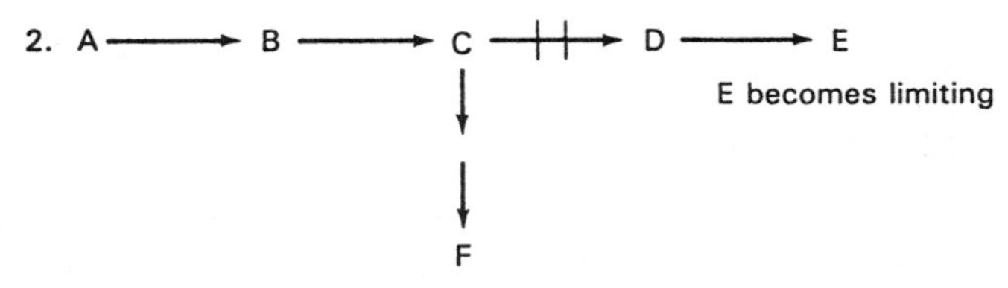

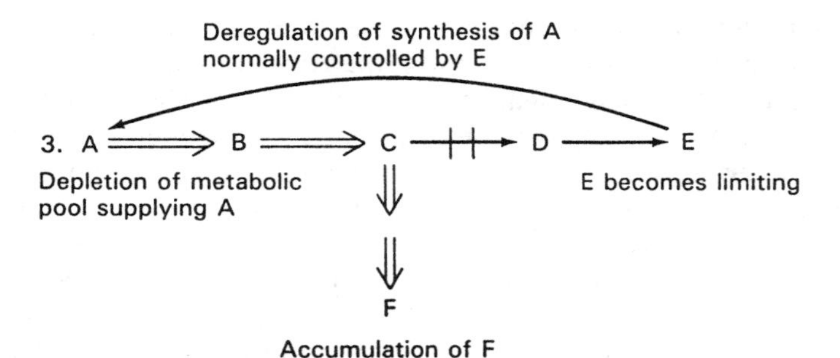

Scheme 1.2 Some possible consequences of herbicidal inhibition of a biosynthetic pathway.

effects on carbon metabolism (Scheme 1.2, reaction 3). This is discussed in detail in reference to glyphosate (Chapter 13).

A major limitation to understanding herbicide action is our lack of detailed knowledge of many plant biochemical and physiological processes. Many biosynthetic pathways are not well characterized, and many aspects of whole-plant physiology are not fully understood. For example, the mechanisms of action of the endogenous plant growth regulating substances (hormones) have not been fully explained. Interestingly, herbicides and herbicide-resistant mutants have become increasingly important in recent years as tools for studying physiological and biochemical processes in plants.

In summary, detailed knowledge of the reasons for plant death is lacking for many herbicide action sites. Consideration of primary target sites remains the most logical physiological or biochemical approach to discussion of herbicide action.

1.3 THE TIME COURSE OF HERBICIDE ACTION

When herbicides are applied under field conditions, the time lag before visible injury can be observed ranges from less than 1 hour to 1 week or longer. The most rapid visible signs of phytotoxicity are desiccation (e.g., bipyridilium herbicides;

see Chapter 8) and epinasty (e.g., auxin-like herbicides; see Chapter 14). However, the absence of visual symptoms of injury is not indicative of a lack of herbicide action within the plant; it is quite possible that a process can be inhibited beyond recovery without any visible effect on the plant.

Absorption of foliage-applied herbicides occurs over a period ranging from several hours to several days; this can be considerably longer for soil-applied herbicides. Because of this slow entry into the plant, translocation and all subsequent steps can occur over extended periods. As a result, considerable time may be required for lethal herbicide concentrations to accumulate, especially in tissues distal to the site of entry into the plant. In contrast, the time course of the final events in Scheme 1.1 is normally much shorter. Herbicide entry into cells from aqueous solutions occurs within seconds, and entry into subcellular organelles occurs equally rapidly. Some biochemical and physiological effects can also be measured very soon after exposure of the sensitive tissue to a herbicide. For example, a rapid rise in chlorophyll fluorescence can be detected very soon after exposure of isolated protoplasts or chloroplasts to photosynthetic inhibitor herbicides (Chapter 7). From the above discussion, we can conclude that the first steps in Scheme 1.1 are normally the rate-limiting steps in herbicide action, and that the interaction between the herbicide and its target site occurs relatively quickly.

1.4 HERBICIDE SELECTIVITY

Rather than devote a separate chapter to herbicide selectivity, this topic is mentioned where relevant throughout this book. Selectivity implies that different plant species do not respond in the same way to a particular herbicide. In other words, one species dies, but the other survives. The margin of selectivity may be small, just enough to provide acceptable weed control without injury to the crop plants. This is not a very desirable situation, because only slight alterations in herbicide activity (under different environmental conditions, for example) may result in poor weed control and/or crop injury. Alternatively, the margin of selectivity may be very wide, as is the case with some of the *s*-triazine- and sulfonylurea-resistant weed biotypes (see Chapters 7 and 13). Recently, this high level of selectivity due to resistance at the site of action has been exploited in the creation of herbicide-resistant crops.

Selectivity is commonly based on one (or more) of the following: differential interception of the herbicide; differential herbicide absorption; differences in rates or pathways of herbicide metabolism; differential sensitivity of target proteins; differential ability to tolerate the toxic effect of the herbicide (Table 1.1). It is very seldom, however, that all of these possible bases of selectivity are considered in studies of herbicide selectivity. Some caution is required in interpreting the results of herbicide selectivity research that describes selectivity without investigation of all of the possible mechanisms.

It should be noted that the selectivity mechanisms included above and in Table 1.1 are presented in order of occurrence in the course of herbicidal action,

TABLE 1.1 MECHANISMS OF HERBICIDE SELECTIVITY. The numbers in brackets are an indication of the relative importance of the different mechanisms; additional details and examples are given in the text.

Process responsible for selectivity	Mechanism
Herbicide interception/ absorption [3]	Leaf angle, leaf area differences; leaf surface characteristics (cuticle composition or leaf hairs); rooting pattern differences
Herbicide leaching in soil [3]	Positional selectivity related to root distribution in soil
Herbicide metabolism [1]	Presence of metabolizing enzymes
Interaction at target site [2]	Different protein structure
Ability to tolerate toxic effects [3]	Presence of "detoxifying" enzyme system; stored reserves to overcome shortages

not in order of importance. In terms of the number of examples that we know of, and of their involvement in crop-weed selectivity, herbicide metabolism is the most important selectivity mechanism (see Chapter 6). Second in importance are differences in the target site in susceptible and resistant species; examples include triazine-resistant weeds (Chapter 7), dinitroaniline-resistant *Eleusine indica* (Chapter 14), and sulfonylurea-resistant weeds (Chapter 13). The remaining mechanisms, although important in some special cases, are generally of less importance as selectivity mechanisms; some examples are discussed in the appropriate chapters throughout the text.

1.5 TESTING FOR HERBICIDE ACTIVITY AND MECHANISM OF ACTION

The routine screening procedures that are commonly used by chemical companies to screen for new and useful herbicides generally include a complement of monocotyledonous and dicotyledonous test species. The visible action discernable at the whole-plant level may often, by comparison with known standards, provide a first indication of the possible mechanism of action of a new compound. Growth regulator-type herbicides may suppress plant growth or stimulate stem elongation into stem curling (epinasty or hyponasty) and other auxin-like effects (Chapter 14). Very severe growth suppression and simultaneous intense anthocyanin accumulation in the tissue may indicate acetolactate synthase-inhibiting herbicides (Chapters 13 and 16). Chlorotic and/or phytotoxic effects, occurring only in the light (but possibly in the dark also) may suggest several possible modes of herbicidal action related to photosynthesis (Sections 8.2, 9.5). Growth suppression and leaf twisting of grass (e.g., maize) seedlings, as typically induced by α-chloroacetamides, may indicate a mechanism of action similar to that of herbicides in this class. Tissue (in particular root tip) swelling caused by thinner cell walls, and cell swelling and enlargement, may be caused by interference with the microtubular system (Chapter 10). Although physiological and/or biochemical mechanisms of herbicide action cannot be determined conclusively from morphological and cytological studies, a close and critical look at the whole-plant effects and at the optimum environmental conditions for damage can provide some valuable hints for further physiological and biochemical investigations.

There have been repeated attempts and suggestions to replace the time-consuming and greenhouse space-consuming routine screening of new compounds by simple and rapid test systems that do not involve whole plants. However, there is no single test that would detect all of the herbicides now known and at the same time reject the inactive structures. Because of the wide ranges of herbicide concentrations required to inhibit different processes, it is impossible to screen all herbicides with one test. Fast and simple test systems can be helpful where greenhouse facilities are not available, or where specific problems or questions that do not require a screening approach are to be studied.

An important application of simple, sensitive, and specific test systems is in mechanism of action studies. Table 1.2 shows a list of some of the more commonly used test systems in such studies. To begin with, the system should be sensitive in the micromolar concentration range (or lower) and should be specific; that is, nonherbicidal analogues should be inactive. Unicellular algae provide good model organisms for the higher plant cell. *Chlamydomonas reinhardtii* and *Chlorella vulgaris* are frequently used, but many others may serve equally well. In principle, all metabolic activities can be monitored in algal cells or cell extracts. The parameters included in Table 1.2 can be measured easily and rapidly, and have been commonly used to compare the action of herbicides or environmental chemicals. Cell count and cell size can be conveniently measured with a Coulter counter. This

TABLE 1.2 TEST SYSTEMS THAT HAVE BEEN SUGGESTED TO BE USEFUL FOR DETECTING HERBICIDAL ACTIVITIES AND FOR STUDYING HERBICIDE MODES OF ACTION. For a collection of methods, see [38].

Test material	Parameter(s) measured	Herbicides detected	Reference
Algae:			
Anabaena variabilis	Cell count, cell size	Photosynthesis inhibitors, diquat, paraquat	1
Chlamydomonas reinhardtii	chlorophyll, O_2-evolution, turbidity	Respiratory uncouplers, peroxidizing herbicides, chlorosis-inducing herbicides, growth inhibitor herbicides	2, 3, 4
Chlorella vulgaris			1, 5, 6, 7, 8
Haematococcus sp.			9
Hormidium sp.			9
Scenedesmus obliquus			9
Dunaliella bioculata			10
Chlorococcum sp.			1

Cell suspension cultures:			
Heterotrophic:			
Soybean Wheat Tomato Potato *Cirsium arvense* Tobacco	Fresh weight, conductivity, ion leakage, fluorescein leakage, radiotracer incorporation	Growth inhibitor herbicides (often with low sensitivity)	11 11 12 13 14 15, 17, 18
Autotrophic:			
Tobacco	O_2-evolution, uptake and metabolism	Most herbicides with few exceptions	15, 16
Isolated mesophyll cells:			
Phaseolus vulgaris Soybean Cotton	Radiotracer incorporation	Photosynthesis inhibitors, diquat, paraquat, fluridone, other herbicides above 10 μM	19, 20 21, 22, 23
Other:			
Leaf pieces:			
Watermelon *Lolium multiflorum*	Sinking leaf disk Starch formation	Photosynthesis inhibitors	24, 25, 26
Cucumber cotyledons	Electrolyte leakage, MDA, ethane, etc.	Photooxidative herbicides	27, 28
Oat	Nitrite formation		29, 30
Sinapis alba			31
Protoplasts, tissues:			
Corn *Allium cepa* *Lemna minor* Oat	Conductivity, uptake and metabolism, membrane rupture, electron microscopy, amino acid leakage	Effects generally only above 100 μM	32 33 34, 35 36
Pollen tubes:			
Tobacco	Turbidity	Diclobenil, organic solvents, environmental chemicals	37
Stems:			
Soybean Oat	Rooting in light and dark	Most herbicides with very few exceptions	4

approach allows the researcher to follow individual cell growth and division in a synchronously growing algal culture. Another advantage of unicellular algae is the possibility of herbicide removal by centrifugation and resuspension of the algal cells. A disadvantage of algal cells is that several herbicidal groups, for example the auxins (Chapter 14), and the arylpropanoic acids and thiocarbamates (Chapter 11), are not inhibitory in this system. The reasons for this could be target insensitivity, absence of the herbicide target site in algal cells, or a requirement for organized tissues for the ultimate expression of activity. A special feature of algal incubation systems is their ability to accumulate lipophilic compounds from an aqueous medium over longer incubation times.

These problems are minimized in higher plant cell suspension cultures. However, only photoautotrophic plant cell suspension cultures provide a good and sensitive test system for herbicidal action. Heterotrophic cell suspension cultures are much less sensitive, or not sensitive at all, while photomixotrophic cell suspension cultures fall between these two groups. Here also, as in the case of the algae, the whole array of the available physiological and biochemical methods of investigation of plant metabolism can be employed. Disadvantages of cell suspension cultures can be the equipment required and the need to maintain sterile conditions when introducing new chemicals, but these problems can be overcome. The advantages of plant cell cultures for herbicide research include their unlimited availability, their homogeneous behavior, and the avoidance of uptake and translocation problems. Cell cultures have, accordingly, been used extensively in herbicide metabolism studies. It must be kept in mind, however, that the absence of tissue and organ formation may lead to profound metabolic differences when compared with the intact plant.

Similar arguments can also be made for the use of freshly isolated mesophyll cells. Their easy access by tissue (leaf) maceration with cellulytic and pectolytic enzymes makes them an attractive study material. The duration of experiments with isolated mesophyll cells is limited, however, because these cultures do not grow. Herbicides that have been studied in isolated cell systems therefore tend to be compounds that interfere with photosynthesis or are rapidly phytotoxic at some other site of action.

The action of photosynthetic inhibitor herbicides can be detected and measured by methods other than complex physical and biochemical methods (Section 7.2). Several simple screening methods are also available, employing, for example, the measurement of starch (decrease), nitrite (large increase), or simply the sinking of leaf discs when shaken in a herbicide-containing buffer. These methods have the advantage of not requiring sophisticated or expensive equipment for detection and quantification of the photosynthesis-inhibiting action. A more general procedure to detect herbicidal action is to measure the result of membrane damage by an increase in conductivity, in a bathing solution of water or low-strength buffer, due to leakage of electrolytes such as K^+ or amino acids. Fluorescein or chromate leakage from preloaded cells or liposomes has also been used to detect membrane damage. However, many membrane effects occur only at concentrations above 100 μM, and are therefore more likely to be unspecific effects of the chemical under study on membrane function.

Several of the test systems listed in Table 1.2 are also useful for environmental monitoring. This is the case particularly for algal growth and pollen tube growth. Both are very sensitive to certain chemicals or organic solvents, but they are also very selective; that is, they can only detect a very limited number of currently used herbicides. Rooting tests with stem sections of oat and soybean, representing the mono- and dicotyledonous plant groups, can be used to detect (with high sensitivity) and classify most known herbicides. These tests use organ (root) regeneration, a process that comprises most of the important metabolic and regulatory steps in higher plants, as a basis for broad herbicidal sensitivity. The ability of algal cells to accumulate lipophilic compounds also makes them useful for environmental monitoring.

Other aspects of herbicide action, such as absorption and translocation, are more commonly studied using whole-plant systems or larger pieces of excised tissue. Some of the specialized techniques used to study these processes are described in Chapters 3, 4, and 5.

1.6 HERBICIDE STRUCTURE-ACTIVITY RELATIONSHIPS

The biological activity of herbicide molecules is a reflection of their ability to penetrate into plant tissues, resist detoxification (or be transformed into toxic products), and interfere with a particular physiological or biochemical process in the plant. Penetration into the plant may include diffusion through the cuticle, followed by movement through cell walls, the plasmalemma, and, in some instances, organelle membranes. For some herbicides, long-distance transport may also be important. Regardless of where the target site is located, and which tissue is most sensitive, sufficient ''active'' molecules must reach and interfere with it for activity to be expressed.

In relation to herbicide discovery, we may consider the above scheme of events and ask the following question: can the behavior of a herbicidal molecule be predicted on the basis of its structural (chemical, physical) properties? If so, can novel molecules be designed with optimized ability to reach the target site and interfere with it? Given the high cost of conventional herbicide screening programs, and the relatively low success rate, interest in the use of structure-activity relationships (SAR) and quantitative structure-activity relationships (QSAR) has increased dramatically in recent years.

In practice, SAR aims to correlate the biological activity of molecules with certain defined physicochemical properties. Two common parameters, relatively easily measured, are log P (logarithm of the 1-octanol/water partition coefficient) and pK_a or pK_b, the dissociation constant of an acid or base (the pH at which 50% of the molecules are dissociated). The importance of these parameters will be highlighted in subsequent chapters; in particular, the influence of log P on membrane transport (Chapter 3), of log P and pK_a on phloem mobility of herbicides (Chapter 5), and of log P on the activity of photosystem II electron transport inhibitors (Chapter 7) will be discussed.

In more detailed analyses, however, additional parameters, used to describe the characteristics of various substituents, can be included [39–41]. These include π, describing the effect of a substituent group on hydrophobicity, σ (the Hammet constant), describing the electronic characteristics of the substituent, and E_s, a steric parameter. The latter parameters are more difficult to determine experimentally, but can be obtained from reference sources [42].

A common approach to optimizing structure involves the determination of biological activity of a homologous series of compounds (often this is determined as an ED_{50} or I_{50}, meaning the dose or concentration required to inhibit growth or a particular process by 50%). The advantages or disadvantages of various substituents, at various positions in the molecule, can then be determined. An example of such data is shown in Table 1.3 [43]. The data are then subjected to regression

TABLE 1.3 SAMPLE STRUCTURE-ACTIVITY DATA BASED ON THE ACTIVITY OF A SERIES OF *N*-(1-METHYL-1-PHENYLETHYL)PHENYLACETAMIDES IN *SCIRPUS JUNCOIDES* AND *ECHINOCHLOA CRUS-GALLI* [43]. Note that this table lists only 14 of the 83 structures tested and used in the correlation analysis.

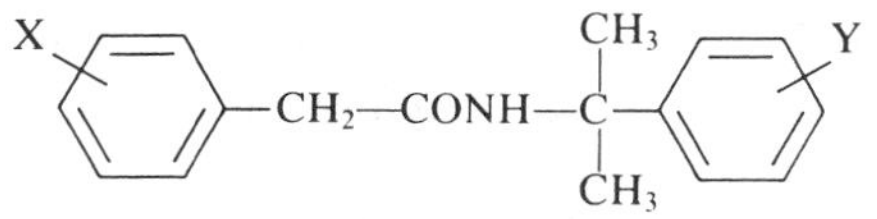

X, Y, CH_3, $—CH_2—CONH—C—$, CH_3

		pI_{50}			
		S. juncoides		*E. crus-galli*	
X	Y	Obsd.	Calcd.	Obsd.	Calcd.
H	H	5.70	5.39	4.95	4.81
2-Me	H	5.43	5.72	5.15	5.16
2-*i*-Pr	H	5.29	5.76	4.67	5.07
2-F	H	5.78	5.49	5.39	4.93
2-Cl	H	5.76	5.77	5.28	5.20
2-OMe	H	5.45	5.37	5.05	4.80
2-NO_2	H	5.15	5.13	4.28	4.54
H	2-Me	5.55	5.72	5.03	5.16
H	2-Cl	5.60	5.77	5.01	5.20
H	3-Me	5.74	5.71	5.13	5.15
H	3-F	6.52	6.42	5.45	5.47
H	3-Cl	6.66	6.58	5.74	5.67
H	3-Br	6.42	6.61	5.42	5.70
H	4-Me	6.03	5.71	5.43	5.15

analysis, in which biological activity is described mathematically in terms of the above parameters. For example, the equation

$$pI_{50} = -a\pi^2 + b\pi + \rho\sigma + \delta E_s + c$$

relates I_{50} to π, σ, and E_s (the other parameters are susceptibility constants, or regression coefficients) [39]. In this way the effect of varying particular physicochemical parameters can be tested by substitution in the above equation, as well as empirically. Both the observed and the predicted I_{50} values are included in Table 1.3. Of course, the better the equation, the more accurate will be the prediction of activity when novel substituents are incorporated into the molecule. More recently, other statistical approaches to QSAR have been introduced [44].

Structure-activity relationships have been used to examine the biological activity of a wide range of herbicide structures [45–49]. In other relevant fields, (Q)SARs have been used to study the activities of insecticides [50, 51], fungicides [52, 53], naturally occurring toxins [54, 55], and environmental pollutants [56, 57].

A serious limitation to the QSAR approach to optimizing activity is that the data are obtained from a single system (e.g., from a plant bioassay or an in vitro assay such as the Hill reaction), and do not take into account differences between experimental systems. Predictions of biological activity based on in vitro assay data fail to take into account the requirements for delivery to the target site, and it is possible that the optimum properties for entry into the tissue may be quite different from those required for inhibitory activity at the target site. In addition, in vitro assays provide no indication of the likelihood of metabolic detoxification of the compounds being tested. Thus, optimization of inhibitor structure based on biochemical assays provides only part of the answer; whole-plant assays are still required to determine the most useful compounds for commercial development.

A further aspect of SAR study involves the examination of the molecular interaction between the herbicide and its target site. However, with the exception of the herbicide binding site in photosystem II, precise details of the topology and electronic characteristics of herbicide binding sites are still unknown. Knowledge of these characteristics of the herbicide binding site in photosystem II allow for prediction of the interaction between various compounds and photosynthetic electron transport; a similar understanding of herbicide interactions with other target sites will come only when those sites have been described in more detail.

An alternative application of SARs, applicable in some instances where knowledge of the binding site is lacking, is to examine the relationship between the herbicide and the reaction substrates or intermediates. In cases where kinetic analysis indicates that the herbicide interacts with one of these products (this could include competitive, noncompetitive, or uncompetitive inhibition), some relationship may be deduced between the herbicide structure and that of a substrate or intermediate. An example is the supposed similarity between sethoxydim and related herbicides and the reaction intermediate formed by acetyl-Coenzyme A carboxylase [58] (see Chapter 11). One group of enzyme inhibitors, the oxalyl hydroxamates, were designed to mimic the structure of the intermediate in the reaction catalyzed by ketol-acid reductoisomerase [59] (Section 13.3). However, although these compounds are extremely potent enzyme inhibitors, their whole-plant activity is relatively low, and they may not be useful as herbicides.

Finally, the interaction of inhibitors with different structures but with activity at the same site can be used to examine the requirement for binding activity (e.g., [60]). Based on this, common structural elements for binding at the target site can be identified, allowing for the design of molecules that may interact more effectively at the target site.

REFERENCES

1. Hawxby, H., B. Tubea, J. Ownby, and E. Basler. 1977. "Effects of various classes of herbicides on four species of algae." *Pestic. Biochem. Physiol.,* 7, 203–09.
2. Hess, F. D. 1980. "A *Chlamydomonas* algal bioassay for detecting growth inhibitor herbicides." *Weed Sci.,* 28, 515–20.
3. Fedtke, C. 1981. "Modes of herbicide action as determined with *Chlamydomonas reinhardii* and Coulter counting." In *Biochemical Responses Induced by Herbicides,* ACS Symposium Series No. 181, D. E. Moreland, J. B. St. John, F. D. Hess, eds., pp. 231–50.
4. Fedtke, C. 1987. "Physiological activity spectra of existing graminicides and the new herbicide 2-(2-benzothiazolyl-oxy)-*N*-methyl-*N*-phenylacetamide (mefenacet)." *Weed Res.,* 27, 221–28.
5. Kratky, B. A., and G. F. Warren. 1971. "A rapid bioassay for photosynthetic and respiratory inhibitors." *Weed Sci.,* 19, 658–61.
6. Lefebvre-Drouet, E., and R. Calvet. 1978. "La detection et le dosage des herbicides à l'aide des chlorelles: recherche sur les conditions expérimentales optimales et application à l'analyse de plusieurs herbicides." *Weed Res.,* 18, 33–39.
7. Chan, Y. T., J. E. Roseby, and G. R. Funnell. 1975. "A new rapid specific bioassay method for photosynthesis inhibiting herbicides." *Soil Biol. Biochem.,* 7, 39–44.
8. Thomas, V. M., L. J. Buckley, J. D. Sullivan, and M. Ikawa. 1973. "Effect of herbicides on the growth of *Chlorella* and *Bacillus* using the paper disc method." *Weed Sci.,* 21, 449–51.
9. Cullimore, R. 1975. "The *in vitro* sensitivity of some species of Chlorophyceae to a selected range of herbicides." *Weed Res.,* 15, 401–06.
10. Felix, H. R., R. Chollet, and J. Harr. 1988. "Use of the cell wall-less alga *Dunaliella bioculata* in herbicide screening tests." *Ann. Appl. Biol.,* 113, 55–60.
11. Harms, H., and C. Langebartels. 1986. "Standardized plant cell suspension test systems for an ecotoxicologic evaluation of the metabolic fate of xenobiotics." *Plant Sci. Lett.,* 45, 157–65.
12. Zilkah, S., and J. Gressel. 1978. "Correlations in phytotoxicity between white and green calli of *Rumex obtusifolius, Nicotiana tabacum,* and *Lycopersicon esculentum.*" *Pestic. Biochem. Physiol.,* 9, 334–39.
13. Zilkah, S., and J. Gressel. 1977. "Cell cultures vs. whole plants for measuring phytotox-

icity. II. Correlations between phytotoxicity in seedlings and calli." *Plant Cell Physiol.*, 18, 657–70.

14. Zilkah, S., and J. Gressel. 1977. "Cell cultures vs. whole plants for measuring phytotoxicity. III. Correlations between phytotoxicities in cell suspension cultures, calli and seedlings." *Plant Cell Physiol.* 18, 815–20.
15. Sato, F., S. Takeda, and Y. Yamada. 1987. "A comparison of effects of several herbicides on photoautotrophic, photomixotrophic, and heterotrophic cultured tobacco cells and seedlings." *Plant Cell Reports*, 6, 401–04.
16. Thiemann, J., A. Nieswandt, and W. Barz. 1989. "A microtest system for the serial assay of phytotoxic compounds using photoautotrophic cell suspension cultures of *Chenopodium rubrum*." *Plant Cell Reports*, 8, 399–402.
17. Talbert, D. M., and N. D. Camper. 1983. "Herbicide and inhibitor effects on cell leakage in suspension cultures of tobacco (*Nicotiana tabacum*)." *Pestic. Biochem. Physiol.* 19, 74–81.
18. Talbert, D. M., and N. D. Camper. 1983. "Herbicide effects on liposome leakage." *Weed Sci.*, 31, 329–32.
19. Rees, R. T., A. H. Cobb, and K. E. Pallett. 1985. "The potential of isolated cells for the metabolic screening of herbicides." Proc. Brit. Crop Prot. Conf.—Weeds, Vol. 1, BCPC Publ., Croydon, pp. 341–48.
20. Ashton, F. M., O. T. De Villiers, R. K. Glenn, and W. B. Duke. 1977. "Localization of metabolic sites of action of herbicides." *Pestic. Biochem. Physiol.*, 7, 122–41.
21. Pallett, K. E., R. T. Rees, P. J. Fitzsimons, and A. H. Cobb. 1986. "The isolation of mesophyll cells and protoplasts for use in herbicide research." Aspects of Applied Biology 11. Biochemical and Physiological Techniques in Herbicide Research, pp. 139–48.
22. Rees, R. T. 1986. "Metabolically active *Glycine max* cells for the study of the activity and movement of herbicides." In Symposium: Movement of Pesticides in Plants. *Pestic. Sci.*, 17, 59.
23. Rafii, Z. E., F. M. Ashton, and R. K. Glenn. 1979. "Metabolic sites of action of fluridone in isolated mesophyll cells." *Weed Sci.*, 27, 422–26.
24. Truelove, B., D. E. Davis, and L. R. Jones. 1974. "A new method for detecting photosynthetic inhibitors." *Weed Sci.*, 22, 15–18.
25. Da Silva, J. F., R. O. Fadayomi, and G. F. Warren. 1976. "Cotyledon disc bioassay for certain herbicides." *Weed Sci.*, 24, 250–52.
26. Kida, T., S. Takano, T. Ishikawa, and H. Shibai. 1985. "A simple bioassay for herbicidal substances of microbial origin by determining de novo starch synthesis in leaf segments." *Agric. Biol. Chem.*, 49, 1299–1303.
27. Kenyon, W. H., S. O. Duke, and K. C. Vaughn. 1985. "Sequence of effects of acifluorfen on physiological and ultrastructural parameters in cucumber cotyledon discs." *Pestic. Biochem. Physiol.*, 24, 240–50.
28. Vanstone, D. E., and E. H. Stobbe. 1977. "Electrolytic conductivity—a rapid measure of herbicide injury." *Weed Sci.*, 25, 352–54.
29. Klepper, L. A. 1975. "Inhibition of nitrite reduction by photosynthetic inhibitors." *Weed Sci.*, 23, 188–90.

30. Klepper, L. A. 1979. "Effects of certain herbicides and their combinations on nitrate and nitrite reduction." *Plant Physiol.*, 64, 273–75.

31. Fedtke, C. 1977. "Formation of nitrite in plants treated with herbicides that inhibit photosynthesis." *Pestic. Sci.*, 8, 152–56.

32. Darmstadt, G. L., N. E. Balke, and L. E. Schrader. 1983. "Use of corn root protoplasts in herbicide absorption studies." *Pestic. Biochem. Physiol.*, 19, 172–83.

33. Beuret, E., and V. Pont. 1987. "Un biotest simple pour mettre en evidence l'action des herbicides sur la membrane plasmique: example du chlorfenprop et de molecules apparantées." *Revue Suisse Vitic. Arboric. Hortic.*, 19, 323–28.

34. O'Brien, M. C., and G. N. Prendeville. 1979. "Effect of herbicides on cell membrane permeability in *Lemna minor*." *Weed Res.*, 19, 331–34.

35. Anderson, J. L., and W. W. Thomson. 1973. "The effects of herbicides on the ultrastructure of plant cells." *Res. Rev.*, 47, 167–89.

36. Schmidt, Th., C. Fedtke, and R. R. Schmidt. 1976. "Synthesis and herbicidal activity of d(+) and l(−) methyl 2-chloro-3-(4-chlorophenyl)-propionate." *Z. Naturforsch.*, 31c, 252–54.

37. Kappler, R., and U. Kristen. 1987. "Photometric quantification of in vitro pollen tube growth: a new method suited to determine the cytotoxicity of various environmental substances." *Envir. Exper. Bot.*, 27, 305–09.

38. The Association of Applied Biologists. 1986. "Biochemical and physiological techniques in herbicide research." *Aspects of Appl. Biol.*, 11, 1–203. Conf. Proc. from: The AAB Office, Natl. Veg. Res. Sta., Wellesbourne, Warwick, U.K.

39. Hansch, C., and T. Fujita. 1964. "ρ-σ-π Analysis. A method for the correlation of biological activity and chemical structure." *J. Amer. Chem. Soc.*, 86, 1616–26.

40. Fujita, T., J. Iwasa, and C. Hansch. 1964. "A new substituent constant, π, derived from partition coefficients." *J. Amer. Chem. Soc.*, 86, 5175–80.

41. Hansch, C., A. Leo, S. H. Unger, K. H. Kim, D. Nikaitani, and E. J. Lien. 1973. "Aromatic" substituent constants for structure-activity correlations." *J. Med. Chem.*, 16, 1207–16.

42. Hansch, C., and A. Leo. 1979. *Substituent Constants for Correlation Analysis in Chemistry and Biology*. New York: Wiley.

43. Kirino, O., C. Takayama, and A. Mine. 1986. "Quantitative structure-activity relationships of herbicidal *N*-(1-methyl-1-phenylethyl)phenylacetamides." *J. Pestic. Sci.* (Japan), 11, 611–17.

44. Livingstone, D. J. 1989. "Multivariate quantitative structure-activity relationship (QSAR) methods which may be applied to pesticide research." *Pestic. Sci.*, 27, 287–304.

45. Mitsutake, K., H. Iwamura, R. Shimizu, and T. Fujita. 1986. "Quantitative structure-activity relationship of photosystem II inhibitors in chloroplasts and its link to herbicidal action." *J. Agric. Food Chem.*, 34, 725–32.

46. Ward, C. E., R. V. Berthold, J. F. Koerwer, J. B. Tomlin, and D. T. Manning. 1986. "Synthesis and herbicidal activity of 1,2,3,4-tetrahydro-1,3,5-triazino[1,2-*a*]benzimidazoles." *J. Agric. Food Chem.*, 34, 1005–10.

47. Macherel, D., M. Tissut, F. Nurit, P. Ravanel, M. Bergon, and J.-P. Calmon. 1986. "Inhibitory action of an isopropyl carbanilate series on mitosis, respiration and photosynthesis." *Physiol. Vég.*, 24, 97–107.
48. Takemoto, I., R. Yoshida, S. Sumida, and K. Kamoshita. 1985. "Quantitative structure-activity relationships of herbicidal *N'*-substituted phenyl-*N*-methoxy-*N*-methylureas." *Pestic. Biochem. Physiol.*, 23, 341–48.
49. Friedman, A. J., and A. J. Hopfinger. 1983. "*N*-(2,6-dihalobenzilidene)arenesulfinamide herbicides and analogous compounds. 2. Structure-activity relationships." *J. Agric. Food Chem.*, 31, 135–37.
50. Tanaka, J., E. Kuwano, and M. Eto. 1986. "Synthesis and pesticidal activities of phosphonate analogs of amino acids." *J. Fac. Agr.*, Kyushu Univ., 30, 209–23.
51. Iwamura, H., K. Nishimura, and T. Fujita. 1985. "Quantitative structure-activity relationships of insecticides and plant growth regulators: comparative studies toward understanding the molecular mechanism of action." *Environ. Health Perspect.*, 61, 307–20.
52. Schwinn, F., and H. Geissbuhler. 1986. "Towards a more rational approach to fungicide design." *Crop Protect.*, 5, 33–40.
53. Snel, M., B von Schmeling, and L. V. Edgington. 1970. "Fungitoxicity and structure-activity relationships of some oxathiin and thiazole derivatives." *Phytopath.*, 60, 1164–69.
54. Wolpert, T. J., V. Macko, W. Acklin, and D. Arigoni. 1988. "Molecular features affecting the biological activity of the host-selective toxins from *Cochliobolus victoriae*." *Plant Physiol.*, 88, 37–41.
55. Edwards, J. V., A. R. Lax, E. B. Lillehoj, and G. B. Boudreaux. 1987. "Structure-activity relationships of cyclic and non-cyclic analogues of the phytotoxic peptide tentoxin." *J. Agric. Food Chem.*, 35, 451–56.
56. Hermens, J. L. M. 1986. "Quantitative structure-activity relationships in aquatic toxicology." *Pestic. Sci.*, 17, 287–96.
57. Clarke, J. U. 1986. "Structure-activity relationships in PCBs: use of principal components analysis to predict inducers of mixed-function oxidase activity." *Chemosphere*, 15, 275–87.
58. Winkler, D. A., A. J. Liepa, J. E. Anderson-McKay, and N. K. Hart. 1989. "A molecular graphics study of factors influencing herbicidal activity of 3-acyltetrahydro-2*H*-pyran-2,4-diones." *Pestic. Sci.*, 27, 45–63.
59. Aulabaugh, A., and J. V. Schloss. 1990. "Oxalyl hydroxamates as reaction-intermediate analogues for ketol-acid reductoisomerase." *Biochem.*, 29, 2824–30.
60. Gardner, G., J. R. Sanborn, and J. R. Goss. 1987. "*N*-Alkylaryltriazine herbicides: a possible link between triazines and phenylureas." *Weed Sci.*, 35, 763–69.

Chapter 2

Reaching the Target

In this chapter, the word "target" is used to refer to the entire plant. This is quite distinct from the term "target site" (discussed in Chapter 1 and in later chapters in the book), referring to the molecular site of action of a herbicide. The first step in reaching the target site is interception by the plant, whether it be on the foliage or via the soil to the roots or basal portion of the shoot. Although delivery to the plant is a physical rather than a physiological process, it is an important first step in herbicide action. The physical, chemical, and microbiological interactions that are involved here will not be discussed in detail; rather, this chapter is included as a general introduction to the topic of herbicide interception by plants.

In conventional herbicide applications, the herbicide is dissolved or suspended in a carrier solution that is sprayed on the soil or plants (or both) under hydraulic pressure. Alternative application methods can also be used, including rope-wick or roller applicators, which apply a more concentrated herbicide solution directly on to the foliage, and a variety of newer droplet-generating devices (e.g., spinning discs and electrostatic sprayers). The design, operating principles, and efficiency of these applicators is the subject of a recent monograph [1], and will not be discussed here. However, the behavior and fate of spray droplets generated by conventional application equipment will be discussed briefly.

2.1 INTERCEPTION AND RETENTION ON LEAVES

Liquid that is forced through narrow nozzles under pressure emerges in the form of "sheets" that quickly atomize to form discrete droplets. In typical agricultural sprays, the droplets generated range in size from approximately 50 μm diameter to

400 μm, the size spectrum being a function of viscosity, surface tension, pressure, and orifice diameter [2]. These droplets are slowed to their terminal velocity very soon after leaving the nozzle orifice (within 10–50 μs), and continue to fall under gravity. If air currents are sufficiently strong, and the droplets sufficiently small, they may move horizontally, or even upward. These small droplets are the source of droplet drift (see following section).

The trajectory of spray droplets is influenced primarily by their mass and by air movement above and within the plant canopy. Larger droplets tend to fall under the influence of gravity, and are little affected by air movement. Consequently, they are likely to impinge on leaves close to the top of the plant canopy, particularly if these leaves are flat. Small droplets, on the other hand, are more likely to be influenced by air movement above and within the canopy. As such, they are more likely to penetrate into a dense canopy than larger droplets, and are more likely to be retained by vertical surfaces (stems, upright leaves) and on fine hairs on lower leaves [3].

Droplets falling vertically can be intercepted by leaves or other plant parts, or may land on the soil. When a droplet lands on a leaf surface, two things can happen: the droplet can be retained on the leaf, or it may bounce and be redirected to another part of the plant or the soil. Whether droplets bounce or are retained on the leaf surface is a function of the kinetic energy and surface tension of the droplet, and the nature of the leaf surface [3]. Other things being equal, droplets with high surface tension are more likely to bounce off leaf surfaces. When droplets strike the leaf, they are distorted laterally, and the energy they possess in falling is either dissipated or retained, according to the surface tension of the droplet and the nature of the leaf surface. If the surface tension of the droplet is very high, the tendency is for the droplet to recoil from surface contact, form a sphere, and bounce off the surface [3]. Conversely, droplets with low surface tension are less likely to recoil in this manner, and are more likely to be retained on the leaf. There are several reports in the literature of increased droplet retention when surfactant

TABLE 2.1 EFFECT OF SURFACTANT ON THE RETENTION OF CHLORMEQUAT BY WHEAT AND BARLEY LEAVES; PLANT GROWTH STAGES WERE NOT SPECIFIED [6].

Growth stage	Surfactant concentration (%)	Spray retention (% of applied dose)	
		Wheat	Barley
Immature	0.03	45	31
	0.1	73	33
	0.3	79	62
Mature	0.03	33	19
	0.1	51	22
	0.3	71	41

is included in the spray solution (e.g., [4–6]); sample data illustrating this are shown in Table 2.1 [6]. In practice, droplets <100 μm in diameter are unlikely to bounce off leaf surfaces. Large droplets, on the other hand, are very likely to bounce on impact. In one study, retention of spray droplets on barley and *Avena fatua* leaves was greatest for 100 μm (diameter) droplets, and considerably less for droplets ranging from 200 to 600 μm [7].

Surface features that promote contact between the spray droplet and the leaf surface enhance herbicide retention. Retention is usually greater on irregular leaf surfaces than on smooth surfaces, and greater on surfaces covered by hairs or trichomes [3]. An illustration of the influence of leaf surface hairiness on retention of MCPA is shown in Figure 2.1 [8]. Spray droplets intercepted by the dense stellate hairs on *Eremocarpus setigerus* leaves break up, with some droplet fragments penetrating to the leaf surface and others remaining on the hairs. Variations in leaf surface characteristics may be responsible for differential retention on the leaves of different species, or on leaves of different ages. Older leaves can be more retentive, due to the effects of weathering on surface morphology [6]; however, by virtue of their position in the canopy, older leaves may intercept less spray than young leaves. The data in Table 2.2 illustrate the difference in retention of spray droplets (applied at different speeds) by two species with different surface characteristics [4]. Radish, by virtue of its rough leaf surface, retained much more of the applied spray than barley.

In some instances, differential spray retention may contribute to herbicide selectivity between species (e.g., [9, 10]). However, for reasons given above, addition of surfactant can decrease the margin of selectivity, by increasing droplet retention on "nonretentive" species [10, 11]. The use of high-speed droplets, and the resulting differences in spray retention by different plant species, has been

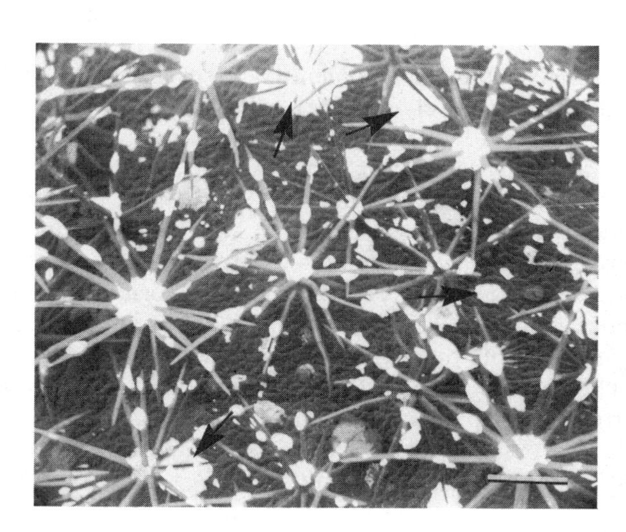

Figure 2.1 Scanning electron micrograph of MCPA (sodium salt) deposits on *Eremocarpus setigerus* leaves. The secondary electron image has been superimposed on a cathodoluminescence image, showing the herbicide as a light-colored deposit. The arrows indicate the herbicide that has penetrated through the leaf hairs to the epidermal surface. Scale bar = 200 μm [8].

TABLE 2.2 SPRAY RETENTION ON BARLEY AND RADISH FOLLOWING APPLICATION AT DIFFERENT DROP SPEEDS; ALL SPRAY SOLUTIONS CONTAINED SURFACTANT [4].

	Spray deposit (μL cm^{-2})	
Drop speed ($m\ s^{-1}$)	Barley	Radish
1.50	0.54	0.84
1.85	0.58	1.02
2.45	0.23	0.91
3.30	0.14	1.30
4.45	0.27	2.49

suggested as a possible way of achieving selectivity where none or an inadequate level existed previously [4].

Droplets are also more likely to be retained on dry leaf surfaces than wet surfaces [3]. Unless surface tension is very low, direct contact between the spray droplet and a water drop on a leaf surface is unlikely. An air film is maintained between the two liquids, aided by the fact that the lower surface can be physically displaced, and this increases the likelihood of the spray droplet rebounding from the leaf surface. Thus, the presence of rain or dew drops on leaf surfaces may lead to decreased herbicide retention; however, other factors, such as enhanced herbicide absorption under conditions of high humidity, may override the effects of decreased retention on herbicide activity under these conditions.

Rebounding droplets frequently break into several smaller droplets that are then distributed according to the processes described above. They may land on leaf, stem, or other plant surfaces, fall to the soil, or be removed from the target zone by droplet drift. Since their volume and momentum are probably lower than those of the original droplet, it is more likely that they will be retained on the next surface they encounter.

Herbicide deposits retained on a leaf surface are available for absorption into the leaf, unless they are removed by leaf abrasion or precipitation. The interval between herbicide application and the timing of the first significant precipitation that does *not* decrease herbicide efficacy is often interpreted as a measure of the time course of herbicide absorption. However, this may not be a correct interpretation; spray droplets dry quickly on leaf surfaces, and the failure of precipitation to reduce herbicide activity may indicate the presence of an insoluble residue (from which absorption may still be occurring) rather than the completion of absorption. This topic is discussed further in Chapter 3.

In addition to its role in droplet deposition and drift, droplet size also influences the biological efficacy of an applied herbicide. However, the literature on this subject is quite confusing and, in some instances, contradictory. There is evidence for enhanced biological activity of small droplets in some reports, and enhanced activity of large droplets in others. For example, the optimum droplet diameter has been reported to range from 100 μm [12] to 400–500 μm [13]. The most likely explanation for the divergence in results reported is that optimum droplet size probably varies with plant species, herbicide, and herbicide concentration in the droplets. Factors such as the role of localized toxicity in reducing translocation [14, 15] and the relative contributions of herbicide deposited on different parts of the plant to overall toxicity must be considered in attempts to explain the observed results.

2.2 HERBICIDE DRIFT

Herbicide drift can occur in either of two ways: by lateral movement of discrete spray droplets, discussed in the previous section, or by vapor drift. Some of the more important factors that influence herbicide drift are illustrated in Scheme 2.1. The complexity of the many possible interactions between these factors makes it very difficult to predict drift accurately, in all but the simplest cases (e.g., small droplets in high wind, or a very volatile herbicide). The obvious consequences of drift are that a lower-than-intended dose reaches the target plants, while a low (but

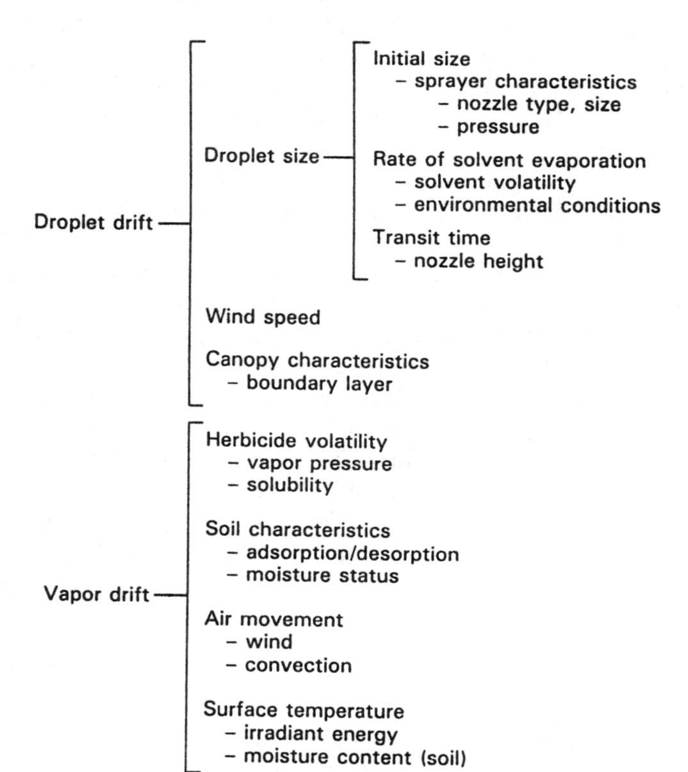

Scheme 2. Factors influencing herbicide drift.

possibly phytotoxic) dose may reach nontarget plants. In addition, herbicide interception by the target plants is often uneven [2]. This uneven spray distribution results in loss of herbicide activity in some places within the target area, due to reduced plant coverage, and enhanced activity (and possible crop phytotoxicity) in others.

The fall speed of spray droplets after they reach their terminal velocity is proportional to the square of their diameter, and evaporation is proportional to the reciprocal of this. A droplet 100 μm in diameter falls 1 m in approximately 3–5 s. The time required for the same droplet to evaporate is also 3–5 s, which means that droplets this size are highly susceptible to drift. Of course, this will vary with environmental factors such as air temperature and relative humidity. Droplet drift can be avoided or reduced by a variety of means, including lowering the spray boom, lowering the spray pressure, and using "thickeners" to reduce evaporation of the droplets. However, all of these approaches usually result in a more uneven spray pattern, which may have detrimental effects on weed control. Adaptations to sprayers to alter air flow beneath the nozzles, mostly aimed at reducing turbulent air flow, can help to reduce droplet drift [2].

Vapor drift is the result of volatilization of herbicide molecules, either from falling droplets or after deposition on plant or soil surfaces. Volatilization, or transfer from solution to the gaseous phase, is a function of the vapor pressure of the herbicide. As a general rule, herbicides with a vapor pressure $> 10^{-2}$ Pa may be subject to volatilization losses [16]. In addition, very low water solubility can promote volatilization of compounds with lower vapor pressures [17]. Losses of foliage-applied, volatile herbicides (e.g., 2,4-D esters) are typically high immediately after application, but decrease rapidly because of absorption into the plant tissue. Losses of herbicide from the soil, on the other hand, can continue for an extended period after application.

Herbicide volatilization from soil is greater under moist soil conditions. In moist soil, more adsorption sites on soil colloids are occupied by water molecules. This results in more herbicide molecules being in the soil solution and available for loss from the soil by volatilization. Consequently, precipitation or dew formation can result in volatilization of herbicide from soil days, or even weeks, after herbicide application. For example, airborne residues of 2,4-D *iso*-octyl ester have been detected following precipitation or deposition of dew on soil treated with the herbicide several days previously [18]. Similar results have been reported for trifluralin, a volatile, soil-applied herbicide [19]. Cumulative losses of unincorporated herbicide from wet soil can be as high as 90 percent of the applied dose, although incorporation can reduce this significantly [20].

2.3 HERBICIDE AVAILABILITY IN SOIL

Herbicide that lands on the soil surface may be subject to a wide variety of physical and chemical processes that make it unavailable to the plant (Figure 2.2). Adsorption to soil constituents (organic matter, clay mineral surfaces, oxides, and so on) is reversible, and can result in fluctuating pools of herbicide being available for plant uptake. Other processes that remove the herbicide from the system permanently will be considered later.

Herbicide adsorption in soil is a function of the chemical properties of the soil and the herbicide molecule, and usually results from weak electrostatic interaction (hydrogen bonding or van der Waals bonds) between the herbicide and the surface of soil particles. Hydrogen bonding may also occur between herbicide molecules and water molecules adsorbed on the soil particles. It is also possible that ionic bonding can occur between charged groups on the soil and permanently charged herbicides, or weak acids or bases in the charged state [21].

Adsorption of herbicide in soil can be described by the Freundlich equation:

$$x = KC^n$$

where x = amount of herbicide adsorbed (μg g^{-1} soil), C = the equilibrium concentration of herbicide in solution (μg ml^{-1}), and K, n = constants for a particular soil. A distribution coefficient, K_d, can be calculated to provide a measure of the relative distribution of herbicide between an adsorbent (e.g., clay, organic matter) and the solvent (usually water):

$$K_d = x/C$$

The amount of herbicide adsorbed in a particular soil depends primarily on the nature of the soil mineral and organic components, the nature of the herbicide, and the soil water status. The interaction between soil and herbicide is strongly influenced by pH, reflecting the effects of pH on both the nature of the soil particle surfaces and on the herbicide (for weak acid and weak base herbicides) [21].

The influence of soil water on herbicide availability to plants is illustrated in Figure 2.3, and is discussed in detail elsewhere [22, 23]. Although adsorption-desorption phenomena largely dictate the amount of herbicide that is in the soil

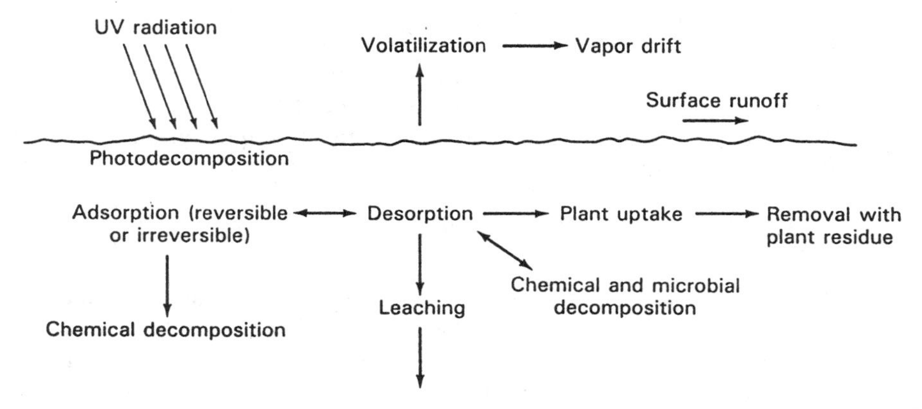

Figure 2.2 Processes resulting in loss of herbicide from soil.

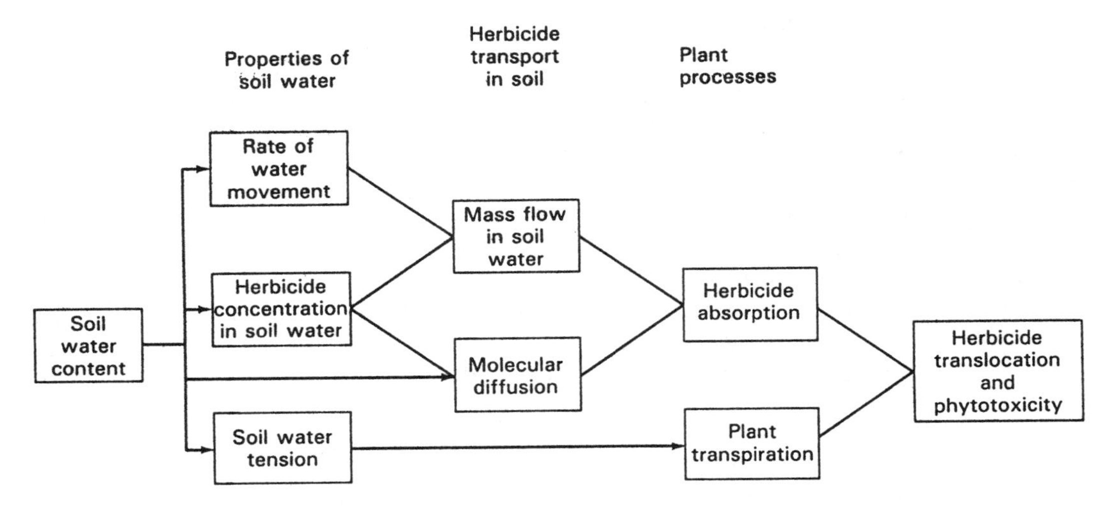

Figure 2.3 Influence of soil water on herbicide availability to plants [23].

solution, this does not necessarily correlate well with biological activity. For example, in a soil with low adsorptive capacity, much of the herbicide will be in the soil solution at low soil water levels; however, mass flow, diffusion, and plant uptake and translocation limitations all act to reduce the phytotoxic effect of the higher herbicide concentration [23]. In more adsorptive soils, herbicide adsorption increases as the soil water level decreases, reflecting competition between water and herbicide for some adsorptive sites. Displacement of herbicide by water was referred to in the previous discussion of herbicide volatilization from soil.

In addition to adsorption-desorption phenomena, herbicide availability in soil is influenced by a number of factors that influence the amount of herbicide present in the soil and available for uptake by plants (Figure 2.2). These processes will not be discussed in detail, but are considered briefly in the context of their contribution to loss of herbicide from the available pool in the soil.

One of the aims of much of the research on herbicide availability in soil has been to construct models that allow prediction of herbicide residues over time. Accumulation of herbicide in soil after repeated application can be calculated by the following equation:

$$R = \frac{AP(1 - P^n)}{1 - P}$$

where R = residue at the end of the n years, A = the dose applied each year, and P = the proportion of the dose that remains after 1 year [24]. However, because of the heterogeneity of most soils, climatic variations, and a variety of other variables, it has been difficult to derive models that are widely applicable to herbicide fate or persistence. More detailed models for predicting herbicide persistence in soil, relating to turnover rates, soil moisture level, temperature, and so on, are available in the published literature [25–27].

Chemical and microbial decomposition are the major processes resulting in herbicide loss from soils. Chemical decomposition is a function of soil pH, and usually starts immediately after the herbicide has entered the soil. First-order reaction kinetics are often used to describe chemical decomposition, providing a half-life value for a particular herbicide-soil combination. The equation

$$dC/dt = kC$$

where C = herbicide concentration after time t, and k is the first-order rate constant, can be solved for k. From this, the half-life of the herbicide in the soil can be determined as:

$$t_{1/2} = 0.693/k$$

However, for a variety of reasons, experimental data often do not fit first-order kinetics, with higher orders of reaction providing better descriptions. Artifacts introduced during experimentation, increased adsorption over time, and different sources of loss from the system may all contribute to this. The kinetics of degradation are discussed in detail by Walker [24].

Microbial decomposition, on the other hand, is a function of the microbial status of the soil, soil temperature, and moisture level. In addition, the herbicide history of the soil can also influence the rate of microbial decomposition of herbicides (see Section 17.3.5). For some herbicides, the list of microorganisms that are capable of degradation is extensive; for example, over 20 species of microorganism can degrade 2,4-D [28].

In soils that have not been exposed previously to a particular herbicide, microbial decomposition is often preceded by a "lag phase" prior to rapid herbicide degradation (Figure 2.4A). The lag phase is interpreted as the time required for microbial population(s) to adapt (either biochemically or in number) so that the herbicide can be degraded. In soils that have been treated previously with a

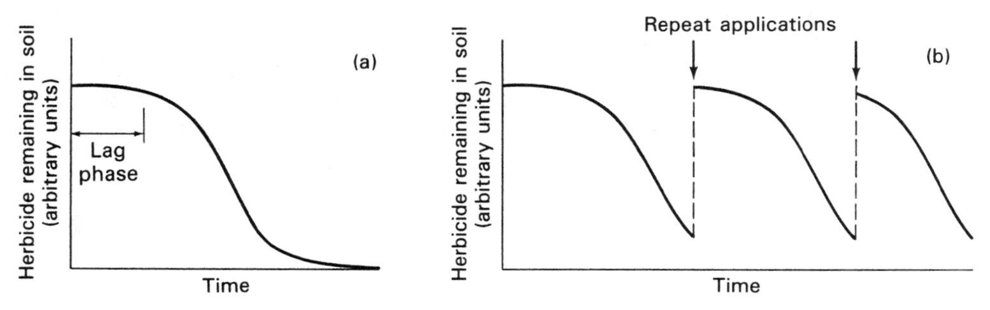

Figure 2.4 Time-course plots of microbial degradation of herbicides. A. First-time application, showing the lag phase prior to rapid degradation. B. Repeat applications, showing adaptation of the soil microflora and more rapid degradation.

particular herbicide, the lag phase can be shortened, reflecting the adaptation of the microflora to that substrate (Figure 2.4B). This can be beneficial, in that the herbicide is degraded quickly and residue problems are avoided. Conversely, accelerated degradation of herbicides can negatively affect their biological activity, to the extent that they no longer provide weed control for a sufficiently long period. Accelerated degradation has been documented for 2,4-D and related phenoxyalkanoic acids and, more recently, for thiocarbamate herbicides [29]. This enhanced degradation can apply to repeat applications of the same herbicide, or to the subsequent application of herbicides in the same chemical group (e.g., butylate in a vernolate-treated soil).

Removal of herbicides in water can occur vertically (by leaching), laterally (by diffusion), and by surface runoff. Leaching of herbicides is a function of water movement through the soil (i.e., amount and intensity of precipitation) and the partitioning of herbicide between the soil solution and the soil matrix (adsorption-desorption). In general, herbicides that are weakly adsorbed to the soil matrix are more likely to be leached than strongly adsorbed herbicides. It must be remembered that downward movement of herbicide can be reversed if evapotranspiration exceeds precipitation, resulting in net upward movement of soil water and herbicide. This can create the situation where a crop may start to grow normally, but be affected later in the season by herbicide that has moved up through the soil with soil water.

Herbicide runoff is a function of herbicide availability at the soil surface and surface water movement (precipitation intensity, soil surface roughness, slope). Herbicide in run-off may be in solution or adsorbed on soil particles that are also removed.

Photodecomposition (or photolysis) can be a significant source of loss of herbicides that are sensitive to uv radiation (> 285 nm) [30]. For soil-applied, uv-sensitive herbicides (e.g., trifluralin), incorporation into the soil immediately after application is required to minimize photodecomposition and maximize biological activity. Some foliage-applied herbicides (e.g., sethoxydim) are also subject to photodecomposition. The only practical approach to minimizing losses in this situation appears to be via formulation, that is, by promoting absorption into the leaf tissue as quickly as possible.

The final source of herbicide loss included in Figure 2.2, volatilization, has been discussed previously in relation to herbicide drift (see Section 2.2). Herbicides subject to volatilization include thiocarbamates, trifluralin and some other dinitroanilines, and short-chain esters of phenoxyalkanoic acids.

REFERENCES

1. McWhorter, C. G., and M. R. Gebhardt. 1988. *Methods of Applying Herbicides*. WSSA Mono. 4. Weed Sci. Soc. Amer., Champaign, Il., 358 pp.
2. Maybank, J. 1987. "Pesticide application technologies". In *Pesticide Science and Biotechnology*, R. Greenhalgh and T. R. Roberts, eds. Oxford, UK: Blackwell, pp. 243–48.

3. Spillman, J. J. 1984. "Spray impaction, retention, and adhesion: an introduction to basic characteristics." *Pestic. Sci.,* 15, 97–106.
4. Taylor, W. A., and G. B. Shaw. 1983. "The effect of drop speed, size and surfactant on the deposition of spray on barley and radish or mustard." *Pestic. Sci.,* 14, 659–65.
5. Tu, Y. Q., Z. M. Lin, and J. Y. Zhang. 1986. "The effect of leaf shape on the deposition of spray droplets in rice." *Crop Prot.,* 5, 3–7.
6. Baker, E. A., and G. M. Hunt. 1985. "Factors affecting the uptake of chlormequat into cereal leaves." *Ann. Appl. Biol.,* 106, 579–90.
7. Lake, J. R. 1977. "The effect of drop size and velocity on the performance of agricultural sprays." *Pestic. Sci.,* 8, 515–20.
8. Hess, F. D., D. E. Bayer, and R. H. Falk. 1974. "Herbicide dispersal patterns: 1. As a function of leaf surface." *Weed Sci.,* 22, 394–401.
9. Sharma, M. P., W. H. Vanden Born, and D. K. McBeath. 1978. "Spray retention, foliar penetration, translocation and selectivity of asulam in wild oats and flax." *Weed Res.,* 18, 169–73.
10. Harper, D. R., and A. P. Appleby. 1984. "Selectivity factors relative to asulam for *Senecio jacobaea* L. control in *Medicago sativa.*" *Weed Res.,* 24, 85–92.
11. Hibbitt, C. J. 1969. "Growth and spray retention of wild oat and flax in relation to herbicidal activity." *Weed Res.,* 9, 95–107.
12. McKinlay, K. S., S. A. Brandt, P. Morse, and R. Ashford. 1972. "Droplet size and phytotoxicity of herbicides." *Weed Sci.,* 20, 450–52.
13. Douglas, G. 1968. "The influence of size of spray droplets on the herbicidal activity of paraquat and diquat." *Weed Res.,* 8, 205–12.
14. Merritt, C. R. 1982. "The influence of form of deposit on the phytotoxicity of difenzoquat applied as individual drops to *Avena fatua.*" *Ann. Appl. Biol.,* 101, 517–25.
15. Taylor, M. J., P. Ayres, and D. J. Turner. 1982. "Effect of surfactants and oils on the phytotoxicity of difenzoquat to *Avena fatua,* barley and wheat." *Ann. Appl. Biol.,* 100, 353–63.
16. Koskinen, W. C., and S. S. Harper. 1988. "Herbicide properties and processes affecting application." In *Methods of Applying Herbicides,* C. G. McWhorter and M. R. Gebhardt, eds. WSSA Mono. 4. Weed Sci. Soc. Amer. Champaign, Il., pp. 9–18.
17. Hance, R. J. 1980. "Transport in the vapour phase," In *Interactions Between Herbicides and the Soil,* R. J. Hance, ed. London, UK: Academic, pp. 59–81.
18. Grover, R., S. R. Shewchuk, A. J. Cessna, A. E. Smith, and J. H. Hunter. 1985. "Fate of 2,4-D *iso*-octyl ester after application to a wheat field." *J. Environ. Qual.,* 14, 203–10.
19. Harper, L. A., A. W. White, Jr., R. R. Bruce, A. W. Thomas, and R. A. Leonard. 1976. "Soil and microclimate effects on trifluralin volatilization." *J. Environ. Qual.,* 5, 236–42.
20. Grover, R. 1990. "Nature, transport, and fate of airborne residues," In *Environmental Chemistry of Herbicides,* R. Grover and A. J. Cessna, eds. Boca Raton, Fl.: CRC Press, pp. 89–117.
21. Calvet, R. 1980. "Adsorption-desorption phenomena," In *Interactions Between Herbicides and the Soil,* R. J. Hance, ed. London: Academic, pp. 1–30.
22. Moyer, J. R. 1987. "Effect of soil moisture on the efficacy and selectivity of soil-applied herbicides." *Rev. Weed Sci.,* 3, 19–34.

23. Green, R. E., and S. R. Obien. 1969. "Herbicide equilibrium in soils in relation to soil water content." *Weed Sci.*, 17, 514–19.

24. Walker, A. 1987. "Herbicide persistence in soil." *Rev. Weed Sci.*, 3, 1–17.

25. Walker, A. 1987. "Evaluation of a simulation model for prediction of herbicide movement and persistence in soil." *Weed Res.*, 27, 143–52.

26. Leistra, M. 1986. "Modelling the behaviour of organic chemicals in soil and ground water." *Pestic. Sci.*, 17, 256–64.

27. Hamaker, J. W., and C. A. I. Goring. 1976. "Turnover of pesticide residues in soil," In *Bound and Conjugated Pesticide Residues*, D. D. Kaufman, G. G. Still, G. D. Paulson, and S. K. Bandal, eds. ACS Symp. Series No. 29. Washington, D.C., pp. 219–43.

28. Sinton, G. L., L. T. Fan, L. E. Erickson, and S. M. Lee. 1986. "Biodegradation of 2,4-D and related xenobiotic compounds." *Enz. Microb. Technol.*, 8, 395–403.

29. Roeth, F. W. 1986. "Enhanced herbicide degradation in soil with repeat application." *Rev. Weed Sci.*, 2, 45–65.

30. Plimmer, J. R. 1976. "Volatility," In *Herbicides: Chemistry, Degradation and Mode of Action*. Vol. 2, P. C. Kearney and D. D. Kaufman, eds. New York: Marcel Dekker, pp. 891–934.

Chapter 3

Foliar Absorption of Herbicides

3.1 INTRODUCTION

Foliar absorption of herbicides is a complex process, involving the passage of herbicide molecules from the outer surface of the plant leaf, through the cuticle, and into the underlying tissue. In this chapter we consider movement as far as the apoplasm only. The final step, penetration into plant cells, will be discussed in Chapter 4. Given the complex interactions between herbicide, plant species, environment, and formulation, the emphasis in this discussion is on the mechanisms controlling the foliar absorption process rather than on particular case studies.

The passage of herbicide molecules into a leaf is a function of the chemical and physical nature of the cuticle, the properties of the herbicide and accompanying formulation ingredients, and the environment in which the leaf has developed and in which absorption is occurring. Ideally, we would like to be able to consider all of these variables and combine them in a useful general model of herbicide absorption. However, experience has shown that there are few rules governing herbicide absorption, and that each herbicide/plant species/formulation/environment combination has its own unique characteristics. Nevertheless, examination of the literature does shed some light on the mechanism and control of foliar uptake of herbicides.

3.2 THE PLANT CUTICLE

All aerial parts of terrestial plants are covered by a cuticle. The primary function of this relatively thin (0.12 to 13.5 μm) [1, 2] membrane is to prevent water loss from plants, so that physiological processes can proceed under water-limiting conditions. In addition, the cuticle acts as an effective barrier to the penetration of xenobiotics and microorganisms into the plant.

The cuticle (often referred to as the cuticular membrane, or CM) is composed of an insoluble cutin framework and soluble waxes (Figure 3.1). "Soluble" here is used to refer to waxes that are soluble in nonpolar solvents (e.g., chloroform), and can be removed relatively easily by immersion of the leaf in such a solvent. The cutin framework is an assembly of single or cross-linked hydroxycarboxylic acids (mostly C_{16} and C_{18}, and particularly rich in the di- and trihydroxyalkanoic acids) that extend from the outer surface of epidermal cell walls into the CM [3–5]. Cross-linking occurs by esterification between carboxylic acid groups and secondary hydroxyl or epoxide groups on the alkyl chains. Some CMs also contain suberin, a combination of cutin-like aliphatic polymers and aromatic moieties, probably shikimate-derived (see Chapter 13). Cutin and suberin are partly hydrophilic, and contain some ionizable groups.

Soluble cuticular lipids are dispersed throughout the CM, but are more predominant toward the outer surface of the CM. The soluble lipids can be divided into two categories, those impregnated in the CM in association with cutin (cuticular waxes), and those forming the outer surface of the CM (epicuticular waxes). Cuticular waxes are mostly intermediate-length fatty acids (C_{16} and C_{18}) and some long-chain hydrocarbons oriented perpendicular to the leaf surface, whereas epi-

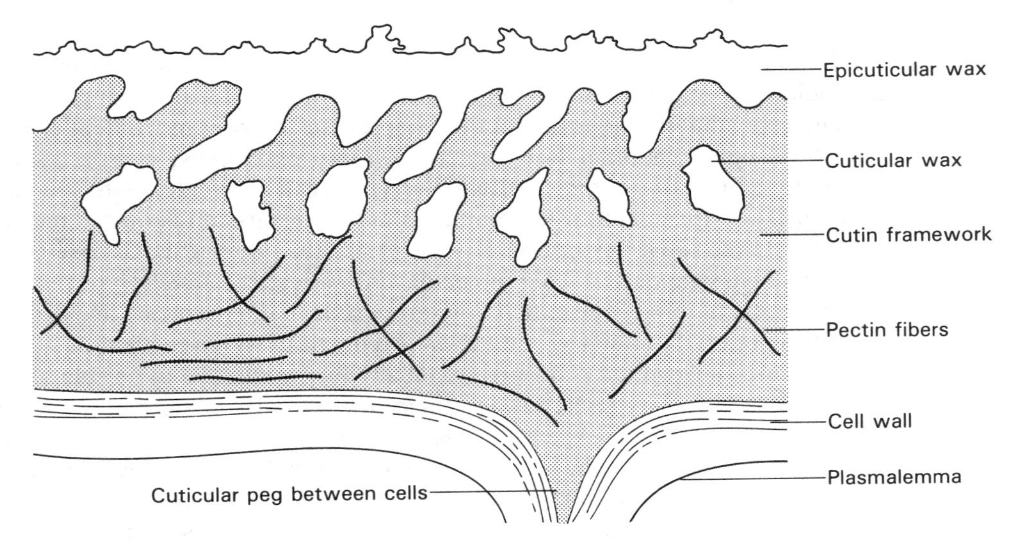

Figure 3.1 Diagrammatic cross section of a plant cuticle.

cuticular waxes are composed of widely varying long-chain (mostly C_{20} to C_{37}, but with some up to C_{50}) aliphatic hydrocarbons with alcohol, ketone, aldehyde, acetate, ketol, β-diketol, and ester substituents [6, 7]. Epicuticular waxes can be deposited in a variety of physical forms; in some species they form a relatively flat layer, whereas in others they are extruded in the form of plates or crystals that range widely in size and shape. These crystals give leaves the characteristic topography that is seen in scanning electron micrographs (Figure 3.2).

The distribution of chemical constituents through the CM dictates that the

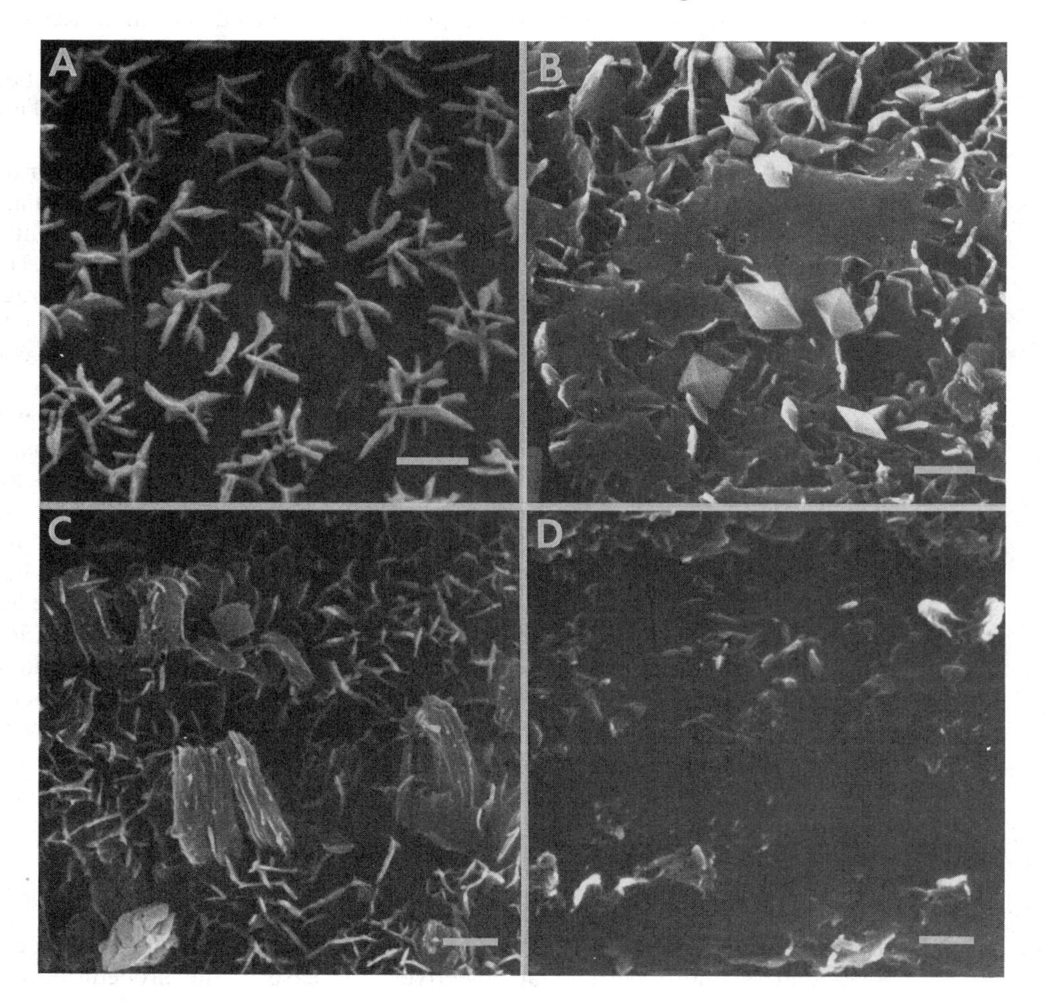

Figure 3.2 Scanning electron micrographs of leaf surfaces. A. *Cassia obtusifolia* L. B. *Setaria glauca* (L.) Beauv.; crystals overlaying was C. *Eriochloa contracta* Hitchc.; wax extrusions from cork cells. D. *E. contracta*; smooth wax. All bars = 1 μm. Photographs courtesy of R. N. Paul.

cuticle is not a homogeneous layer; the outer surface is highly lipophilic, and the environment becomes increasingly hydrophilic as the inner surface of the CM is approached. There has been some speculation in the literature on the presence of hydrophilic and lipophilic channels through the cuticle, and it has been suggested that herbicides move preferentially through these channels, depending on their physicochemical properties. Polar pores have been documented in the cutin of isolated *Citrus* CMs, associated with clusters of carboxyl groups [8], and relatively large channels can be seen in some cuticles by light microscopy [9, 10]. The significance of these in herbicide absorption has not been determined. Regardless, it is likely that there are gradients in polarity within the CM, depending on the relative amounts of soluble cuticular lipids and cutin. The CM might best be described as a "sponge-like" matrix, composed of discontinuous polar and nonpolar regions [11].

At the inner surface of the CM the environment becomes more polar due to the presence of carbohydrate polymers (principally pectin and cellulose) associated with the cell wall. These polymers, often referred to as fibrillae, extend from the cell wall into the more heavily cutinized region of the CM (Figure 3.1). It is unlikely that there is an abrupt gradient between the polar and nonpolar regions, but rather a gradation, depending on how far the fibrillae extend into the CM. In addition, there is evidence that there are some polypeptides located in this region of the CM.

Schönherr and co-workers have described CMs as polyelectrolytes with an isoelectric point about 3.0, and containing three different dissociable groups, at pH 3-6, 6-9, and 9-12 [12, 13]. The first two dissociable groups correspond to carboxylic acid groups, and the third to phenolic hydroxyl groups. The isoelectric point reflects the balance between acidic and basic functional groups; the former are attributed to acidic functions on polygalacturonic acid, free carboxylic acid groups on the cutin matrix, and acidic amino acids, whereas the latter are associated with basic amino acids. Thus the CM acts like a three-dimensional ion exchange matrix, with higher ion exchange capacity at higher pH. At physiological pH the CM is negatively charged, the majority of the exchange sites being occupied by Ca^{2+}. The possible involvement of these exchange sites in herbicide penetration of CMs is discussed in Section 3.4.4.

3.3 TECHNIQUES FOR STUDYING FOLIAR ABSORPTION OF HERBICIDES

A variety of techniques can be used to study foliar absorption, the choice being dependent on the experimental objectives, the ease of the procedure, and the availability of specialized instrumentation. What is measured as foliar absorption is dependent on the experimental procedure employed, and the procedures used vary so widely that often it is impossible to compare data from different experiments.

Foliar absorption is commonly measured by applying radiolabeled herbicide in discrete droplets to a leaf, and measuring the amount remaining on the leaf surface at various time intervals thereafter by use of an arbitrary wash procedure. Caution is required in interpreting the results of such experiments, because solvents vary in their ability to dissolve herbicide deposits [14]. A nonpolar solvent may be required to dissolve a particular herbicide residue, but may also extract some herbicide from within the leaf, leading to underestimation of absorption. Conversely, the same solvent may fail to dissolve a polar herbicide deposit completely, which may result in overestimation of absorption. In practice, it is extremely difficult to verify that a particular leaf surface wash procedure removes only surface-deposited herbicide, and not herbicide that has penetrated the CM.

The importance of the epicuticular wax and cuticular wax as barriers to foliar penetration of herbicides can be assessed by comparing penetration in waxed and dewaxed cuticles. Epicuticular wax can be removed by applying a film of cellulose acetate to the surface, and peeling off the cellulose acetate/epicuticular wax film after the solvent has evaporated [15]. Although this appears to be an effective procedure, it may cause some damage to the leaf, either directly through the effect of acetone, or by the cooling effect as the acetone evaporates. Alternatively, the epicuticular wax can be disrupted by mild abrasion [16]. Cuticular waxes can be removed by treatment of the CM with a nonpolar solvent such as chloroform : methanol [17]. This latter treatment can cause some damage to underlying tissues in intact plants, and its use is more commonly described in isolated CMs.

Much of the detailed research on permeability of CMs to herbicides, and on the relative contributions of epicuticular and cuticular waxes to solute permeation, has been conducted with isolated cuticles. Plant cuticles can be separated from the underlying tissue by incubating leaf discs with cellulase and pectinase (e.g., [18–20]). Pectinase usually contains several degradative enzymes, including pectinesterase, pectintranseliminase, and polygalacturonase. Alternatively, acid digestion or related treatments [1, 21] can be used to remove the CM; some of these procedures also remove epidermal cells beneath the CM. Caution is required in the chemical analysis of isolated cuticles, since it has been shown that they readily absorb lipophilic compounds from the incubation medium [19]. Chamel [22] lists over 25 plant species that have been used in the study of xenobiotic penetration of plant cuticles.

Bioassay methods have also been used to measure herbicide penetration into leaves, based on particular (and, preferably, rapid) physiological effects of the herbicides on the leaf tissue. For example, changes in chlorophyll fluorescence traces (see Chapter 7) can be used to monitor the penetration of photosynthetic inhibitors into a leaf. Other physiological events that can be measured include ethylene evolution and electrolyte leakage. These techniques may be particularly appropriate when uptake is very rapid, for example in the presence of organosilicone surfactants (see Section 3.4.6). One disadvantage of these procedures is that they do not provide quantitative estimates of herbicide penetration, unless standard curves are generated relating herbicide uptake to the physiological parameter being measured.

A relatively recent development in the study of herbicide absorption into leaves is the use of scanning electron microscopy, in conjunction with a range of secondary detection methods, to examine the physical nature of herbicide deposits. Among the secondary detection methods used are cathodoluminescence [23, 24], X-ray fluorescence [25], and energy-dispersive X-ray microanalysis [26]. Some results obtained using these techniques are discussed and illustrated in Section 3.5.2.

3.4 CUTICULAR EFFECTS ON HERBICIDE ABSORPTION

3.4.1 Theoretical Considerations

The movement of herbicides through the CM can be viewed as the sum of three components: partitioning into the CM from the leaf surface, diffusion across the CM, and partitioning into the aqueous medium (apoplasm) at the inner surface of the CM. Cuticular penetration occurs by passive diffusion, meaning that there is no active component involving the expenditure of metabolic energy. Diffusion through the CM can be described by the following equation [2]:

$$J = \frac{k^B \times T \times K}{6\pi r \times n \times dX \times l}(C_o - C_i)$$

where J = flux across the membrane, k^B = the Boltzmann constant, T = absolute temperature, K = partition coefficient, r = radius of the solute molecule, n = viscosity of the solvent, l = tortuosity factor, and C_o and C_i = herbicide concentrations at the outside and inside of the CM, respectively. Unfortunately, there are insufficient data available to permit detailed discussion of all of these factors, the majority of the published reports being concerned with CM thickness and composition, and with physicochemical properties of the herbicide. The following sections are devoted to discussion of the possible roles of these factors in influencing foliar absorption of herbicides.

The driving force for herbicide movement across the CM is the concentration gradient between the inner and outer surface. Permeability of the CM, often referred to in terms of the permeability coefficient, P, can be defined as follows:

$$P = \frac{\delta n}{\delta t} \cdot \frac{1}{C_o - C_i}$$

where C_o and C_i = herbicide concentrations at the outside and inside of the CM, respectively, and $\delta n/\delta t$ = rate of diffusion of the herbicide (mol cm^{-2} sec^{-1}) [22]. Permeability, therefore, can be determined by dividing the flux by the concentration gradient across the CM.

3.4.2 Herbicide Partitioning into the Cuticle

Partition coefficients at the two surfaces of the CM play an important role in herbicide penetration. In the simplest system two separate partition coefficients are involved, one at the outer surface describing partitioning from the spray droplet or deposit into the CM, and a second at the inner surface describing partitioning from the CM into the underlying tissue. This assumes a CM of uniform composition which, as we have seen, does not accurately reflect the nature of most CMs. A more realistic model might include a series of partition coefficients as the herbicide moves from the highly lipophilic epicuticular wax to the more hydrophilic inner surface (Figure 3.3).

Not only do partition coefficients vary through the CM, but also with plant species and with the degree of hydration of the cuticle. In addition, the partition coefficient for weakly acidic herbicides is a function of the pH of the application solution.

In "ideal" membrane transport models, flux is proportional to the solute contact area. However, herbicide deposit area and penetration into the CM are not always positively correlated [27]. Visual examination of herbicide deposits often reveals them to be uneven, with some regions of heavy deposition and others with very little deposition beneath the same droplet [26–28]. Consequently, it is extremely difficult to measure the true deposit contact area.

It might be expected that partitioning into the CM is strongly influenced by the chemical and physical nature of the herbicide and of the epicuticular wax. Sorption of weak acids (e.g., 2,4-D, 2,4,5-T, and NAA) increases at lower pH values, indicating that the nondissociated species preferentially enter the CM [17, 29]. Above the isoelectric point (ca. pH 3), the CM carries a negative charge, and the ionized form of weakly acidic herbicides is subject to Donnan exclusion. However, the heterogeneous nature of CMs, and the presence of polar channels, may provide routes of entry even for dissociated molecules.

In measuring the uptake of ten xenobiotics into the leaves of ten different plant species, Price and Anderson [30] found very little correlation between uptake and the partition coefficients or solubilities of the xenobiotics. Uptake rates varied by a factor of 130, whereas octanol-water partition coefficients varied by 10^{10}. Other reports (e.g., [31, 32]) confirm the lack of a clear relationship between xenobiotic penetration through CMs and their solubilities or partition coefficients.

The role of epicuticular wax in regulating herbicide entry into the CM is also difficult to quantify. A survey of the results of experiments using partially or completely dewaxed CMs indicates that dewaxing does not have uniform effects on all species and herbicides, and each plant species/herbicide combination must be considered separately. For example, brushing wild oat (*Avena fatua* L.) leaves had no effect on the entry of diclofop-methyl, but significantly accelerated the entry of difenzoquat [16]. This was not due to an alteration of the herbicide deposit characteristics on the modified leaf surface. Surprisingly, removal of the epicuticu-

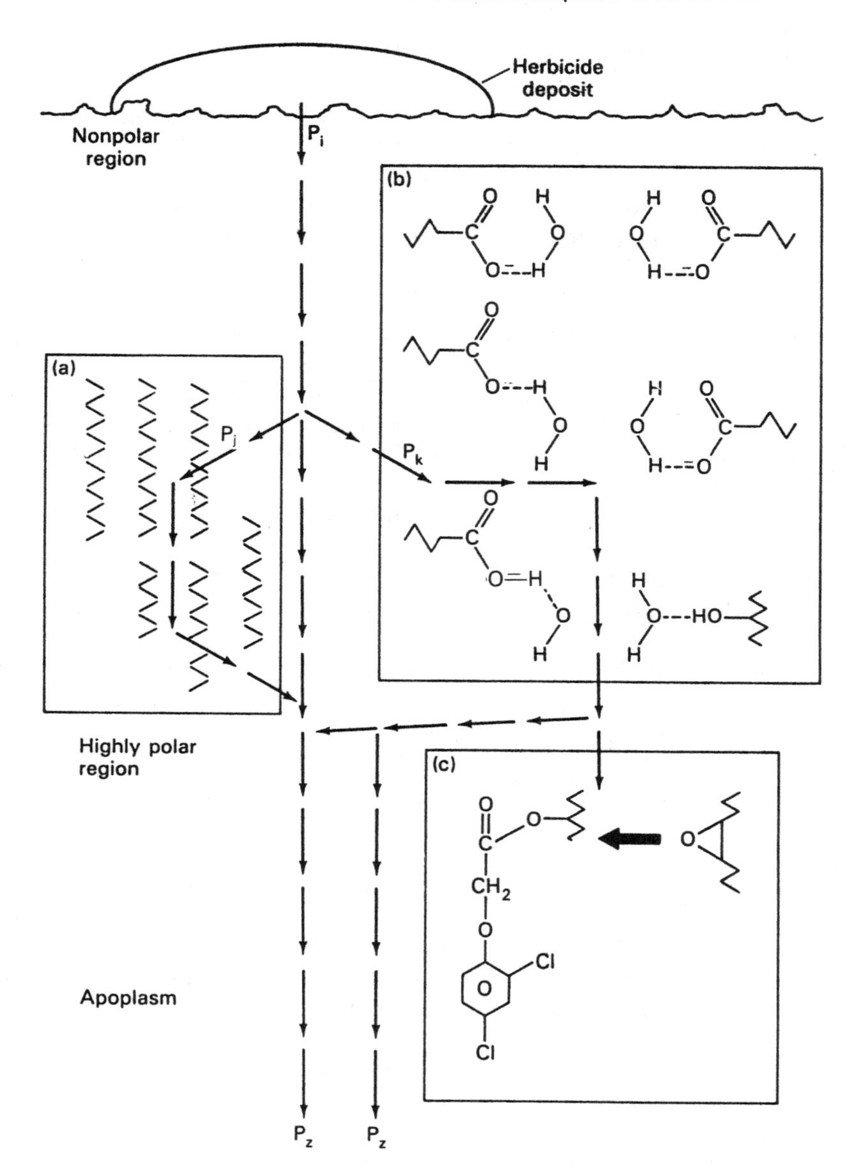

Figure 3.3 Schematic model of herbicide partitioning into the CM, diffusion through the CM, and release into the apoplasm. A. Partitioning into lipophilic regions. B. Passage through a polar pore in the cutin matrix. C. Covalent binding to cutin polymers. "P_n" denotes partition coefficients for herbicide diffusion into the different regions of the CM. Note that the different routes shown are not mutually exclusive.

lar wax with cellulose acetate did not influence entry of either herbicide in this study.

3.4.3 Herbicide Movement through the Cuticle

Although cuticle thickness might be expected to have a major influence on cuticular penetration, there is little evidence for this. In a comparison of 2,4-D movement through isolated CMs of seven species, Norris [33] found no correlation between movement and CM thickness. Indeed, there are several reports in the literature of greater permeability of thicker CMs than thin ones (e.g., [34]). The probable explanation for this is that differences in cuticle structure and composition are much more important than thickness differences, particularly when different species are being compared [2]. In a survey of 10 plant species, Becker et al. [35] found an inverse relationship between cuticle thickness and water diffusion. Although water movement may not provide a reliable estimate of permeability to all herbicides (especially nonpolar herbicides), this result suggests that the relationship between cuticle thickness and permeance is not as clear as one might predict.

Relationships between thickness and herbicide penetration are more likely to be established in leaves of the same species. For example, there are reported cases of greater xenobiotic penetration into younger leaves than older leaves within the same species [36, 37]; in both instances the CM of the younger leaves was thinner than that of the older leaves. Even here, however, some caution is required in interpreting the results. In addition to differences in CM thickness, chemical composition can also vary with ontogeny, and this confounds any assessment of the importance of cuticle thickness.

A further complicating factor is that the CM on any single leaf surface is not of uniform thickness. Examination of attached or isolated CMs by electron microscopy reveals the presence of cuticular "pegs" projecting downward between the anticlinal cell walls [38] (Figure 3.1). However, herbicide deposits often are observed on the depressions overlying the anticlinal cell walls (see Section 3.5.2 and Figure 3.7). Therefore, the shortest path to the nearest cell may not be vertically downward (in the case of a horizontal leaf), but at some angle to the vertical. This, and the heterogeneous composition of the CM, are accounted for by the tortuosity factor. At present this is no more than a theoretical concept, in that there is no satisfactory way of estimating this parameter.

Solute viscosity depends on the degree of hydration and on the relative polarities of the solute and the medium through which it is diffusing. In addition, the viscosity of lipid-rich portions of the CM may be reduced by the solvent and other formulation components [2]. The enhancement of solute penetration following removal of cuticular lipids [33, 39] may be partially due to reduced viscosity.

There is little information on the influence of molecular radius on penetration through the CM. Although pores have been identified and measured in some CM preparations [8], their involvement in herbicide movement through the CM remains to be demonstrated. Molecules larger than the estimated pore size (0.45 nm [8]) are capable of penetrating plant cuticles [2]. This question is complicated by the probable change in molecular radius as polar molecules become hydrated, and further complicated by changes in molecular form of the herbicide (e.g., de-esterification) that may occur within the CM.

The available evidence seems to suggest that the soluble cuticular lipids play a major role in limiting herbicide movement through the CM. Removal of all of the soluble cuticular lipids of isolated CMs has a marked effect on their water permeability, with increases of up to 500-fold reported [40]. Similarly, removal of the soluble waxes increases the penetration of herbicides. For example, penetration of 2,4-D increased by 2.2 to 26-fold when the soluble lipids were removed from isolated cuticles of eight different plant species [33]. However, there was no obvious correlation between the amount of wax removed and the resultant increase in penetration. Retention of pentachlorophenol, but not of 4-nitrophenol, has been demonstrated in isolated *Ficus elastica* cuticles [20]; extraction of the cuticular lipids increased the movement of both compounds through the CM by a factor of over 5,000, illustrating the degree to which the soluble lipids were responsible for limiting permeation. Extraction of cuticular lipids from isolated *Ficus* CMs increased the permeance of 2,4-D by a factor of almost 10,000 [39].

The results of cuticular penetration experiments with intact and dewaxed isolated cuticles are summarized in Table 3.1. It should be noted that the data in Table 3.1 are derived from plants grown under different conditions, and that a variety of procedures was used to extract cuticular lipids from the CM. Consequently, comparison of the data for different compounds may not be valid.

3.4.4 Retention of Herbicides in the Cuticle

Herbicide that enters the CM can either pass through to the underlying tissue, or can be retained in the CM. Retention in the CM can be attributed to high lipophilicity of the herbicide, resulting in partitioning into lipid-rich parts of the CM, or to binding within the CM. For example, more MCPB than MCPA was retained in isolated cuticles of broad bean (*Vicia faba* L.), and this was attributed to the higher oil/water partition coefficient of MCPB [47]. The use of isolated cuticles may not provide an accurate picture of transcuticular movement in vivo, due to chemical alteration of the inner surface of the CM, and the different environment underlying the CM. However, similar results have been obtained with chlorsulfuron and clopyralid, with greater retention of the former attributed to its higher lipophilicity [48]. In a study with intact cabbage (*Brassica oleracea*) leaves, more of the butoxyethanol ester of 2,4-D than the parent acid was retained in the CM. Movement into leaves from which the epidermis had been removed was equal, indicating the differing ability of the two formulations to enter the CM [49]. It appears, therefore, that the CM acts as a temporary reservoir for herbicide as it passes into the plant.

It has been shown recently that 2,4-D and other weak organic acids can be bound covalently by epoxy groups in the polymer matrix following their sorption

TABLE 3.1 EFFECTS OF CUTICULAR LIPIDS ON THE PERMEABILITY OF CUTICULAR MEMBRANES (CM) TO XENOBIOTICS. Data are expressed as the ratio of permeabilities of dewaxed ($P(d)$) and non-dewaxed ($P(i)$) CM from various species [22].

Xenobiotic	$P(d)/P(i)$	Reference
2,4-D	9.5	41
	2.5–120	42
	2.2–25.8	33
	29–9,192	39
2,4-DB	3	22
N-isopropyl-α-chloroacetamide	4	43
NAA	12	41
	23–178	44
	170–700	45
Substituted phenoxyacetic and benzoic acids	8.6–22.9	46
Pentachlorophenol	5,300	20

into the CM [50, 51] (Figure 3.3). This binding is a slow, linear, time-dependent process. Treatment of the CM with HCl inhibits this binding, by alteration of the epoxide groups, but does not alter the sorption characteristics of the CM [17]. Thermodynamics dictate that these bound residues will be released only when the cutin itself is degraded, either chemically or enzymatically.

3.4.5 Herbicide Delivery to the Apoplasm

Partitioning from the inner surface of the CM into the apoplasm or epidermal cells is more difficult to examine, because of the diffuse nature of the CM-apoplasm interface and the structural alteration that almost certainly occurs when CMs are isolated. Initially $C_i = 0$, and movement into the apoplasm is favored. As herbicide accumulates in the apoplasm, however, the concentration gradient may favor retention in the CM. Although this topic has been discussed for many years, there is little experimental evidence to substantiate claims that foliar penetration of herbicides is limited by the rate of removal at the inner surface of the CM. Given the large volume of the apoplasm and epidermal cells relative to the CM, it is unlikely that partitioning into the apoplasm is a limiting step, except perhaps for very lipophilic herbicides. However, rapid physiological damage followed by desiccation (e.g., as induced by paraquat) may limit subsequent herbicide movement from the CM into the apoplasm.

3.4.6 Stomatal Infiltration of Herbicides

The relevance of stomata in herbicide absorption has long been a topic of much debate. Stomata, if they are open, could provide a rapid route of entry into the leaf tissue, without requiring that the herbicide penetrate the CM. Substomatal cavities are lined with a thin cuticle, however, so stomatal entry does not necessarily involve direct access to epidermal cells. The arguments in favor of stomatal entry of herbicides include greater penetration through abaxial leaf surfaces than adaxial surfaces, ostensibly related to the greater number of stomata on most abaxial surfaces. Other factors may be responsible for this uptake difference, however, and they are seldom addressed experimentally. The nature of the cuticle on the two leaf surfaces is generally quite different, due to the different environments above and below the leaf, and to developmental differences.

Critical to stomatal penetration of herbicides is the surface tension of the spray droplets. In order to penetrate stomatal cavities, surface tensions lower than 30 $mN \cdot m^{-2}$ are required [52]. Unfortunately, few agricultural sprays have surface tensions in this range, most being closer to 30–35 $mN \cdot m^{-2}$.

Several other arguments can be offered against the involvement of stomata in foliar penetration of herbicides. In most plant species, stomata are located exclusively or predominantly on the abaxial leaf surfaces. Only very small spray droplets will impact on these surfaces (see Chapter 2), and in any spray operation the fraction of droplets that does so is likely to be very small. In addition, stomata cover approximately 0.1 to 0.5 percent of the leaf surface, and excellent coverage would be required for a significant number of stomata to be contacted. This is unlikely, especially in the case of low-volume applications; there can be exceptions, however, such as in the case of paraffinic oils (see below). Finally, herbicides often can be observed to act under hot, dry conditions in which the spray droplets dry very quickly and the stomata are almost certainly closed.

There are, however, some special cases in which stomatal penetration of herbicides does occur. These are associated with particular herbicide formulations and, although the information is not always provided, probably with low surface tensions. Gudin et al. [53] described very rapid effects on photosynthesis caused by an unidentified low-viscosity oil, and attributed this to stomatal entry; paraffinic oils are known to greatly enhance spreading of herbicide droplets [54]. Similarly, very rapid entry of glyphosate into perennial ryegrass (*Lolium perenne* L.) leaves in the presence of an organosilicone surfactant has been attributed to stomatal penetration [55]. The application solution had a surface tension of 20.7 $mN \cdot m^{-2}$, well below that required for stomatal entry. In addition, the enhancement of penetration was restricted to leaf surfaces in which stomata were present. It seems probable, however, that these instances of stomatal entry of herbicides are exceptions to the rule, and that stomatal entry plays little or no role in the action of most herbicide formulations.

3.5 FORMULATION EFFECTS ON HERBICIDE ABSORPTION

Herbicides are never applied alone or in pure solutions, but in combination with a variety of ingredients that are added to improve the efficacy of the herbicide. These other products include surfactants and oil-surfactant concentrates, used primarily to enhance leaf surface wetting and herbicide penetration, and a number of biologically inert materials that improve the stability and rainfastness of the formulation. In addition, a range of inorganic salts, phosphate esters, and chelating agents have been used to enhance herbicide activity; in many instances, however, data on the effects of these compounds on herbicide absorption are lacking, and there is no clear explanation of how these materials enhance herbicide activity.

3.5.1 Surfactant Effects on Herbicide Absorption

Surfactants are large, heterogeneous organic molecules that are used to facilitate contact between herbicide spray droplets and leaf surfaces. Their structures and physical properties vary widely, although they share the common property of having a hydrophilic and a lipophilic portion, causing them to orient themselves along the droplet-cuticle interface. Although the general function of all surfactants is the same, differences in their physical and chemical properties, and in their compatibility with herbicides and other formulation ingredients, dictate that their action in enhancing herbicide activity is case-specific. For example, there are reports of the activity of a particular herbicide being enhanced by some surfactants, unaffected by others, and decreased by others (although the first two results are most common) [56]. Unfortunately, there are no good explanations for many of the complex interactions that are encountered in this area of research.

The most obvious effects of surfactants are on droplet surface tension, leaf wetting, and contact angle (Figure 3.4). The contact angle is a function of the leaf surface chemistry and the degree of roughness [57]. By facilitating contact between the spray droplet and the leaf surface, the droplets spread and cover a greater area on the cuticle. Contact angle is reduced, making it more difficult to dislodge the droplet. We might expect that surfactant effects on herbicide pentration can be explained entirely on this basis, but experimental evidence suggests that additional factors are involved. In one study in which a wide range of surfactants was tested,

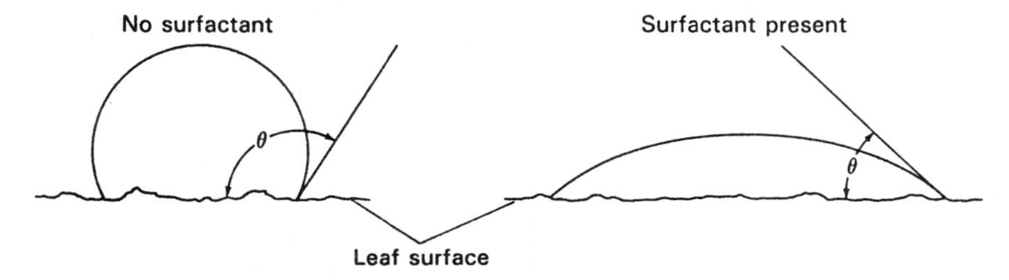

Figure 3.4 Surfactant effects on leaf surface wetting and contact angle. At very low surface tensions, contact angles become almost impossible to measure accurately. Note that decreased contact angle and increased contact area between the droplet and the leaf surface do not necessarily result in increased herbicide absorption.

no relationship was found between contact angle and glyphosate toxicity [56]; although absorption of glyphosate was not measured, it is likely that it was closely related to toxicity. Minimum surface tension and contact angles are reached at an optimum surfactant concentration, usually in the order of 0.1 to 0.5 percent (v/v) [58] (Figure 3.5). However, this does not necessarily coincide with the optimum surfactant concentration for herbicide activity [59]; higher concentrations, in the range of 1 to 5 percent, often provide maximum herbicide activity. Furthermore, deposit area is not clearly related to herbicide penetration (see Section 3.4.2). These results suggest that surfactant effects on herbicide absorption are not related simply to their surface-active properties.

What else do surfactants do? First, by virtue of their physicochemical properties, surfactants may partially solubilize the epicuticular wax on the leaf surface [16], and this may facilitate herbicide penetration into the CM. In addition, surfactant molecules do not necessarily remain on the leaf surface, but can partition into the CM and pass into the underlying tissue [60–62]. In doing so they may affect the chemical or physical properties of the CM, and exert some influence on the path taken by herbicides as they diffuse through the CM. Surfactants themselves are toxic to plant cells [63–65], and may contribute to the overall phytotoxicity of a herbicide treatment. In some instances, a surfactant may promote herbicide absorption, but decrease translocation out of the treated leaf (see Chapter 5). A final possible effect of surfactants is on the nature of the deposit that forms as the spray droplet dries. This is discussed in the following section.

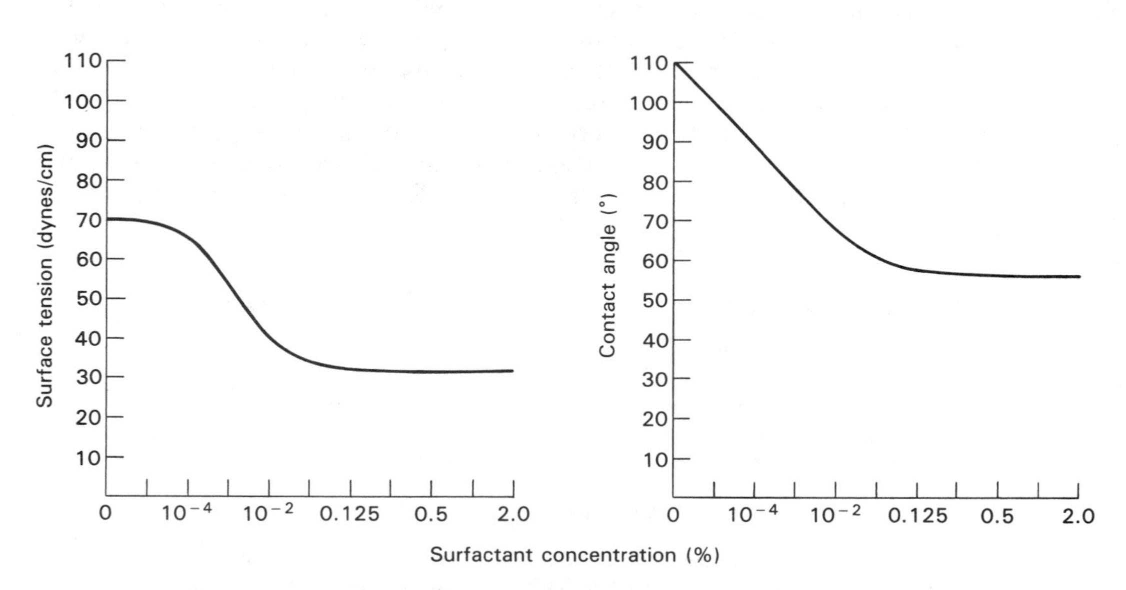

Figure 3.5 Surfactant concentration effects on surface tension and contact angle (mean of three surfactants). Data redrawn from [58].

3.5.2 Influence of Form of Deposit on Herbicide Absorption

When a spray droplet arrives on a leaf surface, the herbicide concentration is relatively low, but increases dramatically as the solvent evaporates. Based on the discussion of theoretical aspects (Section 3.4.1), we would expect this to influence partitioning into the CM, and there is evidence in support of this. For example, Hamilton et al. [66] showed that flamprop-methyl absorption into the CM does not begin until the spray droplet is almost dry. Presumably, when the herbicide is still in the application solution, the concentration (and the driving force) is too low to promote much absorption.

There is an optimum concentration for entry into the CM at which almost all of the original solvent has evaporated and the concentration of herbicide in solution is maximal. On one side of this, the herbicide concentration on the leaf surface is too low to promote much diffusion into the CM, while on the other, saturation is reached and the herbicide starts to precipitate out of solution, leaving a solid or quasi-solid deposit. The herbicide in this deposit may or may not be available for uptake into the leaf. Droplets often appear to be dry very soon after application [67], but it is likely that some bound water is retained in association within or below the herbicide deposit. Subsequent penetration of herbicide into the CM may be facilitated by this bound water (Figure 3.6).

The dried herbicide on the leaf surface may be in crystalline or amorphous form, depending on the other ingredients present and the conditions under which the deposit forms (Figure 3.7). Crystallization is an exothermic process. This implies that if a herbicide crystallizes on a leaf surface, energy (as well as solvent) is required to return the herbicide from the crystal lattice into solution. Although it is

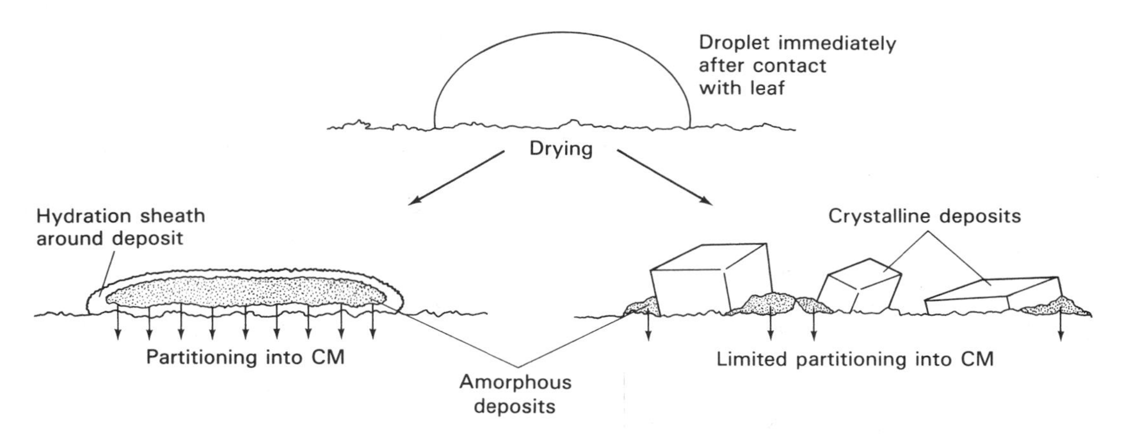

Figure 3.6 Schematic model of herbicide entry into the cuticle as a spray droplet dries, illustrating a possible role of deposit morphology in absorption.

Figure 3.7 Scanning electron micrographs of herbicide deposits on leaf surfaces. A. Clopyralid crystals on *Fagopyrum tataricum* (formulated in distilled water). B. Glyphosate crystals and amorphous deposits on *Fagopyrum tataricum* (formulated in 0.1% Tween 20). C and D. Propanil on sugar beet (*Beta vulgaris* L.) leaves, following the application of two emulsifiable concentrate formulations; the formulation that formed the more amorphous deposit (C) provided greater control of *Echinochloa crus-galli* ([68]). Photographs courtesy of S.A. MacIsaac (A, B) and F.D. Hess (C, D; from reference [68]); used with permission.

possible that a herbicide can become available again for absorption (e.g., upon rewetting), prevention of crystal formation may confer an advantage in terms of maintaining availability for absorption.

Some recent evidence lends support to this hypothesis. Working with propanil, Hess and co-workers [68] observed that different formulations dried to form quite different deposits, from amorphous to highly crystallized. The formulation that formed the amorphous deposit upon drying provided greater control of barnyard grass (*Echinochloa crus-galli*). Differential activity of flamprop formulations also has been related to the nature of the deposit formed upon drying [69]; this was attributed to greater foliar absorption of flamprop from the amorphous deposit than from the crystalline deposit. Recent research on formulation effects on glyphosate absorption has shown similar results [70]. However, the relationship between absorption and deposit form is not consistent; much more glyphosate was taken up from the isopropylamine formulation than from the acid, but the deposits formed by the two formulations were identical. In addition, ammonium sulfate enhanced glyphosate uptake, but the deposit formed by glyphosate plus ammonium sulfate was highly crystalline. Although crystals do form, some glyphosate must remain in a form suitable for absorption into the leaf.

Formulation components may play an important role in the retention of water in herbicide deposits. Some surfactants, by virtue of their hydrophilicity, act as humectants, and this may contribute to their action on herbicide absorption [71, 72]. Herbicide absorption is almost always enhanced at high relative humidity (see Section 3.6), and it seems plausible that this is partly due to the continued presence of water in or around the herbicide deposit. In one study, uptake of 2-deoxyglucose was positively correlated with the hygroscopicity of the surfactant used [62]. In one extreme situation, high relative humidity and the presence of a humectant in the formulation actually decreased amitrole absorption into bean (*Phaseolus vulgaris*) leaves [72]. These deposits remained wet, unlike those that formed under lower relative humidity, from which amitrole absorption was greater. In other words, maintaining the herbicide in a dilute solution on the leaf surface may be detrimental to absorption. Although there have been few reports on the effect of deposit characteristics on herbicide absorption and efficacy, the available evidence suggests that the physical characteristics of herbicide deposits may play an important role in herbicide absorption.

3.6 ENVIRONMENTAL EFFECTS ON HERBICIDE ABSORPTION

Field observations tell us that herbicide action is not consistent, but is related to environmental conditions. Most of the research that has been conducted to explain environmental influences on herbicide action has involved examination of herbicide absorption and translocation, mostly in plants grown in controlled-environment facilities. This allows the manipulation of one parameter (e.g., air temperature) while others are held constant. By conducting a series of experiments under different conditions, the infuence of various factors on herbicide penetration can be assessed. In addition, the chosen environmental factor(s) can be varied before, during, or after herbicide application, providing some insight into the possible mechanism of the environmental effect. The relevance of many such experiments has been questioned, however, since it is almost impossible to extrapolate the results obtained to field conditions [73]. Although there are many reports in the literature of the effects of air temperature, relative humidity (RH), soil moisture level, and irradiance level on herbicide absorption, little information is available on the mechanisms by which absorption is affected.

Environmental effects on herbicide penetration into leaves can be ascribed to effects on the spray droplet and the deposit formed (discussed in Section 3.5.2), or to effects on the plant. The latter includes effects on the thickness, composition, and physical structure of the CM, and effects on water potential and on the physiological status of the leaf tissue. Effects on the cuticle are developmental in nature, and would occur prior to herbicide application; effects on plant water potential may be induced very rapidly, occurring almost immediately after imposition of an environmental stress.

Since the primary function of the cuticle is to prevent water loss from the plant, one might expect cuticles that develop under high moisture deficit conditions to be more impermeable to water and polar solutes. In general, the available evidence on herbicide absorption supports this. Most reports on herbicide absorption in plants grown under varying RH or soil moisture levels confirm that absorption is greater when the atmospheric or soil moisture level is high [74–76]. Some sample data, from a study on NAA uptake into pea (*Pisum sativum*) leaves, are shown in Table 3.2.

Several possible mechanisms can be proposed for the enhancement of absorption at high RH. As discussed in Section 3.5.2, deposits forming under high RH conditions may retain more water, and this may maintain the herbicide in a form that is available for absorption. Alternatively, the hydrophilic pores in the CM (Section 3.2) may swell under high RH conditions, and this may facilitate the movement of herbicide molecules into, and through, the CM. A third possibility, for which there is little experimental support, is that stomatal penetration may be enhanced at high RH because of the effect of high RH on stomatal opening.

Although droplets dry more slowly under high RH conditions, this does not necessarily imply that enhanced absorption is due to the prolonged existence of liquid droplets on the leaf surface. As described previously for flamprop-methyl, penetration into the CM does not begin until the droplets appear to be dry [16], and the same may be true for other herbicides. Increased absorption following rewetting (e.g., [77]) may reflect rehydration of the deposit, rather than the conversion of the dried deposit to a true solution.

TABLE 3.2 INFLUENCE OF ENVIRONMENTAL PARAMETERS (AIR TEMPERATURE, RELATIVE HUMIDITY—RH, IRRADIANCE LEVEL, AND SOIL MOISTURE CONTENT—SMC) ON WAX CONTENT AND PERMEABILITY OF ADAXIAL AND ABAXIAL PEA (*Pisum sativum*) LEAF CUTICLES TO NAA [76].

				Wax content		Penetration	
Temp.	RH	Irradiance	SMC	Adaxial	Abaxial	Adaxial	Abaxial
(C)	(%)	($J \cdot m^{-2} s^{-1}$)	(%)	($\mu g \cdot cm^{-2}$)		($nmol \cdot cm^{-2}$)	
21	87	40	100	7	9	6.4	4.1
			40	10	16	3.3	2.1
21	70	80	100	10	11	2.2	1.7
			40	17	33	0.6	0.3
30	55	80	100	15	19	1.8	1.1
			40	21	27	1.1	0.5

Air temperature effects on herbicide absorption are less consistent than those of RH. Although many reports indicate an increase in foliar absorption at higher temperatures, others indicate no effect of temperature, and yet others, a decrease in absorption at higher temperature (e.g., [75,78,79]). Such plant species-herbicide-environment interactions make it very difficult to predict the effect of air temperature on absorption, or to explain these effects.

Increases in absorption at high temperature have been attributed to increased CM permeability, but little evidence has been offered in support of this. The CM is a partially lipid membrane and, like more typical lipid membranes, can undergo temperature dependent phase transitions, with changes in the orientation of both the cutin matrix and the soluble lipids [80]. Such phase transitions can alter the water permeability of the CM [81], and may influence herbicide penetration, also. Further research is required to determine whether the CM of common crop and weed species is significantly altered by the temperature fluctuations normally encountered under field conditions. Another possible explanation for increased absorption at higher temperature is increased metabolic activity in the plant, resulting in more rapid removal (by translocation or detoxication) of herbicide from the leaf tissue. Again, there is little experimental evidence to support this.

3.7 SUMMARY

In this chapter we have attempted to describe the pathway and mechanism(s) by which herbicides enter leaves. At present, accurate prediction or explanation of herbicide movement through the cuticle is limited primarily by a lack of knowledge of the physical and chemical properties of the cuticle. One reason for this is the tremendous heterogeneity of leaf cuticles. In addition, although the chemical constituents of herbicide sprays can be adequately described, the behavior of the various ingredients of these sprays on the leaf surface is poorly understood. Regardless, some herbicide does penetrate the cuticle, and becomes available for uptake into plant cells. The mechanisms of entry of herbicides into plant cells are discussed in Chapter 4.

REFERENCES

1. Martin, J. T., and B. E. Juniper. 1970. *The Cuticles of Plants*. London: Edward Arnold.
2. Price, C. E. 1982. "A review of the factors influencing the penetration of pesticides through plant leaves." In *The Plant Cuticle,* D. F. Cutler, K. L. Alvin, and C. E. Price, eds. London: Academic, pp. 237–52.
3. Kolattukudy, P. E. 1981. "Structure, biosynthesis, and biodegradation of cutin and suberin." *Annu. Rev. Plant Physiol.* 32, 539–67.

4. Kolattukudy, P. E. 1984. "Biochemistry and function of cutin and suberin." *Can. J. Bot.*, 62, 2918–33.
5. Holloway, P. J. 1982. "The chemical constitution of plant cutins." In *The Plant Cuticle,* D. F. Cutler, K. L. Alvin, and C. E. Price, eds. London: Academic, pp. 45–85.
6. Baker, E. A., M. J. Bukovac, and G. M. Hunt. 1982. "Composition of tomato fruit cuticles as related to fruit growth and development." In *The Plant Cuticle,* D. F. Cutler, K. L. Alvin, and C. E. Price, eds. London: Academic, pp. 33–44.
7. Baker, E. A. 1982. "Chemistry and morphology of plant epicuticular waxes." In *The Plant Cuticle,* D. F. Cutler, K. L. Alvin, and C. E. Price, eds. London: Academic, pp. 139–65.
8. Schönherr, J. 1976. "Water permeability of isolated cuticular membranes: the effect of pH and cations on diffusion, hydrodynamic permeability and size of polar pores in the cutin matrix." *Planta,* 128, 113–26.
9. Miller, R. H. 1985. "The prevalence of pores and canals in leaf cuticular membranes." *Ann. Bot.,* 55, 459–71.
10. Miller, R. H. 1986. "The prevalence of pores and canals in leaf cuticular membranes. 2. Supplemental studies." *Ann. Bot.,* 57, 419–34.
11. Hess, F. D. 1985. "Herbicide absorption and translocation and their relationship to plant tolerances and susceptibility." In *Weed Physiology. Vol II. Herbicide Physiology,* S. O. Duke, ed. Boca Raton, Fla.: CRC Press, pp. 191–214.
12. Schönherr, J., and M. J. Bukovac. 1973. "Ion exchange properties of isolated tomato fruit cuticular membrane: exchange capacity, nature of fixed charges and cation selectivity." *Planta,* 109, 73–93.
13. Schönherr, J., and R. Huber. 1977. "Plant cuticles are polyelectrolytes with isoelectric points around three." *Plant Physiol.,* 59, 145–50.
14. Devine, M. D., H. D. Bestman, J. C. Hall, and W. H. Vanden Born. 1984. "Leaf wash techniques for estimation of foliar absorption of herbicides." *Weed Sci.,* 32, 418–25.
15. Silcox, D., and P. J. Holloway. 1986. "A simple method for the removal and assessment of foliar deposits of agrochemicals using cellulose acetate film stripping." Aspects of Applied Biology 11. Biochemical and Physiological Techniques in Herbicide Research, Wellesbourne: Assoc. Applied Biologists. pp. 13–17.
16. Whitehouse, P., P. J. Holloway, and J. C. Caseley. 1982. "The epicuticular wax of wild oats in relation to foliar entry of the herbicides diclofop-methyl and difenzoquat." In *The Plant Cuticle*. D. F. Cutler, K. L. Alvin, and C. E. Price, eds. London: Academic, pp. 315–30.
17. Bukovac, M. J., P. D. Petracek, R. G. Fader, and R. D. Morse. 1990. "Sorption of organic compounds by plant cuticles." *Weed Sci.* 38, 289–98.
18. Orgell, W. H. 1955. "The isolation of plant cuticle with pectic enzymes." *Plant Physiol.,* 30, 78–80.
19. Schönherr, J., and M. Riederer. 1986. "Plant cuticles sorb lipophilic compounds during enzymatic isolation." *Plant Cell Env.,* 9, 459–66.
20. Kerler, F., M. Riederer, and J. Schönherr. 1984. "Non-electrolyte permeability of plant cuticles: a critical evaluation of experimental methods." *Physiol. Plant.,* 62, 599–606.

21. Jain, K. K. 1976. "Hydrogen peroxide and acetic acid for preparing epidermal peels from conifer leaves." *Stain Technol.,* 51, 202–04.

22. Chamel, A. 1986. "Foliar absorption of herbicides: study of the cuticular penetration using isolated cuticles." *Physiol. Vég.,* 24, 491–508.

23. Ong, B. Y., R. H. Falk, and D. E. Bayer. 1973. "Scanning electron microscope observations of herbicide dispersal using cathodoluminescence as the detection mode." *Plant Physiol.,* 51, 415–20.

24. Leuthold. U., C. Brucher, and E. Ebert. 1978. "The distribution of agrochemicals on leaf surfaces: a methodical study." *Weed Res.,* 18, 265–68.

25. Falk, R. H., F. D. Hess, and D. E. Bayer. 1975. "X-ray fluorescence analysis in weed science." *Weed Sci.,* 23, 373–77.

26. Baker, E. A., G. M. Hunt, and P. J. G. Stevens. 1983. "Studies of plant cuticle and spray droplet interactions: a fresh approach." *Pestic Sci.,* 14, 645–58.

27. Stevens, P. J. G., and E. A. Baker. 1987. "Factors affecting the foliar absorption and redistribution of pesticides. 1. Properties of leaf surfaces and their interactions with spray droplets." *Pestic. Sci.,* 19, 265–81.

28. Hess, F. D., and R. H. Falk. 1990. "Herbicide deposition on leaf surfaces." *Weed Sci.,* 38, 280–88.

29. Riederer, M., and J. Schönherr. 1984. "Accumulation and transport of (2,4-dichlorophenoxy)acetic acid in plant cuticles: 1. Sorption in the cuticular membrane and its components." *Ecotoxicol. Environ. Safety,* 8, 236–47.

30. Price, C. E., and N. H. Anderson. 1985. "Uptake of chemicals from foliar deposits: effects of plant species and molecular structure." *Pestic. Sci.,* 16, 369–77.

31. Davis, D. G., J. S. Mullins, G. E. Stolzenberg, and G. D. Booth. 1979. "Permeation of organic molecules of widely differing solubilities and of water through isolated cuticles of orange leaves." *Pestic. Sci.,* 10, 19–31.

32. Stevens, P. J. G., E. A. Baker, and N. H. Anderson. 1988. "Factors affecting the foliar absorption and redistribution of pesticides. 2. Physicochemical properties of the active ingredient and the role of surfactant." *Pestic. Sci.,* 24, 31–53.

33. Norris, R. F. 1974. "Penetration of 2,4-D in relation to cuticle thickness." *Amer. J. Bot.* 61, 74–79.

34. Cook, G. T., K. E. Carr, and H. J. Duncan. 1979. "The influence of morphological differences in bracken pinnules on the foliar uptake of aminotriazole." *Ann. App. Biol.,* 93, 311–17.

35. Becker, M., G. Kerstiens, and J. Schönherr. 1986. "Water permeability of plant cuticles: permeance, diffusion and partition coefficients." *Trees,* 1, 54–60.

36. King, M. G., and S. R. Radosevich. 1979. "Tanoak (*Lithocarpus densiflorus*) leaf surface characteristics and absorption of triclopyr." *Weed Sci.,* 27, 599–604.

37. Bukovac, M. J., J. A. Flore, and E. A. Baker. 1979. "Peach leaf surfaces: changes in wettability, retention, cuticular permeability, and epicuticular wax chemistry during expansion, with special reference to spray application." *J. Amer. Soc. Hort. Sci.,* 104, 611–17.

38. Holloway, P. J. 1982. "Structure and histochemistry of plant cuticular membranes: an overview." In *The Plant Cuticle,* D. F. Cutler, K. L. Alvin, and C. E. Price, eds. London: Academic, pp. 1–32.

39. Riederer, M., and J. Schönherr. 1985. "Accumulation and transport of (2,4-dichlorophenoxy)acetic acid in plant cuticles. II. Permeability of the cuticular membrane." *Ecotoxicol. Environ. Safety,* 9, 196–208.

40. Schönherr, J. 1976. "Water permeability of isolated cuticular membranes: the effect of cuticular waxes on diffusion of water." *Planta,* 131, 159–64.

41. Bukovac, M. J. 1970. "Movement of materials through the plant cuticle." *Proc. 18th Internat. Hort. Congr.,* 4, 21–42.

42. Wilson, L. A., and R. F. Norris. 1973. "Effect of temperature on penetration of 2,4-D through isolated leaf cuticles." *Plant Physiol.,* 51(Suppl.), 47.

43. Darlington, W. A., and J. B. Barry. 1965. "Effects of chloroform surfactants on permeability of apricot leaf cuticle." *J. Agr. Food Chem.,* 13, 76–78.

44. Norris, R. F., and M. J. Bukovac. 1969. "Some physical-kinetic considerations in penetration of naphthaleneacetic acid through isolated pear leaf cuticle." *Physiol. Plant.,* 22, 701–12.

45. Schönherr, J. 1976. "Naphthaleneacetic acid permeability of citrus leaf cuticle." *Biochem. Physiol. Pflanzen,* 170, 309–19.

46. Bukovac, M. J., J. A. Sargent, R. G. Powell, and M. J. Blackman. 1971. "Studies on foliar penetration. VIII. Effects of chlorination on the movement of phenoxyacetic and benzoic acids through cuticles isolated from the fruits of *Lycopersicon esculentum.*" *J. Exp. Bot.,* 22, 598–612.

47. Kirkwood, R. C., I. McKay, and R. Livingstone. 1982. "The use of model systems to study the cuticular-penetration of ^{14}C-MCPA and ^{14}C-MCPB." In *The Plant Cuticle,* D. F. Cutler, K. L. Alvin, and C. E. Price, eds. London: Academic, pp. 253–66.

48. Devine, M. D., H. D. Bestman, and W. H. Vanden Born. 1990. "Physiological basis for the different phloem mobilities of chlorsulfuron and clopyralid." *Weed Sci.,* 38, 1–9.

49. Bucholtz, D. L., and F. D. Hess. 1987. "A kinetics model for the absorption of 2,4-D acid and butoxyethanol ester into cabbage (*Brassica oleracea* L.) cotyledons." *Pestic. Biochem. Physiol.,* 28, 1–8.

50. Riederer, M., and J. Schönherr. 1986. "Covalent binding of chlorophenoxyacetic acids to plant cuticles." *Arch. Environ. Contam. Toxicol.,* 15, 97–195.

51. Shafer, W. E., and M. J. Bukovac. 1987. "Effect of acid treatment of plant cuticles on sorption of selected auxins." *Plant Physiol.,* 83, 652–56.

52. Schönherr, J., and M. J. Bukovac. 1972. "Penetration of stomata by liquids. Dependence on surface tension, wettability, and stomatal morphology." *Plant Physiol.,* 49, 813–19.

53. Gudin, C., W. J. Syratt, and L. Boize. 1976. "The mechanism of photosynthetic inhibition and the development of scorch in tomato plants treated with spray oils." *Ann. App. Biol.,* 84, 213–19.

54. McWhorter, C. G., and W. L. Barrentine. 1988. "Spread of paraffinic oil on leaf surfaces of johnsongrass (*Sorghum halepense*)." *Weed Sci.,* 36, 111–17.

55. Field, R. J., and N. G. Bishop. 1988. "Promotion of stomatal infiltration of glyphosate by an organosilicone surfactant reduces the critical rainfall period." *Pestic. Sci.*, 24, 55–62.

56. Wyrill III, J. B., and O. C. Burnside. 1977. "Glyphosate toxicity to common milkweed and hemp dogbane as influenced by surfactants." *Weed Sci.*, 25, 275–87.

57. Holloway, P. J. 1970. "Surface factors affecting the wetting of leaves." *Pestic. Sci.*, 1, 156–63.

58. Singh, M., J. R. Orsenigo, and D. O. Shah. 1984. "Surface tension and contact angle of herbicide solutions affected by surfactants." *JAOCS*, 61, 596–99.

59. Foy, C. L., and L. W. Smith. 1965. "Surface tension lowering, wettability of paraffin and corn leaf surfaces, and herbicidal enhancement of dalapon by seven surfactants." *Weeds*, 13, 15–19.

60. Anderson, N. H., and J. Girling. 1983. "The uptake of surfactants into wheat." *Pestic. Sci.*, 14, 399–404.

61. Sherrick, S. L., H. A. Holt, and F. D. Hess. 1986. "Absorption and translocation of MON 0818 adjuvant in field bindweed (*Convolvulus arvensis*)." *Weed Sci.*, 34, 817–23.

62. Stevens, P. J. G., and M. J. Bukovac. 1987. "Studies on octylphenoxy surfactants. Part 2: Effects on foliar uptake and translocation." *Pestic. Sci.*, 20, 37–52.

63. Towne, C. A., P. G. Bartels, and J. L. Hilton. 1978. "Interaction of surfactant and herbicide treatments on single cells of leaves." *Weed Sci.*, 26, 182–88.

64. Norris, R. F. 1982. "Action and fate of adjuvants in plants." In *Adjuvants for Herbicides*, pp. 68–83. Weed Science Society of America. Champaign, Il.

65. McWhorter, C. G. 1985. "The physiological effects of adjuvants on plants." In *Weed Physiology. Vol II. Herbicide Physiology*, S. O. Duke, ed. Boca Raton, Fl. CRC Press, pp. 141–58.

66. Hamilton, R. J., A. W. McCann, and P. A. Sewell. 1982. "Foliar uptake of the wild oat herbicide flamprop-methyl by wheat." In *The Plant Cuticle*, D. F. Cutler, K. L. Alvin, and C. E. Price, eds. London: Academic, pp. 303–13.

67. Hart, C. A., and B. W. Young. 1987. "Scanning electron microscopy and cathodoluminescence in the study of interactions between spray droplets and leaf surfaces." Aspects of Applied Biology 14. Studies of Pesticide Transfer and Performance. Wellesbourne: Assoc. Applied Biologists. pp. 127–40.

68. Hess, F. D., R. H. Falk, and D. E. Bayer. 1981. "Herbicide dispersal patterns: III. As a function of formulation." *Weed Sci.*, 29, 224–29.

69. Hess, F. D., J. R. Goss, D. L. Bucholtz, and R. H. Falk. 1987. "The physical form of flamprop-ethyl herbicide on sprayed leaves influences absorption and subsequent efficacy." In *Pesticide Science and Biotechnology*, R. Greenhalgh and T. R. Roberts, eds. London: Blackwell, pp. 209–14.

70. MacIsaac, S. A., R. N. Paul, and M. D. Devine. 1990. "A scanning electron microscope study of glyphosate deposits in relation to foliar uptake." *Pestic. Sci.* 31, 53–64.

71. Otsuji, K. 1986. "Effects of surfactants on the foliar-absorption of maleic hydrazide." *J. Pestic. Sci.*, 11, 387–92.

72. Cook, G. T., and H. J. Duncan. 1978. "Uptake of aminotriazole from humectant-surfactant combinations and the influence of humidity." *Pestic. Sci.*, 9, 535–44.

73. Devine, M. D. 1988. "Environmental influences on herbicide performance: a critical evaluation of experimental techniques." Proc. EWRS Symp., Factors Affecting Herbicidal Activity and Selectivity. Wageningen, Netherlands: Ponsen & Looijen, pp. 219–26.

74. Gottrup, O., P. A. O'Sullivan, R. J. Schraa, and W. H. Vanden Born. 1976. "Uptake, translocation, metabolism and selectivity of glyphosate in Canada thistle and leafy spurge." *Weed Res.,* 16, 197–201.

75. McWhorter, C. G. 1981. "Effect of temperature and relative humidity on translocation of ^{14}C-metriflufen in johnsongrass (*Sorghum halepense*) and soybean (*Glycine max*)." *Weed Sci.,* 29, 87–93.

76. Hunt, G. M., and E. A. Baker. 1982. "Developmental and environmental variations in plant epicuticular waxes: some effects on the penetration of naphthylacetic acid." In *The Plant Cuticle,* D. F. Cutler, K. L. Alvin, and C. E. Price, eds. London: Academic, pp. 279–92.

77. Babiker, A. G. T., and H. J. Duncan. 1975. "Penetration of bracken fronds by amitrole as influenced by pre-spraying conditions, surfactants and other additives." *Weed Res.,* 15, 123–27.

78. Devine, M. D., J. D. Bandeen, and B. D. McKersie. 1983. "Temperature effects on glyphosate absorption, translocation, and distribution in quackgrass (*Agropyron repens*)." *Weed Sci.,* 31, 461–64.

79. Jordan, T. N. 1977. "Effects of temperature and relative humidity on the toxicity of glyphosate to bermudagrass (*Cynodon dactylon*)." *Weed Sci.,* 25, 448–51.

80. Eckl, K., and H. Gruler. 1980. "Phase transitions in plant cuticles." *Planta,* 150, 102–13.

81. Schönherr, J., K. Eckl, and H. Gruler. 1979. "Water permeability of plant cuticles: the effect of temperature on diffusion of water." *Planta,* 147, 21–26.

Chapter 4

Herbicide Absorption by Roots, Isolated Tissues, and Plant Cells

4.1 INTRODUCTION

This chapter contains a general description of the process of herbicide absorption by roots in intact plants, followed by a more detailed discussion of the mechanisms of herbicide entry into plant cells. The rationale for combining root absorption of herbicides with absorption by isolated tissues and plant cells in general is that, from the available evidence, it appears that diffusion through the free space and movement through the cell wall and plasmalemma involve the same mechanisms, regardless of the tissue type. Root absorption in intact plants will be considered quite briefly, since most of the mechanistic information has come from research with excised tissues or isolated cells.

4.2 HERBICIDE ABSORPTION IN THE SOIL

Herbicide absorption from the soil can take place through any tissue that the herbicide solution contacts, be it root, seed, or shoot tissue. Most of the pertinent information concerns root absorption, but it is likely that it applies equally well to absorption by other plant parts in the soil. Consequently, most of this discussion is devoted to root absorption, with some brief comments on the relevance of shoot and seed absorption in the soil.

4.2.1 Root Absorption

The pathway of root entry of herbicides is quite different from that of foliage-applied herbicides. Root tissue lacks the cuticle of leaf tissue, although older root tissue is covered by a suberized layer. In addition, the root endodermis is lined with a suberized layer (the Casparian strip) which effectively separates the aqueous continuum of the epidermis and cortex from the vascular tissues in the stele [1]. Furthermore, some of the herbicide in soil is maintained in solution, and is therefore available for uptake on a continuing basis (assuming adequate soil moisture levels) whereas foliage-applied herbicides may be available for a considerably shorter period (see Chapters 2 and 3).

To avoid complications of herbicide availability in typical mineral soils, most of the research on root absorption of herbicides has been conducted in nutrient culture, using hydroponics or combinations of nutrient solution and an inert support medium (e.g., coarse silica sand). It should be recognized that while these systems allow more uniformity in herbicide distribution in the root zone, and more uniform availability between experimental units, both root growth and herbicide availability may be very different here than under more natural conditions. The mechanisms of absorption almost certainly remain the same, but the time course may change considerably.

Herbicide absorption can occur along the entire length of the root, but most evidence suggests that it is maximal in the apical portion (e.g., [2,3]). The Casparian strip is least developed in this region, and most water and ion entry into the root occurs here [4]. Since herbicide entry into the root is a result of the combination of mass flow of water in the soil and herbicide diffusion along concentration gradients, it is plausible that root tips constitute the major site of herbicide entry in roots.

The relevance of the Casparian strip in limiting absorption of xenobiotics by root tissue has been considered in several studies over the years. In one such study, it was shown that the fungicide MBC (methyl 2-benzimidazolecarbamate) can readily penetrate to the vascular tissue in intact onion roots, which have no secondary roots, and therefore no breaks in the endodermis [5]. Subsequent research with herbicides and a variety of other compounds has confirmed that the Casparian strip does not pose a significant barrier to the entry of most pesticides. Whereas a direct route of herbicide entry to the vascular tissues (without entering the endodermis) can exist where the endodermis has been broken by the emergence of a secondary root [5] or by mechanical damage, it is more likely that herbicides enter the endodermal cells directly as they diffuse toward to the stele. Strang and Rogers [3] showed that the endodermis was not a significant barrier to the entry of diuron into the xylem in cotton roots; it is likely that this is true for most, if not all, root-absorbed herbicides. In another study, trifluralin uptake into cotton roots was shown to be greater in roots that had been damaged by transplanting [6]. Damage to the epidermis may facilitate entry of trifluralin into the cortex; in

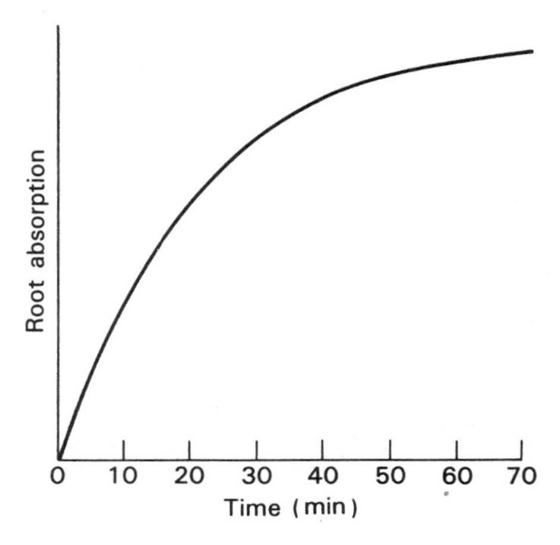

Figure 4.1 Typical time course of root absorption of herbicides.

undamaged cotton roots it is restricted more to the epidermis, where it is largely bound on the root surface.

Herbicide absorption by plant roots is typically characterized by rapid initial entry, over a period of 5 to 30 minutes, followed by a more prolonged period of slow entry (Figure 4.1) [7–11]. The initial phase is often independent of metabolic processes, and reflects permeation of the root tissue. In contrast, the latter phase often is associated with metabolic activity; the mechanisms involved in these two phases will be discussed in detail in subsequent sections.

Root absorption data are often expressed in terms of the root concentration factor (RCF), calculated as [12]:

$$\text{RCF} = \frac{\text{Herbicide concentration in the root tissue}}{\text{Herbicide concentration in the bathing solution}}$$

If RCF = 1.0, the herbicide concentration in the tissue is equal to that in the bathing solution. An RCF value < 1.0 indicates incomplete permeation of the tissue, whereas a value > 1.0 indicates accumulation in the tissue. The mechanisms by which herbicides accumulate in cells beyond the concentration in the external medium are discussed in Section 4.4.

In general, initial absorption rates and RCF are positively correlated with lipophilicity (log P or octanol/water partition coefficient) [12–14] (Figure 4.2). Polar compounds enter the root cells less rapidly, and are initially restricted to the free space, resulting in RCFs of 0.6–1.0 [14, 15]. Lipophilic compounds, on the other hand, enter the root cells rapidly, and can accumulate in lipid-rich domains in the tissue [14]. For weak acids such as 2,4-D and related herbicides, RCF increases as pH decreases (Figure 4.3) [16, 17]. The reasons for this are discussed in Section 4.3.4.

There are conflicting results concerning the role of transpiration in herbicide absorption by roots. In some studies (e.g., [7]), no relationship between transpira-

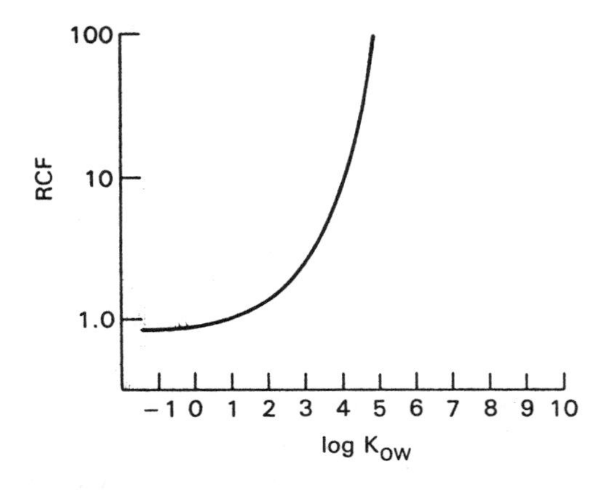

Figure 4.2 Influence of lipophilicity of neutral herbicides on root concentration factor (RCF) [14].

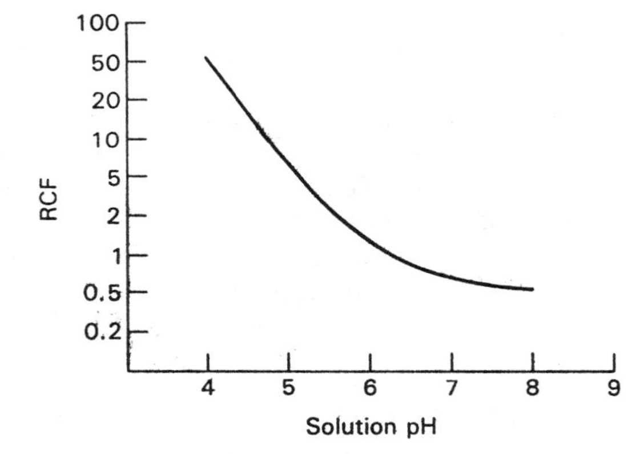

Figure 4.3 Influence of bathing solution pH on root concentration factor (RCF) for weak acid herbicides [17].

tion and the rate of herbicide absorption has been observed. Others, however, have shown herbicide uptake to increase with increased transpiration (see [18]). Briggs et al. [14] found no relationship between the amount of herbicide taken up by roots and that translocated in the xylem. Based on these results, it appears that root absorption of herbicides is relatively independent of water movement into and through the plant, but is related more to physicochemical factors that govern the partitioning of molecules into root tissue.

As might be expected, RCF is negatively correlated to the adsorption coefficient of compounds in the soil [13], indicating that increased adsorption on soil particle surfaces limits the amount of herbicide available for absorption by the plant. Obviously, most herbicides must be in solution in the soil to reach the root surface and enter the tissue. Volatile herbicides may constitute an exception to this, in that they can diffuse through the soil to the root surface in gaseous form. However, the root surface is usually hydrated, and it is likely that herbicides will dissolve in the sheath of water surrounding the root prior to entering the tissue. It is possible, however, that in very dry soil the herbicide could enter directly in the gaseous state [19].

4.2.2 Absorption by Other Tissues in Soil

Although soil-applied herbicides are usually thought of as being absorbed by plant roots, seeds and shoots within the soil can contribute significantly to herbicide uptake. Selective placement of herbicides in the shoot, seed, or root zones (separated by a thin layer of activated charcoal) has demonstrated that seed or shoot uptake can often contribute significantly to phytotoxicity [20–23]. Shoot tissue growing within the soil may be covered by a thin cuticle but, because of the moist environment within the soil, it is unlikely that this is as effective a barrier to herbicide entry as the leaf cuticle, which is exposed to the atmosphere. The relative effectiveness of herbicide placed in different zones in the soil may be related to the sensitivities of different tissues in young seedlings to the herbicide. Thus, if certain shoot tissues are more sensitive than other tissues within the soil, placement of the herbicide close to those shoot tissues may result in maximum phytotoxicity [24, 25]. In some instances, treatment with a root-growth inhibiting herbicide (e.g., trifluralin) has been shown to decrease soil uptake of a second herbicide [26]. While this may imply that the second herbicide is absorbed primarily by roots, it may also reflect a general phytotoxic effect of the first herbicide treatment on the plant.

4.3 HERBICIDE ABSORPTION INTO PLANT CELLS

Since most herbicide target sites are in the cytoplasm, nucleus, or chloroplast, almost all herbicides must enter into plant cells in order to be phytotoxic. At this level absorption involves crossing the cell wall and the cell membrane or plasmalemma. In this regard there appears to be little or no difference between cells from different tissues, and the processes involved are almost certainly the same, regardless of the origin of the tissue or cell. The relevance of the cell wall will be discussed briefly here, and membrane transport will be considered in more detail, reflecting the large amount of research that has been conducted on this aspect of herbicide action.

4.3.1 Methods of Studying Herbicide Absorption into Plant Cells

Herbicide absorption by plant cells is commonly studied using excised tissues or isolated cells or protoplasts. The use of small pieces of tissue immersed in a buffered solution containing herbicide creates a simple experimental system in which various factors (e.g., pH, herbicide concentration) can easily be manipulated (e.g., [27, 28]). The choice of tissue does not appear to be critical, and leaf, stem, root, and tuber tissues have all been used in such experiments. For example, uptake of amitrole has been shown to be equal in bean root and leaf tissue [29]. Similarly, uptake of herbicides into *Fagopyrum tataricum* leaf, stem, and root tissue was shown to be qualitatively similar, with differences in uptake rate related to ease of penetration of the tissue [30]. Root and tuber tissues provide the advantage of more rapid entry, usually because they lack a cuticle on the surface (tuber discs are cut from cores taken through the tuber, and do not include the outer surface). Leaf discs and stem sections can be used effectively, provided the cut edges are clean enough to allow the entry of solutes into the tissue.

After the tissue is prepared, it is normally added to small vessels containing a buffered solution (to maintain cell osmolarity) to which the herbicide is then added. Absorption can be measured directly in the tissue at various time intervals or, more commonly, by removing a small aliquot of the solution and measuring the herbicide content. Herbicide absorption into the tissue is then calculated based on its removal from the bathing solution.

Retention of herbicide in the tissue following a period of absorption can be measured by replacing the herbicide-containing buffer with fresh, herbicide-free buffer, and monitoring efflux of herbicide into the fresh buffer [27, 28]. This can be done on a continuous basis after a single transfer, or by repeated transfers to fresh buffer. Pools of herbicide in the free space, cytoplasm, and vacuole efflux from the tissue at different rates, and efflux profiles can be examined to assess the compartmentalization of herbicide within the tissue [15, 31]. However, these results can be confounded by the slow release of weakly bound herbicide, or by differential efflux of a herbicide and its metabolite(s) from within the tissue.

Similar experiments can be conducted using isolated plant cells or protoplasts [32–34]. Herbicide absorption is measured after addition of herbicide to a suspension of the cells or protoplasts in small vials or tubes. Since absorption occurs very quickly in these experiments, it is usually terminated by high-speed centrifugation of the cells through an oil layer, or by rapid filtration, to separate the cells from the bathing solution. The use of protoplasts removes possible effects of the cell wall on herbicide absorption into the tissue. Since fresh protoplasts start to regenerate new cell walls quickly after preparation, they must be prepared freshly and used soon after isolation.

4.3.2 Movement through the Cell Wall

The cell wall is a highly ordered arrangement of cellulose microfibrils that gives cells and, collectively, whole plants, the mechanical strength that results in their three-dimensional form. Cellulose is the primary component of cell walls; other components include pectins (free and esterified galacturonic acid polymers), xylans, arabans, lignin, suberin, phenolic acids, and some proteins. The cell wall can be regarded, therefore, as a very polar and relatively porous medium surrounding the cell. In addition to its structural role, the cell wall is involved in water and mineral absorption, disease resistance, and in certain extra-protoplastic enzyme activities.

In general, the cell wall appears to offer little resistance to the passage of herbicides into cells. However, there are several reports of herbicides being bound to a constituent of cell walls. The herbicide 2,6-dichlorobenzonitrile, a specific inhibitor of cellulose biosynthesis (see Chapter 15), binds to a protein in cotton cellulose fibers [35]. Presumably, this binding results in reduced entry of the herbicide into the cell proper. There have also been suggestions that paraquat is bound by a cell-wall constituent in paraquat-tolerant lines of *Conyza canadensis*. This binding may prevent phytotoxic quantities of paraquat from reaching the thylakoid membranes in the chloroplast, the site of action of this herbicide (see Chapter 8). Finally, adsorption of trifluralin to components of cell walls has been reported [6]. The component(s) to which the herbicide was bound have not been identified, nor has the significance of this binding in herbicidal activity been determined. Since trifluralin is known to act as a mitotic inhibitor (see Chapter 10), it is unlikely that binding to a cell wall component would contribute to its mechanism of action. However, this could contribute to tolerance in some species, if binding occurred to such an extent that it limited access of the herbicide to the microtubules in the cytoplasm.

4.3.3 Movement through the Plasmalemma

The plasmalemma, which serves as the major barrier between living cells and the external environment, is a highly lipophilic lipid bilayer, very different in structure and properties from the cell wall. The best representation of the cell membrane is perhaps given by Singer's model of a fluid mosaic membrane [36]. In a typical cell membrane, two layers of phospholipids lie against each other, with the hydrocarbon "tails" directed inward, and the hydrophilic "heads" on the outside. Associated with the lipid membrane are proteins that lie on the inner or outer surface, or that span the entire membrane. Within the cytoplasm, an environment is maintained in which the biochemical processes necessary for normal cell function can occur. Much of the internal regulation can be attributed to the action of these membrane-bound proteins (e.g., ATPases, carrier proteins, pumps, and channels). In addition to pH and charge regulation within the cell, these proteins are involved in the transfer of many endogenous solutes across the plasmalemma, tonoplast, chloroplast, and other subcellular membranes. Herbicides can move through the plasmalemma either by simple diffusion or by "active" transport on a carrier protein. The available evidence suggests that the former mechanism is by far the more important for transport of most herbicides.

4.3.4 Mechanisms of Membrane Transport of Herbicides

In an early investigation of uptake into isolated cells, Collander [37] found a highly positive correlation between the permeability of a wide range of exogenous organic compounds into isolated *Nitella* protoplasts and the ether/water partition coefficient of the compounds (comparable to the octanol/water partition coefficient, or log K_{ow} in Figure 4.2). In other words, lipophilic molecules permeate cells more rapidly than hydrophilic molecules. Although Collander did not include any herbicides in his study, more recent results indicate a similar relationship with herbicides [12, 14, 38]. Some of these results were discussed in the section on root absorption of herbicides (Section 4.2.1). Uptake of neutral, lipophilic compounds depends largely on their ability to partition into cell membranes, and is independent of solution pH [12, 39, 40]. Very rapid penetration of isolated corn protoplasts by atrazine [32], and equally rapid uptake of atrazine in live and dead barley roots [41], suggest that the plasmalemma does not pose a significant barrier to atrazine entry into plant cells; the same almost certainly holds true for other compounds with similar physicochemical properties. The lack of involvement of energy-dependent processes in the uptake of neutral, lipophilic molecules suggests that their transport across the plasmalemma occurs by simple diffusion. The possible mechanisms by which herbicides can cross the plasmalemma are illustrated in Figure 4.4.

In contrast to neutral molecules, uptake of weak acids is strongly related to solution pH, with greater uptake at lower pH; a listing of weak acid herbicides for which pH-dependent uptake has been demonstrated is shown in Table 4.1, and a typical uptake-pH response is shown in Figure 4.5. Weak acid herbicides can exist in either the dissociated (anionic) form or the undissociated (neutral) form. The ratio of dissociated to undissociated molecules is governed by the solution pH and the pK_a of the weak acid. These are related by the Henderson-Hasselbach equation:

$$pH = pK_a + \log \frac{[A]}{[HA]}$$

where [A] and [HA] are the concentrations of the dissociated and undissociated species, respectively. At low pH, more herbicide molecules will be in the undissociated form; since the membrane permeability of an undissociated molecule is

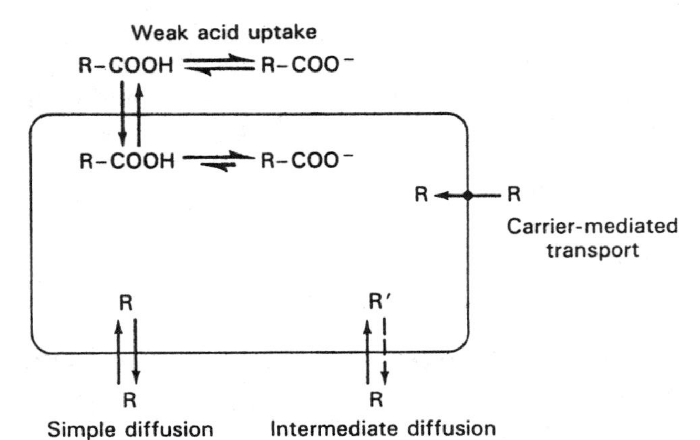

Figure 4.4 Possible mechanisms of herbicide absorption into plant cells.

TABLE 4.1 LIST OF WEAK ACID HERBICIDES THAT SHOW pH-DEPENDENT UPTAKE AND ACCUMULATION IN PLANT TISSUES.

Herbicide	Plant tissue	Reference
2,4-D	Potato tuber discs	27
	Crown gall cells	42
	Isolated corn protoplasts	32
MCPA	*Cyclamen persicum* petioles	43
Picloram	Potato tuber discs	44
	Detached *Psidium cattleianum* leaves	45
Clopyralid	Pea root tips	28
Chlorsulfuron	Pea root tips	28
Metsulfuron-methyl	Corn roots	46
Imazaquin	Soybean leaf discs	47
Imazapyr	Soybean leaf discs	47
Bentazon	*Abutilon theophrasti* cells	48

usually much greater than that of its dissociated ion [17], movement across the plasmalemma is enhanced at low pH.

There have been suggestions in the literature, based on various lines of evidence, that herbicide absorption is an active, or energy-requiring, process. The evidence includes a Q_{10} for absorption greater than 2 [10], inhibition by azide or other metabolic inhibitors [12], or saturable uptake at higher herbicide concentrations [37]. Some authors have concluded, based on this evidence, that herbicide absorption is carrier-mediated. In most instances this conclusion, based on the supposition that the energy requirement is related directly to an energy-requiring transporter for the herbicide, is based on insufficient evidence, and is probably erroneous. Energy is required, however, for the plasmalemma ATPase that is coupled to proton extrusion, which in turn maintains cytoplasmic pH at the desired

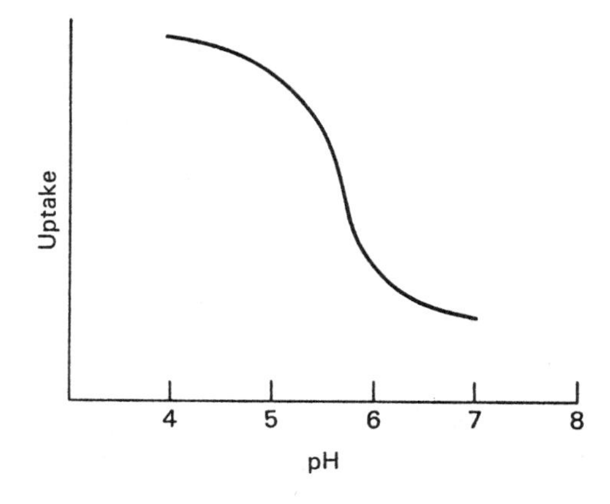

Figure 4.5 Influence of pH on uptake of weak acid herbicides into plant cells. The position of the curve (i.e., left or right of where drawn) will depend on the pK_a of the herbicide.

level. Any interference with normal cell metabolism, whether it be by uncouplers, ionophores, and the like, can affect the pH gradient across the plasmalemma, with corresponding effects on the uptake of weak acid herbicides [43, 47, 48]. Other evidence that endogenous carriers are unlikely to be involved in herbicide transport is that most of these carriers exhibit high substrate specificity, and do not transport, or only poorly transport, structurally related molecules.

There are one or two possible exceptions to the previous statement that carriers are not important in the uptake of herbicides into cells. Kinetic evidence suggests that 2,4-D uptake is saturable (i.e., it is concentration-dependent), implying that a carrier is involved in its transport into plant cells [49]. It may be logical to assume that 2,4-D would be transported on an IAA carrier, since the two molecules are known to compete for binding sites on IAA-binding proteins in cells (see Chapter 14). There is also evidence that glyphosate may be transported on a phosphate carrier in plant [50] and bacterial [51] cells. These results do not indicate the extent of carrier-mediated transport, but it is likely that both these herbicides also enter cells by simple diffusion. Despite the above evidence, diffusion may be the predominant transport mechanism for these herbicides.

4.4 ACCUMULATION AND RETENTION IN CELLS

When plant tissues or cells are exposed to a neutral, lipophilic herbicide, the herbicide quickly permeates the tissue so that the concentration inside and outside the tissue are equal. Following this, more herbicide may enter the tissue or cells, apparently against a concentration gradient, resulting in a slight accumulation in the cells. In the case of lipophilic molecules, the accumulation is usually attributed to partitioning into lipid fractions in the tissue [52, 53]. Weak acid herbicides exhibit similar behavior, but the extent of accumulation can be much higher, with internal concentrations reaching tenfold or more than those in the bathing solution [27, 28].

The mechanism of accumulation of weak acid molecules in the cytoplasm is fundamentally different. Typically, the pH in the apoplast is approximately 5.5, whereas the pH in the cytoplasm is closer to 7.5. Since dissociation of weak acid molecules increases at higher pH, the molecules will tend to ionize in the cytoplasm. Consequently, they are less likely to move out through the plasmalemma, and will be retained in the cytoplasm (Figure 4.4). Accumulation of weak acids in the cytoplasm can be predicted by the following equation [54].

$$\frac{[HA]_i + [A]_i}{[HA]_o + [A]_o} = \frac{(1 + 10^{pHi-pKa})[P_{HA}/P_A + \{(FE/RT)/(1 - e^{-FE/RT})\}.10^{pHo-pKa}]}{(1 + 10^{pHo-pKa})[P_{HA}/P_A + \{(FE/RT)/(1 - e^{-FE/RT})\}.10^{pHo-pKa}.e^{-FE/RT}]}$$

where [HA] and [A] are the concentrations of the undissociated and dissociated species, the subscripts "o" and "i" refer to outside and inside the cell membrane, respectively, P_{HA} and P_A are the permeabilities of the undissociated and dissociated species, F is the Faraday constant, E is the charge on the membrane, T is the temperature (°K), and R is the universal gas constant. For most weak acids, the permeability of the undissociated species is in the order of 10^4 times greater than that of the corresponding anion [17]. As a result, the anion can accumulate significantly in the cytoplasm. This may effectively increase the potency of a compound, by allowing it to accumulate in the vicinity of the site of action. In addition, accumulation in the cytoplasm has major implications for subsequent translocation of the herbicide; this is discussed in more detail in Section 5.4.

There are several other mechanisms by which herbicides can accumulate in plant cells. Binding in the tissue effectively removes the herbicide from the soluble pool, resulting in apparent concentration of the herbicide in the tissue. Many herbicides have been shown to exhibit such nonspecific binding, not related to their mechanism of action [55–57]. Usually this binding is reversible, and the herbicide can be removed from the tissue by transfer to fresh (herbicide-free) solution. A further mechanism that contributes to apparent accumulation of herbicide in tissue is metabolic conversion to more polar products. This is discussed in Section 5.4.4, and estimates of the change in physicochemical properties resulting from some common metabolic conversions are shown in Table 5.2. Reactions such as glycosylation, which result in large decreases in the log K_{ow} of a molecule, decrease membrane permeability of the molecule substantially.

Finally, some polar molecules that do not behave as weak acids (e.g., glyphosate and amitrole) enter the cytoplasm only very slowly, compared to other herbicide molecules [34, 39, 58]. Their uptake is not influenced by pH [29, 34], and they do not accumulate in tissue. However, following entry into plant tissues, they efflux only very slowly. Glyphosate is a zwitterion, and undergoes a change in polarity and net charge upon entering the cytoplasm [59], which may explain its retention in the symplast. The relevance of the "intermediate diffusion" characteristics of these herbicides in relation to phloem translocation is discussed in Chapter 5.

REFERENCES

1. Esau, K. 1977. *Anatomy of Seed Plants*, 2nd ed. New York: Wiley, 550 pp.
2. Jacobson, A., and R. H. Shimabukuro. 1982. "The absorption and translocation of diclofop-methyl and amitrole in wheat and oat roots." *Physiol. Plant.*, 54, 34–40.
3. Strang, R. H., and R. L. Rogers. 1971. "A microautoradiographic study of ^{14}C-diuron absorption by cotton." *Weed Sci.*, 19, 355–62.
4. Tanton, T. W., and S. H. Crowdy. 1972. "Water pathways in higher plants. II. Water pathways in roots." *J. Exp. Bot.*, 23, 600–18.
5. Peterson, C. A., and L. V. Edgington. 1975. "Uptake of the systemic fungicide methyl-2-benzimidazolecarbamate and the fluorescent dye PTS by onion roots." *Phytopathology*, 65, 1254–59.

6. Strang, R. H., and R. L. Rogers. 1971. "A microautoradiographic study of ^{14}C-trifluralin absorption." *Weed Sci.,* 19, 363–69.

7. Moody, K., C. A. Kust, and K. P. Buchholtz. 1970. "Uptake of herbicides by soybean roots in culture solutions." *Weed Sci.,* 18, 642–47.

8. Barrett, M., and F. M. Ashton. 1981. "Napropamide uptake, transport, and metabolism in corn (*Zea mays*) and tomato (*Lycopersicon esculentum*)." *Weed Sci.,* 29, 697–703.

9. Carlson, W. C., E. M. Lignowski, and H. J. Hopen. 1975. "Uptake, translocation, and adsorption of pronamide." *Weed Sci.,* 23, 148–54.

10. Morrison, I. N., and W. H. Vanden Born. 1975. "Uptake of picloram by roots of alfalfa and barley." *Can. J. Bot.,* 53, 1774–85.

11. Balke, N. E., and T. P. Price. 1988. "Relationship of lipophilicity to influx and efflux of triazine herbicides in oat roots." *Pestic. Biochem. Physiol.,* 30, 228–37.

12. Shone, M. G. T., and A. V. Wood. 1974. "A comparison of the uptake and translocation of some organic herbicides and a systemic fungicide by barley. I. Absorption in relation to physicochemical properties." *J. Exp. Bot.,* 25, 390–400.

13. Topp, E., I. Scheunert, A. Attar, and F. Korte. 1986. "Factors affecting the uptake of ^{14}C-labeled organic chemicals by plants from soil." *Ecotoxicol. Environ. Safety,* 11, 219–28.

14. Briggs, G. G., R. H. Bromilow, and A. A. Evans. 1982. "Relationships between lipophilicity and root uptake and translocation of non-ionised chemicals by barley." *Pestic. Sci.,* 13, 495–504.

15. Shone, M. G. T., B. O. Bartlett, and A. V. Wood. 1974. "A comparison of the uptake and translocation of some organic herbicides and a systemic fungicide by barley. II. Relationship between uptake by roots and translocation to shoots." *J. Exp. Bot.,* 25, 401–09.

16. Zsoldos, F., and E. Haunold. 1979. "Effects of pH changes on ion and 2,4-D uptake of wheat roots." *Physiol. Plant.,* 47, 77–80.

17. Briggs, G. G., R. L. O. Rigitano, and R. H. Bromilow. 1987. "Physico-chemical factors affecting uptake by roots and translocation to shoots of weak acids in barley." *Pestic. Sci.,* 19, 101–12.

18. Bukovac, M. J. 1976. "Herbicide entry into plants." In *Herbicides. Physiology, Biochemistry, Ecology,* L. J. Audus, ed. London: Academic, pp. 335–64.

19. Swann, C. W., and R. Behrens. 1972. "Phytotoxicity of trifluralin vapors from soil." *Weed Sci.,* 20, 143–46.

20. Knake, E. L., A. P. Appleby, and W. R. Furtick. 1967. "Soil incorporation and site of uptake of preemergence herbicides." *Weed Sci.,* 15, 228–232.

21. Addala, M. S. A., R. J. Hance, and D.S. Drennan. 1985. "Studies on the site of uptake by plants of chlortoluron and terbutryne." *Weed Res.,* 25, 151–58.

22. Phillips, R. E., D. B. Egli, and L. Thompson, Jr. 1972. "Absorption of herbicides by soybean seeds and their influence on emergence and seedling growth." *Weed Sci.,* 20, 506–10.

23. Rubin, B., and Y. Demeter. 1986. "Dipropetryn absorption during germination by cucurbit seeds and its influence on seedling growth." *Weed Res.,* 26, 333–40.

24. Prendeville, G.N., L. R. Oliver, and M. M. Schreiber. 1968. "Species differences in site of shoot uptake and tolerance to EPTC." *Weed Sci.,* 16, 538–40.

25. Nishimoto, R. K., and G. F. Warren. 1971. "Shoot zone uptake and translocation of soil-applied herbicides." *Weed Sci.,* 19, 156–61.
26. O'Donovan, J. T., and G. N. Prendeville. 1975. "Shoot zone uptake of soil-applied herbicides in some legume species." *Weed Res.,* 15, 413–17.
27. Peterson, C. A., and L. V. Edgington. 1976. "Entry of pesticides into the plant symplast as measured by their loss from an ambient solution." *Pestic. Sci.,* 7, 483–91.
28. Devine, M. D., H. D. Bestman, and W. H. Vanden Born. 1987. "Uptake and accumulation of the herbicides chlorsulfuron and clopyralid in excised pea root tissue." *Plant Physiol.,* 85, 82–86.
29. Lichtner, F. T. 1983. "Amitrole absorption by bean (*Phaseolus vulgaris* L. cv 'Red Kidney') roots". Mechanism of absorption. *Plant Physiol.,* 71, 307–12.
30. Devine, M. D., H. D. Bestman, and W. H. Vanden Born. 1988. "Selection of appropriate tissue for examining herbicide uptake, accumulation, and efflux." *Abstracts, Weed Sci. Soc. Amer.,* 28 (251).
31. Darmstadt, G. L., N. E. Balke, and T. P. Price. 1984. "Triazine absorption by excised corn root tissue and isolated corn root protoplasts." *Pestic. Biochem. Physiol.,* 21, 10–21.
32. Darmstadt, G. L., N. E. Balke, and T. P. Price. 1983. "Use of corn root protoplasts in herbicide absorption studies." *Pestic. Biochem. Physiol.,* 19, 172–83.
33. Burton, J. D., and N. E. Balke. 1988. "Glyphosate uptake by suspension-cultured potato (*Solanum tuberosum* and *S. brevidens*) cells." *Weed Sci.,* 36, 146–53.
34. Singer, S. R., and C. N. McDaniel. 1982. "Transport of the herbicide 3-amino-1,2,4-triazole by cultured tobacco cells and leaf protoplasts." *Plant Physiol.,* 69, 1382–86.
35. Delmer, D. P., S. M. Read, and G. Cooper. 1987. "Identification of a receptor protein in cotton fibers for the herbicide 2,6-dichlorobenzonitrile." *Plant Physiol.,* 84, 415–20.
36. Singer, S. J., and G. L. Nicolson. 1972. "The fluid mosaic model of the structure of cell membranes." *Science,* 175, 720–31.
37. Collander, R. 1954. "The permeability of Nitella cells to non-electrolytes." *Physiol. Plant.,* 7, 420–45.
38. Balke, N. E., and T. P. Price. 1988. "Relationship of lipophilicity to influx and efflux of triazine herbicides in oat roots." *Pestic. Biochem. Physiol.,* 30, 228–37.
39. El Ibaoui, H., S. Delrot, J. Besson, and J.-L. Bonnemain. 1986. "Uptake and release of a phloem-mobile (glyphosate) and of a non-phloem-mobile (iprodione) xenobiotic by broadbean leaf tissues." *Physiol. Vég.,* 24, 431–42.
40. Grimm, E., S. Neumann, and F. Jacob. 1986. "Transport of xenobiotics in higher plants. III. Absorption of 2,4-D and 2,4-dichloroanisole by isolated conducting tissue of *Cyclamen.*" *Biochem. Physiol. Pflanzen,* 181, 69–79.
41. Zhirmunskaya, N. M., and S. S. Kol'tsova. 1973. "Investigation of permeability of barley root cells in relation to the herbicide atrazine." *Soviet Plant Physiol.,* 20, 123–27.
42. Rubery, P. H. 1978. "Hydrogen ion dependence of carrier-mediated auxin uptake by suspension-cultured crown gall cells." *Planta,* 142, 203–06.
43. Grimm, E., S. Neumann, and F. Jacob. 1983. "Uptake of sucrose and xenobiotics into conducting tissue of *Cyclamen.*" *Biochem. Physiol. Planzen,* 178, 29–42.
44. Swanson, C. R., and J. R. Baur. 1969. "Absorption and penetration of picloram in potato tuber discs." *Weed Sci.,* 17, 311–14.

45. Wilson, B. J., and R. K. Nishimoto. 1975. "Ammonium sulfate enhancement of picloram absorption by detached leaves." *Weed Sci.*, 23, 297–301.
46. Milhomme, H., and J. Bastide. 1990. "Uptake and phytotoxicity of the herbicide metsulfuron methyl in corn root tissue in the presence of the safener 1,8-naphthalic anhydride." *Plant Physiol.*, 93, 730–38.
47. Van Ellis, M. R., and D. L. Shaner. 1988. "Mechanism of cellular absorption of imidazolinones in soybean (*Glycine max*) leaf discs." *Pestic. Sci.*, 23, 25–34.
48. Sterling, T. M., N. E. Balke, and D. S. Silverman. 1990. "Uptake and accumulation of the herbicide bentazon by cultured plant cells." *Plant Physiol.*, 92, 1121–27.
49. Minocha, S. C., and P. Nissen. 1985. "Uptake of 2,4-dichlorophenoxyacetic acid and indoleacetic acid in tuber slices of Jerusalem artichoke and potato." *J. Plant Physiol.*, 120, 351–62.
50. Burton, J. D., and N. E. Balke. 1986. "Evidence for carrier-mediated transport of glyphosate into suspension cultured potato cells." 6th Int. Congr. Pestic. Chem. (IUPAC), No. 3D-07.
51. Pipke, R., A. Schulz, and N. Amrhein. 1987. "Uptake of glyphosate by an *Arthrobacter* sp." *Appl. Environ. Microbiol.*, 53, 974–78.
52. Price, T. P., and N. E. Balke. 1983. "Characterization of atrazine accumulation by excised velvetleaf (*Abutilon theophrasti*) roots." *Weed Sci.*, 31, 14–19.
53. Price, T. P., and N. E. Balke. 1983. "Comparison of atrazine absorption by underground tissues of several plant species." *Weed Sci.*, 31, 482–87.
54. Raven, J. A. 1975. "Transport of indoleacetic acid in plant cells in relation to pH and electrical potential gradients, and its significance for polar IAA transport." *New Phytol.*, 74, 163–72.
55. Grimm, E., S. Neumann, and B. Krug. 1987. "Transport of xenobiotics in higher plants. II. Ambimobility of the acidic compounds bromoxynil and pentachlorophenol." *Biochem. Physiol. Pflanzen*, 182, 323–32.
56. Upadhyaya, M. K., and L. D. Noodén. 1987. "Comparison of [^{14}C]oryzalin uptake in root segments of a sensitive and a resistant species." *Ann. Bot.*, 59, 483–85.
57. Barrett, M., and F. M. Ashton. 1983. "Napropamide binding in corn (*Zea mays*) root tissue." *Weed Sci.*, 31, 712–19.
58. Gougler, J. A., and D. R. Geiger. 1981. "Uptake and distribution of *N*-phosphonomethylglycine in sugar beet plants." *Plant Physiol.*, 68, 668–72.
59. Wauchope, D. 1976. "Acid dissociation constants of arsenic acid, methylarsonic acid (MAA), dimethylarsinic acid (cacodylic acid), and *N*-(phosphonomethyl)glycine (glyphosate)." *J. Agric. Food Chem.*, 24, 717–21.

Chapter 5

Herbicide Translocation

5.1 INTRODUCTION

Herbicide action involves the movement of herbicide molecules from the point of entry into the plant to a site of action. In some instances the site of action may be within one or two cell layers from the point of entry. In the case of bipyridilium herbicides, for example, herbicidal action is expressed in the first chlorophyllous cells that the herbicides enter, and long-distance transport is not required for activity (see Section 8.5). For other herbicides, however, transport over longer distances is an important, if not necessary, aspect of their activity.

The most obvious situations in which herbicide translocation over long distances is important are those of postemergence herbicides applied to perennial weeds, in which movement from the shoot to the roots (or rhizomes) is required, and soil-applied photosynthetic inhibitors, which must be transported from the roots to the foliage for their activity to be expressed. Transport over shorter distances can also be important, for example from leaves to apical meristems, or from mature leaves to young leaves.

Translocation over short distances involves diffusion from cell to cell, either via plasmodesmata (the direct cytoplasmic connections between cells) or in the free space. There is almost no information available on the movement of herbicides over such short distances. This can be attributed perhaps to the difficulty of studying such movement, and also to the fact that it is unlikely that movement over short distances poses a significant barrier to herbicide distribution in leaf or root tissue. It is possible that lipophilic molecules partition into the plasmalemma, and diffuse over short distances within the membrane in a manner similar to plastoquinone in the thylakoid membrane (see Chapter 7), whereas more hydrophilic molecules move in the more aqueous medium inside or outside the cell. Regardless, based on the evidence discussed in Chapter 4, it appears that neither the cell wall nor the plasmalemma represents a significant barrier to herbicide movement.

Long-distance transport, on the other hand, has been studied extensively, and much more information is available on the influence of plant and herbicide factors on long-distance translocation of herbicides. In this chapter we will consider briefly the movement of endogenous solutes in the phloem and xylem, and then examine the physicochemical and physiological aspects of herbicide behavior that influence translocation.

Phloem and xylem translocation of herbicides often are discussed as two separate subjects, and herbicides have, in the past, been categorized in broad groups as "phloem-mobile," "xylem-mobile," "ambimobile," or "immobile." However, such groupings are arbitrary, and do not necessarily reflect the true translocation characteristics of herbicides. Some herbicides move much more in the phloem than in the xylem, and vice versa, but all herbicides are capable of movement in both transport systems. Some of the reasons for this will become clear later in this chapter.

The approach to describing herbicide translocation in this chapter is perhaps different from that normally taken in discussions of this subject. Consider the question: why do all herbicides not move in exactly the same manner as sucrose, that is, consistent with our understanding of phloem transport of assimilates? (Alternatively, we could ask why all herbicides do not move exclusively in the xylem, with the transpirational flow of water through the plant.) In attempting to answer the first question, we gain some understanding of the control of herbicide translocation in plants.

Herbicide translocation is inextricably linked to the ability of herbicides to cross the plasmalemma, and to the physicochemical properties of the herbicide molecule that influence its absorption into isolated cells or tissues. It is essential, therefore, that the reader has a thorough understanding of the material covered in Chapter 4 before continuing with this discussion of long-distance transport of herbicides in plants.

5.2 METHODS OF STUDYING HERBICIDE TRANSLOCATION

Herbicide translocation is most conveniently studied using radiolabeled (almost always ^{14}C) herbicide. Usually, small amounts of the labeled herbicide are supplied either to nutrient solution in which the roots are placed, or to one or more individual leaves. Translocation is then monitored by sectioning the plant into different parts at various time intervals after herbicide application, and determining the amounts of radiolabel in those parts. The use of ^{14}C-labeled herbicides has largely replaced the more tedious extraction, cleanup, and separation procedures involved when nonradiolabeled herbicides are used, and allows the processing of many more samples per unit time. However, steps still have to be taken to identify the products being measured in different plant parts, since this procedure does not discriminate between the original material applied and any herbicide metabolites that may be formed. If it is found that the product recovered in some site distal from the site of application is not the original product applied, further work is necessary to determine whether the metabolic transformation occurred in the tissue which the herbicide first entered, or at the site of recovery.

Other important considerations in planning and carrying out experiments on herbicide translocation include the manner in which the herbicide is applied, and the dose applied. Supplying herbicide to plant roots in a small volume of nutrient solution is a convenient method of getting a large amount of uptake in a relatively short time, but is probably not an accurate reflection of root uptake under "natural" conditions. The water status of plants growing under hydroponic conditions is quite different from that in soil-grown plants, and xylem translocation can differ greatly between the two systems. Furthermore, in such experimental systems the entire root often is exposed to herbicide, whereas only a small portion of the root system of soil-grown plants may have access to the herbicide.

Translocation of foliage-applied herbicides is most commonly examined by applying a small volume (e.g., 5 or 10 μL) of herbicide solution to a single leaf, and monitoring the distribution of herbicide at various time intervals. Such experiments often are conducted to explain field observations such as different sensitivities of different plant species, and apparent environmental effects on herbicide action. However, the experiments are often conducted using different herbicide formulations, different application techniques, and on plants grown under such different conditions that extrapolation of results to the field becomes questionable, if not invalid [1]. The use of a commercial formulation blank (i.e., the commercial formulation minus the active ingredient) and of a suitable spray atomizer [2] helps to keep the application procedures consistent with those used in the field. In some instances, the use of radiolabeled herbicide is accompanied by a topical application (usually of the commercially formulated herbicide) to provide overall coverage of the plant. This treatment mimics more closely a "real" herbicide application, in that potential physiological effects are expressed throughout the plant, rather than only at the site at which the radiolabeled herbicide is applied.

Herbicide dose is also important, in that localized toxicity can influence herbicide translocation (see Section 5.5). The dose of herbicide applied should be equal to that applied to a similar area in a conventionally treated plant. This means that for radiolabeled herbicides with high specific activity, some unlabeled herbicide should be added to increase the total herbicide concentration to an appropriate level. For radiolabeled herbicides with low specific activity, care must be taken to ensure that the amount applied is not unrealistically high, that is, that the dose (in

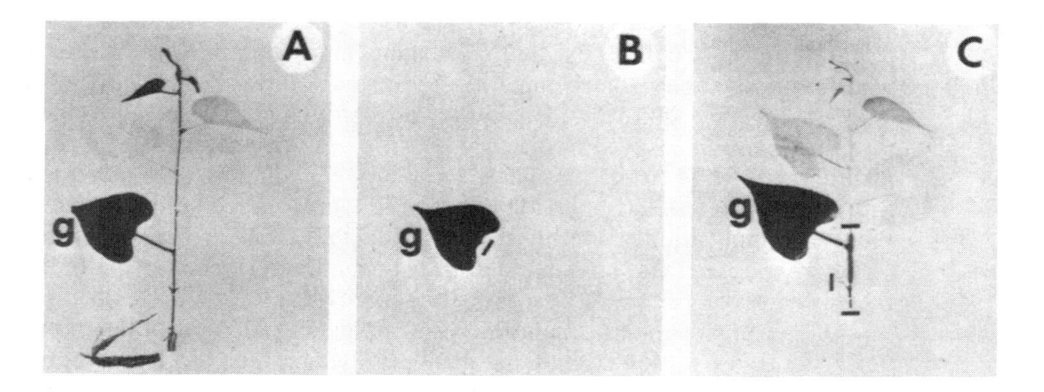

Figure 5.1 Autoradiographs of ^{14}C-glyphosate in *Ipomoea purpurea*. A. Plant treated at leaf *g*. B. Plant treated at leaf *g* but petiole of treated leaf steam-girdled. C. Plant treated at leaf *g* and steam-girdled on stem above and below treated leaf [4].

μg cm^{-2}) is not significantly greater than would be applied in a conventional application. It is particularly important to consider this when using herbicides with intrinsically high levels of activity (e.g., sulfonylureas).

Although experiments using the techniques described above can provide useful information on a gross scale, some of the subtleties of herbicide translocation may be missed. For example, small amounts of localized diffusion away from the site of application may not be detected. Autoradiography, while not ideal for quantitative purposes, can provide more information on the localization of herbicide within plant tissues [3]. Distribution patterns can be visualized quite

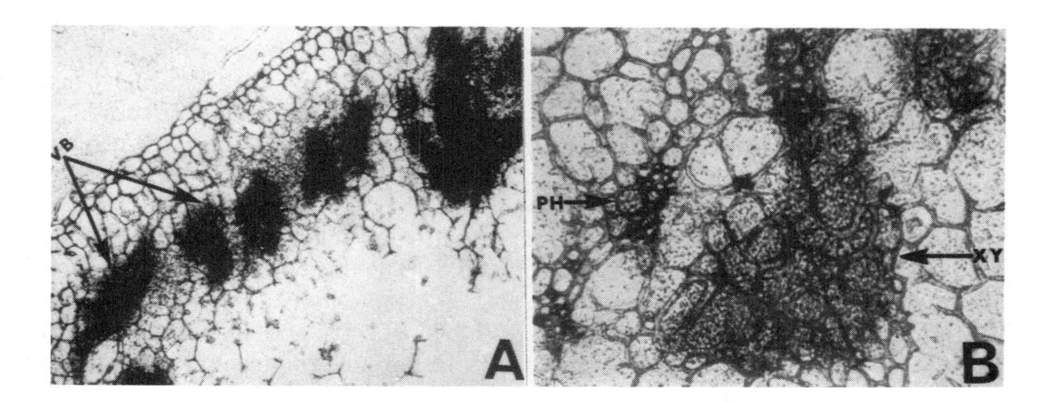

Figure 5.2 Microautoradiographs of soybean stem sections following root application of ^{14}C-picloram. VB = vascular bundle, PH = phloem, XY = xylem [5].

readily, and more information can be gained on the localization of herbicides in different tissues [4] (Figure 5.1). Taking this one step further, microautoradiography can provide a picture of herbicide localization at a cellular or subcellular level [5]. Examples of microautoradiographs of plant tissue treated with ^{14}C-labeled picloram are shown in Figure 5.2. It should be remembered that this technique provides information on the localization of ^{14}C (or radioisotopes), but that the identity of the ^{14}C material (i.e., parent herbicide or metabolite) must be determined by other means.

One complication in studying herbicide translocation, particularly following foliar absorption, is that the results can be confounded by differential herbicide absorption into the leaf or retention in the cuticle. To avoid this problem, Bromilow and co-workers have used a microsyringe to inject small volumes of herbicide into the hollow petioles of *Ricinus communis* leaves [6] (Figure 5.3). The herbicide solution is placed in close proximity to the vascular bundles, and the herbicide has relatively easy access to both phloem and xylem conducting tissue. Arrival of herbicide at different points in the stem below the treated petiole is monitored by collecting phloem exudate through capillary tubes inserted into the stem. The relative herbicide concentrations in upper and lower stem exudates provide a measure of herbicide leakage out of the phloem [6].

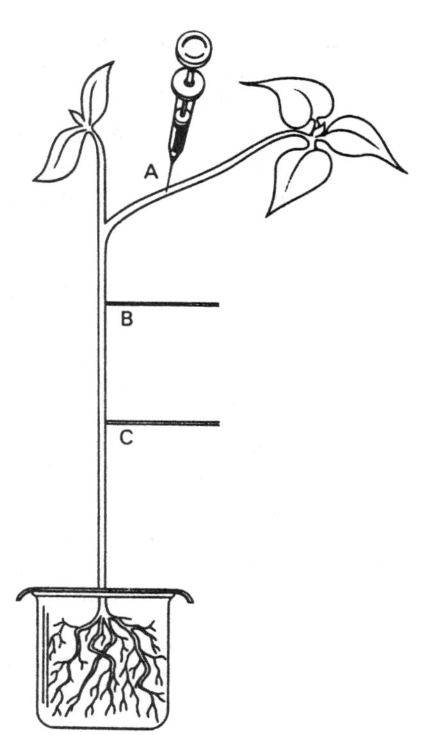

Figure 5.3 Diagram of *Ricinus communis* system used to determine phloem mobility of herbicides and other organic compounds. Herbicide is injected into the hollow petiole at A and phloem sap is collected at B and C. Redrawn from [6].

Finally, translocation experiments can be conducted using excised plant tissues, usually leaves. The herbicide is applied to the leaf, and exudation of herbicide from the cut leaf base or petiole is monitored. This provides a fairly simple experimental system in which translocation can be examined without the "complicating" effects of the remainder of the plant. One of the problems in using excised tissues is that the phloem tissue at the cut end is prone to plugging, due to callose formation at the sieve plate pores. This problem can be overcome by adding a chelating agent (e.g., EDTA) to the nutrient solution in which the leaf base is placed [7]. However, EDTA can cause some phytotoxicity itself, and plant species appear to differ in the concentration of EDTA required to promote exudation without phytotoxic side effects. Consequently, the appropriate concentration of EDTA has to be determined for each species [8].

In some circumstances, movement in the phloem or xylem can be distinguished by steam girdling part of the plant, usually the stem or the petiole of the treated leaf. Since girdling kills the living tissue, translocation through a girdle can occur only in the xylem. The effects of girdling treatments on glyphosate translocation can be seen in Figure 5.1.

5.3 TRANSLOCATION IN THE PHLOEM AND XYLEM

For the purposes of discussing long-distance transport, the plant can be divided into two parts: the symplasm, or continuum of living cells, and the apoplasm, or continuum of nonliving cells (xylem) and the free space. Within the symplasm and apoplasm are well-defined tissues in which long-distance transport occurs. Before considering herbicide movement in the phloem and xylem, the mechanism of translocation of endogenous solutes in the two tissues will be reviewed briefly.

5.3.1 Phloem Translocation

Translocation of solutes (primarily sugars, but also amino acids, inorganic ions, and other compounds) occurs in the phloem, according to osmotic pressure gradients established between "source" and "sink" tissues. The major components of phloem tissue are sieve elements, typically long, narrow cells with open pores at either end through which solutes can pass unimpeded, and companion cells, which are involved in the loading and unloading of sugar into the sieve elements [9]. In addition, specialized parenchyma cells act as storage tissue for organic solutes prior to entry into the phloem proper. Although the sieve elements are highly modified cells that appear to be quite "open," they do contain a functional cytoplasm and exhibit some metabolic activities that are essential to their normal functioning.

Direction of flow in the phloem is dependent on its location relative to carbohydrate sources or sinks. Source tissues, as previously mentioned, are those in which an excess of carbohydrate is available, and some of that excess is made available to sink tissues, in which carbohydrate is utilized for growth, metabolism, or storage. Typical source tissues include mature leaves, and mature thickened roots, rhizomes, and tubers. Sink tissues include young leaves (prior to the stage at which they can produce sufficient carbohydrate for their growth and respiratory needs), inflorescences, and developing seeds, fruits, roots, rhizomes, and so on. Most plant tissues are dynamic, in that they undergo transitions from sink to source (and sometimes back again) depending on their stage of development and the physiological status of other plant parts. Typical source → sink flow is indicated by the "open" lines in Figure 5.4.

The mechanism of phloem transport has been the subject of extensive study for over 50 years, and although much of the process is well understood, there remain some aspects that have yet to be fully explained. Transport sugars—usually, but not always, sucrose—are synthesized in mesophyll cells. From there they follow one of two pathways to the conducting elements in the phloem. The first pathway involves efflux into the apoplasm, followed by loading into the sieve elements or companion cells. It has been shown recently that phloem loading of sucrose is mediated by a plasmalemma-bound carrier protein that is highly specific for sucrose [10, 11]. In some species, however, there is an alternate pathway for sucrose entry into the phloem. Sucrose can be loaded directly into the phloem via plasmodesmal connections between mesophyll cells and the phloem [12]. Even within these continuous connections some selective (unknown) mechanism operates that allows sucrose to pass but prevents many other solutes from reaching the phloem.

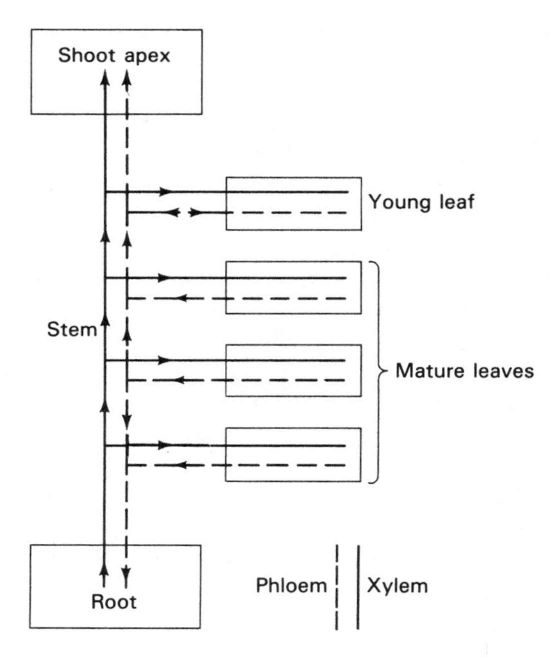

Figure 5.4 Typical distribution patterns in phloem and xylem in a simple annual plant. Phloem translocation is in the direction of carbohydrate sources → carbohydrate sinks, and is maximal near tissues of highest photosynthetic capacity. Xylem translocation is greatest at the base of the shoot, and decreases progressively up through the shoot.

Regardless of the mechanism of sucrose entry into the phloem, it is generally accepted that long-distance flow occurs according to the osmotic pressure flow hypothesis first proposed by Münch [13]. Sucrose is concentrated in the phloem conducting elements by an active loading process or by direct transfer from mesophyll cells; the high osmotic potential of this concentrated solution causes a simultaneous influx of water, and the resultant turgor pressure forces flow toward tissues of low turgor pressure. The latter tissues are sinks, in which sucrose is removed from the system and used for synthesis, energy metabolism, or storage.

5.3.2 Herbicide Translocation in the Phloem

It is generally accepted that herbicide translocation in the phloem occurs by the "passive" movement of herbicides in the direction of solute flow in the phloem. However, there is no clear understanding of the path taken by herbicide molecules to reach the phloem. In plant species in which sucrose effluxes from the mesophyll cells prior to loading into the phloem, it is conceivable that herbicide molecules could enter the phloem from the free space without ever having entered the cytoplasm of leaf cells. However, herbicides that are translocated primarily in the phloem can accumulate in other cells, and it is unlikely that they would fail to enter the epidermal or mesophyll cells of leaves. The herbicide molecules also can efflux from these cells (Chapter 4), and may cycle in and out of leaf cells until they enter a phloem sieve element and are carried out of the leaf [14] (Figure 5.5).

The presence of an apoplasmic step in the transfer from mesophyll cells to the phloem may have some implications for herbicide exchange between phloem and xylem. Sucrose entry into the phloem is assured by the sucrose carrier located on the sieve element (or companion cell) plasmalemma, but no such carrier exists for herbicides. Therefore, herbicide entry into the phloem may be a more random event. There is no information on the relevance of direct plasmodesmal connections between mesophyll and phloem on herbicide translocation; questions of whether or not herbicide molecules can pass through these connections or, if not, by which route they reach the phloem, remain to be addressed.

Observations on the distribution of herbicides in plants following a foliar application have lead to conclusions that some herbicides are "phloem-mobile." That is, their movement appears to parallel that of assimilate. However, only in very few instances have there been detailed comparisons of herbicide and assimilate translocation in the same species under identical conditions, and the results of these studies invariably show some differences in the distribution of herbicide and assimilate. For example, glyphosate is frequently described as being phloem-mobile, but close examination of its movement reveals small but significant quantities moving apoplasmically (i.e., in the xylem) [4]. The movement of 2,4-D in young wheat plants appears to follow that of assimilates closely, but in older plants the distribution patterns of the two differ [15]. Research results with 2,4-D and 2,4,5-T [16], MCPA [17, 18], dalapon [19], and glyphosate [20, 21] in other plant species indicate a close—but not complete—correlation between herbicide and assimilate translocation.

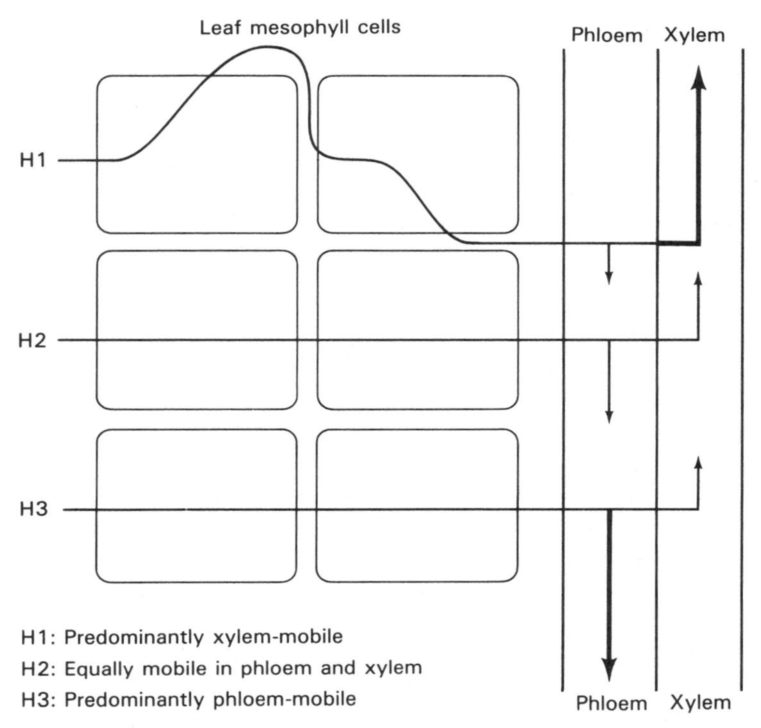

Figure 5.5 Illustration of herbicide movement from cell to cell, and into the phloem and/or xylem.

Although all of these herbicides can be transported in the phloem, small quantities appear to exit from the phloem, enter the xylem, and be distributed with the flow of water in the xylem. Hence, glyphosate has been observed to move through stem girdles in the xylem, following translocation out of a leaf in the phloem [4] (Figure 5.1). Similarly, amitrole and maleic hydrazide have been shown to move quite readily in both phloem and xylem [22, 23]. Picloram and dicamba are translocated primarily in the phloem, but both herbicides can also be translocated in the xylem [24, 25]. This exchange between phloem and xylem is based primarily on physicochemical properties of the herbicides, and is discussed in detail in Section 5.4.

Typical source-sink distribution results in a herbicide being translocated from treated leaves into young leaves and tillers, developing root and shoot buds, and root and rhizome tips [26–28]. Some herbicide that is translocated into roots can be exuded into the rooting medium [29]. Such distribution patterns are influenced by the age of the plant and the location of the treated leaf [30], the location in which the herbicide is applied to the leaf [31], and the conditions under which the plant is growing (see Section 5.6.2).

Distribution of herbicides in a pattern identical to that of assimilate is never achieved, nor is it desirable. If herbicides were to be distributed in a pattern identical to that of assimilate, some tissues would accumulate relatively large amounts, whereas others would receive little or no herbicide; the former tissues would be killed, while the latter would escape the effects of the herbicide, resulting in incomplete control. For example, the uneven distribution of glyphosate and sethoxydim in the rhizome system of *Agropyron repens* (=*Elymus* or *Elytrigia repens*) can result in death of buds toward the rhizome apex, but survival of buds closer to the base of the treated shoots [32, 33]. A pattern of distribution similar to that of assimilate, but with some leakage into other tissues and exchange and redistribution in the xylem, may result in more even herbicide distribution and more complete control.

5.3.3 Xylem Translocation

Xylem, the major water-conducting tissue in higher plants, is usually closely associated with the phloem in the vascular tissue. However, xylem elements do not contain a functional cytoplasm, and can be considered a specialized component of the free space. Xylem elements function as open conduits for the movement of water, inorganic ions, amino acids, and other solutes from tissues of high water potential (e.g., roots) to tissues of lower water potential (e.g., leaves, Figure 5.4). In this way they differ significantly from phloem tissue; the direction of flow in the xylem remains the same throughout the life of the tissue, whereas that in the phloem changes with ontogeny.

As stated above, flow in the xylem is dependent on water potential gradients within the plant. In general, water potential is highest in the soil solution, and decreases in the progression from soil to root to stem to leaves, with lowest levels in the atmosphere. The driving force for water movement, therefore, is the water potential difference between the soil and the air, and water moves into and through the plant as a result of this potential difference. Environmental factors such as soil moisture status and relative humidity that affect the rate of water transport in the xylem exert corresponding effects on the movement of solutes, including herbicides. However, the plant does not function entirely as an open system, wicking water from the soil into the atmosphere. Water loss from the leaves (transpiration) occurs mainly through the stomata, which close under water deficit conditions to maintain plant water status and prevent dehydration of the tissue. Thus, a high water potential difference between soil and air may not give rise to a high rate of translocation in the xylem, or to high amounts of herbicide translocation from root to shoot. However, xylem translocation of herbicides does correlate well in some instances with the total amount of water transpired by the plant [34].

5.3.4 Herbicide Translocation in the Xylem

The entry of herbicides into root tissue was covered in Chapter 4. From the outer surface of the root, herbicide molecules diffuse through the cortex and endodermis into the stele, where the vascular tissue is located. With the exception of passage through the Casparian strip (the suberized layer surrounding the endodermis), movement can occur either in the free space or through the interior of the root cells. However, because diffusion through the free space at the endodermis is prevented by the Casparian strip, herbicides must enter the cytoplasm of the endodermal cells in order to gain access to the vascular tissue. Hence, some of the early suggestions that certain herbicides moved predominantly in the xylem because they were excluded from the living tissue prove to be unfounded. Indeed, most, if not all, of the herbicides in this category (e.g., *s*-triazines and substituted ureas) *must* enter the cytoplasm (and the chloroplast) in leaf tissue to exert their phytotoxic action. This and other evidence discussed in Chapter 4 indicate that entry into cells is not a limiting factor in the movement and action of these herbicides.

In general, distribution of herbicides in the xylem often reflects the movement of water through the plant. Herbicide accumulates in metabolically active leaves (both source and sink leaves in terms of assimilate transport). Within a leaf, herbicide tends to accumulate at the leaf tip and along the leaf margin, ''end points'' of much of the water flow in the leaf.

5.4 PHYSICOCHEMICAL ASPECTS OF HERBICIDE TRANSLOCATION

Herbicide translocation is a function of the entry and retention of herbicide molecules in the conducting elements for sufficiently long that they can move away from the site of entry into the tissue to other tissues in the plant. This retention must be based on factors that maintain the herbicide in an available pool, not on binding or compartmentalization in the tissue. Evidence suggests that herbicides that move in the phloem do so because they are able to enter the phloem, and are retained in the sieve elements while long-distance transport occurs. Translocation in the xylem, on the other hand, occurs because the herbicides are not retained in the symplasm, but freely efflux into the free space and, in particular, into the xylem [35] (Figure 5.5).

The same physicochemical principles that apply to herbicide entry in plant cells in general also apply to entry, retention, and accumulation in the phloem. Weakly acidic herbicides and those that exhibit intermediate permeability are retained because of the low permeability of the absorbed species in the plasmalemma (see Chapter 4). Consequently, a high proportion of the retained herbicide molecules can be translocated in the phloem. It should be noted that these herbicide molecules are not retained selectively in the phloem, but can be retained in the cytoplasm of all living cells. However, retention in the phloem cells results in their translocation. Herbicides that are not retained in plant cells are more likely to be translocated in the xylem.

Because herbicides differ in their physicochemical properties, the degree to which they are retained in the sieve elements varies. Lipophilic molecules may enter the phloem cells more rapidly than more polar molecules, but are also more likely to efflux from the phloem soon after entry. In contrast, other molecules may penetrate the phloem elements more slowly, but are more likely to be retained and translocated in the phloem. These contrasting patterns of uptake and efflux explain the low phloem mobility of neutral, lipophilic herbicides (e.g., most substituted ureas), and the relatively high phloem mobility of weakly acidic herbicides (e.g., clopyralid) [27] and those that form zwitterions (e.g., glyphosate) [21].

In the Introduction to this chapter, the question "why are all herbicides not phloem-mobile?" was posed. As will be shown in the following discussion, identification of the requirements for phloem mobility partially identifies the requirements for xylem mobility. However, the discussion of this topic is divided into two parts, dealing first with phloem translocation and second with xylem translocation. The treatment of these two topics is not equal, but reflects the fact that much more research has focused (especially recently) on physicochemical aspects of phloem translocation of herbicides. It is important to remember, however, that phloem mobility and xylem mobility are not mutually exclusive, and that many compounds are translocated in both systems.

5.4.1 Phloem Translocation

Much of our knowledge of the influence of physicochemical parameters on herbicide translocation comes from studies of the mobility of structurally related series of molecules (e.g., [36–38]). An example is shown in Table 5.1, in which phloem translocation of a series of ω-naphthoxyalkanoic acids is considered [39]. Although

TABLE 5.1 RELATIVE PHLOEM MOBILITY OF A SERIES OF SUBSTITUTED ω-NAPHTHOXYALKANOIC ACIDS. Phloem mobility is given in arbitrary values, relative to ^{3}H-sucrose. Log K_{ow} is the logarithm of the 1-octanol/water partition coefficient [39].

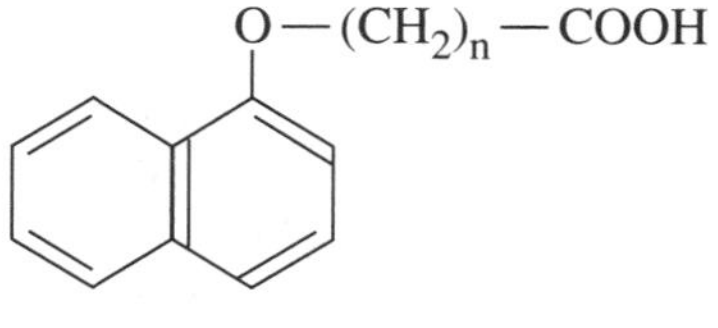

$O-(CH_2)_n-COOH$

n	pK_a	log K_{ow}	Relative phloem mobility
1	3.2	2.6	7.8
2	4.0	3.0	5.6
3	4.4	3.5	2.6
4	4.6	4.0	0.3
5	4.7	4.5	<0.1

this series represents a rather limited combination of pK_a and log K_{ow}, the results clearly indicate that physicochemical parameters influence phloem mobility.

In Figure 5.6, phloem translocation of a series of substituted phenoxyacetic acids is plotted against log K_{ow} [40]. Within this series, the optimum log K_{ow} is slightly greater than 2; below this, uptake into the cells is limited, presumably, and above this, efflux from the phloem increases.

When a herbicide enters and starts to move in the phloem, leakage into the xylem or surrounding tissues can occur at any time, depending on the permeability of the absorbed species in the sieve element membrane, the rate of flow in the phloem, and the size of the plant (i.e., the length of the translocation system). Research with organic chemicals with widely varying physicochemical properties has led to the development of models predicting the optimum permeability for phloem transport (P, equation 1, [41]), or the concentration of herbicide in the phloem relative to that in the leaf apoplasm (C_f, equation 2, [42]). The optimum permeability of a nonionized compound is given by:

$$P = \frac{rV}{2l} \times ln\left(1 - \frac{l}{0.9L}\right)$$

where r is the radius of the sieve elements, V is the average translocation velocity, l is the length of the source tissue along which the herbicide is being loaded, and L is the length of the translocation system of the plant [41].

The second model includes the permeability of weak acids, in addition to nonacidic herbicides, thereby predicting the accumulation of any organic chemicals in the phloem. This equation has the form:

$$C_f = \{([H^+]_i + pK_a)/([H^+]_o + pK_a)\} \times [(a)([H^+]_o P_{AH} + P_{A^-}\, pK_a)/ [H^+]_i(P_{AH} + b) + pK_a(P_{A^-} + b)] \times \exp\{-c([H^+]_i P_{AH} + P_{A^-}\, pK_a)/([H^+]_i + pK_a)\}$$

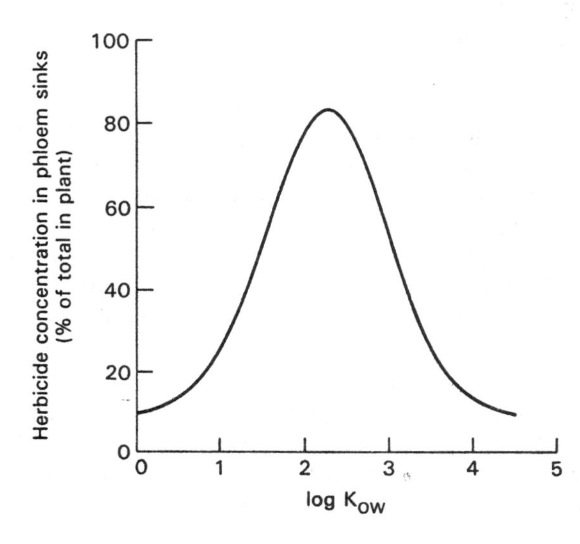

Figure 5.6 Influence of log K_{ow} on the phloem translocation of a series of substituted phenoxyacetic acids. Redrawn from [40].

This equation relates phloem mobility (C_f) to the pH differential across the plasmalemma ($[H^+]_i$, $[H^+]_o$), the permeabilities of the undissociated and dissociated forms of the herbicide through the plasmalemma (P_{HA}, P_{A^-}), the pK_a (in the case of a weak-acid herbicide), and a series of plant parameters that describe the herbicide application zone, sieve tube radius, phloem sap velocity, and the size of the plant (*a*, *b*, and *c*) [42]. Again, it is seen that the rate of flow in the phloem and the length of the translocation system are important parameters in determining phloem mobility. This means that the ''ideal'' phloem-mobile herbicide is not only a physicochemical concept, but is also a function of the size and physiological status of the plant.

The latter model reflects the fact that the two retention mechanisms, weak acid dissociation and intermediate permeability, can confer phloem mobility on herbicide molecules. Phloem mobility can be viewed as a balance between the lipophilicities of the dissociated and nondissociated forms of weak acids (or the corresponding forms of non-weak-acid herbicides) in the phloem conducting elements. This model and the work of Bromilow and co-workers [40] suggest that there is no optimum pK_a or log K_{ow} for herbicide translocation in the phloem, but that the optimum pK_a depends on the log K_{ow} of the herbicide, and vice versa. The relationship between log K_{ow}, pK_a, and phloem mobility (or C_f) can be plotted as a three-dimensional response surface, showing the optimum log K_{ow} and pK_a ranges for phloem mobility (see [8] or [42]). A simplified version of this type of model is shown in Figure 5.7 [40].

Accumulation of weak acids in the phloem is not guaranteed, but is dependent on the ratio of the permeabilities of the dissociated and nondissociated species. Ideally, a P_{AH}/P_{A^-} ratio in the order of 10^4 is desired for high retention and phloem mobility of herbicide molecules. This is not achieved with more lipophilic weak acids, such as members of the arylpropanoic acids. For example, the low phloem mobility of flamprop has been attributed to the low P_{AH}/P_{A^-} ratio [43]. This may also explain the low phloem mobility of related herbicides, such as diclofop [44].

Herbicides that leave source tissue in the phloem can efflux from the phloem at any point along the translocation pathway. Herbicides that are only marginally retained in the phloem are likely to exit the phloem soon after entering it, whereas those that are strongly retained are likely to move longer distances before exiting the phloem and entering the adjoining tissues (including xylem, possibly).

Efflux or leakage from the phloem should not be regarded as a disadvantage. Indeed, it is an essential component of phloem mobility. If retention was complete, very little herbicide would reach the phloem; most of it would enter and be retained in epidermal or mesophyll cells between the point of entry into the leaf and the vascular bundles. The ability to move in the phloem can be regarded as a compromise, therefore, between two opposing forces—retention in cells on one hand, and availability on the other. Hence, there cannot be a herbicide that is completely phloem-mobile; any herbicide that possesses the optimum characteristics for phloem mobility is also able to efflux from the phloem and move in the xylem.

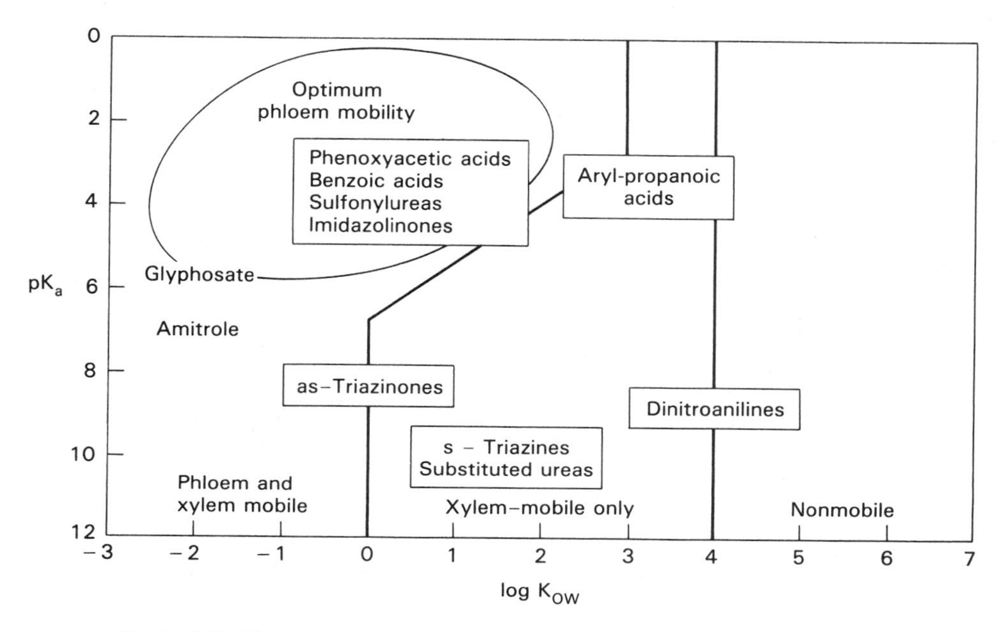

Figure 5.7 Phloem and xylem mobility of organic compounds as influenced by physicochemical parameters. Various groups of herbicides have been superimposed on a model proposed by Bromilow et al. [40].

In some instances, herbicides can be cycled, or retranslocated, through plants several times. For example, dicamba is translocated in *Fagopyrum tataricum* in the phloem, as would be expected, but some efflux and retranslocation in the xylem can occur, resulting in accumulation in immature leaves [25]. Subsequently, however, the herbicide can be exported from these leaves as they mature. Such cycling of herbicide in the plant depends on its remaining available for translocation over an extended period of time.

5.4.2 Xylem Translocation

Xylem translocation following root absorption has been expressed in terms of the transpiration stream concentration factor (TSCF), defined as follows [45]:

$$\text{TSCF} = \frac{\text{Concentration in xylem sap}}{\text{Concentration in uptake solution}}$$

Experiments with a variety of nonionized organic compounds indicate the optimum log K_{ow} for accumulation in the xylem (i.e., maximum TSCF) to be approximately 1.8 [46] (Figure 5.8). The relationship derived has the form of a Gaussian curve with the equation

$$\text{TSCF} = 0.784 \exp - \left[\frac{(\log K_{ow} - 1.78)^2}{2.44}\right]$$

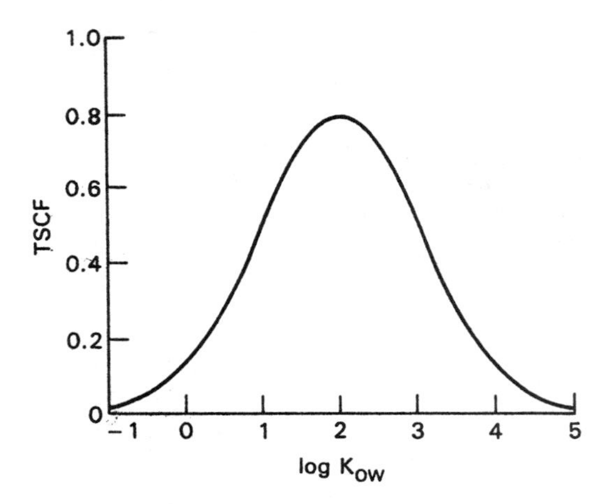

Figure 5.8 Influence of log K_{ow} on xylem translocation (expressed as TSCF) of a wide range of organic compounds in barley. Redrawn from [46].

Decreased uptake into roots limits xylem translocation of more hydrophilic compounds, whereas retention in the roots, by partitioning into lipid materials, limits the translocation of more lipophilic compounds.

For weak acids, the TSCF is closely correlated with the root concentration factor (Chapter 4). Consequently, the TSCF is higher at low solution pH values [47]. However, the TSCF of weak acids is always relatively low (e.g., not more than 10 percent of the concentration in root cortical cell protoplasts), due to retention of the herbicide anions in the root cells.

5.4.3 Design of Phloem Mobile Herbicides

Given our knowledge of the physicochemical characteristics that confer phloem mobility on herbicide molecules, it should be possible to design phloem-mobile molecules by incorporating particular functional groups (e.g., carboxylic acids) into candidate molecules. However, phloem mobility is only one aspect of herbicide activity, and not one that is considered essential in screening programs, in which biological activity and selectivity are considered much more important. Furthermore, the characteristics that confer phloem mobility on herbicides often limit their activity; for example, polar weak acids may be quite mobile in the phloem, but their foliar absorption may be limited. This can be overcome, perhaps, by conversion of the acids to esters to enhance absorption; biological activity is then dependent on hydrolysis of the ester to the free acid, which cannot be guaranteed.

A second approach is to modify molecules with known herbicidal activity to make them more phloem-mobile. For example, the addition of carboxyl groups to lipophilic herbicide molecules can alter their translocation characteristics [48, 49] (Figure 5.9). Unfortunately, this also results in loss of their herbicidal activity. Addition of acidic functions decreases their lipophilicity considerably, which re-

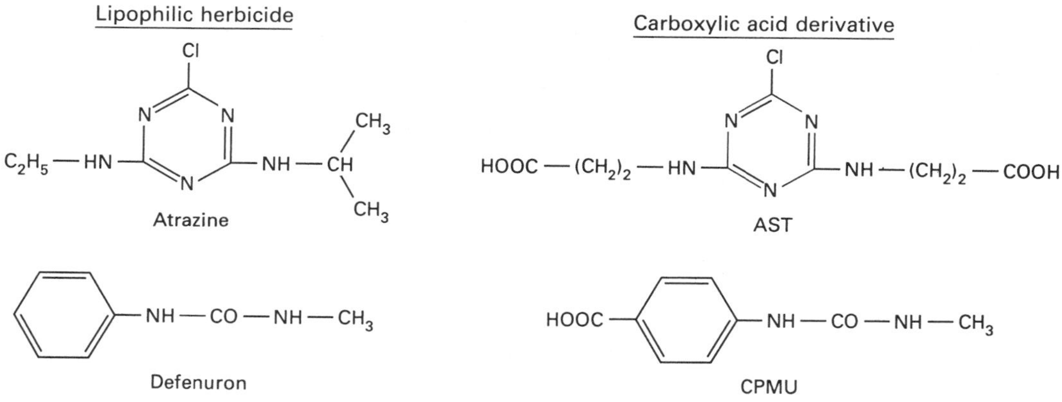

Figure 5.9 Structures of the lipophilic, predominantly xylem-mobile herbicides atrazine and defenuron, and their more hydrophilic, phloem-mobile carboxylic acid derivatives [48, 49].

duces the ability of these inhibitors of photosynthetic electron transport to reach their site of action in thylakoid membranes (Chapter 7).

This latter observation suggests that phloem mobility is not a desirable characteristic for all herbicides. Indeed, it is clear that a high degree of phloem mobility and high herbicidal activity as a photosystem II electron transport inhibitor are mutually exclusive. Given the large number of commercial herbicides that act at this site, it can be assumed that phloem mobility has not been a priority in herbicide screening programs.

It has been suggested that attaching a sugar molecule (glucose, or preferably sucrose) to a herbicide molecule may confer a higher degree of phloem mobility on it. There is no evidence to support this, and it seems likely that the result might be the opposite to that intended. First, markedly changing the physicochemical properties of the herbicide will alter its absorption and retention characteristics in plant cells, resulting in decreased phloem translocation (see following section). Second, the sugar carrier identified on plasma membranes shows very high specificity for sucrose, and does not transport even closely related sucrose analogues [10]. It is unlikely, therefore, that it would load sucrose-herbicide conjugates into the phloem. Third, there is no guarantee that the sugar-herbicide complex would be cleaved to release the active herbicide, either before or after translocation.

5.4.4 Herbicide Metabolism and Herbicide Translocation

The discussion of herbicide translocation thus far has assumed that the herbicide molecules are not chemically altered during the processes of absorption and translocation. However, herbicide metabolism is an important part of their activity and

TABLE 5.2 CHANGES IN PHYSICOCHEMICAL PROPERTIES (pK_a, log K_{ow}) ASSOCIATED WITH SOME COMMON METABOLIC CONVERSIONS [50]. Note that these values are approximate, and will depend on the electronic characteristics of the aromatic and alkyl moieties.

Original formula	Product	Δ log K_{ow}	ΔpK_a
$ArCH_3$	$ArCH_2OH$	−1.8	neutral → neutral
$ArCH_2OH$	ArCOOH	+0.8	neutral → 4
ArCOOH	ArCOO-glucoside	−2.3	4 → neutral
ArOH	ArO-glucoside	−2.3	10 → neutral
$RCOOCH_3$	RCOOH	−0.5	neutral → 4
$RCONH_2$	RCOOH	+1.0	neutral → 4
RCl	$RSCH_2CH(NH_2)COOH$	−4.0	neutral → zwitterion

selectivity. This topic is considered in detail in the following chapter, but is discussed briefly here in the context of herbicide translocation.

Herbicide metabolism invariably results in some change in the pK_a and/or log K_{ow} of the molecule [50] (Table 5.2), and can also result in reversible or irreversible changes in its activity as a metabolic inhibitor. This change in physicochemical properties results in a change in the behavior of the molecule. For example, 2,4-D iso-octyl ester is a neutral, lipophilic molecule that would not be expected to accumulate in plant cells, unless by binding or partitioning into lipids. Conversion to the free acid (i.e., 2,4-D) results in typical weak-acid behavior, with accumulation by an ion-trap mechanism. This species is mobile in the phloem. Further metabolic conversion, however, for example glucose ester formation or ring-glycosylation, is likely to lower the log K_{ow} by at least two orders of magnitude, resulting in a molecule that is neither herbicidal nor phloem mobile.

Xylem translocation of herbicides can be affected in a similar manner, if metabolic conversion changes the properties of the molecules sufficiently. The herbicide metribuzin is normally translocated quite readily in the xylem. Experimental results suggest that reduced translocation of metribuzin from root to shoot in tolerant potato and soybean cultivars is associated with higher rates of formation of polar conjugates than in susceptible cultivars [51, 52]. Presumably, the polar conjugates are strongly retained in the root cells in which they are formed, and relatively little metribuzin is translocated acropetally in the xylem to the leaves.

5.5 HERBICIDE EFFECTS ON PHLOEM TRANSPORT

Any physiological effect of a herbicide on source, sink, or phloem pathway tissue can potentially reduce phloem transport of assimilates and, consequently, phloem transport of the herbicide. Since phloem transport of herbicides is dependent on the production of transport sugars in the leaf, any herbicide effect on carbon metabolism that limits this production may affect herbicide translocation. Other effects on membrane integrity, assimilate loading, long-distance transport, and the metabolic activity of sink tissues may also affect herbicide translocation. A summary of the physiological effects of herbicides that may limit their translocation is presented in Table 5.3.

5.5.1 Effects on Membrane Integrity

Many herbicides affect the structure or function of the plasmalemma or subcellular membranes (see Chapters 11, 15). These can be direct effects on lipid metabolism or on metabolic processes intimately connected with the membrane (e.g., arylpropanoic acid herbicides; see Chapter 11), or indirect effects such as lipid peroxidation (e.g., bipyridilium herbicides; see Chapter 8). For example, paraquat causes very rapid deterioration of all membranes and, consequently, very little translocation can occur before the transport system is disrupted. However, if paraquat is applied in the dark, some (limited) phloem translocation can occur [59]. Prevention of lipid peroxidation allows the plant to maintain translocation long enough that some herbicide is exported from the site of entry out of the leaf.

Amongst other agents that have comparable effects on cell membranes are surfactants. Surfactants can penetrate leaf tissue, and can damage cell membranes [67]. It is not surprising, therefore, that there are reports in the literature of detrimental effects of surfactants on herbicide translocation [65, 66].

More subtle effects on cell membranes may also affect transport processes across those membranes. Although the major physiological/biochemical effect of diclofop in plants appears to be on acetyl co-enzymeA carboxylase (ACCase; Chapter 11), at elevated concentrations this herbicide also depletes the electrogenic potential across the plasma membrane in some species [61]. This effect on the charge and pH gradient across the membrane may affect some of the transport

TABLE 5.3 POSSIBLE SELF-LIMITING EFFECTS OF HERBICIDES OR FORMULATION INGREDIENTS ON HERBICIDE TRANSLOCATION. Note that a direct link between these physiological effects and reduced assimilate or herbicide translocation has not been established in all instances.

Herbicide	Effect	Reference
Glyphosate	Altered carbon metabolism in source and sink tissue	53–55
	Reduced stomatal conductance	56
Chlorsulfuron	Prevention of sucrose entry into phloem	57, 58
Paraquat	Membrane disruption	59
Difenzoquat	Membrane integrity (?)	60
Diclofop	Membrane depolarization	61
Picloram	Swelling of phloem parenchyma	62
Mecoprop, picloram	Reduced carbon fixation (indirect?)	63, 64
Surfactants	Membrane integrity	65, 66

characteristics of the membrane, and may contribute to the low phloem mobility of diclofop.

5.5.2 Effects on Carbon Fixation and Sucrose Production and Loading

Many herbicides block photosynthesis in plants, most commonly via effects on photosynthetic electron transport (see Chapter 7). However, the physicochemical requirements for phloem translocation and for activity on photosynthetic electron transport tend to be mutually exclusive, with relatively low lipophilicity required for the former, but relatively high lipophilicity required for the latter. The result is that many of the herbicides that affect photosynthesis directly have little effect on their own translocation in the phloem, because they are not particularly phloem-mobile (and need not be phloem-mobile). However, many other herbicides exert secondary or indirect effects on carbon fixation, and may thus limit their own transport.

Glyphosate is commonly regarded as being highly phloem-mobile in plants. However, in addition to a small amount of transfer that can occur from phloem to xylem, phloem translocation of glyphosate is also limited by an indirect effect on carbon metabolism in source tissues. Geiger and co-workers have shown that glyphosate affects carbon fixation through an effect on ribulose bisphosphate (RuBP) levels and on starch production [53–55] (Figure 5.10). The effect on RuBP levels is an indirect consequence of inhibition of EPSP synthase by glyphosate (discussed in more detail in Chapter 13); deregulation of the shikimate pathway causes an excess of shikimate-3-phosphate to be produced. The resulting diversion of carbon from the Calvin cycle eventually depletes the amount of RuBP available for carbon fixation. Initially, sugars derived from starch maintain the supply of assimilate for translocation, but eventually starch also becomes depleted. Hence this effect of glyphosate on phloem translocation may not be detectable until one day (or perhaps longer) after application.

Sufonylurea herbicides (e.g., chlorsulfuron) also limit their own translocation via an effect on phloem transport. Photosynthesis and the initial steps in carbon metabolism in treated leaves are not affected directly, but sucrose and starch accumulate in chlorsulfuron-treated leaves soon after herbicide application [57, 58]. This effect can be overcome by supplying the plants with valine, leucine, and isoleucine, the amino acids whose synthesis is inhibited by chlorsulfuron (Chapter 13). It appears that some step in the delivery of transport sugars into the phloem is sensitive to this herbicide, and that the herbicide effectively limits its own translocation in this way.

5.5.3 Effects on Vascular Tissues

A further physiological effect that may limit herbicide translocation can occur along the translocation pathway. Some herbicides, particularly those with auxin-

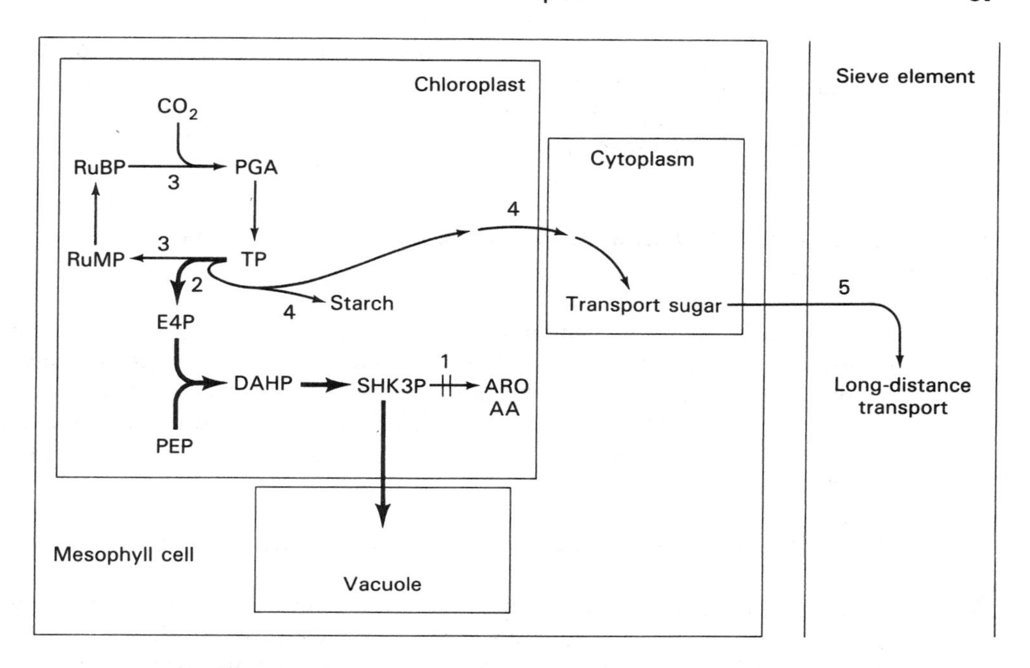

Figure 5.10 Proposed mechanism by which glyphosate limits its own translocation via effects on carbon metabolism in source leaves. 1. Blockage of aromatic amino acid biosynthesis; 2. Diversion of carbon to shikimate-3-phosphate; 3. Reduction of carbon pool available for Calvin cycle; 4. Decreased starch synthesis and decreased export of triose to cytoplasm; 5. Decreased phloem transport of assimilate. Adapted from [55].

like activity, cause swelling of stem or petiole tissue in susceptible species. This is due, presumably, to leakage of herbicide from the phloem into the surrounding vascular parenchyma, and leads to physical constriction and disruption of translocation in both phloem and xylem [62]. While this may contribute to the death of some plants, it may also prevent toxic quantities of herbicide from reaching the roots in perennial species.

5.5.4 Effects on Sink Tissues

In addition to the effects described above on source tissue, Geiger and co-workers have also recorded evidence of an effect of glyphosate on assimilate transport mediated through action in sink tissue [55]. This effect was observed earlier than the reduction in carbon export from source leaves, and has been tentatively attributed to decreased sucrose utilization for protein synthesis in the sink tissue. Other effects of glyphosate and other phloem-mobile herbicides on metabolic activity of sink tissues may also limit the import of assimilate—and herbicide—into those tissues.

One consequence of any of the above effects of herbicides on transport processes in plants is that one herbicide may reduce the translocation of another when the two are applied in combination. This is only a small part of the whole discussion of herbicide interactions (see Chapter 17).

5.6 APPROACHES TO MAXIMIZING PHLOEM TRANSLOCATION OF HERBICIDES

Unfortunately, herbicides are not always translocated to the extent that we might consider adequate or satisfactory. Among the possible approaches to increasing herbicide translocation in plants are the use of additives to alter source-sink relationships within the plant, and the judicious timing of herbicide applications to correspond to the time of maximum photosynthetic activity and/or assimilate translocation.

5.6.1 Use of Additives to Stimulate Phloem Translocation

Attempts have been made to stimulate herbicide translocation (particularly in the phloem) by applying a second chemical either with the herbicide, or some time prior to the herbicide. The rationale for such treatments is that if they result in stimulation of assimilate transport, herbicide translocation will also be increased. Although simultaneous applications would be more convenient, prior treatment with the ''stimulating'' agent would ensure that assimilate translocation will indeed be elevated at the time of herbicide application.

A list of agents used to stimulate herbicide translocation is shown in Table 5.4. The list includes both naturally occurring and synthetic plant growth regulators. One of the problems encountered in this area of research is that assimilate distribution patterns may be altered by an exogenously applied substance, but the total amount of translocation may remain the same. In other words, increased herbicide translocation to one plant part may occur at the expense of translocation to another part.

Some of the compounds listed in Table 5.4 act by stimulating growth in sink tissues. For example, chlorflurenol (a morphactin) and 6-benzyl-aminopurine ap-

TABLE 5.4 CHEMICAL AGENTS USED TO STIMULATE HERBICIDE TRANSLOCATION IN THE PHLOEM.

Chemical agent	Herbicide	Reference
Chlorflurenol	Dicamba	68
6-Benzyl-aminopurine	Glyphosate	69
Gibberellic acid	2,4,5-T	70
Abscisic acid	2,4,5-T	71
Ethephon	Dicamba	72

pear to act by releasing root buds from dormancy, with a resultant increase in both assimilate and herbicide translocation into those buds [68, 69]. Gibberellic acid, on the other hand, is known to stimulate sucrose loading in source tissue [72], and this may contribute to its effect on herbicide translocation [70]. More detailed research is required to explain fully the action of many of these compounds.

5.6.2 Environmental Factors

Assimilate translocation, as discussed earlier, is a function of the synthesis of carbohydrate in source tissues, and its utilization in sinks. Assuming the source tissues to be photosynthetically active leaves, the supply of assimilate for long-distance transport depends on the rate of photosynthesis in those leaves. Conditions that favor high rates of photosynthesis, therefore, also favor high rates of assimilate translocation. This has obvious implications for herbicide translocation, and the influence of environmental parameters on herbicide translocation has been studied extensively. Rather than attempt to summarize all of the available results here, we will consider briefly the underlying principles from a plant physiological viewpoint.

It is well documented in the literature that herbicide translocation is under environmental control. Environmental factors such as air temperature, relative humidity, irradiance level, and soil moisture all have been shown to influence herbicide translocation, although it is not always possible to separate direct effects on translocation from indirect effects mediated through increased herbicide absorption. Nevertheless, timing of herbicide application in relation to environmental conditions is an important aspect of maximizing herbicide efficacy.

The ability of plants to maintain high rates of photosynthesis under different environmental conditions is a function of adaptation (genetic factors) and acclimation (plasticity or ability to respond to changes in environment). Thus, for different plant species there are different optimum temperatures for carbon fixation, as well as for many other processes. There may also be optimum temperature differences within a species, depending on the conditions under which the plants have developed. Photosynthetic response to different temperatures may be a reflection of more obvious biochemical characteristics (e.g., C_3 vs. C_4 plants), or to more subtle differences (e.g., lipid composition and membrane fluidity). Similarly, differences in response to other environmental variables, such as irradiance level and soil moisture levels, reflect species differences in some physiological or biochemical attribute(s).

Without reviewing environmental physiology in detail, it is clear that different species respond in different ways to their environment. Therefore, there is no optimum temperature for assimilate translocation, or for herbicide translocation; the amount of herbicide translocation that occurs in a plant under particular conditions is simply a reflection of that plant's ability to maintain phloem or xylem translocation under those conditions. For example, *Agropyron repens,* a temperate species, is able to maintain relatively high rates of photosynthesis at low temperature, and phloem translocation of a foliage-applied herbicide can occur quite freely under those conditions [28, 73]. The same result would not be expected in a species adapted to subtropical or tropical environments.

Similar reasoning can be applied to effects of other environmental variables on herbicide translocation in different plant species. Unfortunately, in most of the research that has been conducted on environmental influences on herbicide translocation, only the movement of the herbicide is monitored, and not assimilate production and translocation, or water entry and translocation in the plant. The interpretation of the results, therefore, is often by inference rather than by direct comparison of herbicide and assimilate or water movement data. If we assume that herbicide translocation and distribution patterns are a reflection of the physiological status of the plant, herbicide translocation data can provide some insight into plant response to varying environments. It must be remembered, however, that the nature of herbicide translocation can be obscured by binding in tissues, exchange between phloem and xylem, and so on.

Finally, it must be recognized that higher rates of assimilate translocation in the phloem, or water translocation in the xylem, do not necessarily result in more herbicide translocation, or in more herbicide activity. Although translocation rates may differ, control ultimately depends on the total amount of herbicide reaching different tissues. Other factors being equal, it may take longer for a critical herbicide concentration to be reached in plants growing under "adverse" conditions, but if that critical concentration can be reached after an extended period, equal control may result [74]. Thus, initial translocation rate differences do not necessarily lead to different levels of control.

REFERENCES

1. Devine, M. D. 1988. "Environmental influences on herbicide performance: a critical evaluation of experimental techniques." Proc. EWRS Symposium, Factors Affecting Herbicidal Activity and Selectivity. Wageningen, Netherlands: Ponsen and Looijen, pp. 219–26.
2. Bucholtz, D. L., and F. D. Hess. 1988. "An atomizer for application of very low volumes of herbicide solutions." *Weed Sci.*, 36, 406–09.
3. Crafts, A. S., and S. Yamaguchi. 1964. *The Autoradiography of Plant Materials*. Calif. Agric. Exp. Stn. Manual 35. 143 pp.
4. Dewey, S. A., and A. P. Appleby. 1983. "A comparison between glyphosate and assimilate translocation patterns in tall morningglory (*Ipomoea purpurea*)." *Weed Sci.*, 31, 308–14.
5. O'Donovan, J. T., and W. H. Vanden Born. 1981. "A microautoradiographic study of ^{14}C-labelled picloram distribution in soybean following root uptake." *Can. J. Bot.*, 59, 1928–31.
6. Bromilow, R. H., R. L. O. Rigitano, G. G. Briggs, and K. Chamberlain. 1987. "Phloem translocation of non-ionised chemicals in *Ricinus communis*." *Pestic. Sci.*, 19, 85–99.

7. Groussol, J., S. Delrot, P. Caruhel, and J.-L. Bonnemain. 1986. "Design of an improved exudation method for phloem sap collection and its use for the study of phloem mobility of pesticides." *Physiol. Vég.*, 24, 123–33.
8. Devine, M. D. 1989. "Phloem translocation of herbicides." *Rev. Weed Sci.*, 4, 191–213.
9. Esau, K. 1977. *Anatomy of Seed Plants,* 2nd ed. New York: Wiley.
10. Hitz, W. D. 1986. "Molecular determinants of sugar carrier specificity." In *Phloem Transport,* J. Cronshaw, W. J. Lucas, and R. T. Giaquinta, eds. New York: Alan R. Liss, Inc., pp. 27–40.
11. Gallet, O., R. Lemoine, C. Larsson, and S. Delrot. 1989. The sucrose carrier of the plant plasma membrane. I. Differential affinity labeling. Biochim. Biophys. Acta 978, 56–64.
12. van Bel, A. J. E. 1987. "The apoplast concept of phloem loading has no universal validity." *Plant Physiol. Biochem.*, 25, 677–86.
13. Münch, E. 1930. *Die Stoffbewegungen in der Pflanze.* Jena: Gustav Fischer.
14. Devine, M. D., and L. M. Hall. 1990. "Implications of sucrose transport mechanisms for the translocation of herbicides." *Weed Sci.*, 38, 299–304.
15. Olunuga, B. A., P. H. Lovell, and G. R. Sagar. 1977. "The influence of plant age on the movement of 2,4-D and assimilates in wheat." *Weed Res.*, 17, 213–17.
16. Kühbauch, W., A. Suss, and V. Lang. 1975. "Wanderung von ^{14}C-Assimilaten und ^{14}C-Herbiziden in Barenklaupflanzen (*Heracleum sphondylium*)." *Angew. Botanik,* 49, 253–62.
17. Fykse, H. 1974. "Untersuchungen über *Sonchus arvensis* L. I. Translokation von C-14-markierten Assimilaten." *Weed Res.*, 14, 305–12.
18. Fykse, H. 1975. "Untersuchungen über *Sonchus arvensis* L. II. Translokation von ^{14}C-MCPA unter verschiedenen Bedingungen." *Weed Res.*, 15, 165–70.
19. Hull, R. J. 1969. "Translocation of assimilates and dalapon in established johnsongrass." *Weed Sci.*, 17, 314–20.
20. McAllister, R. S., and L. C. Haderlie. 1985. "Translocation of ^{14}C-glyphosate and $^{14}CO_2$-labeled photoassimilates in Canada thistle (*Cirsium arvense*)." *Weed Sci.*, 33, 153–59.
21. Gougler, J. A., and D. R. Geiger. 1981. "Uptake and distribution of *N*-phosphonomethylglycine in sugarbeet plants." *Plant Physiol.*, 68, 668–72.
22. Crafts, A. S. 1959. "Further studies on comparative mobility of labeled herbicides." *Plant Physiol.*, 34, 613–20.
23. Crafts, A. S. 1967. "Bidirectional movement of labeled tracers in soybean seedlings." *Hilgardia,* 37, 625–38.
24. Gaudiel, R., and W. H. Vanden Born. 1979. "Picloram translocation and redistribution in soybean plants following root uptake." *Pestic. Biochem. Physiol.*, 11, 129–34.
25. Chang, F. Y., and W. H. Vanden Born. 1971. "Translocation and metabolism of dicamba in Tartary buckwheat." *Weed Sci.*, 19, 107–12.
26. Chandrasena, N. R., and G. R. Sagar. 1986. "Uptake and translocation of ^{14}C-fluazifop by quackgrass (*Agropyron repens*)." *Weed Sci.*, 34, 676–84.
27. Devine, M. D., and W. H. Vanden Born. 1985. "Absorption, translocation, and foliar activity of clopyralid and chlorsulfuron in Canada thistle (*Cirsium arvense*) and perennial sowthistle (*Sonchus arvensis*)." *Weed Sci.*, 33, 524–30.

28. Devine, M. D., J. D. Bandeen, and B. D. McKersie. 1983. "Temperature effects on glyphosate absorption, translocation, and distribution in quackgrass (*Agropyron repens*)." *Weed Sci.*, 31, 461–64.
29. Coupland, D., and J. C. Caseley. 1979. "Presence of ^{14}C activity in root exudates and guttation fluid from *Agropyron repens* treated with ^{14}C-labelled glyphosate." *New Phytol.*, 83, 17–22.
30. Lund-Hoie, K., and A. Bylterud. 1969. "Translocation of aminotriazole and dalapon in *Agropyron repens* (L.) Beauv." *Weed Res.*, 9, 205–10.
31. Chandrasena, N. R., and G. R. Sagar. 1987. "The effect of site of application of ^{14}C-fluazifop on its uptake and translocation by quackgrass (*Agropyron repens*)." *Weed Sci.*, 35, 457–62.
32. Stoltenberg, D. E., and D. L. Wyse. 1986. "Regrowth of quackgrass (*Agropyron repens*) following postemergence applications of haloxyfop and sethoxydim." *Weed Sci.*, 34, 664–68.
33. Harker, K. N., and J. Dekker. 1988. "Effects of phenology on translocation patterns of several herbicides in quackgrass, *Agropyron repens*." *Weed Sci.*, 36, 463–72.
34. Shone, M. G. T., and A. V. Wood. 1972. "Factors affecting absorption and translocation of simazine by barley." *J. Exp. Bot.*, 23, 141–51.
35. Peterson, C. A., P. P. Q. de Wildt, and L. V. Edgington. 1978. "A rationale for the ambimobile translocation of the nematicide oxamyl in plants." *Pestic. Biochem. Physiol.*, 8, 1–9.
36. Lichtner, F. T. 1986. "Phloem transport of agricultural chemicals." In *Phloem Transport*, J. Cronshaw, W. J. Lucas, and R. T. Giaquinta, eds. New York: Alan R. Liss, Inc., pp. 601–08.
37. Crisp, C. E., and J. E. Larson. 1983. "Effect of ring substituents on phloem transport and metabolism of phenoxyacetic acid and six analogues in soybean (*Glycine max*)." In *Pesticide Chemistry: Human Welfare and the Environment*, J. Miyamoto and P. C. Kearney, eds. New York: Pergamon, pp. 213–22.
38. Crisp, C. E., and M. Look. 1978. "Phloem loading and transport of weak acids." In *Advances in Pesticide Science*. Vol. 3, H. Geissbuhler, G. T. Brooks, and P. C. Kearney, eds. New York: Pergamon, pp. 430–37.
39. Chamberlain, K. D., D. N. Butcher, and J. C. White. 1986. "Relationships between chemical structure and phloem mobility in *Ricinus communis* var. Gibsonii with reference to a series of ω-(1-naphthoxy)alkanoic acids." *Pestic. Sci.*, 17, 48–52.
40. Bromilow, R. H., K. Chamberlain, and A. A. Evans. 1990. "Physico-chemical aspects of phloem translocation of herbicides." *Weed Sci.*, 38, 305–14.
41. Tyree, M. T., C. A. Peterson, and L. V. Edgington. 1979. "A simple theory regarding ambimobility of xenobiotics with special reference to the nematicide, oxamyl." *Plant Physiol.*, 63, 367–74.
42. Kleier, D. A. 1988. "Phloem mobility of xenobiotics. I. Mathematical model unifying the weak acid and intermediate permeability theories." *Plant Physiol.*, 86, 803–10.
43. Rigitano, R. L. O., R. H. Bromilow, G. G. Briggs, and K. Chamberlain. 1987. "Phloem translocation of weak acids in *Ricinus communis*." *Pestic. Sci.*, 19, 113–33.
44. Hall, C., L. V. Edgington, and C. M. Switzer. 1982. "Translocation of different 2,4-D,

bentazon, diclofop, or diclofop-methyl combinations in oat (*Avena sativa*) and soybean (*Glycine max*)." *Weed Sci.*, 30, 676–82.

45. Shone, M. G. T., and A. V. Wood. 1974. "A comparison of the uptake and translocation of some organic herbicides and a systemic fungicide by barley. I. Absorption in relation to physicochemical properties." *J. Exp. Bot.*, 25, 390–400.
46. Briggs, G. G., R. H. Bromilow, and A. A. Evans. 1982. "Relationships between lipophilicity and root uptake and translocation of non-ionised chemicals by barley." *Pestic. Sci.*, 13, 495–504.
47. Briggs, G. G., R. L. O. Rigitano, and R. H. Bromilow. 1987. "Physico-chemical factors affecting uptake to roots and translocation to shoots of weak acids in barley." *Pestic. Sci.*, 19, 101–12.
48. Neumann, S., E. Grimm, and F. Jacob. 1985. "Transport of xenobiotics in higher plants. I. Structural prerequisites for translocation in the phloem." *Biochem. Physiol. Pflanzen*, 180, 257–68.
49. Grimm, E., S. Neumann, and F. Jacob. 1985. "Transport of xenobiotics in higher plants. II. Absorption of defenuron, carboxyphenylmethylurea, and maleic hydrazide by isolated conducting tissue of *Cyclamen*." *Biochem. Physiol. Pflanzen*, 180, 383–92.
50. Bromilow, R. H., K. Chamberlain, and G. G. Briggs. 1986. "Techniques for studying the uptake and translocation of pesticides in plants". Aspects of Applied Biology. 11. Biochemical and Physiological Techniques in Herbicide Research. pp. 29–44. Wellesbourne: Assoc. Applied Biologists.
51. Abusteit, E. O., F. T. Corbin, D. P. Schmitt, J. W. Burton, A. D. Worsham, and L. Thompson, Jr. 1985. "Absorption, translocation, and metabolism of metribuzin in diploid and tetraploid soybean (*Glycine max*) plants and cell cultures." *Weed Sci.*, 33, 618–28.
52. Gawronski, S. W., L. C. Haderlie, R. H. Callihan, and R. B. Dwelle. 1985. "Metribuzin absorption, translocation, and distribution in two potato (*Solanum tuberosum*) cultivars." *Weed Sci.*, 33, 629–34.
53. Geiger, D. R., S. W. Kapitan, and M. A. Tucci. 1986. "Glyphosate inhibits photosynthesis and allocation of carbon to starch in sugarbeet leaves." *Plant Physiol.*, 82, 468–72.
54. Servaites, J. C., M. A. Tucci, and D. R. Geiger. 1987. "Glyphosate effects on carbon assimilation, ribulose bisphosphate carboxylase activity, and metabolite levels in sugarbeet leaves." *Plant Physiol.*, 85, 370–74.
55. Geiger, D. R., and H. D. Bestman. 1990. "Self-limitation of herbicide mobility by phytotoxic action." *Weed Sci.*, 38, 324–29.
56. Shaner, D. L., and J. L. Lyon. 1979. "Stomatal cycling in *Phaseolus vulgaris* L. in response to glyphosate." *Plant Sci. Lett.*, 15, 83–87.
57. Vanden Born, W. H., H. D. Bestman, and M. D. Devine. 1988. "The inhibition of assimilate translocation by chlorsulfuron as a component of its mechanism of action." Proc. EWRS Symposium, Factors Affecting Herbicidal Activity and Selectivity. Wageningen, Netherlands: Ponsen and Looijen, pp. 69–74.
58. Bestman, H. D., M. D. Devine, and W. H. Vanden Born. 1990. "Herbicide chlorsulfuron reduces assimilate transport out of treated leaves of field pennycress (*Thlaspi arvense*) seedlings." *Plant Physiol.*, 93, 1441–48.

59. Calderbank, A., and P. Slade. 1966. "The fate of paraquat in plants." *Outl. Agric.*, 5, 55–59.

60. Merritt, C. R. 1982. "The influence of form of deposit on the phytotoxicity of difenzoquat applied as individual drops to *Avena fatua*." *Ann. Appl. Biol.*, 101, 517–25.

61. Wright, J. P., and R. H. Shimabukuro. 1987. "Effects of diclofop and diclofop-methyl on the membrane potentials of wheat and oat coleoptiles." *Plant Physiol.*, 85, 188–93.

62. Peterson, R. L., G. R. Stephenson, and B. F. J. Mitchell. 1974. "Effects of picloram on shoot anatomy of red maple and white ash." *Weed Res.*, 14, 227–29.

63. Whipps, J. M., and M. P. Greaves. 1986. "Effect of mecoprop on plant growth and distribution of photosynthate in wheat (*Triticum aestivum* L.) seedlings." *Weed Res.*, 26, 227–32.

64. Sharma, M. P., and W. H. Vanden Born. 1971. "Effect of picloram on $^{14}CO_2$-fixation and translocation of ^{14}C-assimilates in Canada thistle, soybean, and corn." *Can. J. Bot.*, 49, 69–74.

65. Sherrick, S. L., H. A. Holt, and F. D. Hess. 1986. "Effects of adjuvants and environment during plant development on glyphosate absorption and translocation in field bindweed (*Convolvulus arvensis*)." *Weed Sci.*, 34, 811–16.

66. Stevens, P. J. G., and M. J. Bukovac. 1987. "Studies on octylphenoxy surfactants. Part 2. Effects on foliar uptake and translocation." *Pestic. Sci.*, 20, 37–52.

67. St. John, J. B., P. G. Bartels, and J. L. Hilton. 1974. "Surfactant effects on isolated plant cells." *Weed Sci.*, 22, 233–37.

68. Baradari, M. R., L. C. Haderlie, and R. G. Wilson. 1980. "Chloflurenol effects on absorption and translocation of dicamba in Canada thistle (*Cirsium arvense*)." *Weed Sci.*, 28, 197–200.

69. Waldecker, M. A., and D. L. Wyse. 1985. "Chemical and physical effects of the accumulation of glyphosate in common milkweed (*Asclepias syriaca*) root buds." *Weed Sci.*, 33, 605–11.

70. Basler, E. 1974. "Abscisic acid and gibberellic acid as factors in the translocation of auxin." *Plant Cell Physiol.*, 15, 351–61.

71. Carson, A. G., and J. D. Bandeen. 1975. "Influence of ethephon on absorption and translocation of herbicides in Canada thistle." *Can. J. Plant Sci.*, 55, 795–800.

72. Aloni, B., J. Daie, and R. E. Wyse. 1986. "Enhancement of [^{14}C]sucrose export from source leaves of *Vicia faba* by gibberellic acid." *Plant Physiol.*, 82, 962–66.

73. Devine, M. D., and J. D. Bandeen. 1983. "Fate of glyphosate in *Agropyron repens* (L.) Beauv. growing under low temperature conditions." *Weed Res.*, 23, 69–75.

74. Coupland, D. 1983. "Influence of light, temperature and humidity on the translocation and activity of glyphosate in *Elymus repens* (=*Agropyron repens*)." *Weed Res.*, 23, 347–55.

Chapter 6

Herbicide Metabolism

6.1 INTRODUCTION

The metabolism of herbicides in plants constitutes the most important mechanism of herbicide selectivity among weed and crop plants. In general, the crop plant or tolerant weed can detoxify the herbicide fast enough to avoid an accumulation to phytotoxic levels in the tissue. More precisely, the herbicide molecules are removed from the active cellular pool by conjugation, detoxification, deposition, and so on, more rapidly than they are replenished from the outside. Alternatively, a "pro-herbicide" (precursor or "protrac") molecule may be toxified faster (e.g., by hydrolysis) in a sensitive than in a tolerant plant. Other important mechanisms of herbicide tolerance or resistance include differences in uptake, transport, and/or at the site of action. It should be noted, however, that herbicide translocation differences between species often reflect differences in metabolism of the herbicide in those species (see Section 5.4.4).

Herbicide metabolism in plants has been studied extensively, for two major reasons: (a) to understand the mechanism of selectivity among weed and crop plants and between different crop cultivars [1, 2], and (b) to fulfill herbicide registration requirements [3]. In the physiological context of this book, those metabolic herbicide conversions that detoxify active compounds are the most interesting. Several recent review papers and comprehensive discussions of herbicide metabolism in plants are available [1–5]. The treatment of herbicide metabolism in plants usually presents schemes showing the conversion, degradation, and conjugation of different herbicidal structures and structural groups. The approach used here is to classify and discuss the different known types of metabolic conversions and conjugations of xenobiotics in plants that are important for herbicide detoxification. No attempt is made to cover all herbicidal groups or to discuss selectivity mechanisms in detail.

As mentioned above, the focus of this discussion is the physiological or biochemical basis of herbicide selectivity resulting from the primary detoxification reaction(s) that confer tolerance on particular plant species or cultivars/biotypes. "Tolerance" here is meant to define a gradual tolerance, which may be low or high, depending on the activity of the detoxifying/metabolizing enzyme. "Resistance" would, by contrast, be much more pronounced and usually be caused by an insensitive target site/binding protein. Examples are the mutant D-1 protein in triazine-resistant weeds (Section 7.4) and the acetyl-CoA carboxylase (ACCase) in dicotyledonous plants resistant to aryl-propanoic acids (Section 11.2). In these instances the plants are completely insensitive (resistant) to the respective herbicides. This definition (tolerance by detoxification, resistance by site of action differences) reflects the frequent practical situation that site-of-action resistance is much more pronounced than the gradual tolerance caused by elevated activities of detoxifying enzymes. In the herbicide field, the terms "resistance" and "tolerance" are most commonly used as described above, that is, resistance and tolerance for high and low levels, respectively, of insensitivity to herbicides. However, it may not always be possible to differentiate between "low-level resistance" or "cross-resistance" (caused by a partial loss of binding specificity at the site of action) and "high-level tolerance" (caused by very high rates of enzymatic herbicide detoxification), unless the mechanism of the resistance/tolerance is known. According to this definition, a particular plant could conceivably tolerate higher levels of herbicide under conditions of "high-level tolerance" than under conditions of "low-level resistance." This example serves to illustrate the ambiguity of the present use of the terms "tolerance" and "resistance." In the fungicide and insecticide fields, the term tolerance has been abandoned after similar discussions. The question of crop tolerance is, however, restricted to herbicides.

The metabolic behavior of herbicides in plants is summarized in Scheme 6.1. The detoxification of herbicides in plant tissues can be roughly divided into three phases, namely conversion, conjugation, and deposition [3]. This crude division should be seen mainly as a guide; individual compounds may behave differently in that (a) not all three phases need be involved, (b) the parent structure may be an inactive pro-herbicide that must first be converted to the active herbicide, or (c) reversible conjugation and/or binding or partitioning into lipid bodies may further decrease the pool of "free," intracellular herbicide. The bioactivation of pro-herbicides holds potential for the introduction of new safe and selective herbicides [6]. Bioactivation reactions include several types of metabolic conversions, including: oxidations (fatty acid β-oxidation, Section 14.1; sulfoxidation of EPTC, Sections 6.2.2 and 11.4; *N*-dealkylation of metflurazon, Sections 6.2.2 and 8.1.1); reductions (bipyridylium herbicides, Section 8.5); hydrolyses (phenoxypropionic acid esters, Section 11.1; pyridate, Section 6.3.4); and reversible conjugations

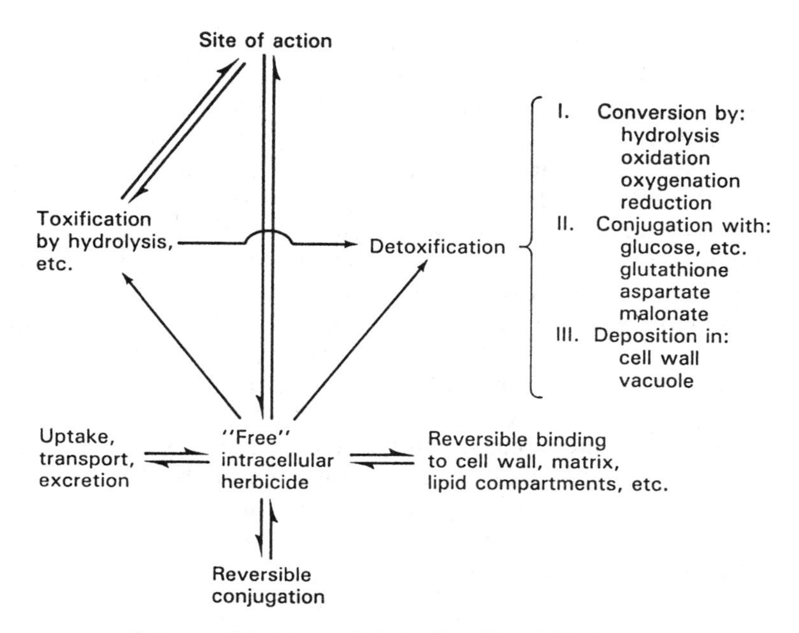

Scheme 6.1 Metabolic behavior of herbicides in plants.

(auxins, Section 14.3, aryl-propanoic acid glucosides, Section 6.3.2). Pro-herbicides may penetrate into the plant more readily (e.g., esters, Chapter 4, Sections 11.1, 14.1) and may move differently in the plant than the herbicidally active structure (e.g., acid vs. ester, Chapter 5). However, in in vitro studies, pro-herbicides may not be active because the required bioactivation reactions usually do not occur in cell-free test systems.

6.2 CONVERSION OF HERBICIDE MOLECULES

Some herbicides may be conjugated directly, for example by nucleophilic displacement of a chlorine substituent, or by amide bond formation at a free carboxyl group (see Section 6.3). Many herbicides have, however, no substituents available in the parent molecule that can react to form conjugates with cellular constituents. These herbicides must first be converted into (a) metabolite(s) that can serve as substrate(s) for one of the conjugating enzymes. The possible conversions are hydrolysis, oxidation, and reduction.

6.2.1 Hydrolysis

Hydrolysis can be expected if the herbicide is a carboxylic acid ester, a phosphate ester, or an amide. The hydrolyzing enzymes can accordingly be classified as esterases, phosphatases, or amidases. For example, esterases are known to split and thereby activate the aryl-propanoic acid esters of ACCase inhibitors described in Section 11.2. Two esterases have been studied in vitro: those for chlorfenprop-methyl [7] and for benzoylprop-ethyl [8]. These esterases produce the herbicidally active molecules and may control the expression of tolerance in some plants. However, the "chlorfenprop-methyl esterase" is highly active in both tolerant and sensitive plants and tissues and does not, therefore, provide a basis for the observed tolerance. The "benzoylprop-ethyl esterase," on the other hand, is considerably more active in wild oat (sensitive) than in wheat (tolerant). The substrate specificities of these esterases have not been tested. However, carboxyl esterases, phosphatases, and amidases that hydrolyze xenobiotic compounds in plants and animals in most cases have low substrate specificities. The K_m is 2 mM for chlorfenprop-methyl and 3.8 μM for benzoylprop-ethyl. Most of the herbicide esters in this group, including the most intensely studied compound, diclofop-methyl [9, 10], are rapidly hydrolyzed in tolerant as well as in sensitive plants. The respective "carboxyl esterases" might therefore be of the "chlorfenprop-methyl esterase" type, that is, unspecific with high activity.

Another example where the activity of an esterase leads to the bioactivation of a pro-herbicide is the case of pyridate, of which the hydrolysis product 6-chloro-3-phenyl-pyridazine-4-ol (CPP) is an inhibitor of photosynthetic electron transport [11]:

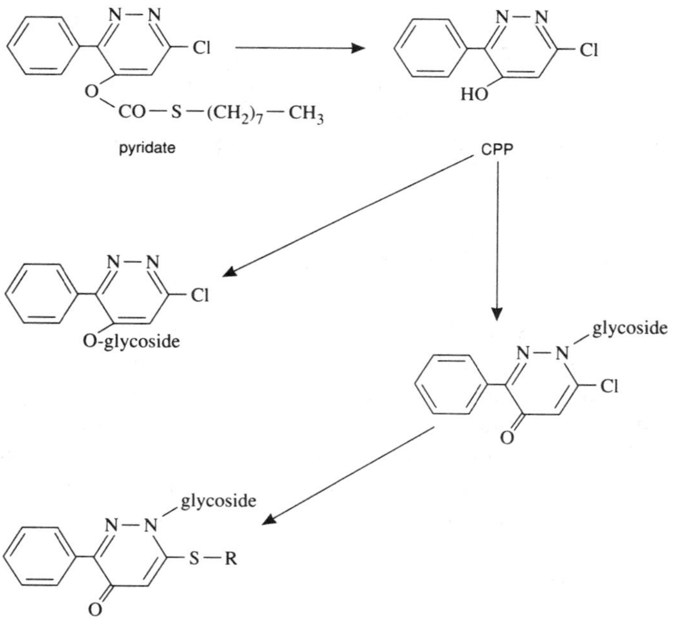

N—N
Cl
O
CO—S—$(CH_2)_7$—CH_3
pyridate

N—N
Cl
HO
CPP

N—N
Cl
O-glycoside

glycoside
N—N
Cl
O

glycoside
N—N
S—R
O

This scheme also indicates further metabolism of CPP by conjugation [*O*- and *N*-glycosylation of CPP, and glutathione (?) conjugation of the *N*-glycoside]. These conjugation reactions are discussed in more detail in Section 6.3.4. A carboxyl esterase is also responsible for the de-esterification of chlorimuron-ethyl and similar sulfonylurea ester molecules [12]. In this case the product of the esterase is herbicidally inactive.

A nonenzymatic hydrolysis catalyzed by benzoxazinone (2,4-dihydroxy-3-*keto*-7-methoxy-1,4-benzoxazine) in (primarily) corn roots results in conversion of atrazine to the herbicidally inactive product 2-hydroxy-atrazine [13, 14]. The reaction mechanism is a nucleophilic displacement of the chlorine, similar to glutathione conjugation, but in this case Cl^- is displaced by OH^- (via benzoxazinone catalysis).

Propanil is hydrolyzed to propionic acid and 3,4-dichloroaniline by an amidase that has been studied extensively [15–17]. The enzyme appears to be ubiquitous; it has been found in many plant species and also in soil bacteria. The amidase activity is particularly high in rice extracts, which correlates well with the tolerance of rice plants to propanil. The "aryl acylamidase" from red rice has been purified 4.8 times [17]. The K_m is 170 μM, the optimum pH 7.4–7.8, and the enzyme is inhibited by SH-reagents and heavy metals. The amidase is also inhibited by a number of insecticidal and fungicidal phosphorous and carbamate compounds, thereby causing synergistic herbicidal interactions in rice plants (Section 17.3.4). A comparable situation exists in the case of the herbicide mefenacet, which is also hydrolyzed by an amidase. This amidase is similarly inhibited by the thiophosphate fungicide ediphenfos, and a synergistic interaction is observed in *Echinochloa crus-galli*.

Very few herbicides contain a phosphate group. The phosphate bond is often of the phosphonate type (e.g., in fosamine, glufosinate, and glyphosate), but phosphate ester, thioester, and amide bonding also occur (e.g., in amiprophos-methyl, anilophos, butamiphos, and piperophos). The hydrolyzing reactions and the respective enzymes that might split the phosphate esters, thioesters, and amides have not been studied.

There are two examples of the creation of herbicide-resistant plants by introduction of a detoxifying gene from an unrelated species. A specific bromoxynil nitrilase from the soil bacterium *Klebsiella ozaenae* has been cloned and transferred into tobacco plants. It is expressed under the control of the light-regulated ribulose-bisphosphate carboxylase small subunit promoter [18]. The enzyme hydrolyzes bromoxynil to the acid form (3,5-dibromo-4-hydroxy benzoic acid), thereby conferring tolerance to the transgenic tobacco plants. The second example is the incorporation of a gene that encodes for acetylation of glufosinate; this example is discussed in more detail in Section 13.4.

Hydrolysis is also important in situations of reversible conjugate formation such as have been proposed for auxin herbicides (Section 14.3) and the aryl-propanoic acid ACCase inhibitors (Section 6.3.2). The respective conjugates (amides and acid glycosides) have clearly been shown to be herbicidally active in vivo and must, therefore, be activated by hydrolysis in sensitive tissue.

6.2.2 Oxidation, Oxygenation, and Hydroxylation

Hydroxylations are among the most common herbicide metabolic conversions observed in plants. Detoxification by hydroxylation and subsequent glycoside formation is especially important as a mechanism of herbicide selectivity in wheat and other cereals. The first hydroxylation or oxidative demethylation may not always lead to a herbicidally inactive structure (e.g., in chlortoluron and metoxuron), but subsequent hydroxylations and conjugations will eventually detoxify the herbicides [19]. In the special case of metflurazon, *N*-demethylation to norflurazon results in bioactivation of the herbicide (Section 8.1.1) [20].

The enzymes that catalyze hydroxylation reactions are of the monooxygenase type (formerly referred to as mixed function oxygenases). They are known to introduce one atom from molecular oxygen into the substrate molecule and, simultaneously, to reduce the second oxygen atom to water using NADPH. The reaction can lead to an oxidation, for example, a sulfoxidation of the type

$$-S- + 1/2\ O_2 \rightarrow -S(O)-$$

or (more commonly) to a hydroxylation (see below). Several monooxygenases have recently been studied in vitro (21–24). From comparative in vivo metabolism studies it is known that a spectrum of several different monooxygenases with individual and differing substrate specificities may exist in any one plant tissue. With respect to herbicides, first and second *N*-demethylation, *O*-demethylation, ring-methyl hydroxylation, aliphatic chain hydroxylation, and phenyl-ring hydroxylation can be distinguished, and these reactions may be catalyzed by different monooxygenases. As mentioned above, thiol sulfur oxygenation is also suggested to be dependent on monooxygenation.

An example for successive *N*-demethylation is the oxygenation of chlortoluron and related *N*-methylated phenylureas [19, 25, 26]:

$CH_3 \rightarrow -CH_2OH \rightarrow -H$

$HOH_2C- \leftarrow H_3C-$ (ring, Cl) $-NH-CO-N<$

$CH_3 \rightarrow -CH_2OH \rightarrow -H$

chlortoluron

In addition, chlortoluron can be oxygenated in the ring-methyl position, as indicated on the left side of the above scheme.

Several observations support the interpretation that distinct monooxygenase isoenzymes may be responsible for different hydroxylations. These include: (a) the ratio of *N*-demethylation of chlortoluron to 4-methylphenyl hydroxylation increases in older tissue cultures [25], (b) different hydroxylations are differently affected by the monooxygenase inhibitor 1-aminobenzotriazole (ABT) [27], and (c) different plants and tissues show different spectra of hydroxylation [26].

Studies on in vitro *N*-demethylation of aminopyrine (chosen as a nonphytotoxic xenobiotic model substrate) in microsomes from *Helianthus tuberosus* have shown the reaction to be inhibited by CO and by inhibitors of monooxygenases (e.g., ABT), and to require NADPH as a reductant [28]. The enzymatic activity is very low in dormant tubers but can be increased by pretreatment with phenobarbital and clofibrate. This phenomenon of induction is frequently seen with monooxygenase enzymes [29]. It is also the basis for the safening action of dichlormid, naphthalic anhydride, and other safeners on chlorsulfuron (Section 17.4).

Both ring-methyl hydroxylation and phenyl-ring hydroxylation have been observed for chlorsulfuron [30, 31]:

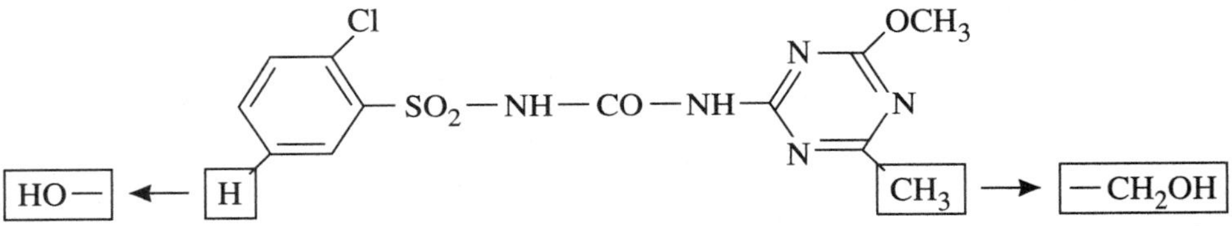

Cl
OCH$_3$
N
SO$_2$—NH—CO—NH—
N
N
HO— ← H
CH$_3$ → —CH$_2$OH

chlorsulfuron

In cisanilide, phenyl-ring hydroxylation and hydroxylation in the pyrrolidine ring have both been observed in carrot and cotton leaves and cell suspension cultures [32]:

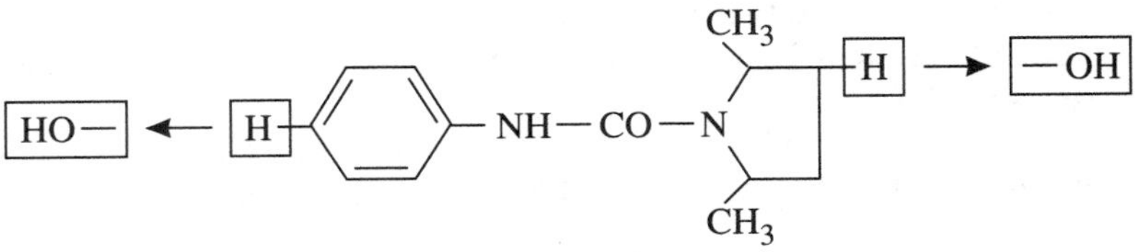

CH$_3$
H → —OH
HO— ← H
NH—CO—N
CH$_3$

cisanilide

Phenyl-ring hydroxylation at position 6 is the first step in bentazon (Figure 7.1) metabolism in rice [33], followed by conjugation with glucose to the β-D-glucoside. Extensive phenyl-ring hydroxylation occurs in the phenoxyacetic acid auxin herbicides (Scheme 14.2). In these and other similar cases, a halogen (chlorine) atom is subject to intramolecular transfer to one of the adjacent positions, and the position previously occupied by the halogen is hydroxylated. This type of intramolecular rearrangement is known as an NIH-shift reaction. Another herbicide for which phenyl-ring hydroxylation initiates the detoxification pathway is diclofop [9, 24] (e.g., in wheat). However, NIH-shift reactions have not been reported for this herbicide.

Two-step *O*-demethylation has also been reported, for example for metoxuron in wheat, carrot, and parsnip; in these species, the rates of *O*-demethylation correlate with the observed tolerance to metoxuron [34]:

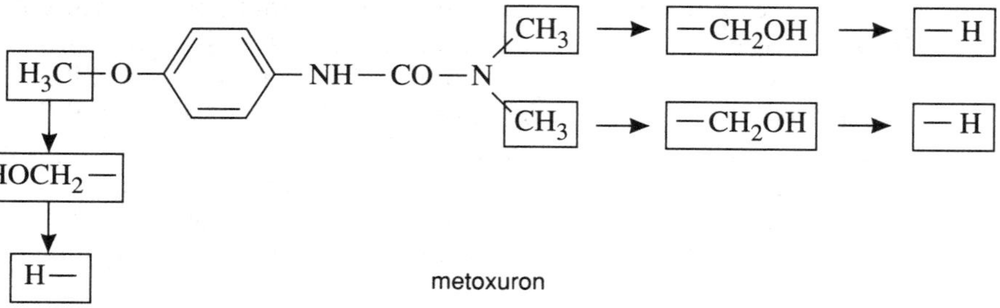

$$H_3C{-}O{-}C_6H_4{-}NH{-}CO{-}N(CH_3)_2$$

H_3C- → $HOCH_2-$ → $H-$; CH_3 → $-CH_2OH$ → $-H$ (each N-methyl group)

metoxuron

The metabolism of EPTC in plants is initiated by an oxygenation of the thiol-bridge sulfur [35]:

EPTC $O{=}C(N(CH_2{-}CH_2{-}CH_3)_2){-}S{-}CH_2{-}CH_3$ → $-S(=O)-$

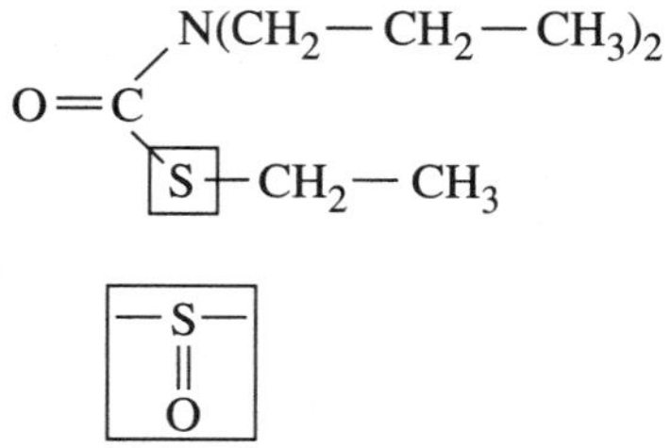

The resulting EPTC-sulfoxide may be conjugated with glutathione, or may be further oxidized to the EPTC-sulfone (see Scheme 11.2). The implications for the mode of action of EPTC and for the mechanism of safening action are discussed in Section 17.4.

Finally, peroxidases have been reported to oxidize the hydrolysis product of propanil, 3,4-dichloroaniline (DCA), to tetrachloroazobenzol in plants and in soil [36].

6.2.3 Reduction

Reductive metabolism of herbicides occurs only rarely in plants. A well-documented case is the reductive deamination of the *as*-triazinone herbicides metamitron and metribuzin in tolerant crop plants and in some weeds [37–39]:

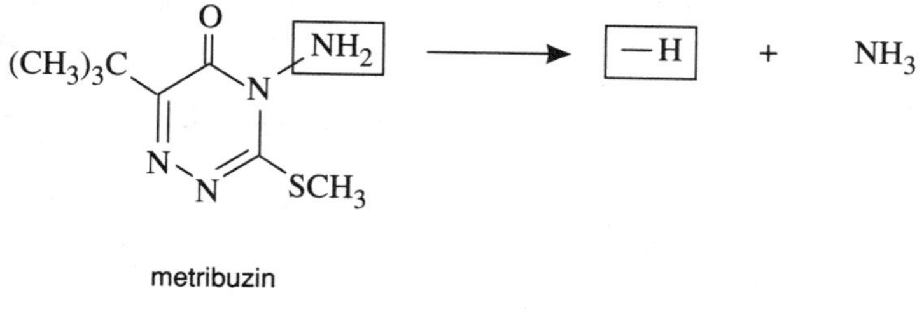

$(CH_3)_3C$, O, N, NH_2, N, N, SCH_3 (NH_2 → $-H$ + NH_3)

metribuzin

The deamination can be measured in vitro in isolated leaf peroxisomes and results in the stoichiometric formation of the deaminated herbicide and ammonia. The herbicides as well as the deaminated metabolite are conjugated with sugars and/or homoglutathione, depending on the particular crop plant (Sections 6.3.2, 6.3.4). Many of the conjugates have not, however, been sufficiently characterized. The mechanism of action of the metribuzin synergist picolinic acid *tert*-butylamide has been explained by an inhibition of the in vivo deamination of metribuzin (Section 17.3.3).

6.3 CONJUGATION

The terminal metabolites of herbicides in plants are usually conjugates. Many different conjugates can be formed, primarily with sugars, amino acids, peptides, and lignin building blocks as conjugation partners. The type of chemical bond formed may be ester, ether, thio ether, glycoside (semiacetalic), or amide. Only a fraction of the conjugates is usually "soluble" and can be extracted from the tissue with water or other solvents. Much of the herbicide may form an "insoluble" fraction that cannot be extracted by standard extraction procedures. The insoluble fraction may contain the herbicide and/or its conversion products chemically bound to insoluble matrix molecules such as lignin, hemicelluloses, or proteins. Also, adsorption to cell wall components may be very strong. The formation of insoluble and/or unextractable metabolites constitutes the third phase of herbicide metabolism (introduced in Section 6.1 as the "deposition" phase), and will be discussed separately in Section 6.4.

6.3.1 Conjugation with Glutathione

Glutathione conjugation is a very important detoxification mechanism in many plant tissues, and is particularly prominent in corn. Glutathione conjugation is also of great importance in the mechanism of herbicidal safening for thiocarbamate and chloroacetamide herbicides (see Sections 11.4 and 17.4, and Scheme 11.2), and in the mechanism of synergism between atrazine and tridiphane (Section 17.3.1 and Figure 17.5).

The chemical mechanism of glutathione conjugation is a nucleophilic displacement in the herbicide molecule. The glutathione anion GS^- serves as the nucleophile; chlorine, *p*-nitrophenol, or an alkyl-sulphoxide are possible leaving groups in herbicide molecules. Examples are the chlorine in propachlor and the *p*-nitrophenol in fluorodifen [4] (see next page).

The accompanying metabolism schemes show two additional pecularities: (a) in some cases, particularly in soybean, homoglutathione rather than glutathione is transferred to the herbicide, and (b) the (homo)glutathione conjugate is frequently further metabolized in the tissue by peptide hydrolysis, sulfur oxygenation, and *N*- or *O*-malonylation (see Section 6.4). A homoglutathione conjugate is also formed in soybean plants by chlorine displacement in chlorimuron-methyl [12], by 3-

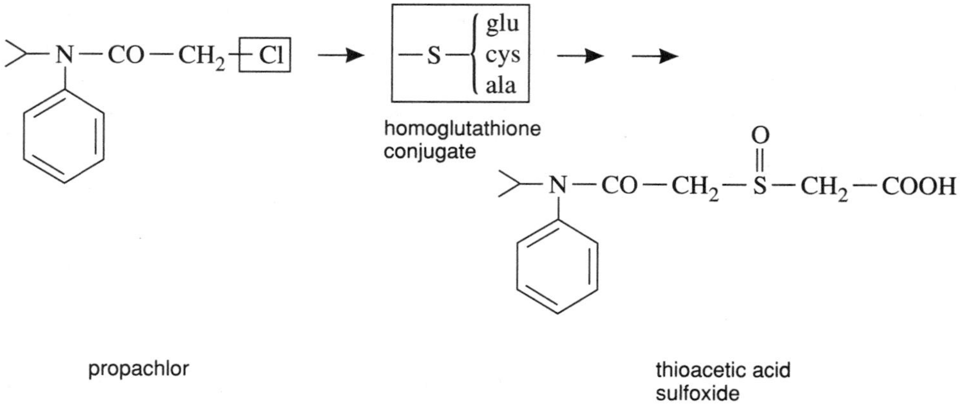

homoglutathione conjugate

propachlor

thioacetic acid sulfoxide

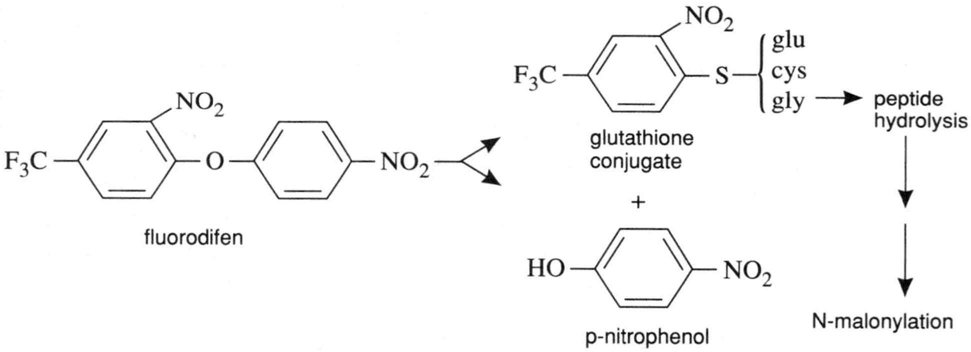

glutathione conjugate

peptide hydrolysis

fluorodifen

p-nitrophenol

N-malonylation

carboxy-4-nitrophenol displacement in acifluorfen [40], and after oxidation of the bridge sulfur to the sulfoxide, by S(O)-methyl displacement in metribuzin [41].

The reaction with glutathione is catalyzed by more or less specific glutathione-*S*-transferases (GSTs). The involvement of specific GST induction in herbicide-safening effects is described in Section 17.4. It appears that there is a family of constitutive and inducible GST isoenzymes in many plant tissues, in addition to the family of constitutive and inducible monooxygenases (Sections 6.2.2 and 17.4). These enzymes may constitute a plant defense system against naturally occurring xenobiotics and also against metabolic waste products from the plant's own catabolism.

A "GST isoenzyme III" from corn has been found to metabolize alachlor in vitro, with a K_m of 8.9 mM [42]. The Michaelis constants and the enzymatic activities compare favorably with the tolerance exhibited by the intact plants, strongly suggesting that this is the primary mechanism of tolerance. Similarly, a

GST from pea conjugates fluorodifen with a K_m of 12 μM [43]. The corn GST isoenzymes involved in herbicide detoxification have been reviewed recently [44].

6.3.2 Conjugation with Sugars

β-D-glucopyranoside is one of the most common glycoside conjugates formed in plants, and *N*-glycosides, *O*-glycosides, and glucose esters have been described in the literature. In addition to glucose, other sugars can form the glycoside linkage, or can be added to the already formed glycoside, thereby causing growth of the sugar moiety with time [45].

An example of *N*-glucoside formation is metribuzin-β-D-(*N*-glucoside), which is subsequently malonylated to metribuzin-6-*O*-malonyl-β-D-(*N*-glucoside) [41, 46]:

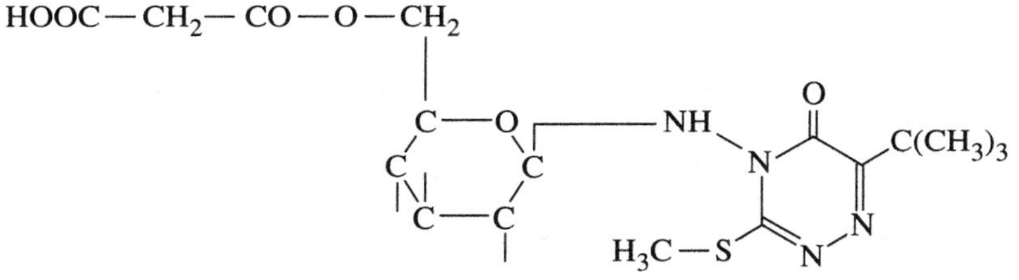

The glycosylating enzyme has been partially purified from tomato plants, and has been found to utilize UDP-glucose as the glucose donor [46]. It may, accordingly, be called a UDP-glucose : arylamine *N*-glucosyl transferase.

The formation of *O*-glucosides is more common and usually follows the initial introduction of hydroxyl groups into the herbicide molecule by monooxygenation (Section 6.2.2). As an example, the structure of 2,5-dichloro-4-(6-*O*-malonyl-β-D-glucosyl)-phenoxyacetic acid (see Section 14.3) is shown:

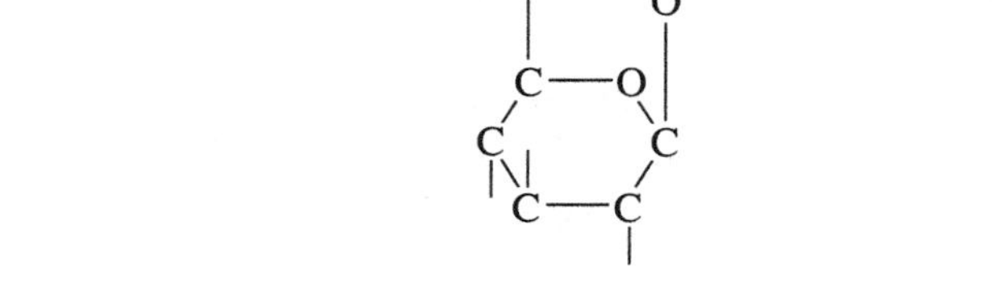

The formation of glucose esters has been described with aryl-propanoic acid herbicides such as diclofop in (sensitive) wild oat [10]:

$$Cl-C_6H_3(Cl)-O-C_6H_4-O-CH(CH_3)-CO-O-\text{glucose}$$

The exact structure of the diclofop glucose ester is not known. Interestingly, diclofop is aryl-hydroxylated and subsequently *O*-glucosylated in (tolerant) wheat plants [9, 24]. It has been suggested that the sensitivity of wild oat reflects the reversible nature of the ester conjugate, which therefore does not represent a permanent detoxification product. However, irreversible detoxification by ring hydroxylation and *O*-glycosylation confers tolerance on wheat.

6.3.3 Conjugation with Amino Acids

The *N*-aspartyl and *N*-glutamyl conjugates of the natural auxin IAA and of auxin herbicides are described in Section 14.3 as "slow release" forms of these compounds. This formulation indicates the reversible character of the ester bond:

$$C_6H_3Cl_2-O-CH_2-CO-NH-CH(COOH)-CH_2-COOH$$

N-aspartyl-2, 4-D

The glutathione and homoglutathione conjugates described in Section 6.3.1 can of course also be considered as amino acid conjugates, but the type of binding and the enzymatic biosynthetic mechanism (nucleophilic displacement) is very different from the carboxylic acid amide formation considered here.

The further breakdown of the glutathione and homoglutathione conjugates by peptide bond hydrolysis has, in several instances, been shown to yield the cysteine conjugate, which is eventually malonylated. In the case of chlorfenprop, only the cysteine conjugate, but not the glutathione conjugate, has been described in wheat [47]:

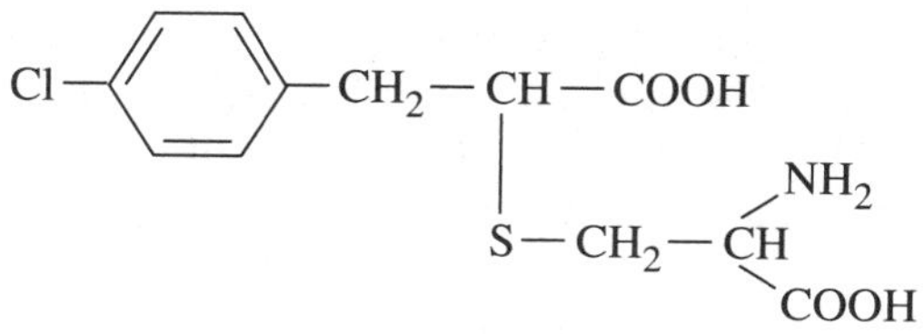

$$Cl-C_6H_4-CH_2-CH(COOH)-S-CH_2-CH(NH_2)COOH$$

2-cysteinyl chlorfenprop

It is possible, therefore, that cysteine could also react as a nucleophile in a conjugation reaction similar to the glutathione conjugation (Section 6.3.1).

6.3.4 Conjugation Reactions: a Synopsis

The initial conjugates described above (Sections 6.3.1–6.3.3) are in many instances further metabolized by many different metabolic conversions and conjugation reactions occurring in the plant. The early and the later metabolites are subject to oxygenation, hydrolysis (with a possibility of restoring the active herbicide), and further conjugation. The sugar moieties of metribuzin conjugates have been shown to grow in size as they move up in the plant in the transpiration stream [45]. In addition to mono- and disaccharides, ninhydrin-reactive conjugation components and lipids are also incorporated during this phase, yielding complex conjugates. Metribuzin conjugates have been reported to contain the sugar components glucose, galactose, mannose, arabinose, and rhamnose, the amino acids alanine, leucine, glutamate, α-aminobutyrate, phenylalanine, asparagine, and proline, and several unidentified lipids. It is generally believed that these complex metabolites/conjugates are eventually deposited in the cell wall matrix and/or the vacuole. The amino acid conjugates of auxin herbicides have been shown to be targeted into the cell wall, whereas the glucosides are predominantly targeted into the vacuole (Section 14.3).

An example of the complex metabolism and conjugation pattern that might occur is illustrated by the herbicide pyridate (see Section 6.2.1). After initial hydrolysis to the active herbicidal structure CPP, *O*-glycosylation (slow) and *N*-glycosylation (fast) both occur in corn and peanut. In the *N*-glycoside the chlorine atom is a much better leaving group for nucleophilic displacement by glutathione or cysteine than it was in the CPP molecule. Fast conjugate formation by chlorine displacement is observed in corn, but not in peanut. However, both crop plants are herbicide-tolerant, because both the *N*-glycosylation and the *O*-glycosylation effectively detoxify the photosynthetic inhibitor CPP.

Malonylation of herbicide conjugates has already been mentioned several times in this chapter. *N*-malonylation (of the fluorodifen-cysteine conjugate, Section 6.3.1) and *O*-malonylation (of metribuzin-*N*-glucoside and of 2,4-D-*O*-glucoside, Section 6.3.2) have been described. Hydrolysis and breakdown of glutathione in the glutathione conjugate goes beyond the cysteine conjugate to the thioacetic acid conjugate, and the thioacetic acid sulfoxide conjugate, with propachlor and EPTC. For EPTC, the additional terminal conjugate EPTC-*O*-malonyl-3-thiolactic acid has been described [35]:

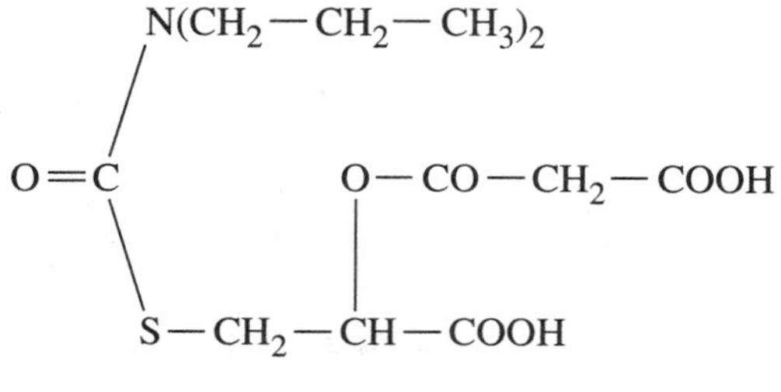

$$
\begin{array}{l}
\quad\; N(CH_2-CH_2-CH_3)_2 \\
\;\;\;/ \\
O=C \qquad\qquad O-CO-CH_2-COOH \\
\;\;\;\backslash \qquad\qquad\quad | \\
\quad\; S-CH_2-CH-COOH
\end{array}
$$

The suggested sequence of reactions leading from the cysteine conjugate to the malonyl thiolactic acid conjugate is transamination → reduction → malonylation. The malonylation reaction of course requires malonyl-CoA. Another example of *N*-malonylation is the formation of the *N*-malonyl conjugate of metribuzin via the exocyclic amino group [41]. Lamoureux and Rusness concluded that "it seems likely that malonylation may be a reaction utilized by plants to compartmentalize and/or terminate metabolism" [35]. Whether plants "use" this as a terminating step is open to debate; however, malonylation is frequently the final step in herbicide metabolism pathways.

The characterization and structure determination of herbicide conjugates in plants has been greatly facilitated by the recently developed method of FABS (*F*ast *A*tom *B*ombardment mass *S*pectroscopy) [4]. The use of this technique will undoubtedly contribute greatly to a more complete understanding of the formation of complex herbicide conjugates. However, contrasting this tremendous improvement in conjugate structure elucidation, the enzymatic mechanisms and cofactor requirements responsible for the biosynthesis of these conjugates remain largely unknown.

6.4 TERMINAL CONJUGATE FORMATION AND DEPOSITION

The pathway of metabolism of a herbicide may largely determine the eventual targeting of the terminal metabolites and conjugates. As mentioned above, glycosides are predominantly deposited in the vacuole, while amino acid conjugates are mainly excreted into the cell wall. For example, the auxin herbicide 2,4-D is metabolized and deposited by the sequence

2,4-D → 2,5-Cl-4-OH-D → 2,5-Cl-4-β-D-glucosyl-D
→ 2,5-Cl-4-(6-*O*-malonyl-β-D-glucosyl)-D

(Sections 6.2.2, 6.3.2, and 14.3). The detoxification and conjugation pathways may not be entirely irreversible, but re-entry of the aglycone herbicide or herbicide conversion product into the active cytoplasmic pool occurs very slowly, if at all. Eventually the herbicide or other xenobiotic is removed from the active cytoplasmic pool by the cooperating metabolic processes of conversion, conjugation, and deposition (Scheme 6.1). Another example of this "cooperation" of metabolic processes is the deposition of PCP in the cell wall [3, 48]:

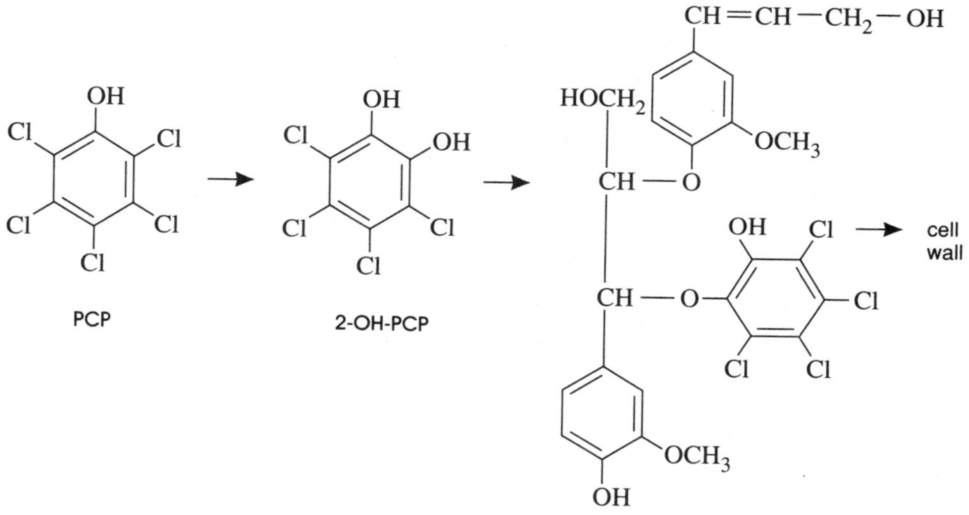

PCP

2-OH-PCP

cell wall

The xenobiotic in this case is chemically integrated into the lignin component of the cell wall and forms an "insoluble" residue. Of course, the term insoluble is dependent to some extent on the type of solvent and on a possible pretreatment of the plant material by chemical or enzymatic hydrolysis before the extraction. Specific extraction procedures for pectin, lignin, hemicellulose, and protein fractions have shown that all of these cell wall components can bind herbicides and other xenobiotics [48, 49].

REFERENCES

1. Cole, D., R. Edwards, and W. J. Owen. 1987. "The role of metabolism in herbicide selectivity." In *Herbicides. Progress in Pesticide Biochemistry and Toxicology,* Vol. 6, D. H. Hutson, T. R. Roberts, eds. New York: Wiley, pp. 57–104.
2. Owen, W. J. 1987. "Herbicide detoxification and selectivity." In Proc. Brit. Crop Prot. Conf.—Weeds, Vol. 1. Thornton Heath, U.K.: BCPC Publ., pp. 309–18.
3. Sandermann Jr., H. 1987. "Pestizid-Rückstände in Nahrungspflanzen. Die Rolle des pflanzlichen Metabolismus." *Naturwiss.,* 74, 573–78.

4. Lamoureux, G. L., and D. S. Frear. 1987. "Current problems, trends, and developments in pesticide metabolism in plants." In *Pesticide Science and Biotechnology*, R. Greenhalgh, T. R. Roberts, eds. Oxford: Blackwell, pp. 455–62.
5. Hathway, D. E. 1989. *Molecular Mechanisms of Herbicide Selectivity*. Oxford: Oxford University Press, 214 pp.
6. Hutson, D. H. 1987. "The bioactivation of herbicides." In Proc. Brit. Crop Prot. Conf.—Weeds, Vol. 1. Thornton Heath, U.K.: BCPC Publ., pp. 319–28.
7. Fedtke, C., and R. R. Schmidt. 1977. "Chlorfenprop-methyl: its hydrolysis *in vivo* and *in vitro* and a new principle for selective herbicidal action." *Weed Res.*, 17, 233–39.
8. Hill, B. D., E. H. Stobbe, and B. L. Jones. 1978. "Hydrolysis of the herbicide benzoylprop-ethyl by wild oat esterase." *Weed Res.*, 18, 149–54.
9. Dusky, J. A., D. G. Davis, and R. H. Shimabukuro. 1980. "Metabolism of diclofop-methyl (methyl-2-[4-(2′,4′-dichlorophenoxyl)phenoxy] propionate in cell suspensions of diploid wheat (*Triticum monococcum*)." *Physiol. Plant.*, 49, 151–56.
10. Shimabukuro, R. H., W. C., Walsh, and R. A. Hoerauf. 1979. "Metabolism and selectivity of diclofop-methyl in wild oat and wheat." *J. Agric. Food Chem.*, 27, 615–23.
11. Zohner, A. 1987. "Mode of crop tolerance to pyridate in corn and peanuts." In Proc. Brit. Crop Prot. Conf.—Weeds, Vol. 3. Thornton Heath, U.K.: BCPC Publ., pp. 1083–90.
12. Brown, H. M., and S. M. Neighbors. 1987. "Soybean metabolism of chlorimuron-ethyl: physiological basis for soybean selectivity." *Pestic. Biochem. Physiol.*, 29, 112–20.
13. Hamilton, R. H. 1964. "Tolerance of several grass species to 2-chloro-*s*-triazine herbicides in relation to degradation and content of benzoxazinone derivatives." *J. Agric. Food Chem.*, 12, 14–17.
14. Jensen, K. I. N., G. R. Stephenson, and L. A. Hunt. 1977. "Detoxification of atrazine in three graminae subfamilies." *Weed Sci.*, 25, 212–20.
15. Hoagland, R. E., and G. Graf. 1972. "Enzymatic hydrolysis of herbicides in plants." *Weed Sci.*, 20, 303–05.
16. Hoagland, R. E., G. Graf, and E. D. Handel. 1974. "Hydrolysis of 3′,4′-dichloropropionanilide (propanil) by plant aryl acylamidases." *Weed Res.*, 14, 371–74.
17. Hoagland, R. E. 1978. "Isolation and some properties of an aryl acylamidase from red rice, *Oryza sativa* L., that metabolizes 3′,4′-dichloropropionanilide." *Plant Cell Physiol.*, 19, 1019–27.
18. Stalker, D. M., K. E. McBride, and L. D. Malyj. 1988. "Herbicide resistance in transgenic plants expressing a bacterial detoxification gene." *Science*, 242, 419–23.
19. Cabanne, F., P. Gaillardon, and R. Scalla. 1985. "Phytotoxicity and metabolism of chlortoluron in two wheat varieties." *Pestic. Biochem. Physiol.*, 23, 212–20.
20. Tantawy, M. M., T. Braumann, and L. H. Grimme, 1984. "Uptake and metabolism of the phenylpyridazinone herbicide metflurazon during the bleaching and regeneration process of the green alga, *Chlorella fusca*." *Pestic. Biochem. Physiol.*, 22, 224–31.
21. Frear, D. S., H. R. Swanson, and F. S. Tanaka. 1969. "*N*-Demethylation of substituted 3-(phenyl)-1-methylureas: isolation and characterization of a microsomal mixed function oxidase from cotton." *Phytochemistry*, 8, 2157–69.

22. Moreland, D. E., F. T. Corbin, and W. P. Novitzky. 1990. "Metabolism of metolachlor by a microsomal fraction isolated from grain sorghum (*Sorghum bicolor*) shoots." *Z. Naturforsch.*, 45c, 558–64.
23. Mougin, C., F. Cabanne, M.-C. Canivenc, and R. Scalla. 1990. "Hydroxylation and N-demethylation of chlortoluron by wheat microsomal enzymes." *Plant Sci.*, 66, 195–203.
24. Zimmerlin, A., and F. Durst. 1990. "Xenobiotic metabolism in plants: aryl hydroxylation of diclofop by a cytochrome-P-450 enzyme from wheat." *Phytochemistry*, 29, 1729–32.
25. Cole, D. J., and W. J. Owen. 1988. "Metabolism of chlortoluron in cell suspensions of *Lactuca sativa:* a qualitative change with age of culture." *Phytochemistry*, 27, 1709–11.
26. Gonneau, M., B. Pasquette, F. Cabanne, and R. Scalla. 1988. "Metabolism of chlortoluron in tolerant species: possible role of cytochrome P-450 monooxygenases." *Weed Res.*, 28, 19–25.
27. Canivenc, M. C., B. Cagnac, F. Cabanne, and R. Scalla. 1988. "Manipulation of chlortoluron fate in wheat cells." In Proc. EWRS Symp. Factors Affecting Herbicidal Activity and Selectivity. Wageningen, Netherlands: Ponsen & Looijen, pp. 115–20.
28. Fonne-Pfister, R., A. Simon, J.-P. Salaun, and F. Durst. 1988. "Xenobiotic metabolism in higher plants: Involvement of microsomal cytochrome P-450 in aminopyrine *N*-demethylation." *Plant Sci.*, 55, 9–20.
29. O'Keefe, D. P., J. A. Romesser, and K. J. Leto. 1988. "Identification of constitutive and herbicide inducible cytochromes P-450 in *Streptomyces griseolus*." *Arch. Microbiol.*, 149, 406–12.
30. Sweetser, P. B., G. S. Schow, and J. M. Hutchison. 1982. "Metabolism of chlorsulfuron by plants: biological basis for selectivity of a new herbicide for cereals." *Pestic. Biochem. Physiol.*, 17, 18–23.
31. Hutchison, J. M., R. Shapiro, and P. B. Sweetser. 1984. "Metabolism of chlorsulfuron by tolerant broad-leaves." *Pestic. Biochem. Physiol.*, 22, 243–47.
32. Frear, D. S., and H. R. Swanson. 1975. "Metabolism of cisanilide (cis-2,5-dimethyl-1-pyrrolidinecarboxanilide) by excised leaves and cell suspension cultures of carrot and cotton." *Pestic. Biochem. Physiol.*, 5, 73–80.
33. Mine, A., M. Miyakado, and S. Matsunaka. 1975. "The mechanism of bentazon selectivity." *Pestic. Biochem. Physiol.*, 5, 566–74.
34. Vassiliou, G., and F. Müller. 1978. "Metabolism of metoxuron in umbelliferae of varying selectivity." *Meded. Fac. Landbouwwet. Rijksuniv. Gent*, 43, 1181–91.
35. Lamoureux, G. L., and D. G. Rusness. 1987. "EPTC metabolism in corn, cotton, and soybean: identification of a novel metabolite derived from the metabolism of a glutathione conjugate." *J. Agric. Food Chem.*, 35, 1–7.
36. Lieb, H. B., and C. C. Still. 1969. "Herbicide metabolism in plants: specificity of peroxidases for aniline substrates." *Plant Physiol.*, 44, 1672–73.
37. Schmidt, R. R., and C. Fedtke. 1977. "Metamitron activity in tolerant and susceptible plants." *Pestic. Sci.*, 8, 611–17.
38. Fedtke, C., and R. R. Schmidt. 1979. "Characterization of the metamitron deaminating enzyme activity from sugar beet (*Beta vulgaris* L.) leaves." *Z. Naturforsch.*, 34c, 948–50.

39. Fedtke, C. 1983. "Leaf peroxisomes deaminate *as*-triazinone herbicides." *Naturwiss.*, 70, 199.

40. Frear, D. S., H. R. Swanson, and E. R. Mansager. 1983. "Acifluorfen metabolism in soybean: diphenylether bond cleavage and the formation of homoglutathione, cysteine, and glucose conjugates." *Pestic. Biochem. Physiol.*, 20, 299–310.

41. Frear, D. S., H. R. Swanson, and E. R. Mansager. 1985. "Alternative pathways of metribuzin metabolism in soybean: formation of *N*-glucoside and homoglutathione conjugates." *Pestic. Biochem. Physiol.*, 23, 56–65.

42. O'Connell, K. M., E. J. Breaux, and R. T. Fraley. 1988. "Different rates of metabolism of two chloroacetanilide herbicides in Pioneer 3320 corn." *Plant Physiol.*, 86, 359–63.

43. Frear, D. S., and H. R. Swanson. 1973. "Metabolism of substituted diphenylether herbicides in plants. I. Enzymatic cleavage of fluorodifen in peas." *Pestic. Biochem. Physiol.*, 3, 473–82.

44. Timmerman, K. P. 1989. "Molecular characterization of corn glutathione S-transferase isozymes involved in herbicide detoxication." *Physiol. Plant.*, 77, 465–71.

45. Frear, D. S., L. N. Falb, and A. E. Smith. 1987. "Metribuzin metabolism in soybeans: partial characterization of the polar metabolites." *Pestic. Biochem. Physiol.*, 27, 165–72.

46. Mansager, E. R., H. R. Swanson, and F. S. Tanaka. 1983. "Metribuzin metabolism in tomato: isolation and identification of *N*-glucoside conjugates." *Pestic. Biochem. Physiol.*, 19, 270–81.

47. Pont, V., and G. F. Collet. 1980. "Métabolisme du chloro-2-(*p*-chlorophényl)-3 propionate de méthyle et problème de sélectivité." *Phytochemistry*, 19, 1361–63.

48. Langebartels, C., and H. Harms. 1985. "Analysis for nonextractable (bound) residues of pentachlorophenol in plant cells using a cell wall fractionation procedure." *Ecotox. Environ. Safety*, 10, 268–79.

49. Pillmoor, J. B., J. K. Gaunt, and T. R. Roberts. 1984. "Examination of bound (nonextractable) residues of MCPA and flamprop in wheat straw." *Pestic. Sci.*, 15, 375–81.

Chapter 7

Herbicidal Inhibition of Photosynthetic Electron Transport

7.1 INTRODUCTION

The inhibitors of photosynthetic electron transport constitute a very important group of herbicides, historically as well as with respect to their numbers. Together with many of the auxin-like phenoxyacetic acids (Chapter 14), they were among the first organic herbicides to be developed (early 1950s). Their photosynthesis-inhibiting activity was detected several years after their herbicidal activity [1], which in turn was found in a normal routine screening for herbicidally active organic chemicals. Today we know that these compounds inhibit the photosynthetic electron transport system at the reducing side of photosystem II (Figure 7.2) by binding to the photosystem II reaction center protein D-1 at the native binding site for plastoquinone (Figures 7.5, 7.6, 7.7) [2–4]. With this knowledge, which includes the amino acid sequence and tertiary structure of the herbicide-binding protein and the binding niche, the photosystem II herbicidal target is the best characterized of all pesticide targets [5].

Inhibition of photosynthetic electron flow leads to excessive radiative excitation in the blocked photosynthetic pigment system. The consequences include maximum fluorescence emission (Figure 7.4), energy spillover to oxygen and other nearby molecules (Chapter 9), photooxidation and, eventually, phytotoxicity at the organelle, cell, and tissue level (see Section 7.6).

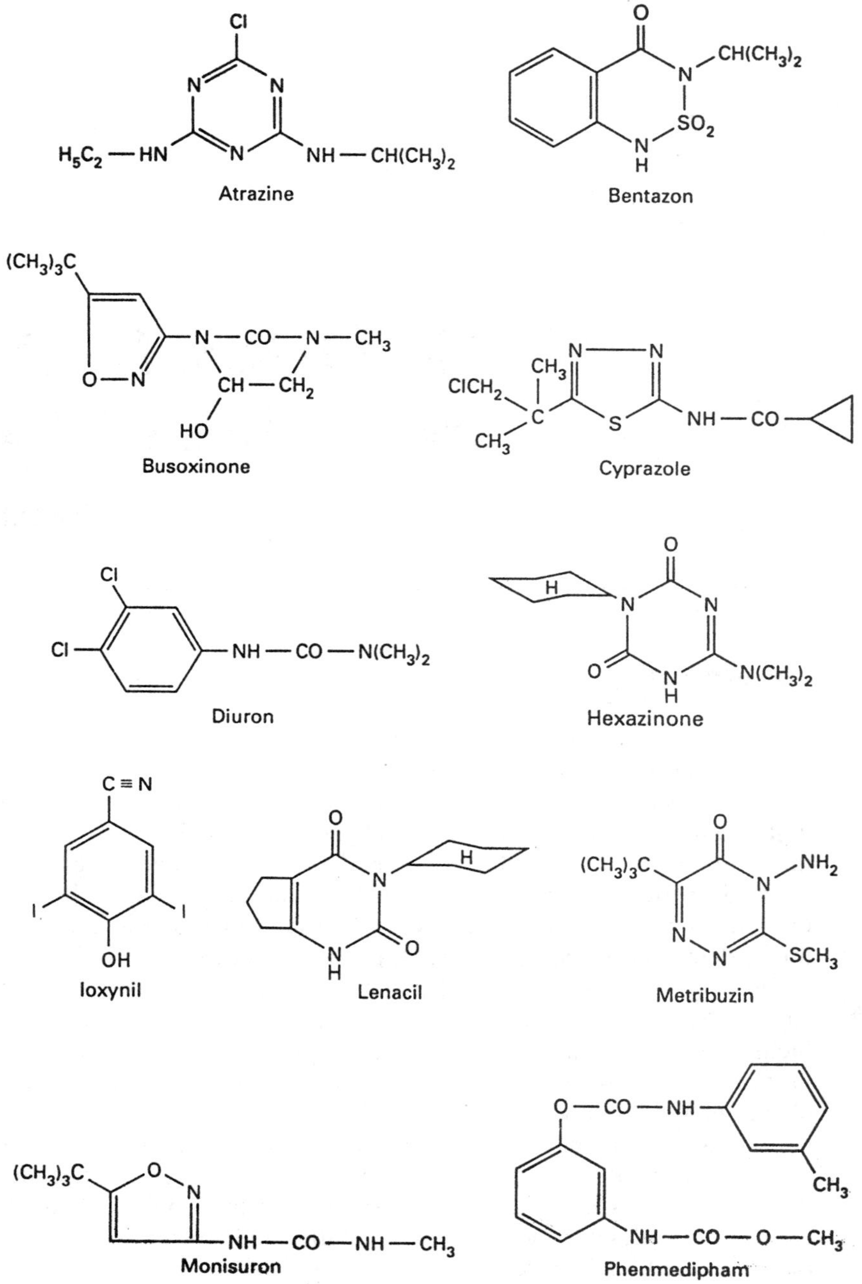

Cl
N N
H_5C_2 — HN N NH — $CH(CH_3)_2$
Atrazine

O
N — $CH(CH_3)_2$
SO_2
N
H
Bentazon

$(CH_3)_3C$
N — CO — N — CH_3
O — N CH — CH_2
HO
Busoxinone

N — N
CH_3
$ClCH_2$ C S NH — CO
CH_3
Cyprazole

Cl
Cl — NH — CO — $N(CH_3)_2$
Diuron

O
H N N
O N $N(CH_3)_2$
H
Hexazinone

C ≡ N
I I
OH
Ioxynil

O
N H
N O
H
Lenacil

O
$(CH_3)_3C$ N NH_2
N N SCH_3
Metribuzin

$(CH_3)_3C$ O N
NH — CO — NH — CH_3
Monisuron

O — CO — NH
CH_3
NH — CO — O — CH_3
Phenmedipham

Figure 7.1 A selection of herbicides that inhibit photosynthetic electron transport. These examples are representatives of chemical subgroups listed in Table 7.1.

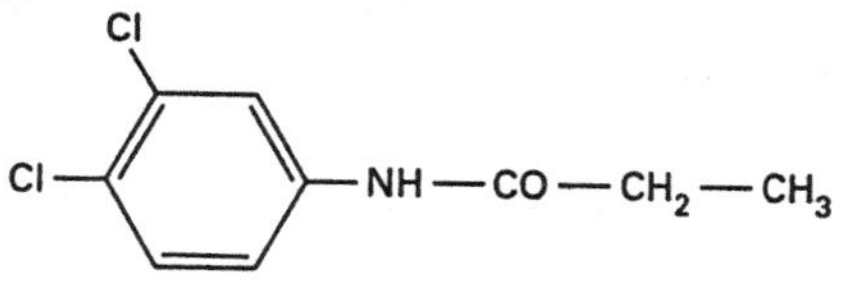

Cl
Cl
NH—CO—CH_2—CH_3
Propanil

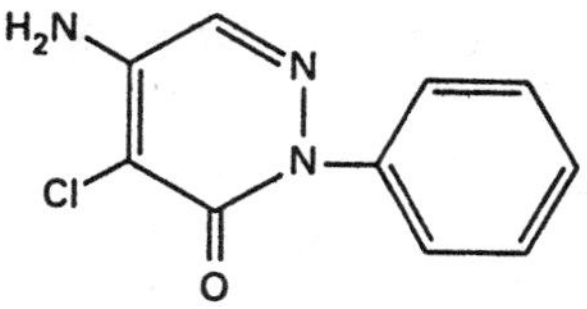

H_2N
N
N
Cl
O
Pyrazon

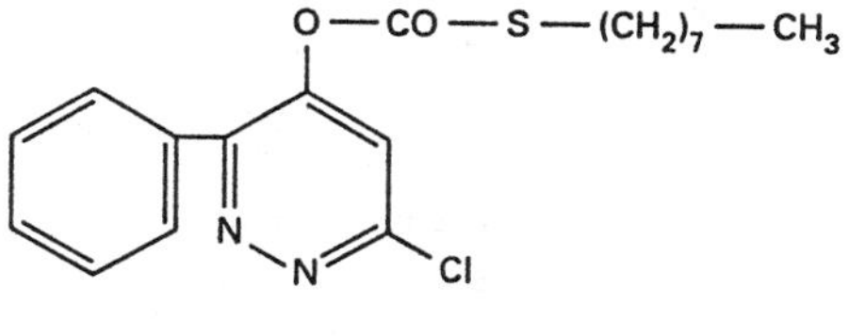

O—CO—S—$(CH_2)_7$—CH_3
N
N
Cl
Pyridate

Figure 7.1 (Continued).

7.2 HERBICIDES THAT INHIBIT PHOTOSYNTHETIC ELECTRON TRANSPORT

In the early years of the development of organic herbicides, inhibitors of photosystem II formed the largest group; around 1970 they constituted about 50 percent of the commercially available herbicides. This proportion has now (early 1990s) fallen to about 30 percent, reflecting the growing share and importance of new herbicidal compounds with other modes of action. However, new photosystem II-inhibiting herbicides are still being found and developed (e.g., [6–10]). A selection from the known photosystem II inhibitor structures is given in Figure 7.1. As far as possible, representatives of important subgroups are presented in Figure 7.1, for example, atrazine for the *s*-triazines, busoxinone, diuron, and monisuron for the substituted ureas, lenacil for the uracils, and so on (see also Table 7.1). The diversity of structures presented in Figure 7.1 appears at a first glance to reflect an enormously low specificity of the binding site. However, many of the structures presented in Figure 7.1 contain the element —CO—NH—, which consequently was first thought to be the main structural element for inhibitory action. In the cyclic molecules, the similar element would be —N=C—NH—, pointing to the importance of the underlying electronic configuration. However, the H on the nitrogen is obviously not required. This has been demonstrated in the substituted urea series with benzothiazole as the aromatic substituent: in this series methylation of the three available free urea hydrogens does not lead to loss of the inhibitory activity on chloroplast electron transport [11].

Therefore, the minimum structural element in the above functional groups does not include hydrogen. Instead, two atoms with a partial negative charge and a

TABLE 7.1 ALLOCATION OF 117 HERBICIDAL PHOTOSYSTEM II ELECTRON TRANSPORT INHIBITORS TO STRUCTURAL GROUPS. Sample compounds from Figure 7.1 are included for reference.

Structural group	Examples	Number of compounds
Substituted ureas	Busoxinone, diuron, monisuron	43
s-Triazines	Atrazine	29
as-Triazinones	Metribuzin	5
s-Triazinediones	Hexazinone	2
Anilides	Propanil (compare cyprazole)	7
Uracils	Lenacil	4
Biscarbamates	Phenmedipham	4
Unclassified heterocycles	Bentazon, pyrazon, pyridate	9
Benzimidazoles	Fluromidine	3
Hydroxybenzonitriles	Ioxynil	6
Quinones	Chloranil, quinonamid	5

fixed distance between them form the important core element [12]:

$$—CO—N\lesssim \text{ or } —N{=}C—N\lesssim$$

These two atoms should be complementary to specific amino acids in the binding niche of the binding protein D-1 (Section 7.4) where they can form hydrogen bridges. In addition, a positive π-charge at a particular atom considered to be essential for binding is common to all these molecules, except for the phenols [13]. A similar electronic core element can also be found in the other molecules given in Figure 7.1, again with the possible exception of the hydroxybenzonitriles (e.g., ioxynil, bromoxynil) and nitrophenols.

With the exception of ioxynil, phenol and quinone structures are not given in Figure 7.1, but are mentioned in Table 7.1. Whereas phenols are inhibitors of photosystem II at the D-1 binding site, quinones inhibit photosynthetic electron transport at different sites, in particular at a plastohydroquinone oxidizing site of the Rieske iron-sulfur protein (Figure 7.2).

There are additional substructures in the inhibitor molecules that bind to specific amino acids in the D-1 binding niche and that differ among the various inhibitors. These additional molecular binding interactions contribute to binding strength and specificity, and exert an influence on the binding constant. The remainder of the molecule appears, however, to fulfill largely steric and lipophilic requirements. Considering the large variability of the inhibitor structures, and taking into account our knowledge of the herbicide-binding protein structure, it can be concluded that the herbicide-binding niche in this protein is rather large. Most of the inhibitors discussed here require only part of the available space in the binding niche, and thereby make use of only some of the many possible electronic and lipophilic interactions. These multiple interaction sites have also been termed "subreceptors," but there is no such thing as a clearly defined subreceptor; many types of intermediate binding interactions appear to be possible. Regardless, it is clear from a number of studies that all of the compounds discussed here (Figure 7.1 and Table 7.1) compete freely for the same binding site [13].

Finally, there is a strict steric requirement in the active inhibitor molecules that enables them to fit into the binding niche. A guide to the correct steric and electronic requirements and to the correct "fit" of the inhibitor molecule into the herbicide binding niche can be obtained by comparison of herbicide molecules with the native binding substrate, plastoquinone (Section 7.3). For example, this comparison reveals that the lipophilic substituent, taking the place of the isoprenoid side chain of the plastoquinone, can be much smaller but must form the correct angle (120°) with the hydrophilic head of the inhibitor molecule.

The number of compounds given in Table 7.1 for the different structural groups differs widely between the different groups. The discovery of active compounds has been particularly successful in the substituted urea and the *s*-triazine groups. Within these groups, however, only a few of the many compounds developed have gained commercial importance. Still, the possibility of large variations of a herbicidally active basic structure, without loss of herbicidal activity (such as in the substituted urea or the *s*-triazine molecule), greatly increases the chances of finding commercially useful (selective) representatives in these series. A more recent illustration of this is the wide range of herbicidally active molecules based on the sulfonylurea core (Chapter 13).

7.3 METHODS OF STUDYING HERBICIDAL INHIBITION OF PHOTOSYNTHESIS

One of the most popular and convenient methods to test inhibitors of photosystem II is with isolated chloroplast membranes or, more precisely, with isolated thylakoid membranes. It does not really matter which plant species is used as a source material, but availability, ease of procedure, and yield of photosynthetically active thylakoid membrane preparations have favored spinach, sugar beet, and lettuce leaves. Though the inhibition constant obtained does not depend on the plant species, as long as triazine-resistant plants are avoided, the conditions of growth of the plant and of storage of the leaves can clearly influence the value obtained. It is most convenient to extract fresh leaves and store the chloroplast suspensions in 10 percent glycerol buffers in liquid nitrogen, in which they maintain their activity for several months, or longer. This procedure ensures material of constant quality over relatively long time spans.

The present view of the chloroplast electron transport system from water to $NADP^+$ is presented in Figure 7.2 [14–16]. Three integral membrane protein complexes can be distinguished: photosystem II with the attached water splitting system (Mn) and cytochrome b_{559} (cyt. b_{559}), photosystem I, and, between these

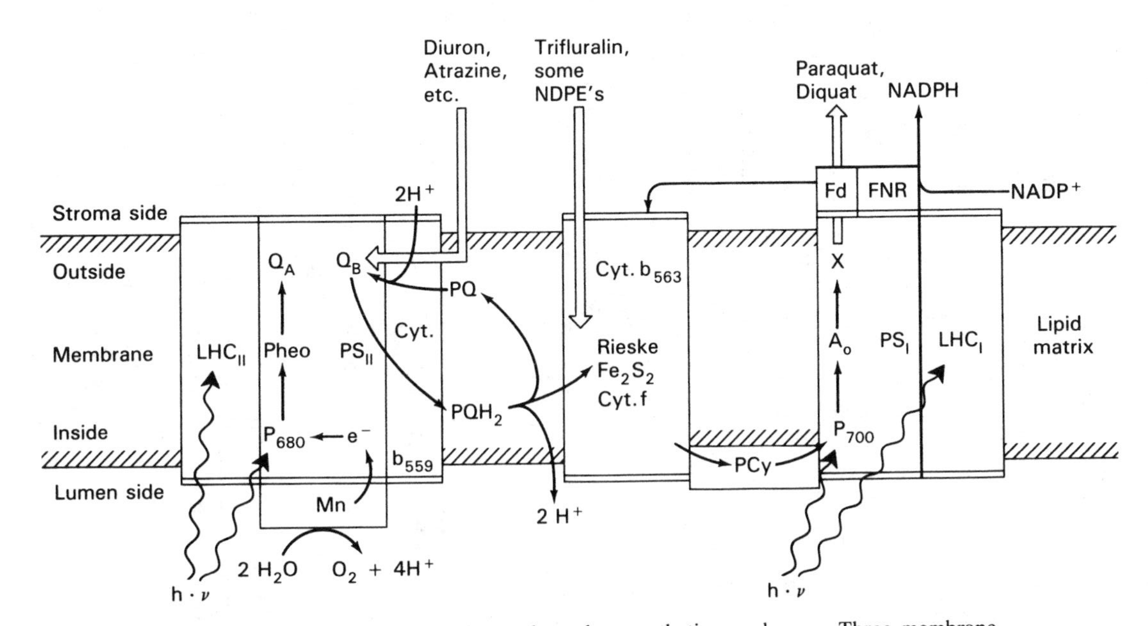

Figure 7.2 Electron flow in higher plant photosynthetic membranes. Three membrane integrated complexes (Photosystem II, Rieske iron-sulfur protein, and Photosystem I) are connected by mobile electron carriers (plastoquinone, plastocyanin). Inhibition and interference sites of herbicidal compounds are indicated by broad arrows.

two, the Rieske iron-sulfur protein with the cytrochromes f and b_{563}. Electron flow between these different complexes is mediated by plastoquinone (PQ) and plastocyanin (PCy). Light harvesting complexes (LHCII and LHCI) are associated with their respective photosystems; they contain the pigments that absorb light energy ($h \cdot \nu$), and they also transfer the absorbed energy to the respective photosynthetic reaction center pigments P_{680} (PS II) and P_{700} (PS I).

For measuring photosynthetic electron flow in isolated thylakoids, assay of the Hill reaction with potassium ferricyanide (FeCy) is perhaps the most convenient method. The segment of photosynthetic electron flow measured thereby includes photosystem II. However, since FeCy is not always reduced at a single defined site in the electron transport system and may, depending on the intactness of the chloroplast membranes, be reduced at several locations between PQ and ferredoxin (Fd), inhibition sites other than photosystem II are not always included when FeCy is used as a Hill reagent. As is indicated in Figure 7.2, some dinitroanilines (e.g., trifluralin) [17–19] and nitrodiphenylethers (NDPEs) [20], at elevated concentrations (i.e., in excess of 10 μM), inhibit photosynthetic electron flow at the reducing side of the plastoquinone cycle. Also indicated in Figure 7.2 is cyclic electron flow mediated by Fd, the reduction of $NADP^+$ mediated by Fd and ferredoxin-NADP-oxidoreductase (FNR), and the liberation of protons (acidification) in the thylakoid interior (lumen side) by the active electron transport system. The substructure of the photosystem II reaction center protein will be discussed in the next section.

The inhibition constant used in electron flow measurements is usually the molar inhibitor concentration required for 50 percent inhibition (I_{50} concentration), expressed as the negative logarithm (base 10, pI_{50} value). The pI_{50} value of a "good" inhibitor of photosynthetic electron transport should be around 6.5, representing a concentration of 0.3 μM. However, there is a particular problem with the interpretation of pI_{50} values that should be discussed here. The pI_{50} value is a composite value containing the contributions of both (a) the hydrophilic-lipophilic distribution of the inhibitor between the aqueous phase of the test buffer and the lipophilic phase of the thylakoid membranes and (b) the actual binding constant at the binding site. Table 7.2 gives an example of the accumulation or "concentration" of atrazine in the unicellular alga *Scenedesmus acutus* [21]. This system differs from the isolated thylakoids considered before, but still demonstrates the magnitude of the concentration increase in intact cells.

In isolated suspended thylakoid membranes, the lipophilic-hydrophilic distribution can be compared with the log octanol-water partition coefficient (log P or log K_{ow}), which can be measured easily by comparative reversed phase HPLC [22]. A second parameter of importance, which does not strictly parallel the log P value, is the water solubility. A correlation of pI_{50} with water solubilities of 13 phenylurea compounds is shown in Figure 7.3 [23]. For discussion, let us first ignore the possibility of differences in the binding constant at the binding site. It is clear, however, that two structural series emerge for which there is a clear dependency of the pI_{50}-value on water solubility. These may be series of compounds with very

TABLE 7.2 ACCUMULATION OF ATRAZINE IN *Scenedesmus acutus* FROM CULTURE MEDIUM CONTAINING DIFFERENT ATRAZINE CONCENTRATIONS [21].

Atrazine concentration ($M \times 4.65$)	Accumulation factor	
	Living cells	Dead cells
10^{-9}	90	12
10^{-8}	60	9
10^{-7}	35	9
10^{-6}	10	9

similar binding constants, and for which a relationship between pI_{50} and water solubility (lipophilic-hydrophilic distribution behavior) clearly emerges. The result, as expected, is that the more lipophilic compounds (which have lower water solubilities) are the more potent inhibitors in this system.

At the same time, if the reasoning given above is correct, the compounds in Figure 7.3 that are connected by a line should all have similar binding constants. The variation in the measured pI_{50} value would then largely be caused by differences in the lipophilicity of the compounds. For the development of a highly active preemergence, root-absorbed photosystem II-inhibiting herbicide, an optimum combination of water solubility and log P (for high uptake and mobility in the plant;

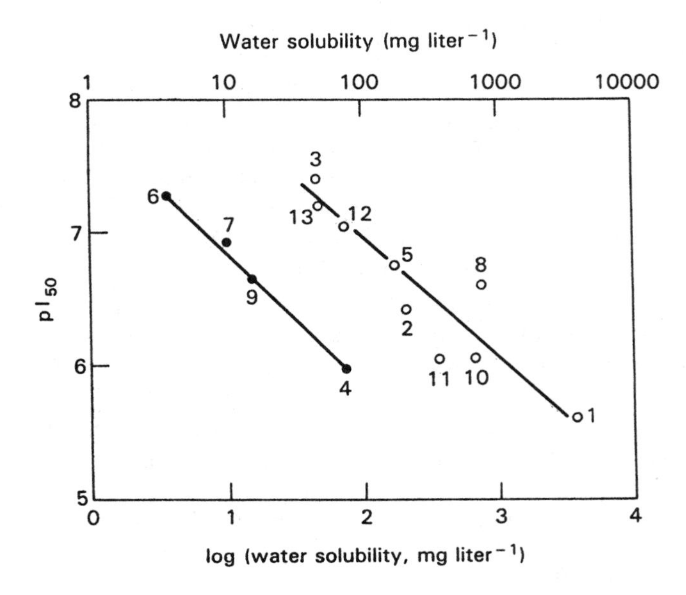

Figure 7.3 Relationship between pI_{50} and log (water solubility) for 13 phenylurea herbicides [23]. (1) fenuron, (2) monuron, (3) diuron, (4) fluometuron, (5) isoproturon, (6) chloroxuron, (7) chlortoluron, (8) metoxuron, (9) dimefuron, (10) monolinuron, (11) metobromuron, (12) linuron, (13) chlorbromuron.

see Chapters 4 and 5), optimum partitioning into the thylakoid membranes, and of course a high binding constant (i.e., an optimum fit in the binding niche) should all be realized. Metribuzin is one compound that comes close to meeting these requirements.

In the thylakoid membrane suspension test system, the herbicide is distributed between the aqueous test buffer and the lipophilic membrane phase according to its hydrophilic-lipophilic distribution behavior. Since the buffer-membrane system differs profoundly from the water-octanol system, the log P values can be used only for a comparative estimation. A delayed hydrophilic-lipophilic equilibrium can also occur (e.g., with bentazon and ioxynil) [24]. The binding niche in the herbicide binding D-1 protein is accessible only from the interior lipid membrane matrix, where the natural substrate plastoquinone also moves by two-dimensional diffusion (Section 7.4). The herbicide must move in a similar manner in the membrane to reach its binding site. In this situation the physiology of the plant is also of importance. Leaves may grow in the shade or in the sun, they may be young or old, and they respond to water supply and temperature, only to name some of the more important influences. Thylakoid membranes prepared from such differently adapted tissues differ in chlorophyll and lipid contents and can, via altered hydrophilic-lipophilic distribution behavior, lead to clearly different pI_{50} values for the same herbicidal compound. Since thylakoid electron transport test systems are usually pipetted to a constant chlorophyll concentration, the variability of this parameter in different thylakoid membranes can also be a source of variation.

From the previous discussion of the problems and peculiarities concerning the pI_{50} value obtained in isolated thylakoid membranes, it is clear that in vivo measurements of herbicidal photosynthesis inhibition are very important. Infrared gas analysis of CO_2 uptake measures the final result of photosynthesis and is therefore less specific than other parameters more closely related to the photosynthetic electron transport system.

The most important method of measuring in vivo herbicide interference with photosynthetic electron transport involves measurement of fluorescence when dark-adapted leaves are irradiated with photosynthesis-saturating light (Figure 7.4) [25–29]. A very rapid rise within less than 0.1 μs leads to a fluorescence level called F_0. At this time electrons start to flow through photosystem II into the plastoquinone pool, which is quite large (Figure 7.2). If the transfer of electrons to Q_B and into the plastoquinone pool is inhibited by a herbicide, the excitation energy builds up in photosystem II and leads to increased fluorescence emission (Figure 7.4, traces 1, 2, and 3). Increasing herbicide concentrations therefore lead to increasing F_0 and F_{max}; the rapid fluorescence rise is increased because of the closed Q_B sites (no electron flow possible), and the induction kinetics are changed. "Variable fluorescence" (F_{var}) is usually calculated as follows:

$$F_{var} = \frac{F_{max} - F_o}{F_{max}}$$

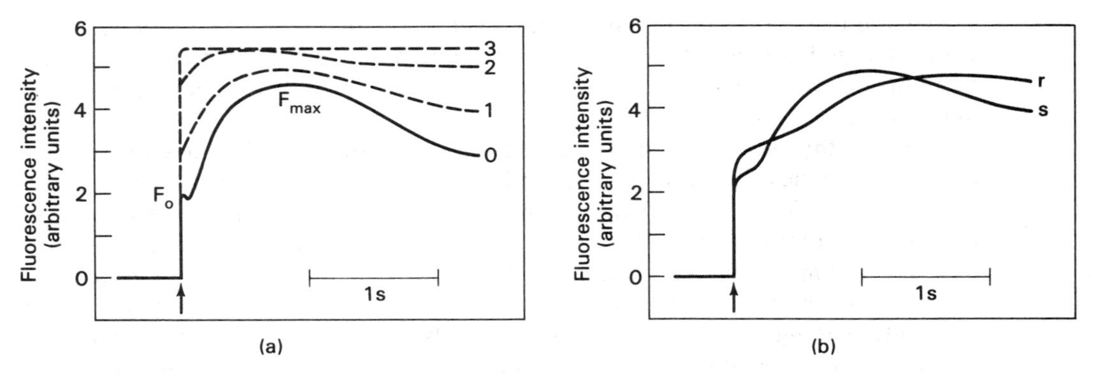

Figure 7.4 (a) Prompt fluorescence kinetics of undisturbed (0) and progressively inhibited (1, 2, 3) whole-leaf photosynthesis after dark adaptation. (b) Similar curves for wild-type (*s*) and triazine-resistant (*r*) *Brassica napus* leaves [28].

F_{var} is usually 0.5 to 0.6 in resting leaf tissue and is reduced to zero with increasing herbicidal inhibition. The exact shape of the fluorescence curve depends very much on the pretreatment and physiology of the tissue and on the conditions of measurement. In some cases (low excitation light intensity) the leveling at F_0 can be very minor, and F_0 is therefore sometimes not easily recognized. The time lapse to F_{max} can be 0.4 to several seconds. With increasing intensity of the triggering light F_0 and F_{max} both rise, and the shape of the curve changes considerably [30]. New developments in fluorescence measurement techniques allow the separate measurement of photochemical and non-photochemical quenching in continuous light [30a,b].

Two other related procedures have been proposed for the study of in vivo photosynthetic capacity: thermoluminescence [31, 32] and photoacoustic spectroscopy [33, 34]. In thermoluminescence, a thylakoid preparation is frozen to −195°C in white light (trapping of excitation energy) and subsequently warmed up at a constant rate, for example, 10° C/minute, to yield the thermoluminescence "glow curve." This glow curve has several peaks that are related to different charge recombinations. The shape of the glow curve and the position of the bands are affected by various factors, among them photosynthesis-inhibiting herbicides. The consensus, based on these observations, is that thermoluminescence is generated in photosystem II. It has been claimed that the glow curves are specific for specific photosynthesis inhibitors.

In photoacoustic spectroscopy (PAS), the heat emitted due to nonradiative de-excitation is measured. The leaf material is illuminated with chopped light in a closed compartment, and the pulsed heat emissions that create pressure changes in the gas phase are measured by a sensitive microphone. After treatment with an inhibitor of photosynthesis, the PAS signal is strongly increased, similar to the increase in fluorescence described above.

7.4 THE HERBICIDE BINDING PROTEIN

The dramatic increase in our understanding of the structure of the chloroplast protein responsible for herbicide binding and the action of the photosystem II-inhibiting herbicides has been made possible by a number of different and, in part, unrelated lines of research. Scheme 7.1 gives an overview of some new approaches, developments, and results that have contributed to our present understanding and model of the "herbicide binding protein." These are described here in some more detail:

1. Mutants of weed plants with very high resistance to *s*-triazines were found to have an altered 32 kD thylakoid membrane protein [35].

2. This same protein could be labeled by an azido-atrazine which, when activated by UV light in the presence of thylakoid membranes, was bound to met-214 of the 32 kD protein (D-1; Figures 7.5, 7.6, Table 7.3) [36, 37]. At the same time it became clear that the 32 kD protein is identical to the previously well-known "rapidly turning over protein" that is constantly formed and replaced in the light.

3. Displacement of plastoquinone in the thylakoid membrane by herbicides.

4. Sequencing of the now well-defined protein. The sequences that have been obtained clearly show that the psbA gene is highly conserved in higher plants and algae (Table 7.3). The understanding of the precise structure and function of the encoded D-1 protein was eventually made possible after it was discovered that there is both a structural and functional homology with the bacterial photosyn-

1. Triazine-resistant mutants (higher plants, algae, bacteria).
2. Labeling of the binding protein with azido-atrazine at met-214.
3. Displacement of functional plastoquinone from its binding site by herbicides inhibiting at photosystem II.
4. Base and amino acid sequences of psbA genes and gene products from wild type and mutant plants.
5. Hydropathy plotting of membrane-integrated proteins.
6. X-ray structure of the *Rhodopseudomonas viridis* photosynthetic reaction center protein at 3 Å resolution.

Scheme 7.1 A summary of research lines and areas of research progress that have contributed to the development of the herbicide-binding site model described in the text and in Figures 7.5, 7.6, and 7.7 and Table 7.4.

TABLE 7.3 NUCLEOTIDE BASE AND AMINO ACID SEQUENCES OF THE psbA GENES, CODING FOR THE CHLOROPLAST D-1 PROTEIN, FROM DIFFERENT PLANT SPECIES. Only amino acids 207–288, which form the herbicide/plastoquinone binding niche and part of the spans IV and V, are presented. Amino acid differences in the *Anacystis nidulans* sequence are given in the last row. (Compare with Figures 7.5, 7.6.)

207	208	209	210	211	212	213	214	215	216	217	218	219	220	221	222
gly	gly	ser	leu	phe	ser	ala	met	his	gly	ser	leu	val	thr	ser	ser
GGC	GGC	TCC	CTA	TTC	AGT	GCT	ATG	CAT	GGT	TCC	TTG	GTA	ACT	TCT	AGT
GGC	GGC	TCC	CTA	TTC	AGT	GCT	ATG	CAT	GGT	TCC	TTG	GTA	ACT	TCT	AGT
GGC	GGC	TCC	CTA	TTC	AGT	GCT	ATG	CAT	GGT	TCC	TTG	GTA	ACT	TCT	AGT
GGC	GGC	TCC	CTA	TTT	AGT	GCT	ATG	CAT	GGT	TCC	TTG	GTA	ACT	TCT	AGT
GGC	GGC	TCC	CTA	TTT	AGT	GCT	ATG	CAT	GGT	TCC	TTG	GTA	ACT	TCT	AGT
GGT	GGT	TCA	TTA	TTC	TCA	GCT	ATG	CAC	GGT	TCT	TTA	GTT	ACT	TCA	TCT
GGT	GGT	TCG	CTG	TTC	TCG	GCA	ATG	CAC	GGT	TCG	TTG	GTG	ACC	AGC	TCG
—	—	—	—	—	—	—	—	—	—	—	—	—	—	—	—

223	224	225	226	227	228	229	230	231	232	233	234	235	236	237	238
leu	ile	arg	glu	thr	thr	glu	asn	glu	ser	ala	asn	glu	gly	tyr	arg
TTG	ATC	AGG	GAA	ACC	ACA	GAA	AAT	GAA	TCT	GCT	AAT	GAA	GGT	TAC	AGA
TTG	ATC	AGG	GAA	ACC	ACA	GAA	AAT	GAA	TCT	GCT	AAT	GAA	GGT	TAC	AGA
TTG	ATC	AGG	GAA	ACC	ACA	GAA	AAT	GAA	TCT	GCT	AAT	GAA	GGT	TAC	AGA
TTG	ATC	AGG	GAA	ACC	ACA	GAA	AAT	GAA	TCT	GCT	AAC	GAA	GGT	TAC	AGA
TTG	ATC	AGG	GAA	ACC	ACA	GAA	AAT	GAA	TCT	GCT	AAT	GAA	GGT	TAC	AGA
TTA	ATC	CGT	GAA	ACA	ACT	GAA	AAC	GAA	TCA	GCT	AAC	GAA	GGT	TAC	CGT
TCG	GTG	CGT	GAG	ACG	ACC	GAG	ACC	GAG	AGC	CAA	AAC	TAC	GGC	TAC	AAA
—	val	—	—	—	—	—	thr	—	—	gln	—	tyr	—	—	lys

239	240	241	242	243	244	245	246	247	248	249	250	251	252	253	254
phe	gly	gln	glu	glu	glu	thr	tyr	asn	ile	val	ala	ala	his	gly	tyr
TTC	GGT	CAA	GAG	GAA	GAA	ACT	TAT	AAC	ATC	GTA	GCC	GCT	CAT	GGT	TAT
TTC	GGT	CAA	GAG	GAA	GAA	ACT	TAT	AAC	ATC	GTA	GCC	GCT	CAT	GGT	TAT
TTC	GGT	CAA	GAG	GAA	GAA	ACT	TAT	AAT	ATC	GTA	GCC	GCT	CAT	GGT	TAT
TTC	GGT	CAA	GAG	GAA	GAA	ACT	TAT	AAC	ATC	GTA	GCT	GCT	CAT	GGT	TAT
TTC	GGT	CAA	GAG	GAA	GAA	ACT	TAT	AAT	ATC	GTA	GCT	GCT	CAT	GGT	TAT
TTC	GGT	CAA	GAA	GAA	GAA	ACT	TAC	AAC	ATT	GTA	GCT	GCT	CAT	GGT	TAC
TTT	GGT	CAA	GAG	GAA	GAG	ACC	TAC	AAC	ATC	GTG	GCA	GCC	CAC	GGT	TAC
—	—	—	—	—	—	—	—	—	—	—	—	—	—	—	—

255	256	257	258	259	260	261	262	263	264	265	266	267	268	269	270
phe	gly	arg	leu	ile	phe	gln	tyr	ala	ser	phe	asn	asn	ser	arg	ser
TTT	GGC	CGA	TTG	ATC	TTC	CAA	TAT	GCT	AGT	TTC	AAC	AAC	TCT	CGT	TCG
TTT	GGC	CGA	TTG	ATC	TTC	CAA	TAT	GCT	AGT	TTC	AAC	AAC	TCT	CGT	TCG
TTT	GGC	CGA	TTG	ATC	TTC	CAA	TAT	GCT	AGT	TTC	AAC	AAC	TCT	CGT	TCG
TTT	GGT	CGA	TTG	ATC	TTC	CAA	TAT	GCT	AGT	TTC	AAC	AAC	TCT	CGT	TCT
TTT	GGT	CGA	TTG	ATC	TTC	CAA	TAT	GCT	AGT	TTC	AAC	AAC	TCT	CGT	TCT
TTT	GGT	CGT	CTA	ATC	TTC	CAA	TAC	GCT	TCT	TTC	AAC	AAC	TCT	CGT	TCA
TTC	GGT	GCG	TTG	ATC	TTC	CAA	TAC	GCA	TCG	TTC	AAC	AAC	AGC	CGT	TCG
—	—	—	—	—	—	—	—	—	—	—	—	—	—	—	—

271	272	273	274	275	276	277	278	279	280	281	282	283	284	285	286
leu	his	phe	phe	leu	ala	ala	trp	pro	val	val	gly	ile	trp	phe	thr
TTA	CAC	TTC	TTC	CTA	GCT	GCT	TGG	CCT	GTA	GTA	GGT	ATC	TTG	TTT	ACC
TTA	CAC	TTC	TTC	CTA	GCT	GCT	TGG	CCT	GTA	GTA	GGT	ATC	TTG	TTT	ACC
TTA	CAC	TTC	TTC	CTA	GCT	GCT	TGG	CCT	GTA	GTA	GGT	ATC	TTG	TTT	ACC
TTA	CAC	TTC	TTC	TTA	GCT	GCT	TGG	CCG	GTA	ATC	GGT	ATT	TTG	TTT	ACT
TTA	CAC	TTC	TTC	TTA	GCT	GCT	TGG	CCT	GTA	GTA	GGT	ATT	TTG	TTT	ACT
TTA	CAC	TTC	TTC	TTA	GCT	GCT	TGG	CCG	GTA	ATC	GGT	ATT	TTG	TTC	ACT
CTG	CAC	TTC	TTC	CTG	GGT	GCA	TGG	CCG	GCT	GTG	GGC	ATC	TGG	TTT	ACC
—	—	—	—	—	gly	—	—	—	—	—	—	—	—	—	—

287	288
ala	leu
GCT	TTA
GCT	TTA
GCT	TTA
GCT	TTG
GCT	TTA
GCT	TTA
TCC	ATG
ser	met

Key: Row

1 *Nicotiana*
2 *Nicotiana tabacum*
3 *Nicotiana debnei*
4 *Solanum nigrum*
5 *Amaranthus hybridus*
6 spinach
7 *Chlamydomonas reinhardtii*
8 *Anacystis nidulans*
9 *Anacystis nidulans*

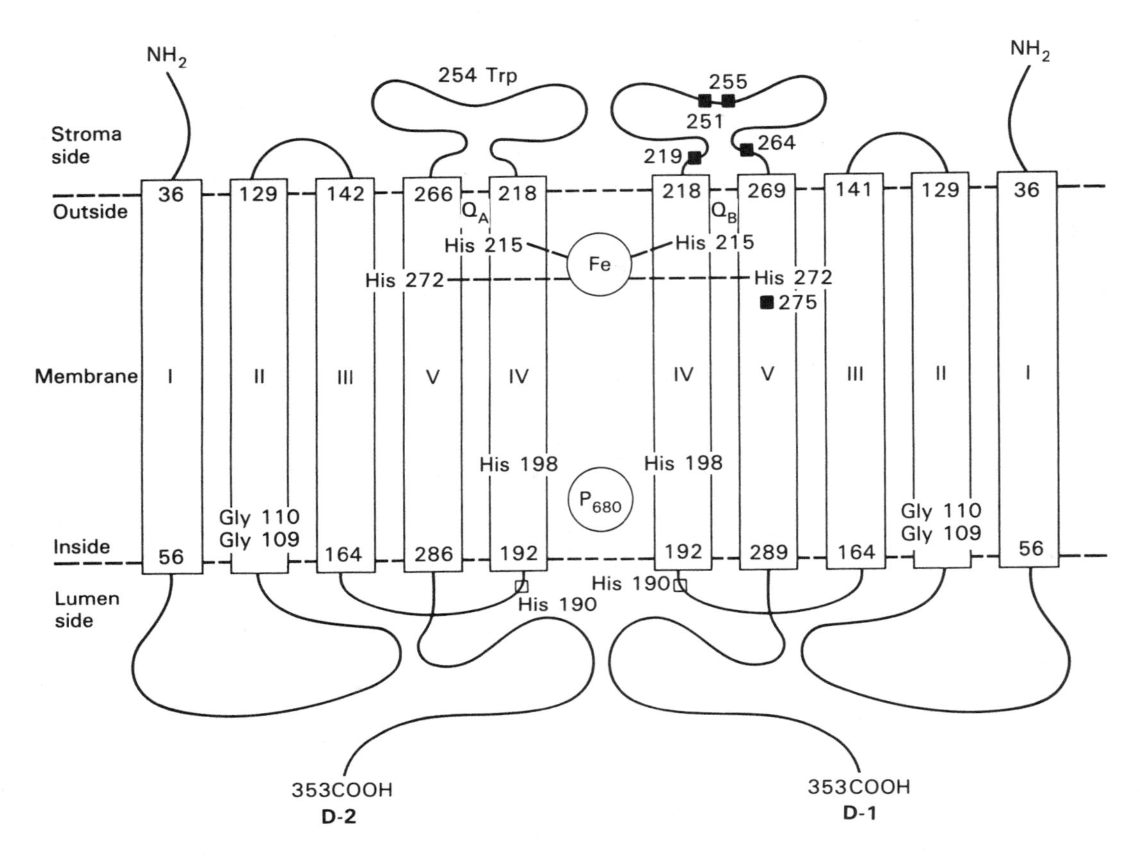

Figure 7.5 Schematic representation of the herbicide-binding photosynthetic reaction center proteins D-1 and D-2. Solid squares indicate mutation sites that confer herbicide resistance [2, 3]. The membrane spanning helices I to V are represented by open rectangles.

thetic reaction center protein. The important leap for the conversion of structure to function therefore eventually came from the elucidation of the bacterial tertiary protein structure [2–4].

5. Membrane-integrated proteins are now generally considered to consist of "spans" and "loops" (Figure 7.5). The membrane-crossing spans contain about 20–25 hydrophobic amino acids which form α-helices (3.6 amino acids per turn), whereas the loops contain the more hydrophilic amino acids and connect the spans [14]. In order to locate the spans in an amino acid chain, a hydropathy index plot is generated: the hydrophobicity index of a window of 4–7 amino acids is moved through the chain, and the indices are plotted against the amino acid number. Sometimes shorter or longer hydrophobic stretches than the normal 20–25 amino

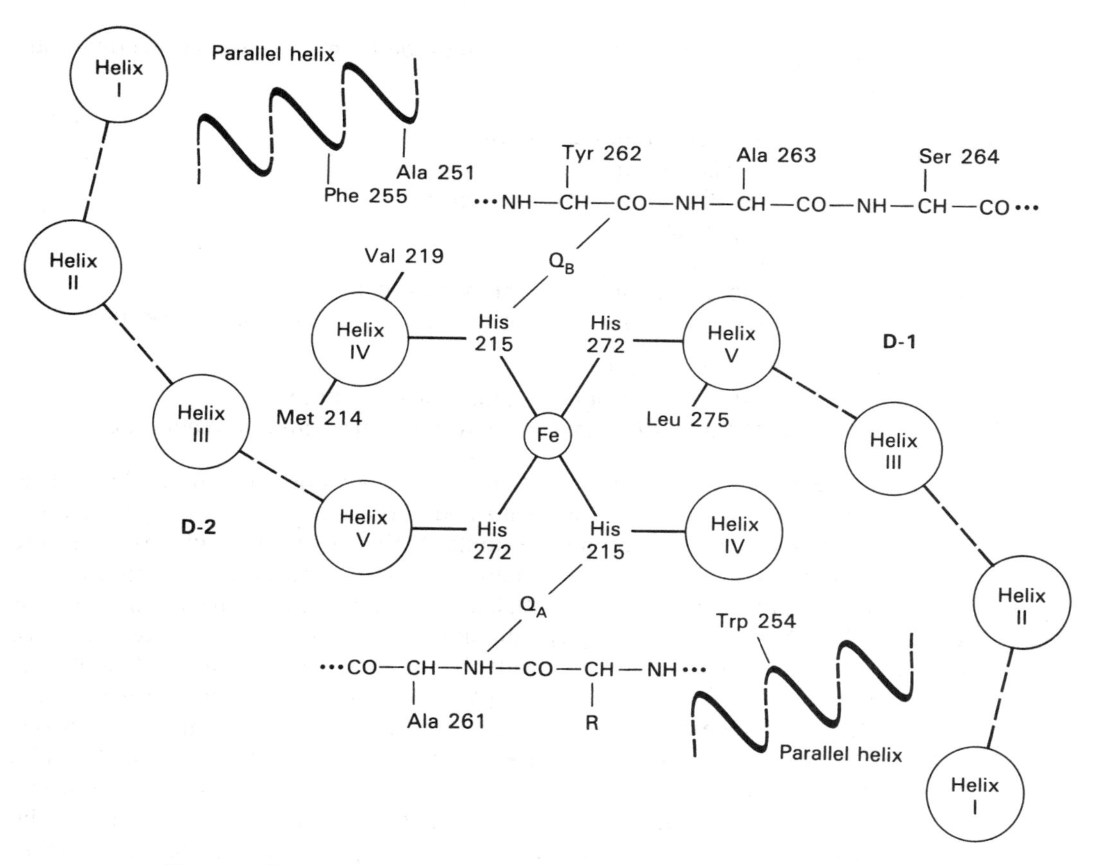

Figure 7.6 Top view of the photosynthetic reaction center proteins D-1 and D-2. The helical membrane spans I to V are represented by circles. The amino acid positions important for herbicide (plastoquinone) binding and resistance are marked [4].

acids required for a membrane span are found, which makes the interpretation difficult. This was also the case with the 32 kD herbicide binding protein.

6. Elucidation of the X-ray structure of the bacterial *Rhodopseudomonas viridis* photosynthetic reaction center protein complex provided answers to many of the remaining questions. This was a formidable achievement, since the bacterial photosynthetic reaction center is the largest and most complex structure yet determined. (In 1988, Drs. H. Michel, J. Deisenhofer, and R. Huber were awarded the Nobel Prize in chemistry for this work.) Some similarities between bacterial photosynthesis and photosystem II of higher plants were already known, but the homology of the bacterial L (light) chain with D-1 and the M (medium) chain with D-2 has only recently been detected. Many different names have been used for the

different proteins considered here, and it may be helpful to list the synonymous designations:

D-1 protein = herbicide binding protein
= 32 kD protein
= rapidly turning over protein
= Q_B protein
= B protein
= psbA gene product (chloroplast genome)
≅ L (light) chain in bacterial photosynthetic reaction center

D-2 protein = 34 kD protein
= psbD gene product (chloroplast genome)
≅ M (medium) chain in bacterial photosynthetic reaction center

The bacterial photosynthetic reaction center contains an additional H (heavy) chain for which (at present) no counterpart is known in higher plants. However, the D-1 and D-2 proteins are similar to the bacterial L and M polypeptide chains if the structurally and functionally important aspects are compared: both form 5 spans; span II always starts with a glycine pair; the histidines, important for iron, plastoquinone, and chlorophyll binding, are conserved; lysine residues are absent or very scarce; and the mutations that confer resistance to photosynthesis inhibitors cluster in the loop between spans IV and V of D-1 or L, respectively (Figure 7.5) [2]. The extensive loop between spans IV and V, which also carries most of the herbicide resistance mutations (see Section 7.5), forms a ''parallel helix'' that covers the Q_B/herbicide binding site on the stroma side of the membrane (Figure 7.6). In the native state, two plastoquinones are bound to the D-1/D-2 complex, Q_A in a nondissociative way and Q_B in a reversible way. Further examination of the structure of the bacterial photosynthetic reaction center shows that spans IV and V are tilted to the plane of the membrane by up to 38° (Figure 7.7) and that spans I to V, when viewed from the top, form a curved line (Figure 7.6). One difference in the plant D-1 and D-2 proteins is that their loops are considerably longer than in L and M, but the functional significance of this deviation is not known.

Turning to the Q_B binding site, it was mentioned previously that herbicide resistance mutations occur only in this part of the molecule (Figures 7.5, 7.6, and Table 7.4). Native Q_B binding is suggested to include quinone carbonyl interactions with his-215 and a peptide bond close to ser-264 (the triazine-resistance mutation site in higher plants) and ring-ring overlap with phe-255 (a *Chlamydomonas* mutation site, Table 7.4) [38,38a]. The complementary nondissociative Q_A binding could similarly include trp-254 on D-2. Much detailed information on proteins D-1 and D-2 and the herbicide/plastoquinone binding site has been accumulated, and the homology of the bacterial photosynthetic reaction center with the plant photosystem II reaction center has been confirmed by recent preparations (Figure 7.7) that contain chlorophyll, pheophytin (Pheo), β-carotene, and nonheme iron. The

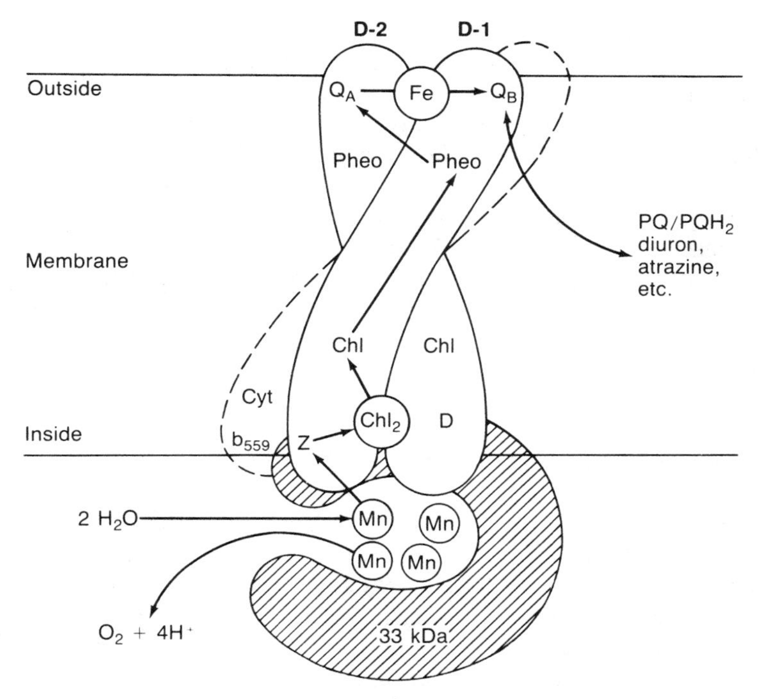

Figure 7.7 Membrane (cross-sectional) view of electron flow from water through the D-1/D-2 reaction center pigment-protein complex to plastoquinone [2].

TABLE 7.4 CHARACTERISTICS OF SOME MUTANT LINES SELECTED BY EXPOSURE TO SUBLETHAL HERBICIDE DOSES.

Species	Herbicide used in selection	Amino acid		Resistance Factor	
		Posn.	Change	Atrazine	Diuron
Higher plants	Atrazine	264	ser → gly	1000	1
Chlamydomonas	Diuron	264	ser → ala	100	10
Anacystis	Diuron	264	ser → ala	10	100
Synechococcus	Atrazine	211	phe → ser	7	2
Chlamydomonas	Diuron	219	val → ile	2	15
Chlamydomonas	Metribuzin	251	ala → val	25	5
Chlamydomonas	Atrazine	255	phe → tyr	15	0.5
Chlamydomonas	Metribuzin	275	leu → phe	1	5
Rhodopseudomonas	Terbutryn	(229)	ile → met	7	—
Rhodobacter	Terbutryn	(222)	tyr → gly	—	—
Rhodobacter		(223)	ser → pro	—	—
Rhodobacter		(229)	ile → met	—	—

complex also contains cytochrome b_{559}. Figure 7.7 indicates electron flow from water to plastoquinone in the plant photosystem II reaction center, again on the basis of information obtained from the bacterial system. The water-splitting system feeds electrons through only one branch (the D-1) via Z and the reaction center chlorophyll dimer (P_{680}) to chlorophyll a and pheophytin, crossing from there to the firmly bound Q_A on D-2. Q_A can only form a semiquinone, but can reduce Q_B to the hydroquinone. For this to occur, Q_B must acquire two protons that come from the stroma side (top of Figure 7.7), possibly aided by arginine residues in the D-1 loop above the binding site. Plastohydroquinones can then leave the D-1 binding niche through an open protein channel oriented toward the membrane's lipid interior. From there another plastoquinone molecule or, by competitive interaction, a herbicide molecule, can enter the vacant binding niche.

7.5 OCCURRENCE AND SELECTION OF MUTANTS

Herbicide-resistant mutants have aided greatly in the development of our present thinking and our understanding of the herbicide-binding protein (D-1). The selection and detection of *s*-triazine-resistant weed plants after repeated field applications over several years dates back to the early 1970s (Table 7.5) [39]. The locations were mainly in maize fields, but resistance also developed after *s*-triazine applications in orchards and along railroad tracks. The list of resistant weeds has grown considerably since then. Table 7.6 shows the results of a recent European survey,

TABLE 7.5 LIST OF EARLY REPORTS OF *s*-TRIAZINE-RESISTANT WEEDS IN NORTH AMERICA [39].

Species	Year of first report	Location of first report
Amaranthus hybridus	1972	Maryland
Amaranthus arenicola	1978	Colorado
Amaranthus powellii	1968	Washington, Oregon
Amaranthus retroflexus	1979	Ontario
Ambrosia artemisiifolia	1977	Ontario
Brassica campestris	1978	Quebec
Bromus tectorum	1977	Nebraska, Kansas, Oregon, Montana, Washington
Chenopodium album	1973	Ontario
Chenopodium missouriense	1978	Pennsylvania
Chenopodium strictum	1978	Ontario
Echinochloa crus-galli	1978	Maryland
Kochia scoparia	1976	Nebraska, Washington
Panicum capillare	1976	Michigan
Poa annua	1977	California
Senecio vulgaris	1976	Washington

TABLE 7.6 LIST OF EUROPEAN WEEDS REPORTED TO BE HIGHLY RESISTANT TO *s*-TRIAZINES [40].

Species	A	B	BG	DK	F	H	I	NL	CH	GB	D
Amaranthus bouchonii					+				+		
Amaranthus cruentus							+				
Amaranthus hybridus					+	+			+		
Amaranthus lividus									+		
Amaranthus retroflexus	+		+		+	+			+		+
Arenaria serpyllifolia					+						
Atriplex patula	+										+
Bidens tripartita	+										
Bromus tectorum					+						
Chenopodium album	+	+			+	+	+	+	+		+
Chenopodium ficifolium								+	+		+
Chenopodium polyspermum					+	+			+		
Chenopodium strictum						+					
Digitaria sanguinalis					+						
Echinochloa crus-galli					+						
Epilobium ciliatum	+	+									
Epilobium tetragonum					+						
Erigeron canadensis					+	+			+		
Galinsoga ciliata											+
Myosoton aquaticum											+
Poa annua	+	+			+			+	+		+
Polygonum convolvulus	+										
Polygonum lapathifolium					+						
Polygonum persicaria					+						
Senecio vulgaris		+	+	+	+	+		+	+	+	+
Setaria glauca					+						
Setaria viridis					+						
Setaria viridis major					+						
Solanum nigrum	+	+			+	+	+	+	+		+
Sonchus asper					+						
Stellaria media	+								+		+

A = Austria, B = Belgium, BG = Bulgaria, DK = Denmark, F = France, H = Hungary, I = Italy, NL = Netherlands, CH = Switzerland, GB = Great Britain, D = Germany but not all of these resistant plants have been studied in sufficient detail to ensure that resistance is based on a point mutation in the psbA gene (AGT → GGT, replacing glycine for serine, position 264, Table 7.3) [38]. This point mutation has so far been confirmed for *Amaranthus hybridus, Solanum nigrum, Brassica napus/campestris,* and *Senecio vulgaris*.

The reason for grouping these resistant plants together is primarily the high level of resistance (more than 10 times at the whole plant level and more than 1,000 times at the chloroplast level), which sets them aside from examples of increased tolerance (for which the level of tolerance is only 2–3 times, and develops slowly over several years; Section 6.1). Increased tolerance, where studied, is usually associated with higher herbicide degradation rates, and represents the result of a population drift caused by herbicide application over several years. Typical *s*-triazine resistance occurs spontaneously within two generations, with a probability close to 10^{-8}, and does not change its level once it has occurred. Since one chloroplast contains 150–300 plastome (chloroplast genome) copies, and since the behavior of the mutation in the chloroplast is recessive, there must be an extrachloroplastic mechanism that favors a homozygous chloroplast population. This mechanism leads to homozygous resistance in the second generation when the herbicide pressure continues [41–43]. The intermediate generation shows intermediate characteristics.

Selection for resistance in the field by *s*-triazines (mainly simazine and atrazine) is accelerated by the long half life (several months) of these herbicides in the soil. One method of studying the level of resistance in intact leaves is by fluorescence measurements (Figure 7.4b). *s*-Triazine-resistant tissue shows a herbicide-insensitive fluorescence signal, but with altered kinetics. The higher intermediate fluorescence level during the transient rise (from F_0 to F_{max}) has been interpreted to reflect a lower electron transfer rate from Q_A to Q_B.

The pattern of cross-resistance in *s*-triazine-resistant chloroplasts isolated from resistant weeds has been studied extensively [44, 45]. The pattern is similar in chloroplasts isolated from different resistant weed species, but different in herbicide-resistant algae (Table 7.4), adding support to the suggestion that the same mutation is responsible for *s*-triazine resistance in the different resistant weed species. Approximate cross-resistance values of triazine-resistant weeds to other classes of photosynthetic inhibitors are shown in Table 7.7.

It is clear that the groups listed in Table 7.7 do not really behave as groups. Exceptions and intermediate cases are possible, and an even better fit into the mutant binding niche is possible, leading to increased sensitivity (values < 1.0). This is not only the case for bentazon and for phenols but also for some very lipophilic *as*-triazinones, confirming the notion that each inhibitor molecule must be considered individually [44]. The types of binding have tentatively been classified as "serine type" or "histidine type," reflecting binding to different "subreceptors" in the binding niche. Binding of herbicidal inhibitors may occur either closer to serine (e.g., *s*-triazines) or closer to histidine (e.g., ioxynil), leading to the different behavior of these inhibitors in the ser-264 mutant.

TABLE 7.7 CROSS RESISTANCE OF CHLOROPLASTS FROM TRIAZINE—RESISTANT WEEDS TO OTHER CLASSES OF PHOTOSYNTHETIC ELECTRON TRANSPORT INHIBITORS.

Herbicide group	Resistance factor I_{50} resistant/susceptible
s-Triazines	100–1000
as-Triazinones	10–100
Uracils	10–50
Biscarbamates	10–50
Heterocyclic ureas	10–100
Phenylureas	1–50
Hydroxybenzonitriles	1–3
Bentazon	0.5
Phenols	0.1–0.5

Selection of herbicide-resistant mutants in unicellular algae and photosynthetic bacteria has been very successful (Table 7.4) [46–55]. Of the many resistant selections obtained, however, many have been found to be unstable after herbicide withdrawal, and many have other resistance mechanisms. For example, altered herbicide uptake appears to be a common resistance mechanism in these mutants.

Point mutations in the D-1 protein that confer herbicide resistance are listed in Table 7.4, and are also indicated in Figures 7.5 and 7.6. Additional mutation sites are located around the plastoquinone binding niche, and some of these mutation sites are also known from several algal species [52–55]. The known herbicide-resistant mutants all clearly point to the plastoquinone Q_B binding niche. However, slowed electron transfer from Q_A to Q_B, which appears to be a consequence of the mutation at the ser-264 site, has not been found in association with other mutation sites. It may be possible, therefore, to find higher plant mutants with an unaltered electron transport rate.

Although it is clear from field work with triazine-resistant *Brassica napus* "Canola" cultivars, which have been made resistant by backcrossing with resistant *Brassica campestris,* that triazine resistance leads to a 10–20 percent "yield penalty" [56, 57], the physiological and bioproductive consequences of this mutation have been repeatedly questioned. The slower $Q_A \rightarrow Q_B$ electron transfer has already been mentioned, but physiological conditions apparently exist under which factors other than the electron transfer rate through Q_B limit growth, so that the mutation does not necessarily result in measurable differences in growth. For example, whole-chain electron transport is unaltered in some experimental systems [58, 59]. Under normal growth conditions, however, chlorophyll a/b decreases [60], chloroplast lipid composition changes [61, 62], and changes occur in temperature sensitivity [63–65]. Decreased stability of the oxygen-evolving system and decreased abscisic acid concentrations [66] have also been reported. Some, if not all, of these changes can be interpreted as consequences of decreased photosynthate production, leading to adaptive responses that are at least in part related to the "shade adaptation response" [67] (see Chapter 16).

7.6 PHYTOTOXIC EFFECTS OF PHOTOSYNTHETIC ELECTRON TRANSPORT INHIBITORS

In high light, that is, at irradiance levels of 1,000–1,700 $\mu E \cdot m^{-2} \cdot s^{-1}$ photosynthetically active radiation, the first necrotic effects of photosynthetic electron transport inhibitors occur approximately 5–10 hours after the cessation of photosynthesis [68]. Some of the physiological and biochemical damage indicators that can be measured during this time (and thereafter) are presented in Figure 7.8 [69]. The most sensitive process is CO_2 fixation, which is lost within a few hours after herbicidal inhibition of photosynthetic electron transport. Also decreasing from the onset of herbicidal action are (a) β-carotene and (b) in vitro photosystem II activity in thylakoid membranes isolated from treated tissue and tested with either ferricyanide (reduction after the herbicide inhibition site) or silicomolybdate (reduction before and after the herbicide inhibition site). Photosystem I activity, measured as photooxidation of ascorbate, is temporarily increased, either because of increased ascorbate accessibility or because of more efficient energy transfer from the damaged photosystem II to the still-intact photosystem I, and only decreases in later stages of herbicidal action. Increases in lutein, and more specifi-

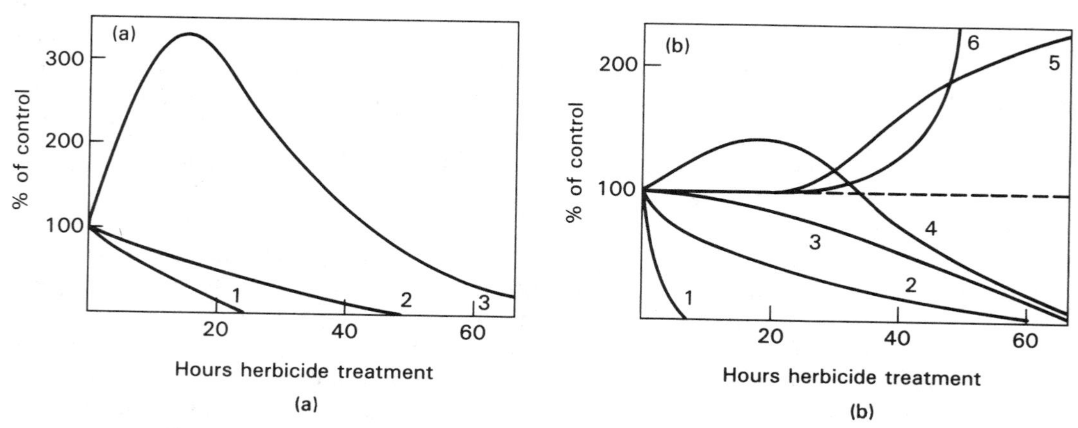

Figure 7.8 Progressive damage in cucumber cotyledons after treatment with a photosystem II inhibiting herbicide (monuron). Measurements were made either in chloroplasts isolated from treated leaves (a: 1—reduction of ferricyanide, 2—reduction of silicomolybdate, 3—photooxidation of ascorbate) or in intact treated leaves (b: 1—CO_2-fixation, 2—content of β-carotene, 3—total chlorophyll, 4—lutein, 5—malondialdehyde, 6—ethane evolution) [69].

cally in malondialdehyde levels, indicate photooxidation of membrane lipids, as does the evolution of ethane. Ethane evolution is close to zero in control tissues and rises sharply after a delay of several hours in herbicide-treated tissues. β-carotene is known to be a quencher of singlet oxygen ($^1O_2$) and protects chlorophyll from photooxidation for several hours. The interpretation of singlet oxygen as one of the primary toxicants generated in herbicidally inhibited chloroplasts is supported by the protective action of the $^1O_2$ quencher DABCO (1,4-diazabicyclo(2,2,2)octane) and also by the protection of diuron-treated chloroplasts by a bypass electron transfer system for the inhibition site (TMPD = N,N,N',N'-tetramethyl-*o*-phenylenediamine/ascorbate). The general background and mechanisms of oxidative membrane damage are discussed in Chapter 9. The mode of toxic action, however, also includes damage by more direct radical/pigment/lipid interactions.

On an ultrastructural level, chloroplast swelling and membrane rupture can be observed after 5–10 hours, depending on the irradiance level. At the highest irradiance levels, membrane rupture occurs prior to any other visible effects, indicating the high sensitivity of the membranes, and the chloroplast envelope in particular, to damage by oxidants [69].

In the in vivo situation in the greenhouse or in the field, photosynthesis-inhibiting herbicides are most commonly applied to the soil and must be taken up by the roots and transported to the leaves. Numerous experiments have shown that the amount of herbicide reaching the leaf and the herbicidal site in the chloroplast is linearly correlated with the amount of water moving through the plant (Chapter 5). Herbicidal action, therefore, is strongly increased by high transpiration rates [70]. Once in the leaf, and more precisely in the chloroplast, the activity of the herbicide is very much determined by its hydrophilic-lipophilic distribution behavior, as discussed in Section 7.3.

Besides inducing different kinds of energy spill (fluorescence, heat, singlet oxygen) in the blocked but otherwise active chloroplast, photosynthesis-inhibiting herbicides also block light-dependent processes other than photosynthetic CO_2-fixation, such as the reduction of nitrite to ammonia. This results in nitrite accumulation in these tissues [71, 72] (Section 16.2). The toxic herbicidal effects can, however, be adequately explained by the photooxidative and photoradical damage; the toxicity of nitrite is considerably lower, and does not appear to play any significant role in the action of these herbicides.

It has already been mentioned that the herbicide-binding D-1 protein has a very high turnover rate. It is constantly being replaced in the thylakoid membrane in the light, at a rate that is much higher than for any other chloroplast protein [3]. For example, the rate is 50 times the turnover rate of the ribulosebisphosphate carboxylase large subunit [73, 74]. Furthermore, it has been observed that the rate of degradation increases with irradiance level. It has been suggested, therefore, that the D-1 protein is itself subject to photooxidation or to free radical attack. The damaged protein would subsequently be removed by a specific protease and replaced by a newly synthesized D-1. There may even be a preformed photooxidation site, and the constant removal may be part of the chloroplast protective system against photooxidations. However, after a herbicide binds to the D-1 protein and photooxidation occurs, removal of the damaged protein from the membrane is inhibited. This would provide a mechanism for irreversible blocking of the photosynthetic electron transport system after extended illumination times in herbicidally inhibited tissues.

REFERENCES

1. Good, N. E. 1961. "Inhibitors of the Hill reaction." *Plant Physiol.*, 36: 788–803.
2. Barber, J. 1987. "Photosynthesis reaction centers: a common link." *TIBS*, 12, 321–26.
3. Mattoo, A. K., J. B. Marder, and M. Edelman. 1989. "Dynamics of the photosystem II reaction center." *Cell*, 56, 241–46.

3a. Trebst, A. 1991. "The molecular basis of resistance of photosystem II herbicides." In *Herbicide Resistance in Weeds and Crops*, J. C. Caseley, G. W. Cussans, R. K. Atkin, eds. Oxford: Butterworth-Heineman Ltd., pp. 145–164.

4. Trebst, A. 1986. "The topology of the plastoquinone and herbicide binding peptides of photosystem II in the thylakoid membrane." *Z. Naturforsch.*, 41c, 240–45.
5. Arntzen, C. J., P. Böger, D. E. Moreland, and A. Trebst, eds. 1987. "Herbicides affecting chloroplast functions." *Z. Naturforsch.*, 41c, 661–844.
6. Draber, W. 1987. "Can quantitative structure activity analyses and computer graphics assist in developing new inhibitors of photosystem II?" *Z. Naturforsch.*, 41c, 713–17.
7. Chamberlain, H. E., W. L. Kurtz, D. A. Addison, J. R. Beck, P. E. Brewer, R. B. Cooper, M. D. Hammond, M. P. Lynch, B. Miller, D. G. Ortega, S. J. Parka, C. R. Salhoff, R. D. Schultz, and H. L. Webster. 1987. "EL-177: A new pre-emergence herbicide for control of annual broadleaf and grass weeds in field corn." Proc. Brit. Crop Prot. Conf.—Weeds, Vol. 1. Thornton Heath, U.K.: BCP Publ., pp. 35–39.
8. Ducruet, J.-M., and R. Scalla. 1985. "Action of methylthiopyrimidine experimental herbicides as diuron-like inhibitors of photosynthesis." *Z. Naturforsch.*, 40c, 388–90.
9. Asami, T., S. Yoshida, and N. Takahashi. 1986. "Photosynthetic electron transport inhibition by pyrones and pyridones: structure-activity relationships." *Agric. Biol. Chem.*, 50, 469–74.
10. Asami, T., N. Takahashi, and S. Yoshida. 1987. "Synthesis of conjugated enamino compounds inhibiting photosynthetic electron transport." *Agric. Biol. Chem.*, 51, 205–10.
11. Trebst, A., and E. Harth. 1974. "Herbicidal *N*-alkylated and ring-closed *N*-acylamides as inhibitors of photosystem II." *Z. Naturforsch.*, 29c, 232–35.
12. Trebst, A., W. Donner, and W. Draber. 1984. "Structure activity correlations of herbicides affecting plastoquinone reduction by photosystem II: electron density distribution in inhibitors and plastoquinone species." *Z. Naturforsch.*, 39c, 405–11.
13. Mitsutake, K., H. Iwamura, R. Shimizu, and T. Fujita. 1986. "Quantitative structure-activity relationship of photosystem II inhibitors in chloroplasts and its link to herbicidal action." *J. Agric. Food Chem.*, 34, 725–32.

14. Murphy, D. J. 1986. "The molecular organization of the photosynthetic membranes of higher plants." *Biochim. Biophys. Acta,* 864, 33–94.

15. Anderson, J. M. 1986. "Photoregulation of the composition, function and structure of thylakoid membranes." *Annu. Rev. Plant Physiol.,* 37, 93–136.

16. Cramer, W. A., W. R. Widger, R. G. Hermann, and A. Trebst. 1985. "Topography and function of thylakoid membrane proteins." *TIBS,* 10, 125–29.

17. Moreland, D. E., and S. C. Huber. 1972. "Inhibition of photosynthesis and respiration by substituted 2,6-dinitroaniline herbicides. I. Effects on chloroplast and mitochondrial activities." *Pestic. Biochem. Physiol.,* 2, 342–53.

18. Droppa, M., S. Demeter, and G. Horvath. 1981. "Two sites of inhibition of the photosynthetic electron transport chain by the herbicide trifluralin." *Z. Naturforsch.,* 36c, 853–55.

19. Trebst, A. 1980. "Inhibitors of electron flow: tools for the functional and structural localization of carriers and energy conservation sites". In *Methods in Enzymology,* Vol. 69, A. San Pietro, ed. New York: Academic, pp. 675–715.

20. Böger, P. 1984. "Multiple modes of action of diphenyl ethers." *Z. Naturforsch.,* 39c, 468–75.

21. Böhm, H.-H., and H. Müller. 1976. "Model studies on the accumulation of herbicides by microalgae." *Naturwiss.,* 63, 296.

22. Eadsforth, C. V. 1986. "Application of reverse-phase H.P.L.C. for the determination of partition coefficients." *Pestic. Sci.,* 17, 311–25.

23. Dicks, J. W. 1978. "Inhibition of the Hill reaction of isolated chloroplasts by herbicidal phenylureas." *Pestic. Sci.,* 9, 59–62.

24. Mine, A., and S. Matsunaka. 1975. "Mode of action of bentazon: effect on photosynthesis." *Pestic. Biochem. Physiol.,* 5, 444–50.

25. Trebst, A., and H. Wietoska. 1975. "Hemmung des photosynthetischen Elektronentransports von Chloroplasten durch Metribuzin." *Z. Naturforsch.,* 30c, 499–504.

26. Voss, M., G. Renger, and C. Kötter. 1984. "Fluorimetric detection of photosystem II herbicide penetration and detoxification in whole leaves." *Weed Sci.,* 32, 675–80.

27. Habash, D., M. P. Percival, and N. R. Baker. 1985. "Rapid chlorophyll fluorescence technique for the study of penetration of photosynthetically active herbicides into leaf tissue." *Weed Res.,* 25, 389–95.

28. Ali, A., E. P. Fuerst, C. J. Arntzen, and V. Souza Machado. 1986. "Stability of chloroplastic triazine resistance in Rutabaga backcross generations." *Plant Physiol.,* 80, 511–14.

29. Ali, A., and V. Souza Machado. 1981. "Rapid detection of triazine resistant weeds using chlorophyll fluorescence." *Weed Sci.,* 21, 191–97.

30. Franck, U. F., N. Hoffmann, H. Arenz, and U. Schreiber. 1969. "Chlorophyllfluoreszenz als Indikator der photochemischen Primärprozesse der Photosynthese." *Ber. der Bunsenges. für physikalische Chemie,* 73, 871–79.

30a. Schreiber, U., U. Schliwa, and W. Bilger. 1986. "Continuous recording of photochemical and non-photochemical quenching with a new type of modulation fluorimeter." *Photosynth. Res.,* 10, 51–62.

30b. Krause, G. H., and E. Weis, 1991. "Chlorophyll fluorescence: the basics." *Annu. Rev. Plant Physiol. Mol. Biol.,* 42, 313–49.

31. Horvath, G. 1986. "Usefulness of thermoluminescence in herbicide research." *CRC Critical Reviews in Plant Sciences,* 4, 293–310.

32. Demeter, S., and Govindjee. 1989. "Thermoluminescence in plants." *Physiol. Plant.,* 75, 121–30.

33. Buschmann, C., and H. Prehn. 1986. "Photosynthetic parameters as measured via non-radiative de-excitation." In *Biological Control of Photosynthesis,* R. Marcelle, H. Clijsters, M. van Poucke, eds. Dordrecht, Netherlands: Martinus Nijhoff Publishers, pp. 83–91.

34. Buschmann, C. 1986. "Fluoreszenz- und Wärmeabstrahlung bei Pflanzen. Anwendung in der Photosyntheseforschung." *Naturwiss.,* 73, 691–99.

35. Arntzen, C. J., C. L. Ditto, and P. E. Brewer. 1979. "Chloroplast membrane alterations in triazine-resistant *Amaranthus retroflexus* biotypes." *Proc. Natl. Acad. Sci. U.S.A.,* 76, 278–82.

36. Pfister, K., K. E. Steinback, G. Gardner, and C. J. Arntzen. 1981. "Photoaffinity labeling of a herbicide receptor protein in chloroplast membranes. *Proc. Natl. Acad. Sci. U.S.A.,* 78, 981–85.

37. Oettmeier, W., K. Masson, H.-J. Soll, and W. Draber. 1984. "Herbicide binding at photosystem II. A new azido-triazinone photoaffinity label." *Biochim. Biophys. Acta,* 767, 590–95.

38. Trebst, A., B. Depka, B. Kraft, and U. Johanningmeier. 1988. "The Q_B-site modulates the conformation of the photosystem II reaction center polypeptides." *Photosynthesis Res.,* 18, 163–77.

38a. Tietjen, K. G., J. F. Kluth, R. Andree, M. Haug, M. Lindig, K. H. Müller, H. J. Wroblowsky, and A. Trebst. 1991. The herbicide binding niche of photosystem II - a model." *Pestic. Sci.,* 31, 65–72.

39. LeBaron, H. M., and J. Gressel., eds. 1982. *Herbicide Resistance in Plants.* New York: Wiley, 401 pp.

40. Barralis, G., and J. Gasquez. 1987. "Biology and distribution of weeds: studies on the distribution of triazine resistant weeds." European Weed Research Society, Basel, Newsletter No. 38, 4–9.

41. Gasquez, J., A. Al Monemar, and H. Darmency. 1985. "Triazine herbicide resistance in *Chenopodium album* L.: occurrence and characteristics of an intermediate biotype." *Pestic. Sci.,* 16, 392–96.

42. Bettini, P., S. McNally, M. Sevignac, H. Darmency, J. Gasquez, and M. Dron. 1987. "Atrazine resistance in *Chenopodium album.* Low and high levels of resistance to the herbicide are related to the same chloroplast psbA mutation." *Plant Physiol.,* 84, 1442–46.

43. Robertson, D. 1985. "Origin and phenotypic expression of recessive chloroplast mutations in higher plants: atrazine resistance as a model system." *Plant Molec. Biol. Reporter,* 3, 99–106.

44. Oettmeier, W., K. Masson, C. Fedtke, J. Konze, and R. R. Schmidt. 1982. "Effect of different photosystem II inhibitors on chloroplasts isolated from species either susceptible or resistant towards *s*-triazine herbicides." *Pestic. Biochem. Physiol.,* 18, 357–67.

45. Fuerst, E. P., C. J. Arntzen, K. Pfister, and D. Penner. 1986. "Herbicide cross-resistance in triazine resistant biotypes of four species." *Weed Sci.,* 34, 344–53.

46. Hirschberg, J., and L. McIntosh. 1983. "Molecular basis of herbicide resistance in *Amaranthus hybridus*." *Science,* 222, 1346–49.
47. Erickson, J. M., M. Rahire, J.-D. Rochaix, and L. Mets. 1985. "Herbicide resistance and cross-resistance: changes at three distinct sites in the herbicide binding protein." *Science,* 228, 204–07.
48. Golden, S. S., and R. Haselkorn. 1985. "Mutation to herbicide resistance maps within the psbA gene of *Anacystis nidulans*." *Science,* 229, 1104–07.
49. Johanningmeier, U., U. Bodner, and G. F. Wildner. 1987. "A new mutation in the gene coding for the herbicide binding protein in Chlamydomonas." *FEBS Letters,* 211, 221–24.
50. Brown, A. E., C. W. Gilbert, R. Guy, and C. J. Arntzen. 1984. "Triazine herbicide resistance in the photosynthetic bacterium *Rhodopseudomonas sphaeroides*." *Proc. Natl. Acad. Sci. U.S.A.,* 81, 6310–14.
51. Astier, C., I. Meier, C. Vermotte, and A. L. Etienne. 1986. "Photosystem II electron transfer in highly herbicide resistant mutants of *Synechocystis* 6714." *FEBS Letters,* 207, 234–38.
52. Brusslan, J., and R. Haselkorn. 1988. "Molecular genetics of herbicide resistance in cyanobacteria." *Photosynthesis Res.,* 17, 115–24.
53. Paddock, M. L., S. H. Rongey, E. C. Abresch, G. Feher, and M. Y. Okamura. 1988. "Reaction centers from three herbicide-resistant mutants of *Rhodobacter sphaeroides*. 2.4.1: Sequence analysis and preliminary characterization." *Photosynthesis Res.,* 17, 75–96.
54. Hirschberg, J., A. B. Yehuda, I. Pecker, and N. Ohad. 1988. "Mutations resistant to photosystem II herbicides." In *Plant Molecular Biology,* D. von Wettstein, N.-H. Chua, eds. New York: Plenum, pp. 357–66.
55. Gingrich, J. C., J. S. Buzby, V. L. Stirewalt, and D. A. Bryant. 1988. "Genetic analysis of two new mutations resulting in herbicide resistance in the cyanobacterium *Synechococcus* sp. PCC 7002." *Photosynthesis Res.,* 16, 83–99.
56. Forcella, F. 1987. "Herbicide resistant crops: yield penalties and weed thresholds for oilseed rape." *Weed Res.,* 27, 31–34.
57. Ireland, C. R., A. Telfer, P. S. Covello, N. A. Baker, and J. Barber. 1988. "Studies on the limitations to photosynthesis in leaves of the atrazine-resistant mutant of *Senecio vulgaris* L." *Planta,* 173, 459–67.
58. Ahrens, W. H., and E. W. Stoller, 1983. "Competition, growth rate, and CO_2-fixation in triazine resistant smooth pigweed." *Weed Sci.,* 31, 438–44.
59. Jansen, M. A. K., J. H. Hobe, J. C. Wesselius, and J. J. S. van Rensen. 1987. "Comparison of photosynthetic activity and growth performance in triazine-resistant and susceptible biotypes of *Chenopodium album*." *Physiol. Vég.,* 24, 475–84.
60. Holt, J. S., and D. P. Goffner. 1985. "Altered leaf structure and function in triazine-resistant common groundsel (*Senecio vulgaris*)." *Plant Physiol.,* 79, 699–705.
61. Chapman, D. J., J. De Felice, and J. Barber. 1985. "Characteristics of chloroplast thylakoid lipid composition associated with resistance to triazine herbicides." *Planta,* 166, 280–85.
62. Tremolieres, A., H. Darmency, J. Gasquez, M. Dron, and A. Connan. 1988. "Variation of trans-hexadecanoic acid content in two triazine resistant mutants of *Chenopodium album* and their susceptible progenitor." *Plant Physiol.,* 86, 967–70.

63. Darmency, H., and J. Gasquez. 1982. "Differential temperature-dependence of the Hill activity of isolated chloroplasts from triazine resistant and susceptible biotypes of *Polygonum lapathifolium* L." *Plant Sci. Lett.*, 24, 39–44.

64. Ducruet, J.-M., and Y. Lemoine. 1985. "Increased heat sensitivity of the photosynthetic apparatus in triazine resistant biotypes from different plant species." *Plant Cell Physiol.*, 26, 419–29.

65. Ducruet, J.-M., and D. R. Ort. 1988. "Enhanced susceptibility of photosynthesis to high leaf temperature in triazine-resistant *Solanum nigrum* L. Evidence for photosystem II D-1 protein site of action." *Plant Sci.*, 56, 39–48.

66. Tournaud, G., M. T. Leydecker, and H. Darmency. 1987. "Abscisic acid in triazine-resistant and susceptible *Poa annua*." *Plant Sci.*, 49, 81–83.

67. Vaughn, K. C. 1986. "Characterization of triazine-resistant and susceptible isolines of Canola (*Brassica napus* L.)." *Plant Physiol.*, 82, 859–63.

68. Potter, J. R., and W. P. Wergin. 1975. "The role of light in bentazon toxocity in cocklebur: physiology and ultrastructure." *Pestic. Biochem. Physiol.*, 5, 458–70.

69. Pallett, K. E., and A. D. Dodge. 1980. "Studies into the action of some photosynthetic inhibitor herbicides." *J. Exp. Bot.*, 31, 1051–66.

70. van Oorschot, J. L. P. 1970. "Influence of herbicides on photosynthetic activity and transpiration rate in intact plants." *Pestic. Sci.*, 1, 33–37.

71. Fedtke, C. 1977. "Formation of nitrite in plants treated with herbicides that inhibit photosynthesis." *Pestic. Sci.*, 8, 152–56.

72. Churchill, K., and L. Klepper. 1979. "Effect of ametryn (2-(ethylamino)-4-(isopropylamino)-6-(methylthio)-*s*-triazine) on nitrate reductase activity and nitrite content of wheat (*Triticum aestivum* L.)." *Pestic. Biochem. Physiol.*, 12, 156–62.

73. Gaba, V., J. B. Marder, B. M. Greenberg, A. K. Mattoo, and M. Edelmann. 1987. "Degradation of the 32 KD herbicide binding protein in far red light." *Plant Physiol.*, 84, 348–52.

74. Ikeuchi, M., and Y. Inone. 1987. "Specific ^{125}I labeling of D1 (herbicide-binding protein)." *FEBS Letters*, 210, 71–76.

Chapter 8

Other Herbicidal Interactions with Photosynthesis

8.1 INHIBITORS OF CAROTENOID BIOSYNTHESIS

8.1.1 Inhibitors of Desaturase Reactions

Carotenoids are present in large amounts in the thylakoid membranes in close vicinity to the light harvesting chlorophylls and also to the photosynthetic reaction centers (Sections 7.4, 7.6). Carotenoids are synthesized from the C-5 building block isopentenyl pyrophosphate (IPP) which, in turn, is synthesized from mevalonic acid (MVA, Figure 8.1). The important role of carotenoids as protectants of the photosynthetic pigments is discussed in detail in Section 9.4.

The carotenogenic enzymes include the condensing synthases, the desaturases, and the cyclases. They have been found in the chloroplast envelope but seem to be absent in thylakoid membranes. The desaturation reactions require the presence of molecular oxygen. It is possible, therefore, that desaturation is initiated by a hydroxylation, but the involvement of a monooxygenase-type enzyme reaction has not been demonstrated (see Section 6.2.2). The double bond would then be formed concomitantly with the release of a water molecule. This interpretation is supported by the presence of hydroxy-phytoene and hydroxy-phytofluene following treatment with the herbicidal desaturase inhibitors diflufenican [1] and norflurazon [2, 3]. The presence of epoxy-phytoene might indicate a photooxidation step [4].

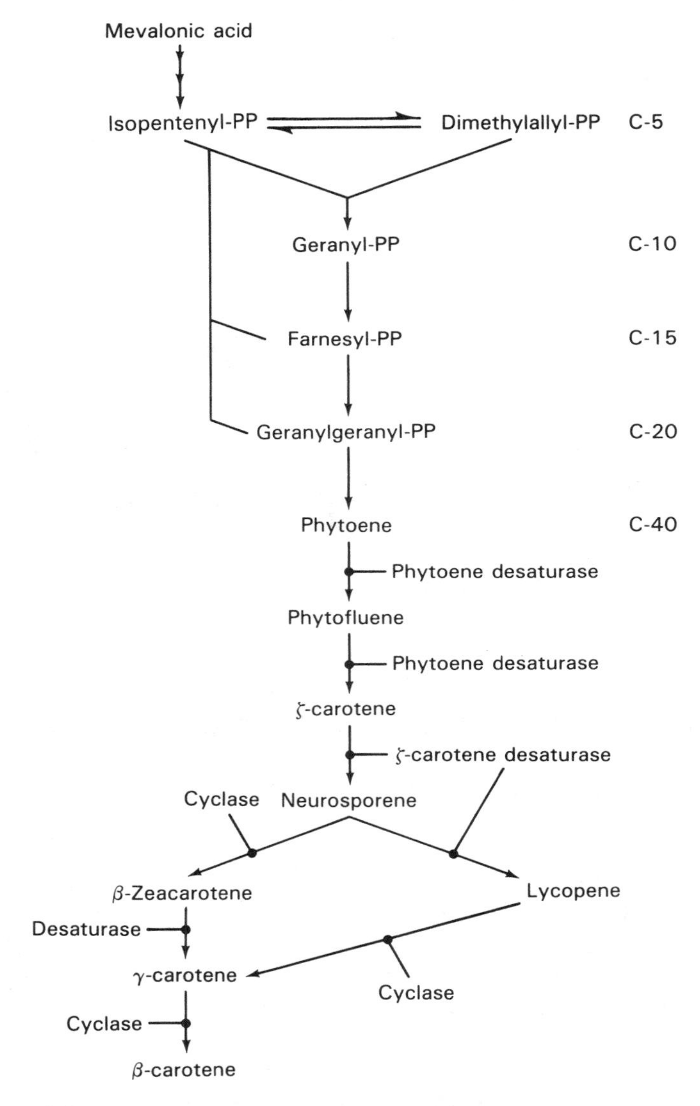

Figure 8.1 Pathway of plastidic carotenoid biosynthesis from mevalonic acid (MVA). The condensation phase from C-5 to C-40 is catalyzed by synthases. The reactions catalyzed by desaturases and by the cyclase are specifically marked.

The most important herbicidal target enzymes in the biosynthesis of carotenoids are the desaturases. The structures of some herbicides that inhibit either phytoene desaturase or ζ-carotene desaturase, or both, are shown in Figure 8.2. Two groups are presented: Group 1 herbicides lead to the in vivo accumulation of phytoene, indicating inhibition of phytoene desaturase [5–7]. Group 2 herbicides

Group 1:

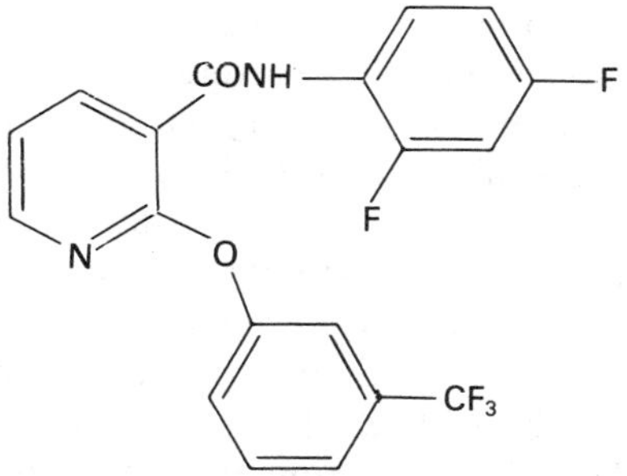

Diflufenican

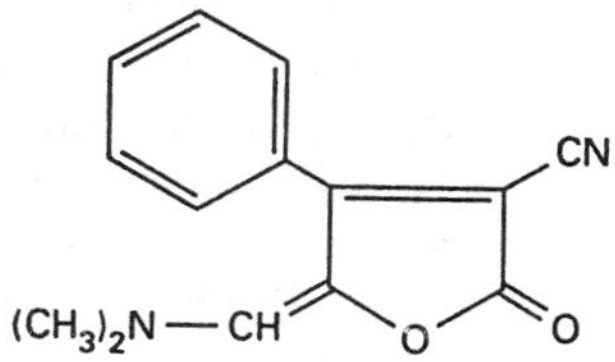

Difunon

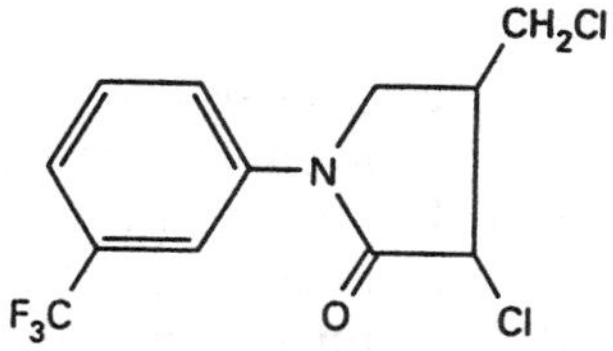

Fluorochloridone

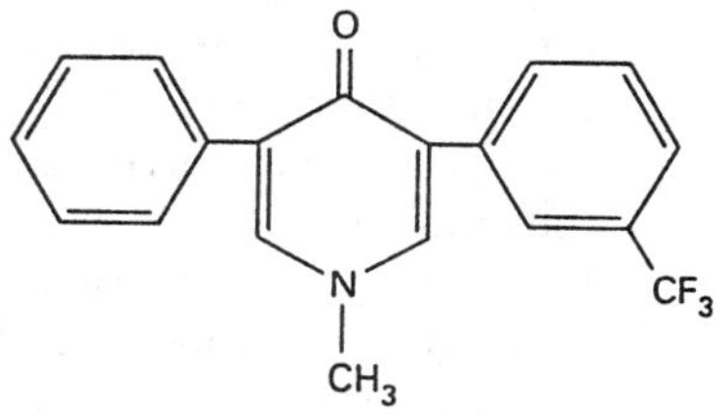

Fluridone

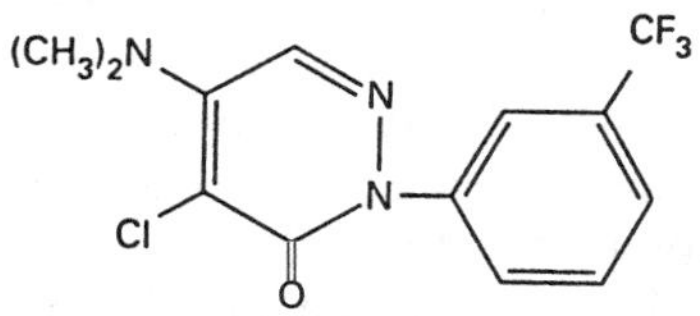

Metflurazon

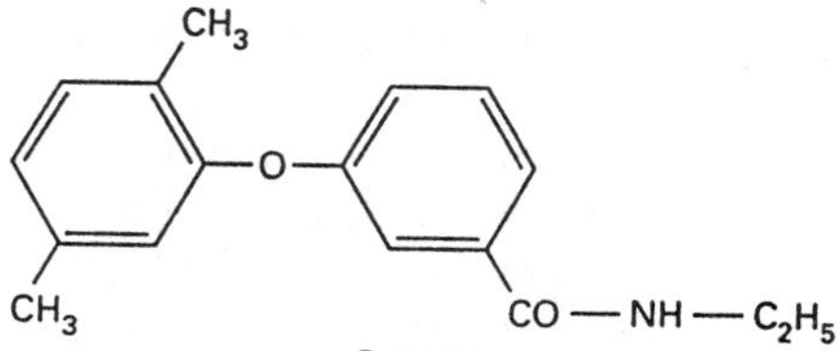

S-3422

Group 2:

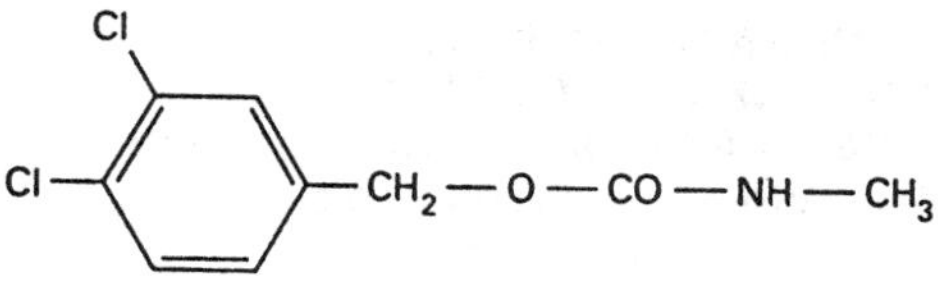

Dichlormate

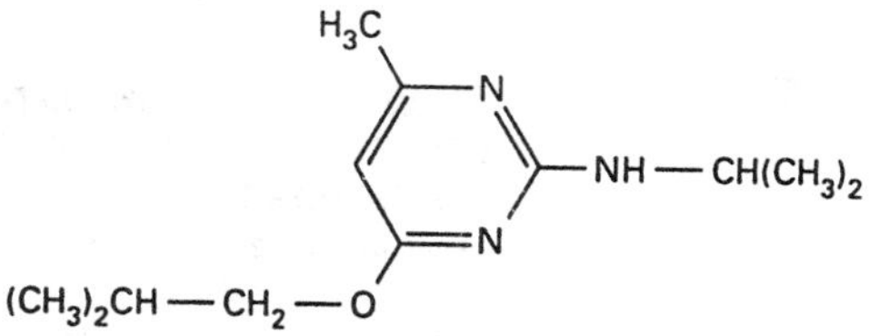

J-852

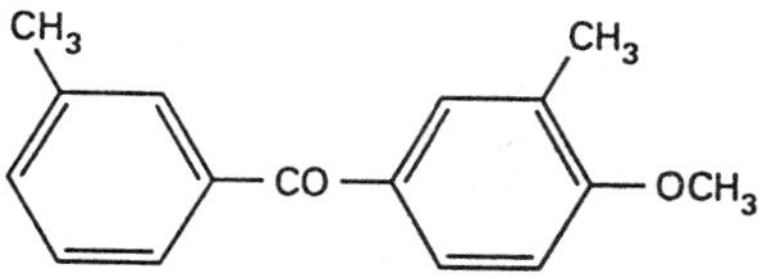

Methoxyphenone

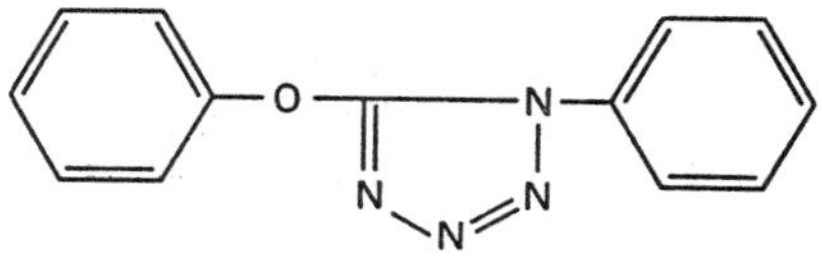

WL-110547

Figure 8.2 Herbicidal inhibitors of desaturases in carotenoid biosynthesis. Group 1 herbicides preferentially inhibit phytoene desaturase, whereas Group 2 herbicides inhibit both desaturases.

are less specific and lead to the accumulation of phytoene, phytofluene, and ζ-carotene in varying amounts and proportions, indicating preferential inhibition of ζ-carotene desaturase or of both desaturase enzymes [8–11]. Other compounds leading to the accumulation of phytoene are also likely to be inhibitors of the desaturase enzymes [12]. Phytoene desaturase and ζ-carotene desaturase differ in the position of the double bond they introduce, the first enzyme introducing the inner two double bonds, and the second the outer two double bonds. The end product is an open chain tetraterpenoid (C-40) with 11 conjugated double bonds, namely the red pigment lycopene (Figure 8.1). The precursors phytoene and phytofluene accumulate in plastoglobuli [13]. The development of the etioplast in the dark is not affected by these herbicides, but extensive photobleaching occurs in the light (Section 8.3).

The most convenient method for quantifying chlorosis and for elucidating the mechanism by which it is caused is by HPLC [1]. The pigments are extracted from the tissue with ethanol and are subsequently analyzed on a reversed-phase HPLC column. This procedure allows for the separation and quantification of the colored pigments (chlorophylls and carotenoids) and, more importantly, the detection and quantification of the precursor molecules, such as phytoene, phytofluene, ζ-carotene (Figure 8.1) that accumulate after herbicide application (Tables 8.1 and 8.2). Hydroxy-phytoene and hydroxy-phytofluene have also been identified by HPLC [1].

A total of about 25 in vitro carotenoid biosynthesis systems have been described in the literature. One problem encountered in the establishment of these systems has been the lack of suitable, commercially available substrates. ^{14}C-IPP and ^{14}C-MVA are now commonly used. Some of the in vitro systems use a *Phycomyces blakesleeanus* preparation, which synthesizes ^{14}C-phytoene from ^{14}C-MVA, with a chloroplast or thylakoid preparation that catalyzes the subsequent steps [14–17]. Among the other in vitro systems described are red pepper

TABLE 8.1 CONCENTRATIONS OF CAROTENOIDS AND CHLOROPHYLLS IN *SCENEDESMUS* CELLS TREATED WITH 10 μM NORFLURAZON OR OXYFLUORFEN FOR 15 H IN THE LIGHT. Standard deviations were ±7%. Contents are given as μg/ml packed cell volume [3].

Pigments	Control	Norflurazon	Oxyfluorfen
Carotenoids:			
α-Carotene	22.6	6.1	2.2
β-Carotene	34.6	8.2	3.9
Phytoene	0.0	79.0	0.0
Xanthophylls	79.8	70.4	10.9
Chlorophylls:			
Chlorophyll a	5380	4050	1740
Chlorophyll b	1920	1690	660

TABLE 8.2 CONVERSION OF ^{14}C-PHYTOENE TO DESATURATED AND CYCLYZED CAROTENOIDS IN CELL-FREE PREPARATIONS FROM *APHANOCAPSA* AND *CHENOPODIUM*[18].

Species:	Conc.	Radioactivity incorporated (dpm)			
Herbicide		Phytoene	ζ-Carotene	Lycopene	β-Carotene
Aphanocapsa:					
Control		3924	1013	816	1070
Difunon	1 μM	<u>7354</u>	397	227	399
J-852	50 μM	4035	<u>2180</u>	399	561
CPTA	50 μM	3649	941	<u>1432</u>	411
Chenopodium:					
Control		4125	712	316	355
Difunon	1 μM	<u>5331</u>	276	147	123
J-852	50 μM	4291	<u>1688</u>	171	231
CPTA	50 μM	3846	744	<u>622</u>	200

and *Narcissus pseudonarcissus* chromoplasts and thylakoid membranes from *Aphanocapsa* and *Chenopodium* (Table 8.2). The in vitro system allows for more precise differentiation betwen the different enzymatic steps. Table 8.2 indicates that, in the particular system employed, difunon preferentially inhibits phytoene desaturase, J-852 inhibits ζ-carotene desaturase, and CPTA inhibits the cyclase (Section 8.1.2). The sequential biosynthetic pathway clearly leads to accumulation of the intermediate compound that is the substrate for the inhibited enzyme. In vivo and in vitro systems might, however, yield different results. For example, the compound WL-110547 leads to ζ-carotene accumulation in vivo, but to phytoene accumulation in vitro [11]. Phytoene desaturase therefore appears to be more sensitive in vitro, whereas ζ-carotene desaturase is more sensitive in vivo. In these different systems, however, steady state equilibrium concentrations of the intermediates will not be alike. Amitrole (Figure 8.4) also leads to ζ-carotene accumulation in vivo, but does not lead to any accumulation in an in vitro carotenogenic system [15].

It can be concluded that the bleaching herbicides considered here are in many situations not entirely specific for one of the two carotenogenic desaturases, but instead inhibit both to varying degrees, depending on the molecular environment. On the other hand, the desaturases themselves have not been sufficiently characterized, and it is not known whether two clearly different enzymes exist in the many different systems that have been studied.

The rather low specificity of desaturase inhibition is further illustrated by the inhibition of linoleic acid desaturase by several of the herbicides included in Figure 8.2. WL-110547 [11], flurochloridone [19], and in particular many pyridazinones (e.g., metflurazon in Figure 8.2) [20] have been found to inhibit the desaturation of linoleic acid (C18 : 2) to linolenic acid (C18 : 3). Structure-activity studies have, however, shown that linoleic acid desaturase and phytoene desaturase are affected differently by different pyridazinone compounds. The derivative 4-chloro-5-dimethylamino-2-phenyl-3(2H)pyridazinone (BAS-13-338 = SAN-9785) leads to accumulation of linoleic acid but does not affect the synthesis of carotenoids, and seems, therefore, to be highly specific for linoleic acid desaturase [21, 22]. Consequently, it can be expected that the herbicidal desaturase inhibitors presented in Figure 8.2, and also similar or similarly acting compounds, affect different cellular desaturases with variable specificities. The precise inhibition(s) must be studied individually for each compound and for each plant species or tissue/system.

Linoleic acid is very prominent among the fatty acids in the galactolipids of chloroplast membranes. Moreover, there is a clear relationship between the extent of unsaturation (C18 : 3 to C18 : 2 ratio), fluidity of the membrane, and temperature adaptation. Cold-acclimated membranes contain higher amounts of C18 : 3, and are therefore more fluid at low temperatures. Consequently, herbicides that inhibit linoleic acid desaturase also inhibit cold acclimation in treated tissues [21, 23]. Whereas lipid unsaturation appears to be a clear and necessary adaptive metabolic change under low temperature conditions, it is not sufficient to explain all of the adaptive changes that lead to cold tolerance. Additional factors (e.g., the induction of "stress" proteins) are also involved [24–26]. Linoleic acid and also oleic acid (C18 : 1) accumulate, and linolenic acid decreases, under conditions of water stress [27].

The specific inhibitor of linoleic acid desaturase, SAN-9785, can be used to study the effects of a decrease in linolenate in the presence of normal carotenoid levels. Greening and thylakoid membrane formation appear to be normal; however, chloroplast development is slowed down [28], and subtle changes can be observed in the chlorophyll a/b ratio and in the proportion of appressed thylakoid membranes [29]. Fluorescence kinetics indicate an altered lipid environment in these thylakoid membranes.

The essentially white tissue produced by inhibitors of carotenoid synthesis in the light has been employed for measuring phytochrome levels in vivo. The facilitated spectrophotometric investigation of albino tissues allows for a more detailed study of photomorphogenesis [30]. However, in a comparison of norflurazon-bleached (white) and tentoxin-treated (yellow) mung bean primary leaves, it has been shown that less phytochrome accumulates and phytochrome-controlled processes are slowed down in the presence of norflurazon [31].

An indirect effect (supposedly unrelated to photobleaching) is a decrease in abscisic acid (ABA) concentrations after application of norflurazon [32, 33] or fluridone [34]. Decreased cellular ABA concentration is presumably the basis for the delayed stomatal closure that occurs after leaves are subjected to rapid water stress (Figure 8.3) [35]. Stomatal closure induced by a light-dark transition is similarly delayed in norflurazon-treated plants. Greatly decreased ABA concentrations in the presence of desaturase inhibitors suggest, but do not prove, a synthetic route to ABA via carotenoids [36]. The concentration of xanthoxine, structurally

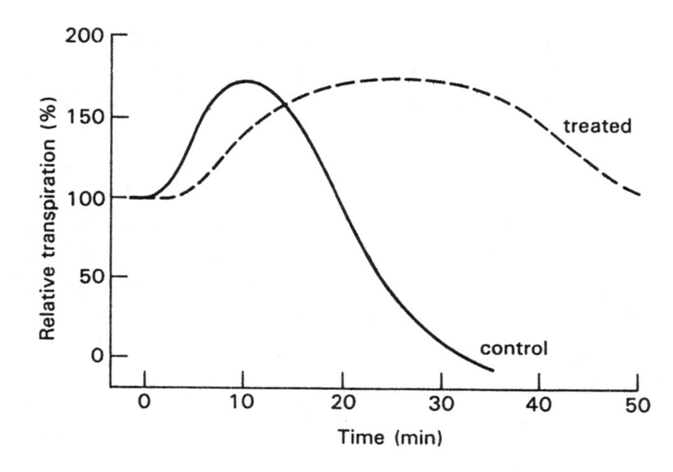

Figure 8.3 Transpiration rate of wheat leaves after induction of a rapid water stress by lowering of the water potential (time 0). The transient stomatal opening was delayed and subsequent stomatal closure was inhibited by 10 μM norflurazon [35].

similar to ABA, increases in the presence of norflurazon. Thus, xanthoxine might be synthesized, at least in part, via a different metabolic pathway than ABA.

8.1.2 Inhibitors of Other Carotenogenic Reactions

Three compounds are dealt with in this section: MPTA, clomazone, and amitrole (Figure 8.4). MPTA and the structurally similar CPTA (2-(4-chlorophenylthio)-triethylamine) are not used as commerical herbicides, although they do have herbicidal properties. However, they are used as growth regulators. These compounds inhibit the carotenogenic cyclase reactions (Figure 8.1), leading to the accumulation of lycopene, mainly, but also of some phytofluene in citrus fruits [37], in *Myxococcus fulvus* [38], in *Phycomyces blakesleeanus* [39, 40], and also in in vitro systems (Table 8.2) [18]. In cotton seedlings, phytoene, phytofluene, ζ-carotene, and lycopene all accumulate, with lycopene as the main product [41]. Lycopene also accumulates markedly in citrus fruits [42]. The primary inhibition of the cyclase induces a general increase of the biosynthetic capacity of the carotenogenic pathway. This increase can be inhibited by cycloheximide or actinomycin D, but not by chloramphenicol. Therefore, it appears that transcription of

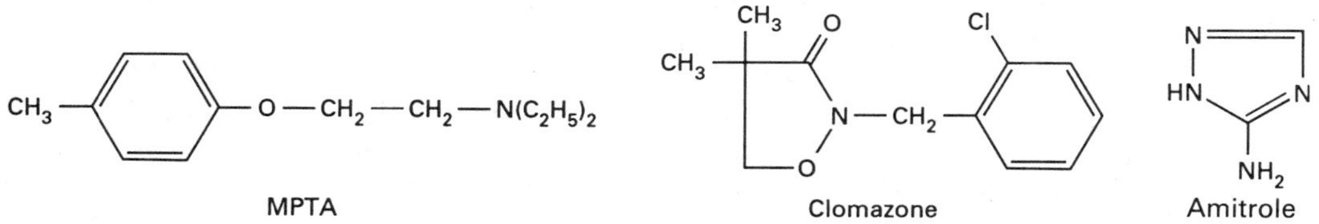

MPTA Clomazone Amitrole

Figure 8.4 Structures of herbicidal inhibitors of carotenoid biosynthesis that do not affect the desaturase enzymes.

nuclear genes and subsequent cytoplasmic protein synthesis are involved in this induction response. The result is accumulation of massive amounts of acyclic carotenoids and their precursors [43]. The accompanying precursor accumulation (phytoene, etc.) may be caused by feedback inhibition of desaturase reactions. In the in vitro system only lycopene accumulates (Table 8.2), supporting the interpretation of a specific primary inhibition of the cyclase.

The second compound in Figure 8.4, clomazone, is a recent development in the group of the bleaching herbicides, and the mode of action of this herbicide is not yet precisely known. Bleaching of treated tissue is observed, but phytoene and phytofluene do not accumulate. Carotenoids are not formed in the presence of clomazone but, contrasting with the development of etiolated white seedlings in the presence of most carotenoid synthesis inhibitors, clomazone also inhibits seedling growth [44–46]. More detailed analyses of the terpenoid synthesis pathway in the presence of clomazone have shown that inhibition by this herbicide occurs after farnesyl-PP; the synthesis of sterols, which originate from the triterpenoid squalene, is not affected [47] (the triterpenoid squalene (C-30) is synthesized by head-to-head condensation of two farnesyl-PP (C-15) units). However, diterpene (C-20) and tetraterpene (C-40) syntheses are both inhibited in the presence of clomazone (Figure 8.1) [45, 46, 48, 49]. The block in the in vivo synthesis of terpenoids by clomazone may therefore be localized between farnesyl-PP and geranylgeranyl-PP (Figure 8.1). The synthesis of the diterpenoid compound phytol is also affected, resulting in decreased chlorophyll phytylation and membrane integration [44, 46]. Chlorophyll accumulation is reduced, consequently, in the presence of clomazone. Another very important diterpene derivative is the plant growth hormone gibberellic acid (GA_3). Interestingly, 100 μM GA_3 reverses the growth inhibition induced by 100 μM clomazone in *Pisum sativum* [48].

Noncompetitive inhibition of IPP-isomerase by clomazone has been noted in an enzymatic cell-free system from spinach leaves [47]. A substrate analogous enzyme inhibitor for IPP-isomerase has also been described [50]. However, an independent study found no inhibition of the in vitro and in vivo enzyme activities leading from IPP to phytoene [50a]. Future research will have to address the following questions: which enzyme of the carotenogenic pathway is inhibited in vivo? How does inhibition of the enzyme in vivo relate to the primary effect(s) of the herbicide? Could the inhibition of the terpenoid pathway be a secondary/regulatory consequence of the unknown primary clomazone effect(s)?

Unlike clomazone, amitrole (aminotriazole) (Figure 8.4) has been an established herbicide for more than 30 years. Consequently, many more studies have been published on it. However, except for the obvious fact that chlorosis is induced through inhibition of carotenoid biosynthesis [8, 51, 52], little is known about the exact site of molecular interference. Carotenoid biosynthesis is not very strongly inhibited [52], but phytoene, phytofluene, ζ-carotene, and lycopene all accumulate in vivo. In one study, no accumulation of carotenoid precursors was found in the presence of amitrole in an in vitro system from *Phycomyces blakesleeanus,* although carotene biosynthesis was strongly inhibited [15].

Fluridone and difunon lead to phytoene accumulation in this system, whereas ζ-carotene accumulates in the presence of J-852 (Section 8.1.1). On the other hand, amitrole stimulates squalene accumulation. In conclusion, amitrole is an inhibitor of an unknown (early) enzymatic step in carotenoid/terpenoid biosynthesis. The accumulation of the triterpenoid squalene points to inhibition of a condensing synthase, not unlike clomazone. However, the diversity of observed effects of amitrole could also be explained by indirect inhibition or by several simultaneous independent inhibitions.

Amitrole is known to have two additional independent sites of inhibition. Elevated concentrations (in excess of 0.1 mM) inhibit the enzyme catalase (see Section 9.4) [53]. This does not lead to measurable increases in the cellular hydrogen peroxide concentrations, presumably because H_2O_2 is destroyed by other enzymatic mechanisms. A large accumulation of oxidized glutathione (GSSG) can, however, be measured after amitrole application [54, 55]. The synthesis of GSSG increases greatly, and the reduced/oxidized ratio (GSH/GSSG) decreases, indicating regulatory responses and cellular redox potential changes after catalase inhibition by amitrole. A contribution of these effects to the herbicidal action is questionable because of the high concentrations required.

Early studies of amitrole activity in plant tissues and plant cell cultures led to the suggestion of primary effects on amino acid and purine metabolism [56–60]. An inhibition site in the histidine biosynthetic pathway has been implicated (Chapter 13). In addition, interactions found in microorganisms could be taken to support an inhibitory action of amitrole on some step(s) in basic metabolism [61, 62]. Inhibition of purine biosynthesis has been reported [63], but the concentrations employed (0.1–1 mM) are higher than those required for the induction of chlorotic effects in higher plants. It is unclear, therefore, whether or not the effects on nucleic acid and protein/amino acid biosynthesis seen in higher plants are primary effects. Some, at least, may be consequences of the inhibition of carotenoid biosynthesis, catalase activity (partially), and/or some step in amino acid or purine biosynthesis.

8.2 UNCLASSIFIED BLEACHING HERBICIDES AND MULTIFUNCTIONAL BLEACHING

The molecular basis for the bleaching activity of the herbicidal compounds presented in Figure 8.5 is not known, as most have not yet been sufficiently studied. Some may induce bleaching by so far unknown mechanisms, whereas others could induce bleaching by an already well-known mechanism. Eventually, some of these herbicides may be relocated into one of the more defined mechanism groups described in the other parts of this chapter.

Pyrazolate, a rather new herbicide that has been the subject of only limited mode of action studies, can be compared with the similar herbicides benzophenap, pyrazoxyfen, and NC-310 [64]. With these it shares the partial structure DTP

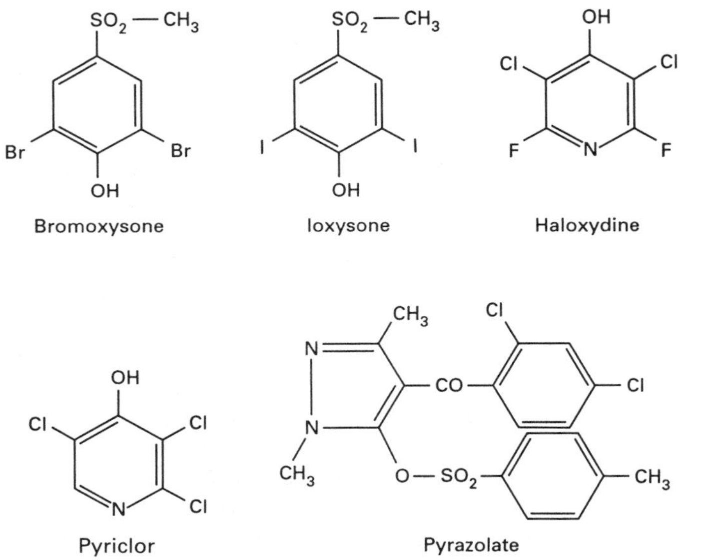

Bromoxysone Ioxysone Haloxydine

Pyriclor Pyrazolate

Figure 8.5 Structures of bleaching herbicides that possibly inhibit carotenoid biosynthesis.

(1,3-dimethyl-4-(2,4-dichlorobenzoyl)-5-hydroxypyrazole), or the closely similar partial structure, which supposedly is the active herbicide [65, 66]. It has been shown that DTP does not inhibit carotenoid biosynthesis and has only a minor effect on chlorophyll biosynthesis in the dark [16, 67, 68]. Photobleaching occurs in the light, but malondialdehyde (an indicator of unsaturated lipid peroxidation) is not produced (compare Section 8.4.3). In another study, inhibition of carotenoid and chlorophyll biosynthesis, and concurrent phytoene accumulation, have been reported [12].

Bromoxysone and ioxysone [69], haloxydine [52, 70], and pyriclor [71] exert strong bleaching activity of green tissue in the light. The ultrastructural and biochemical effects are very similar to those that have been observed with amitrole and, as for amitrole, the primary mechanism of herbicidal interference is unknown. Albino leaves that develop in the light following treatment with these herbicides are similar to those developing in the presence of herbicidal inhibitors of carotenoid biosynthesis. For example, accumulation of phytoene, phytofluene, and ζ-carotene has been detected [8]. However, in vitro inhibition of carotenoid biosynthesis has not been reported.

In a more detailed study, haloxydine and amitrole have been grouped together, but separate from herbicides representing the desaturase inhibitors [52]. Both compounds only partially inhibit carotenoid and chlorophyll biosynthesis; tissue bleaching occurs more slowly and is less efficient than that caused by desaturase inhibitors. Photoinactivation of NADP-glyceraldehyde-3-phosphate-dehydrogenase, which occurs rapidly after application of difunone, does not occur in the presence of haloxydine or amitrole.

It should be possible to study those bleaching herbicides whose mechanisms of action are still obscure with present-day techniques, in particular with HPLC analysis of chloroplast pigments and with in vitro carotenoid biosynthesis systems. It is possible that hitherto unknown molecular inhibition and interference sites may be found using these approaches.

Some bleaching herbicides may have more than one primary mechanism of interference in the induction of bleaching. The inhibitors of desaturase enzymes illustrate the possibility of multifunctional interaction: several desaturases with similar or closely related mechanisms/substrates are inhibited to varying degrees, producing variable and complex modes of herbicidal action (Section 8.1.1). Amitrole may also have several modes of action, but a contribution to herbicidal action is clear only for the inhibition of carotenoid biosynthesis (Section 8.1.2). The nitrodiphenylether fomesafen (Section 8.4.1) has been shown to inhibit phytoene desaturation at 80 μM in an in vitro system, but to induce lipid peroxidation in vivo at 20 μM [18]. The contribution of the former mechanism to its herbicidal action is probably minor because of the higher concentration needed. More interesting is a shift in the type of herbicidal action by substitution variation in the nitrodiphenylether series [72] and in the cyclic amide series [73] (Section 8.4.1). The highly active herbicides of both groups require light for their herbicidal action, and lead to malondialdehyde (MDA) accumulation. In both groups, however, molecular variations have been found that do not require light for their herbicidal activity and do not cause MDA to accumulate in damaged tissue. The possibility of multiple (related?) mechanisms contributing to varying degrees to the herbicidal action should therefore be considered in these and other cases.

8.3 PHOTOBLEACHING IN THE ABSENCE OF COLORED CAROTENOIDS

Compounds that inhibit the synthesis of colored carotenoids generally do not affect protochlorophyll(ide) synthesis in etiolated tissue. The amount of carotenoids present in untreated, etiolated stems and leaves is sufficient to protect the chlorophyll that is synthesized after light induction, and thylakoid membranes develop normally under these conditions. Only when the synthesis of carotenoids has been inhibited during the development and growth of *new* tissue will this new tissue eventually become white. If the tissue was grown in the dark in the presence of an inhibitor of carotenoid biosynthesis, photobleaching will occur in strong light. In weak light, however, the tissue will turn green. This comparison demonstrates the protective role of carotenoids against photooxidation and also shows that, besides their primary effect on carotenoid biosynthesis, these herbicides generally do not inhibit other metabolic reactions or pathways [74].

When carotenoid-free green plants are transferred to strong light, severe photooxidation occurs. However, when plants treated with carotenoid biosynthesis inhibitors are grown in strong light from the beginning, products of membrane lipid peroxidation or other photooxidation products are not observed. The reason why photooxidation does not occur is that chloroplast development is arrested at a very early stage. Obviously a signal is produced in the light by the carotenoid-free membranes that inhibits the synthesis of nucleic acids and proteins needed for normal, light-dependent chloroplast development [75]. The m-RNA for two major nuclear gene products destined for the chloroplast (those for the light harvesting complex and for the ribulosebisphosphate carboxylase small subunit) are lacking from the cytosolic m-RNA pool. Other minor m-RNAs are also absent. These deficiencies can be observed in carotenoid-free mutants, and also in herbicide-treated plants free of carotenoids. In low light, however, m-RNA synthesis and chloroplast development are normal. These results suggest that the signal produced by illuminated, carotenoid-free chloroplast membranes is a photooxidation product. The specificity of this regulatory/coordinating mechanism is demonstrated by the uninhibited nuclear synthesis of another light-regulated m-RNA that encodes a cytosolic protein.

The bleached chloroplasts are free of 70S ribosomes and of endogenous membrane systems [52, 70, 71, 76–79]. Although nitrate reductase is not a chloroplast enzyme, it is usually also lost from bleached tissues [80, 81]. Certain isoenzymes of nitrate reductase may, however, be superinduced in some bleached tissues [81–83]. Simultaneously, the peroxisomal enzymes catalase, glycolate oxidase, and hydroxypyruvate reductase are lost [52, 84, 85]. This observation reflects the close cooperation of chloroplasts and peroxisomes in the actively photosynthesizing cell. It is not clear whether the described effects are due to regulatory interactions, that is, to altered gene expression, or to direct phytotoxic (photooxidative) effects, or both. The actual mechanism of bleaching will often be a mixture of these molecular mechanisms, depending on the concentration and distribution of the herbicide, on the type of herbicide, and on the environmental conditions.

In practical situations, herbicide-treated tissue does occasionally turn green, either because of low light levels or because of only partial inhibition of carotenoid biosynthesis. Of course, if such tissue is transferred to high light, photooxidation of the pigments, membrane lipids, and further chloroplast constituents then occurs. The mechanism of photooxidation is discussed in Section 9.3. In specific cases of slow damage, ethylene may be synthesized as a secondary stress response [86].

8.4. INHIBITORS OF PROTOPORPHYRINOGEN-OXIDASE

8.4.1 Structures and Inhibition at the Enzyme Level

The compounds to be discussed in this section have been investigated intensively during the last three decades, but only very recently has the chloroplast and mitochondrial enzyme protoporphyrinogen-oxidase (PPG-oxidase) been identified as the molecular target. The inhibition of this enzyme in the two organelles (chloroplasts and mitochondria) blocks chlorophyll and heme synthesis and leads to excessive formation of the singlet oxygen-generating protoporphyrin IX (PPIX) and other tetrapyrrole structures. PPIX accumulates in the tissue when the herbicides are applied in the dark, but in light, photooxidation can rapidly destroy newly formed tetrapyrroles, leading to less accumulation.

A selection of important inhibitors of PPG-oxidase is shown in Figure 8.6. Group 1 contains the "nitrodiphenylether (NDPE) herbicides," which have been investigated intensively. The possible structural variation in the *ortho* position to the nitro group is quite extensive, and includes carboxylic acids, carboxylic esters, amines, and hydrogen. It therefore allows for a large number of highly active compounds that have been exploited chemically and biologically over the past 30 years. The group name nitrodiphenylether is not particularly fitting since diphenylethers with a halogen (chlorine) instead of the nitro group have the same mode of action (Table 8.3) [87], whereas diphenylethers with a *para*-phenoxypropionic acid substituent have a completely different mode of action (Section 11.1). On the other hand, several structurally unrelated compounds are known which, by all standards and comparisons, have the same mode of action as the NDPEs (Figure 8.6, Group 2) [73, 88–93].

PPG-oxidase is the enzyme that converts protoporphyrinogen IX to protoporphyrin IX by an oxidative aromatization requiring molecular oxygen (Figure 8.7). (The rather complicated porphyrin nomenclature is a special field with many traditional names, and can be looked up elsewhere [94]). PPG-oxidase is extremely sensitive to the NDPEs and similarly acting herbicides, with I_{50}-concentrations mostly in the nanomolar range (Table 8.3). From the data in Table 8.3 it can also be deduced that isoenzymes with different herbicide sensitivities appear to exist in the two different organelles, the etioplast/chloroplast and the mitochondrion. This is not yet clear, however, since the comparison in Table 8.3 is based on organelles from different plant species; future experiments will have to address this and other unanswered questions.

PPG-oxidase was identified as the sensitive enzyme only after prolonged research efforts. Misleading conclusions from earlier experiments repeatedly hindered or delayed progress in determining the target site of these herbicides. Three such examples are given here: first, the insensitivity of white (carotenoid-free) tissue led researchers to speculate about a carotenoid sensitizer pigment. Second, antagonism with diuron and other inhibitors of photosynthetic electron transport initiated speculation that NDPEs might act as redox catalysts (Section 8.4.2). Third, accumulation of PPIX in herbicide-treated tissue in the dark was taken as an indication that Mg- and/or Fe-chelatases might be inhibited (Figure 8.7). However, chelatases are not sensitive to the herbicides considered here, or are only inhibited by very high concentrations (in excess of 100 μM) [95–97].

We are therefore left with the apparent contradiction that the accumulating PPIX is the *product* of the inhibited reaction, the enzymatic conversion of protoporphyrinogen IX to PPIX. In the native chloroplast or mitochondrion, the

Group 1:

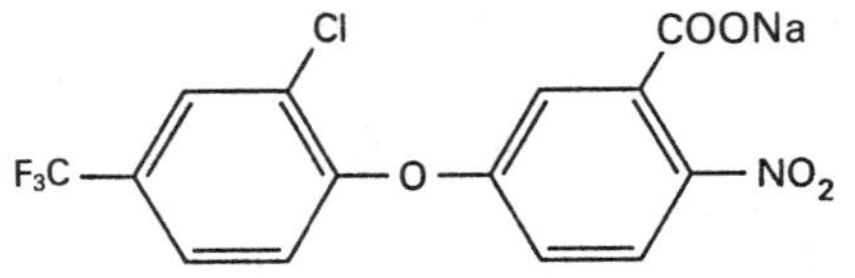

Acifluorfen

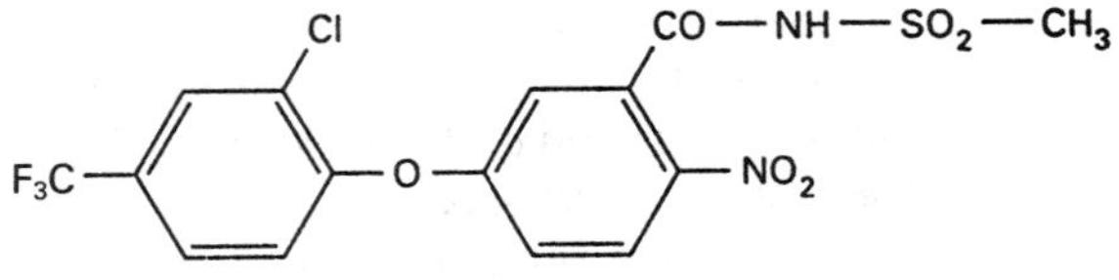

Fomesafen

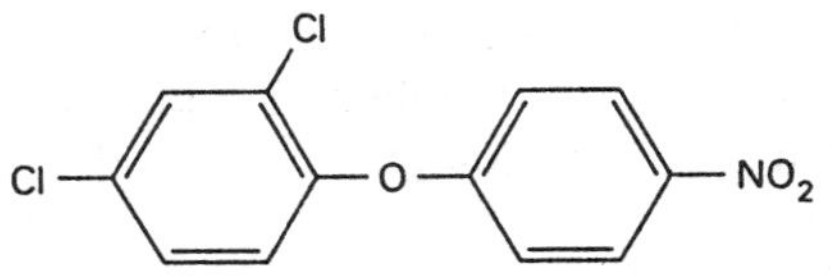

Nitrofen

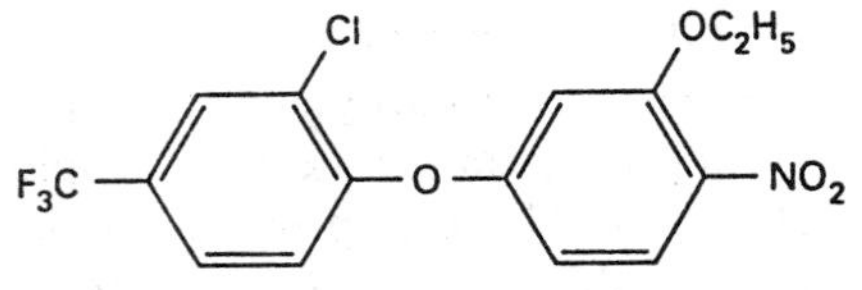

Oxyfluorfen

Group 2:

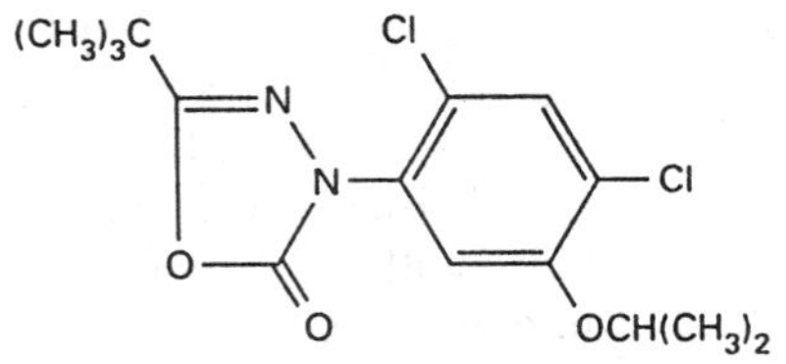

Oxadiazon

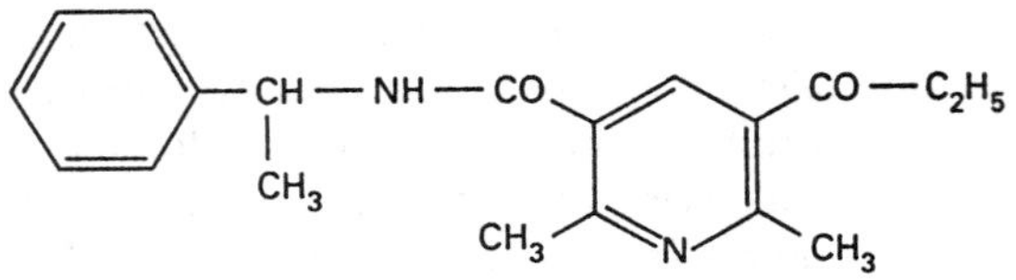

LS-82-556

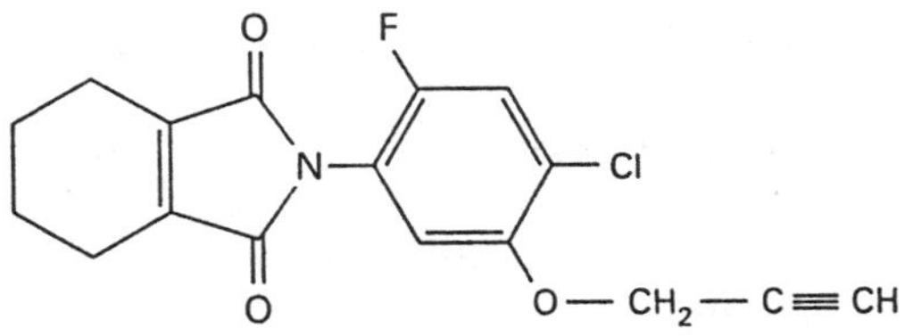

S-23142

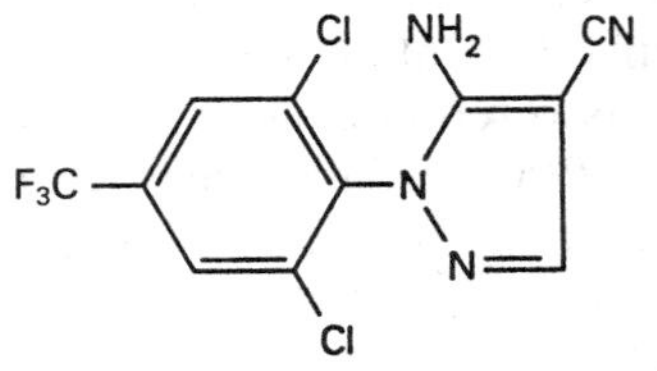

M & B-39279

Figure 8.6 Selected structures of nitrodiphenylether herbicides (Group 1) and of similarly acting herbicides (Group 2).

sequentially acting biosynthetic enzymes are supposedly organized in a multienzyme complex from which the intermediates do not become free (i.e., they are not released from the enzyme complex). After herbicidal inhibition of PPG-oxidase, the substrate protoporphyrinogen IX apparently diffuses out of the enzyme site and is then subject to nonenzymatic oxidative aromatization to PPIX. However,

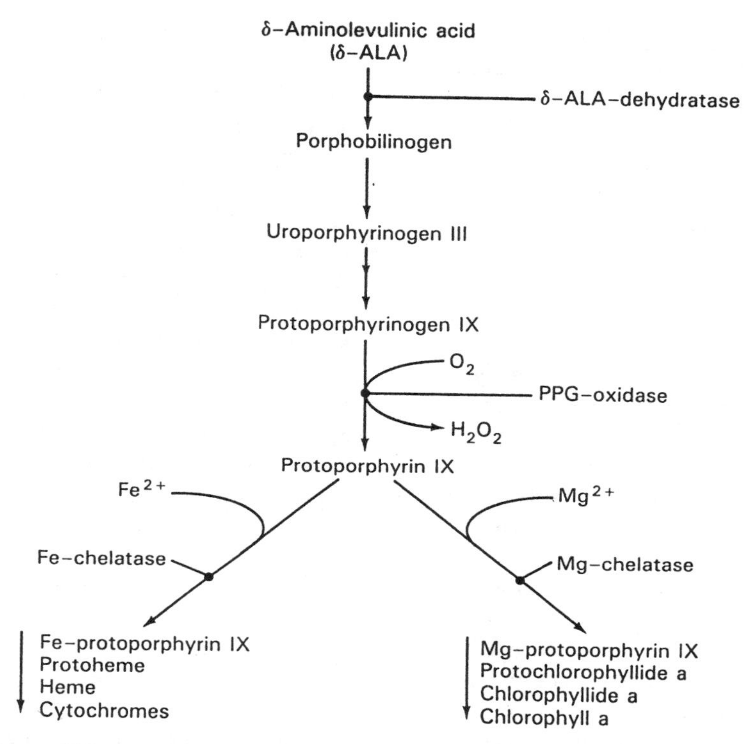

Figure 8.7 Pathway of tetrapyrrole synthesis in higher plants. Enzymes that are involved in the pathway and mentioned in the text are labeled.

TABLE 8.3 INHIBITION OF PROTOPORPHYRINOGEN-OXIDASE (PPG-OXIDASE) BY SEVERAL OF THE HERBICIDAL COMPOUNDS SHOWN IN FIGURE 8.6, BY THE CHLORO-SUBSTITUTED (INSTEAD OF NITRO) ANALOGUE OF ACIFLUORFEN, LS-820340, AND BY ITS M-TRIFLUOROMETHYL-ANALOGUE, RH-5348. I_{50}-values (concentrations required for 50% inhibition of enzyme activity) are listed for the enzymes in corn etioplasts and potato mitochondria [95, 96].

	PPG-oxidase inhibition (I_{50})	
Herbicide	Corn etioplasts (nM)	Potato mitochondria (nM)
Acifluorfen-methyl	4.0	0.43
LS-820340	10.0	3.0
RH-5348	180.0	19.0
Oxadiazon	11.5	9.0
LS-82-556	4,000.0	40,000.0
M&B-39279	80.0	15.0

the free PPIX is obviously not a substrate for the chelatase enzymes in the multienzyme complex, and either accumulates in the dark or reacts as a singlet oxygen generator in the light (Section 9.2). PPIX has been shown to accumulate after application of NDPEs [98–102] and oxadiazon [93]. Tetrapyrroles other than PPIX have not been intensively searched for, but they appear to be formed, also; the nonenzymatic oxidation and aromatization of protoporphyrinogen IX also yields more hydrophilic and more acidic tetrapyrroles, in addition to PPIX, that are not found with the usual basic extraction procedures used for PPIX [103].

In many cases it is possible to identify the pigment responsible for a light-dependent process or effect by studying the action spectrum. In the case of the NDPEs, however, the situation is complicated by the diversity of different systems: both green and yellow (etiolated and carotenoid containing) tissue is sensitive. White tissue is often, but not always, insensitive (when caused by mutation) [72], or partially insensitive (after treatment with an inhibitor of carotenoid biosynthesis) [104, 105]. The action spectra published with different tissues usually reflect the prevailing pigment situation: no really inactive wavelengths are obtained, but yellow light and intermediate wavelengths are often very effective [105–107]. With the knowledge that PPIX and other tetrapyrroles are the photosensitizer pigments that cause herbicidal action, it can now be seen in suitable systems that the action spectrum conforms with the tetrapyrrole absorption spectrum. For example, nonchlorophyllous soybean cells are particularly useful because of their low pigment background [92, 99]. The action spectrum obtained with these cells implies the participation of a pigment system that strongly absorbs in the blue region (400 nm), corresponding to PPIX and other tetrapyrroles.

8.4.2 Interactions with Other Herbicides and Inhibitors

Before the molecular enzyme target of the NDPEs and similar herbicides was known, attempts were made to obtain mode of action information by studying the interactions of the herbicides with other inhibitors with known metabolic inhibition sites. A list of the interactions observed is presented in Table 8.4. The possible involvement of photosynthetic electron transport in NDPE action has been studied in detail. The question of whether photosynthetic electron transport reduces the NDPE molecule to a radical was a matter of controversy for several years because diuron, an inhibitor of photosynthetic electron transport (Section 7.1), antagonized the phytotoxic action of NDPEs in some systems [108, 109], but not in others [110–113]. The same is true for other herbicides that inhibit photosynthetic electron transport through photosystem II (Table 8.4). Although it should be clear from the results obtained with white, yellow, or green tissue (see descriptions above and below) that photosynthesis is not an absolute requirement for the herbicidal action of the NDPEs, it was considered possible for some time that the photosynthetically active pigment system interacted in a special way with NDPEs. The antagonistic interactions that were observed in some algal and higher plant photosynthetic

TABLE 8.4 REPORTED INTERACTIONS BETWEEN NITRODIPHENYLETHERS AND INHIBITORS OF KNOWN BIOSYNTHETIC PATHWAYS OR BIOCHEMICAL REACTIONS.

Subcellular location: Compound	Inhibition	Interaction	Reference
Chloroplast:			
Prometryne	Photosystem II	Antagonism in	111
Atrazine	Photosystem II	algae, occasional	112, 113
Bentazon	Photosystem II	partial	112
Diuron	Photosystem II	antagonism in	111, 112
Monuron	Photosystem II	higher plants	109
Norflurazon	Carotenoid biosynthesis	Partial antagonism	104, 112, 121
Tentoxin	Chloroplast biogenesis	Strong synergism	112
Gabaculine	Chlorophyll biosynthesis	Strong antagonism	100, 101, 103, 122, 123
4,6-Dioxo-heptanoic acid	δ-ALA-Dehydratase	Strong antagonism	92, 100
Rifampicin	Organelle RNA biosynthesis	Antagonism	103
Chloramphenicol	Organelle protein biosynthesis	No interaction	103
Ethanol	Scavenging of	Partial antagonism	87,
α-Tocopherol	activated oxygen		110,
Cu-penicillamin			112, 120
Mitochondria:			
Antimycin A	Electron transport		
Rotenon	Electron transport	Antagonism	103, 112
CCCP	Electron transport		
2,4-Dinitro-phenol	Electron transport		
Cytosol:			
Puromycin	Protein synthesis	Antagonism	103,
Cycloheximide	Protein synthesis	Antagonism	111

systems can now be explained by an oxygen effect: the inhibitor of photosynthetic electron transport reduces the concentration of molecular oxygen in the chloroplast, thereby lowering the rate of lipid peroxidation. This interpretation is strongly supported by experiments in which different oxygen pressures were applied; under conditions of high aeration the antagonizing effect of diuron or prometryne disappeared [111].

The antagonism between NDPEs and inhibitors of photosystem II described above suggested for some time a possible mode of action of NDPEs involving the shuttling of an electron to molecular oxygen. A similar mode of action has been established for the bipyridylium herbicides (Section 8.5), leading in the latter case to the production of superoxide anions. However, no indications of superoxide anion formation can be found after NDPE application [110, 112]. The concept of an electron transfer from the photosynthetic electron transport system to oxygen via an NDPE molecule would require the transient reduction of the NDPE. A reduction spin signal has been found with oxyfluorfen under artificial conditions [114]; however, extensive voltametric studies of NDPEs [115–117], and the comparison with similarly acting diphenylethers with a chlorine (which cannot accept an electron) instead of the nitro substituent [87, 116], eliminates the possibility that the NDPEs may react by redox catalysis in vivo.

Another concept suggested a herbicide-pigment interaction leading to destabilization and subsequent photooxidation and bleaching. A series of experiments has been conducted on the interference of NDPEs with β-carotene in organic solutions and in micelles [118]. In these artificial preparations, β-carotene was bleached and reductions did occur in UV-A light. However, the artificial conditions employed and the inactivity of visible light and of the chloro-diphenylether analogue suggest that the results obtained are not relevant to in vivo conditions.

Further interactions listed in Table 8.4 include those with inhibitors of different pigment biosyntheses. Compounds that inhibit the pathway leading to tetrapyrroles (e.g., gabaculin and 4,6-dioxoheptanoic acid) are strong antagonists of NDPE herbicidal action; 4-amino-5-hexynoic acid also acts in this manner [119]. The antagonistic action of norflurazon and similar inhibitors of carotenoid biosynthesis can probably be explained by the inhibition of chloroplast biogenesis (Section 8.3), which then indirectly inhibits the biosynthesis of tetrapyrroles (chlorophylls). Oxygen scavengers and antioxidants antagonize the herbicidal action more directly by interfering with the mechanism of photooxidation. Secondary plant constituents also contain a large number of antioxidants [120].

Some inhibitors of RNA and protein biosynthesis are also antagonistic with PPG-oxidase inhibitors. The reason for their antagonistic interaction can be seen in the inhibition of an induction process that normally leads to increased rates of tetrapyrrole biosynthesis after inhibition of PPG-oxidase. As discussed earlier, PPIX is formed during herbicidal action, but chelation to Mg- or Fe-PPIX (heme) no longer occurs. As a consequence, the normal feedback inhibition of tetrapyrrole biosynthesis by heme is no longer in effect; specific mRNAs are induced, and enzymes of this pathway are produced in higher amounts, leading to even greater PPIX accumulation. This induction of RNA and protein synthesis that takes place during herbicidal action is inhibited by rifampicin, puromycin, and cycloheximide. The antagonistic interaction with inhibitors of mitochondrial electron transport indicates that herbicidal inhibition of mitochondrial PPG-oxidase also contributes to the overall herbicidal action. The reason for this antagonism is not clear; an indirect effect on mitochondrial growth/biogenesis might be involved. The long list (Table 8.4) and extensive discussion of interactions illustrates the difficult (and sometimes erroneous) route to elucidating a molecular site of herbicide action. In spite of being partially outdated, the detailed description of the many different interactions might be helpful in understanding special facets of the mode of action of the PPG-oxidase inhibitors.

8.4.3 Herbicidal Action after Inhibition of PPG-Oxidase

It has long been known that the NDPEs require light, molecular oxygen, and a pigment system for their herbicidal action [72]. Whereas green or yellow (etiolated) tissue is sensitive, white tissue is usually insensitive. Precursors of carotenoid biosynthesis (phytoene, phytofluene, ζ-carotene) do not accumulate in treated tissues (Section 8.1, Table 8.1) [73], but existing photosynthetic pigments (chlorophylls and carotenoids) are bleached in the course of herbicidal action in the presence of light. Simultaneously, malondialdehyde and other lipid peroxidation products are formed (Section 9.3). Table 8.5 gives a list of the sequence of events during herbicidal action. Electrolyte leakage, an indication of membrane damage by lipid peroxidation, is among the earliest effects that can be detected. Lipid peroxidation starts immediately after illumination when the tissue is preincubated with the herbicide in the dark (as was the case in Table 8.5). Simultaneously, the loss of ascorbate and reduced glutathione (GSH) indicate (per)oxidative actions in the chloroplast stroma and in the cytosol [126]. The respective oxidized species, dehydroascorbate and GSSG, do not accumulate, indicating that the redox buffer molecules are excessively oxidized, then destroyed. Glutathione reductase, dehydroascorbate reductase, superoxide dismutase, ascorbate oxidase, peroxidase and catalase activities decrease rapidly. Ultrastructural destruction follows after a short lag period [124, 125]. Figure 8.8 shows the early effects of acifluorfen at the ultrastructural level under conditions similar to those in Table 8.5. Proliferation of

TABLE 8.5 SEQUENTIAL DETECTION OF DAMAGE INDICATORS IN ACIFLUORFEN TREATED GREEN CUCUMBER COTYLEDONS EXPOSED TO STRONG LIGHT (420 μE m^{-2} s^{-1} PHOTOSYNTHETICALLY ACTIVE RADIATION). The herbicide had been applied 20 hours earlier, and the plants were kept in darkness prior to the strong illumination [124–126].

Hours in light	Damage indicators detected
1	Electrolyte leakage, decrease in ascorbate and glutathione, decrease in enzyme activities
1.5	Ultrastructural damage: chloroplast thylakoids, mitochondria, tonoplast, plasmalemma
2	Evolution of ethane and ethylene, accumulation of malondialdehyde
5	Decrease in carotenoids and chlorophylls

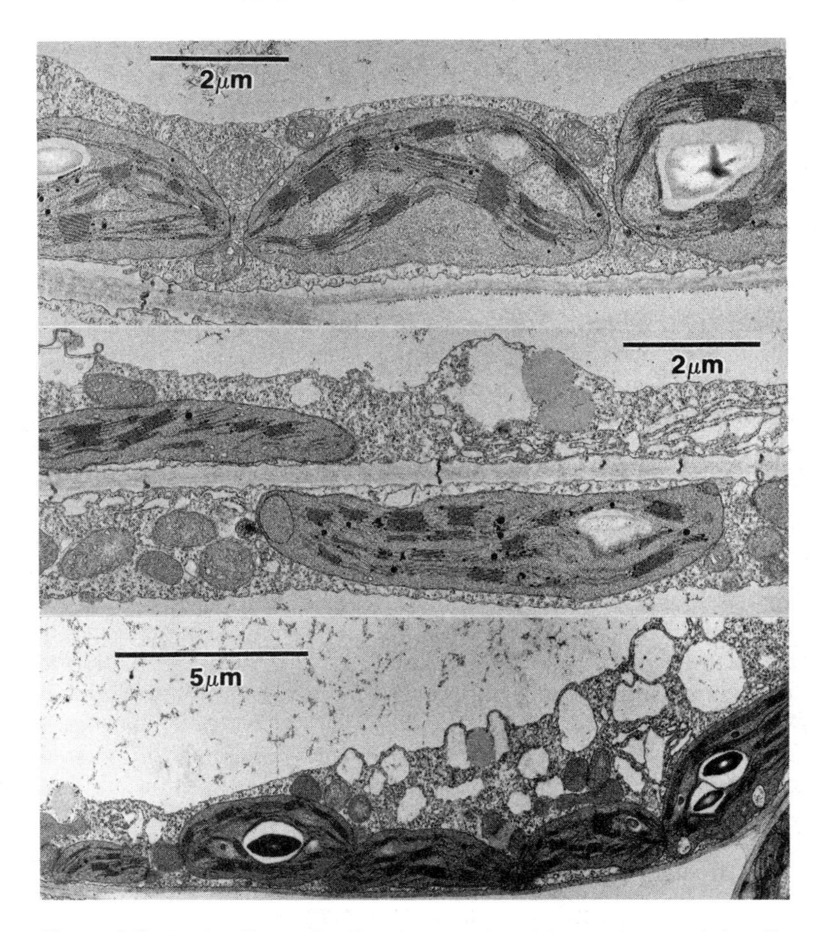

Figure 8.8 Early effects of acifluorfen on 7-day-old cucumber cotyledon discs that have been incubated for 20 h in darkness on 30 μM herbicide and then exposed to light (300 μE m^{-2} s^{-1}) for 0 (top), 1 (middle), or 2 (bottom) h. Photographs courtesy of R.N. Paul.

cytoplasmic vesicles starts after 1 h and becomes very marked after 2 h. Damage, as judged by a loss of clear membrane staining, can also be seen after 1 h in some mitochondria and some chloroplast grana. At the same time, lipid droplets accumulate outside the grana. Cytoplasmic vesicle proliferation indicates that damage is not confined to the organelles containing the sensitive enzyme PPG-oxidase, but occurs simultaneously in the cytoplasm. After 2 h the symptoms become more obvious, but they are not basically different. The chloroplast grana are among the most resistant membrane systems in the cell interior, possibly because of their massive structure and their high antioxidizing potential. Early damage in mitochondria has been paralleled by the cytochemical localization of hydrogen peroxide and the superoxide radical in these organelles, but not in chloroplasts [111]. Chloroplasts can remove these activated oxygen species more effectively, and therefore accumulate them more slowly and/or after a longer lag phase.

During the peroxidation of unsaturated membrane lipids (Section 9.3) malondialdehyde accumulates up to 20–30 nmoles/g fresh weight. The production of ethylene can be inhibited by AVG (aminoethoxyvinylglycine), again providing evidence for its synthesis as an induced stress response [103]. Pigment bleaching occurs rather late, apparently reflecting the special importance of these pigments: because of their high sensitivity to photobleaching, they are more effectively protected than other cell constituents.

The kinetics of herbicidal damage described above have been obtained with tissue preloaded with herbicide in the dark and then exposed to high light levels. Detailed kinetic studies without a herbicide preincubation time have shown that a light-independent lag phase of 5–7 h precedes the onset of lipid peroxidation [103]. This lag phase is also independent of herbicide concentration and cannot, therefore, be explained by the time required for the herbicide to reach the site of action. Similarly, the lag phase cannot be explained by the time-dependent loss of redox buffers like ascorbate, glutathione, and α-tocopherol during light-dependent action, since this loss occurs much more rapidly (within 1–2 h) [126]. Moreover, the lag phase is always 5–7 h long, whether studied in light or darkness, light being required only after the end of the lag phase for the induction of photooxidation. As was mentioned above when discussing the interactions of PPG-oxidase inhibitors with inhibitors of RNA and protein synthesis (Table 8.4), a derepression of the tetrapyrrole biosynthesis pathway occurs after herbicidal inhibition of PPG-oxidase. The lag phase may reflect the time required for derepression and induction of increased rates of tetrapyrrole biosynthesis.

Light-dependent phytotoxicity induced by the massive synthesis of tetrapyrroles is also the mode of action of δ-aminolevulinic acid which, when aided by the "activator" 2,2′-dipyridyl, can be applied as a herbicide [127, 128]. δ-Aminolevulinic acid is the basic building block for tetrapyrrole biosynthesis (Figure 8.7); its excessive availability leads to an accumulation of tetrapyrroles in a manner similar to that of the NDPEs and related herbicides and, ultimately, to a similar mode of herbicidal action.

8.4.4 Other Modes of Action of Diphenylether Herbicides

Early studies with NDPE herbicides concentrated largely on their interference with mitochondrial and chloroplast electron transport systems and with respiratory and photosynthetic phosphorylation systems [111, 129–134]. A small number of NDPEs act as inhibitors of photosynthetic electron transport, either at the diuron site or at the DBMIB site (Section 7.3) [131]. Some NDPEs, for example nitrofen, nitrofluorfen, and fomesafen, bind to the chloroplast coupling factor and then act as inhibitors of energy transfer in the process of photophosphorylation (Figure 8.9) [135, 136]. However, the concentrations required for this effect are generally in the 10 to 100 μM range, about three orders of magnitude higher than the concentrations required for the inhibition of PPG-oxidase (Table 8.3). Therefore, the NDPE inhibitions seen in different electron transport and phosphorylation systems can be concluded to be independent mechanisms at higher concentrations and have little or nothing to do with the herbicidal action. The different interference mechanisms of diphenylether herbicides are positioned in a "concentration diagram" in Figure 8.10. The inhibitors of carotenoid biosynthesis include some diphenylether structures, the *m*-phenoxybenzamides [7]. Some aryl-substituted propanoic acids (including the "aryl-propanoic acids," Section 11.1) are occasionally also included in presentations of diphenylethers, but their mode of action is completely different. Figure 8.10 is an illustration of the frequent observation that chemicals/herbicides have several different modes of interaction with plant metabolism at different concentration ranges. For herbicidal considerations, only those inhibitions or interferences that occur in the nanomolar (roughly corresponding to g per ha) to micromolar (kg per ha) range are relevant. The stress response, which fills the same concentration range as photooxidation in Figure 8.10, appears to be an indirect (secondary) metabolic response following herbicidal action.

One stress response induced by NDPEs is the induction of a number of isoflavonoid glucosides and their key enzymes chalcone synthase, phenylalanine ammonia lyase, and UDP-glucose : isoflavone-7-*O*-glucosyltransferase in soybean leaves by acifluorfen [138]. In general, the phenylpropanoid biosynthesis pathway is induced in a number of crop species, leading to increased synthesis of phytoalexins (glyceollins and glyceofuran in soybean, phaseollin in bean, pisatin in pea) [139,

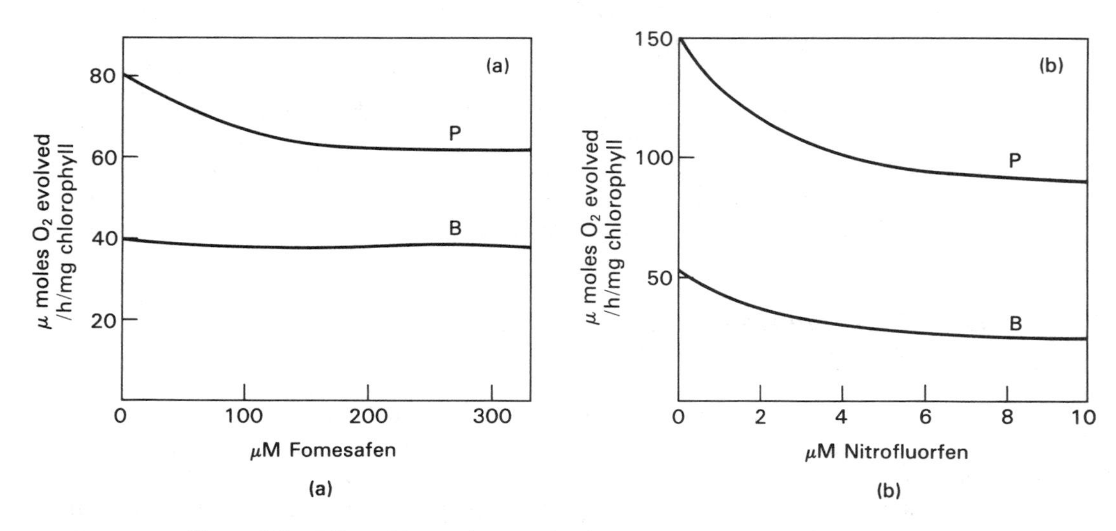

Figure 8.9 Effect of fomesafen and nitrofluorfen on oxygen evolution in basal (B) and state 3 phosphorylating (P) chloroplast electron transport [136].

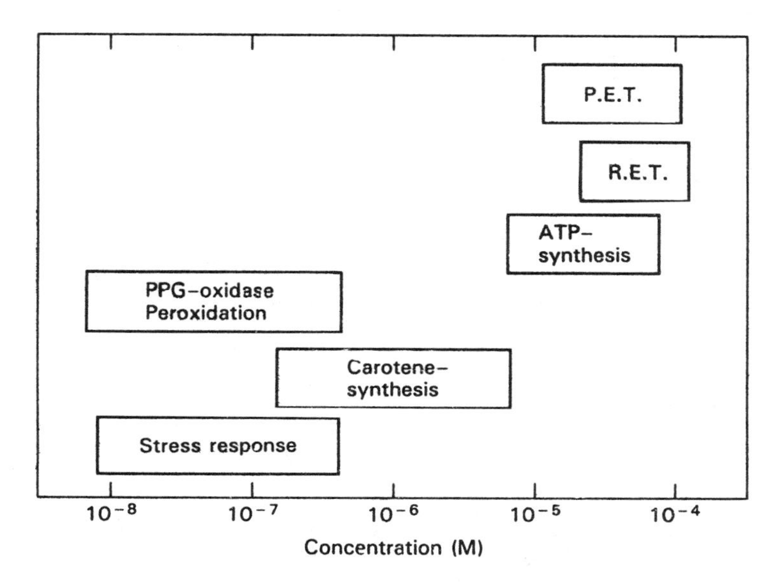

Figure 8.10 Modes of action of diphenyl ethers applied at different concentrations. The sensitive metabolic pathways or reactions are represented by boxes positioned in the respective concentration ranges needed for 50% inhibition [137].

140]. In cotton, the sesquiterpenoid hemigossypol is induced (see Chapter 16 for further elaboration of these effects).

8.5 PHOTOOXIDATION VIA PHOTOSYSTEM I REDOX INTERACTION

The herbicides to be dealt with under the heading "redox interaction with photosystem I" are the fastest acting herbicides, faster even than the NDPEs (Section 8.4.3). Under strong illumination, the first leaf damage, consisting of areas showing symptoms of wilting, can be observed after 20–30 min. These are the leaf areas with the highest photosynthetic electron transport rates. After a foliar application, the light-dependent herbicidal action is so fast that herbicide export from the treated tissue rarely occurs because of rapid desiccation (see also Section 5.5). After treatment in the dark, the herbicides are more mobile, resulting in even more tissue damage throughout the plant if the plants are exposed to light later.

The two compounds from this group that are most commonly used are the bipyridylium herbicides diquat and paraquat (Figure 8.11). They are reduced by photosystem I in a one-electron transfer reaction (Figure 7.2). The resulting radical

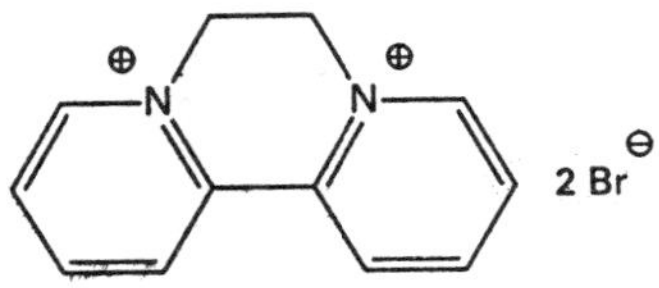

Figure 8.11 Structures of important bipyridylium herbicides and of a representative heteropentalene.

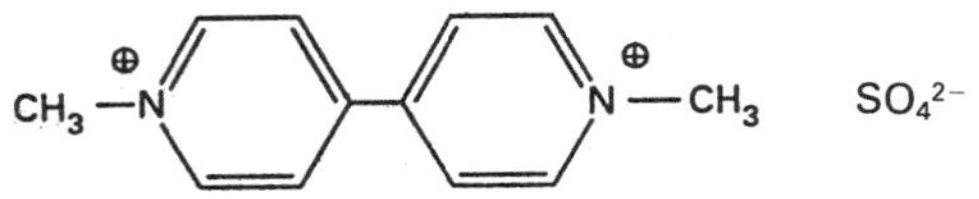

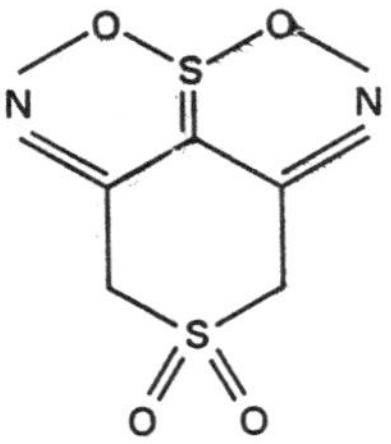

anion reacts spontaneously with molecular oxygen to produce the superoxide anion, $O_2^{\cdot -}$ (Section 9.2) [141]. The superoxide anion is considered to initiate the reaction sequence that eventually causes herbicidal action (Figures 9.2, 9.3). By a series of interactions, hydrogen peroxide and the highly reactive hydroxyl radical are produced. These toxic oxygen species affect unsaturated membrane lipids most strongly, resulting in fatty acid peroxidation, loss of membrane semipermeability and, eventually, wilting and desiccation.

In order to serve as an autooxidizable redox catalyst at the reducing side of photosystem I, a suitable molecule should have a redox potential around -400 to -500 mV and should form a stable, water-soluble radical upon accepting the single electron. The radical must of course not be *too* stable but must react with O_2 to produce $O_2^{\cdot -}$ and the original (now oxidized) redox catalyst. A number of quaternary ammonium compounds are known to fulfill these requirements, and about six of these serve as herbicides. Other structures that meet these requirements, such as the heteropentalenes (Figure 8.11), can also act as herbicides [142, 143]. The latter compounds have lower herbicidal activity and also lower mammalian toxicity than the bipyridiliums.

As may be speculated from the elevated mammalian toxicity, the bipyridylium herbicides do not only react with photosystem I in the light. Low-potential electrons from other cellular sources besides photosynthesis can reduce diquat and paraquat and initiate the same sequence of oxidations and peroxidations, albeit at a much lower rate. It has been stated repeatedly that the slow type of damage occurring at low concentrations or in the dark closely resembles the membrane alterations observed during normal cell senescence [144–146]. This conclusion implies that normal cell senescence is to a large extent caused by the action of molecular oxygen on membrane lipids (compare Section 9.3). The slow type of bipyridylium herbicide action could therefore be characterized as an "accelerated senescence."

In high light, however, the much higher rate of low-potential electron transfer from photosystem I to the bipyridylium herbicides leads to rapid lipid peroxidation. The type of damage and the sequence of damage indicators that occur is not very different from those produced by inhibitors of photosystem II (Figure 7.8) or by NDPEs (Table 8.5), but the time scale can be different. The irradiance level and the age of the tissue greatly influence the speed of the herbicidal action, with younger tissues and higher irradiance levels producing the fastest action. Under comparable conditions, bipyridylium herbicides induce the most rapid phytotoxicity (within 0.2 to 1 h in high light), closely followed by NDPEs and similarly acting compounds (Section 8.4.1, 0.5 to 2 h). The inhibitors of photosystem II act more slowly (1 to 5 h).

Pigment photooxidation, loss of photosynthetic activities, membrane leakage, and pheophytin and malondialdehyde production have all been recorded after paraquat action [147]. Reduced glutathione and ascorbate are rapidly lost without the simultaneous appearance of dehydroascorbate [148]. Propane, ethane, and ethylene are evolved [149]. In algal cultures, elevation of the iron concentration in the culture solution increases the production of the above gases, indicating that Fenton-type reactions are involved in the photooxidation process (Sections 9.2, 9.3). Diuron decreases the herbicidal activity of paraquat and diquat, presumably by inhibiting photosynthetic electron transport [150, 151]. Not surprisingly, a nitrogenous atmosphere also decreases the herbicidal damage [150]. Ultrastructural damage can be seen within a few hours. It includes membrane rupture at the level of the thylakoids, mitochondria, tonoplast, and plasmalemma, and is accompanied by the accumulation of plastoglobuli [151, 152]. Many different aspects of bipyridylium herbicides, including synthesis, chemistry, application, fate in plants, animals, soil, and water, analysis, physiology of action, environmental behavior, and toxicology are discussed in a book dedicated to these herbicides [153].

Several paraquat-resistant plant species are known. The level of resistance is usually quite high; for example, 100-fold higher herbicidal concentrations often are required for the same level of activity. The resistance is also frequently termed tolerance since in the herbicide field tolerance and resistance are not strictly defined (see Section 6.1 for a discussion of this topic). Because of the usually much higher herbicide concentrations required for measurable toxicity in the herbicide-surviving plants, the term resistance will be used here. Paraquat resistance has been reported in *Lolium perenne* [154–157], three *Conyza* (=*Erigeron*) species [158–164], *Hordeum glaucum* [165, 166], *Nicotiana tabacum* [167, 168], and the fern *Ceratopteris richardii* [169]. The mechanism of resistance remains a matter of some controversy; two different mechanisms have been suggested, one based on increased levels of those enzymes involved in the detoxification of activated oxygen species (catalase, superoxide dismutase, peroxidase) [155, 157, 160, 161,

168], and a second based on the rapid sequestering of paraquat in the apoplast (veinal tissue and/or cell walls) [158, 159, 162, 166]. In some cases none of these resistance mechanisms has been demonstrated, and the physiological or biochemical basis of the resistance remains unknown [164, 169]. The strategies underlying these mechanisms are of course fundamentally different, one resting on the more rapid destruction of toxic oxygen species, and the second avoiding the movement of the herbicide to its site of action. The mechanism of resistance may well be different in different resistant species. In *Conyza bonariensis,* sequestration of paraquat provides a very effective mechanism of herbicide removal. Furthermore, since other similar herbicidal acceptors of electrons from photosystem I are not or only slightly affected by this resistance (i.e., the tissue is sensitive to them), the paraquat-sequestering mechanism seems to be the relevant mechanism of resistance in this species [163]. The occasionally reported increased levels of superoxide dismutase and other enzymes might result indirectly from the altered cell wall structure that has been observed in these mutants.

In a study of the superoxide dismutase variation in different *Conyza bonariensis* selections, enzyme activity did not correlate with the segregation of paraquat resistance [158]. However, increased levels of superoxide dismutase can convey higher tolerance levels, as has been shown for *Spirodela oligorrhiza* pretreated with benzyl viologen [170], for *Chlorella sorokiniana* treated with sulfite [171], and for *Escherichia coli* [172]. Increased levels of superoxide dismutase, catalase, and peroxidase have been reported for paraquat-resistant lines of *Lolium perenne* [155, 157, 161] and tobacco [168]. Whether or not this mechanism alone can convey a high level of resistance on a whole-plant basis remains to be determined.

REFERENCES

1. Britton, G., P. Barry, and A. J. Joung. 1987. "The mode of action of diflufenican: its evaluation by HPLC." In Proc. Brit. Crop Prot. Conf.—Weeds, Vol. 3. Thornton Heath, U.K.: BCPC Publ., pp. 1015–22.
2. Britton, G. 1987. "Carotenoid biosynthesis—an overview." In 8th International Symposium on Carotenoids, Boston, Mass.
3. Sandmann, G., and P. Böger. 1983. "Comparison of the bleaching activities of norflurazon and oxyfluorfen." *Weed Sci.,* 31, 338–41.
4. Ben-Aziz, A., and E. Koren. 1974. "Interference in carotenogenesis as a mechanism of action of the pyridazinone herbicide Sandoz 6706." *Plant Physiol.,* 54, 916–20.
5. Wightman, P., and C. Haynes. 1985. "The mode of action and basis of selectivity of diflufenican in wheat, barley, and selected weed species." In Proc. Brit. Crop Prot. Conf.—Weeds, Vol. 1. Croydon, U.K.: BCPC Publ., pp. 171–78.
6. Lay, M.-M. 1983. "The herbicidal mode of action of R-40244 and its absorption by plants." *Pestic. Biochem. Physiol.,* 19, 337–43.
7. Lambert, R., and P. Böger. 1983. "Inhibition of carotenogenesis by substituted di-

phenylethers of the m-phenoxybenzamide type." *Pestic. Biochem. Physiol.,* 20, 183–89.

8. Burns, E. R., G. A. Buchanan, and M. C. Carter. 1971. "Inhibition of carotenoid synthesis as a mechanism of action of amitrole, dichlormate, and pyriclor." *Plant Physiol.,* 47, 144–48.
9. Fujii, Y., T. Kurokawa, Y. Inone, I. Yamaguchi, and T. Misato. 1977. "Inhibition of carotenoid biosynthesis as a possible mode of herbicidal action of 3,3′-dimethyl-4-methoxybenzophenone (NK-049)." *J. Pestic. Sci.,* 2, 431–37.
10. Fujii, Y. 1979. "Studies on the mode of herbicidal activity of methoxyphenone." *J. Pestic. Sci.,* 4, 391–99.
11. Kerr, M. W., and D. P. Whitaker. 1987. "The mode of action of the herbicide WL-110547." In Proc. Brit. Crop Prot. Conf.—Weeds, Vol. 3. Thornton Heath, U.K.: BCPC Publ., pp. 1005–13.
12. Soeda, T., and T. Uchida. 1987. "Inhibition of pigment synthesis by 1,3-dimethyl-4-(2,4-dichlorobenzoyl)-5-hydroxypyrazole, norflurazon, and new herbicidal compounds in radish and flatsedge plants." *Pestic. Biochem. Physiol.,* 29, 35–42.
13. Dahlin, C., and H. Ryberg. 1986. "Accumulation of phytoene in plastoglobuli of SAN-9789 (norflurazon)-treated dark-grown wheat." *Physiol. Plant.,* 68, 39–45.
14. Sandmann, G., and P. M. Bramley. 1985. "Carotenoid biosynthesis by *Aphanocapsa* homogenates coupled to a phytoene-generating system from *Phycomyces blakesleeanus.*" *Planta,* 164, 259–63.
15. Bramley, P. M., I. E. Clarke, G. Sandmann, and P. Böger. 1984. "Inhibition of carotene biosynthesis in cell extracts of *Phycomyces blakesleeanus.*" *Z. Naturforsch.,* 39c, 460–63.
16. Sandmann, G., I. E. Clarke, P. M. Bramley, and P. Böger. 1984. "Inhibition of phytoene desaturase—the mode of action of certain bleaching herbicides." *Z. Naturforsch.,* 39c, 443–49.
17. Beyer, B., K. Kreuz, and H. Kleinig. 1980. "β-carotene synthesis in isolated chromoplasts from *Narcissus pseudonarcissus.*" *Planta,* 150, 435–38.
18. Sandmann, G., and P. Böger. 1987. "Herbicides affecting plant pigments." In Proc. Brit. Crop Prot. Conf.—Weeds, Vol. 1. Thornton Heath, U.K.: BCPC Publ., pp. 139–48.
19. St. John, J. B. 1985. "Action of R-40244 on chloroplast pigments and polar lipids." *Pestic. Biochem. Physiol.,* 23, 13–18.
20. St. John, J. B. 1976. "Manipulation of galactolipid fatty acid composition with substituted pyridazinones." *Plant Physiol.,* 57, 38–40.
21. St. John, J. B., M. N. Christiansen, E. N. Ashworth, and W. A. Carter. 1979. "Effects of BASF-13-338, a substituted pyridazinone, on linolenic acid levels and winterhardiness of cereals." *Crop Sci.,* 19, 65–69.
22. Wang, X.-M, D. F. Hildebrand, H. D. Norman, M. L. Dahmer, J. B. St. John, and G. B. Collins. 1987. "Reduction of linolenate content in soybean cultures by a substituted pyridazinone." *Phytochemistry,* 26, 955–60.
23. Ashworth, E. N., M. N. Christiansen, J. B. St. John, and G. Patterson. 1981. "Effect of temperature and BASF 13-338 on the lipid composition and respiration of wheat roots." *Plant Physiol.,* 67, 711–15.

24. de la Roche, A. I. 1979. "Increase in linolenic acid is not a prerequisite for development of freezing tolerance in wheat." *Plant Physiol.*, 63, 5–8.
25. Quinn, P. J., and W. P. Williams. 1978. "Plant lipids and their role in membrane function." *Prog. Biophys. Molec. Biol.*, 34, 109–73.
26. Quinn, P. J. 1981. "The fluidity of cell membranes and its regulation." *Prog. Biophys. Molec. Biol.*, 38, 1–104.
27. Ferrari-Ilion, R., A. T. Pham Thi, and J. V. da Silva. 1984. "Effect of water stress on the lipid and fatty acid composition of cotton (*Gossypium hirsutum*) chloroplasts." *Physiol. Plant.*, 62, 219–24.
28. Davies, A. D., and J. L. Harwood. 1983. "Effect of substituted pyridazinones on chloroplast structure and lipid metabolism in greening barley leaves." *J. Exp. Bot.*, 34, 1089–100.
29. Leech, R. M., C. A. Walton, and N. R. Baker. 1985. "Some effects of 4-chloro-5-(dimethylamino)-2-phenyl-3(2H)-pyridazinone (SAN 9785) on the development of chloroplast thylakoid membranes in *Hordeum vulgare*." *Planta*, 165, 277–83.
30. Jabben, M., and G. F. Deizer. 1979. "Effects of the herbicide SAN 9789 on photomorphogenetic responses." *Plant Physiol.*, 63, 481–85.
31. Duke, S. O., and A. D. Lane. 1984. "Phytochrome control of its own accumulation and leaf expansion in tentoxin- and norflurazon-treated mung bean seedlings." *Physiol. Plant.*, 60, 341–46.
32. Feldman, L. J., and P. S. Sun. 1986. "Effects of norflurazon, an inhibitor of carotenogenesis, on abscisic acid and xanthoxin in the caps of gravistimulated maize roots." *Physiol. Plant.*, 67, 472–76.
33. Stegink, S. J., and K. C. Vaughn. 1988. "Norflurazon (SAN-9789) reduces abscisic acid levels in cotton seedlings: a glandless isoline is more sensitive than its glanded counterpart." *Pestic. Biochem. Physiol.*, 31, 269–75.
34. Moore, R., and J. D. Smith. 1984. "Growth, graviresponsiveness and abscisic acid content of *Zea mays* seedlings treated with fluridone." *Planta*, 162, 342–44.
35. Hoglund, H.-D., and R. Klockare. 1987. "Stomatal responses to rapidly imposed water stress and light/dark transition in norflurazon-treated wheat leaves." *Physiol. Plant.*, 69, 477–80.
36. Creelman, R. A. 1989. "Abscisic acid physiology and biogenesis in higher plants." *Physiol. Plant.*, 75, 131–36.
37. Yokoyama, H., C. W. Caggins, and G. L. Henning. 1971. "The effect of 2-(4-chlorophenylthio)-triethylamine hydrochloride on the formation of carotenoids in Citrus." *Phytochemistry*, 10, 1831–34.
38. Kleinig, H. 1974. "Inhibition of carotenoid synthesis in *Myxococcus fulvus*." *Arch. Mikrobiol.*, 97, 217–26.
39. Yokoyama, H., W. J. Hsu, E. Hayman, and C. R. Benedict. 1985. "The role of 2-(4-methylphenoxy)triethylamine (MPTA) on tetraterpene biosynthesis in carotenogenic mold *Phycomyces blakesleeanus*." *Plant Physiol. Suppl.*, 79, 298i.
40. Murillo, F. J. 1980. "Effect of CPTA on carotenogenesis by *Phycomyces* car A mutants." *Plant Sci. Lett.*, 17, 201–05.
41. Benedict, C. R., C. L. Rosenfield, J. Miller, and H. Yokoyama. 1985. "The regulation of isoprenoid synthesis by 2-(4-methylphenoxy)triethylamine (MPTA)." *Plant Physiol. Suppl.*, 79, 297i.

42. Benedict, C. R., C. L. Rosenfield, J. R. Mahan, S. Madhavan, and H. Yokoyama. 1985. "The chemical regulation of carotenoid biosynthesis in Citrus." *Plant Sci.,* 41, 169–73.

43. Greenblatt, G. A., C. L. Rosenfield, H. Yokoyama, and C. R. Benedict. 1986. "The mode of action of 2-(4-methylphenoxy)triethylamine (MPTA) on carotenoid biosynthesis in cotton cotyledons." *Plant Physiol. Suppl.,* 80, 349i.

44. Duke, S. O., W. H. Kenyon, and R. N. Paul. 1985. "FMC-57020 effects on chloroplast development in pitted morningglory (*Ipomoea lacunosa*) cotyledons." *Weed Sci.,* 33, 786–94.

45. Duke, S. O., and R. N. Paul. 1986. "Effects of dimethazone (FMC 57020) on chloroplast development. I. Ultrastructural effects in cowpea (*Vigna unguiculata* L.) primary leaves." *Pestic. Biochem. Physiol.,* 25, 1–10.

46. Duke, S. O., and W. H. Kenyon. 1986. "Effects of dimethazone (FMC 57020) on chloroplast development. II. Pigment synthesis and photosynthetic function in cowpea [*Vigna unguiculata* L.) primary leaves." *Pestic. Biochem. Physiol.,* 25, 11–18.

47. Sandmann, G., and P. Böger. 1987. "Inhibition of prenyl-lipid biosynthesis by dimethazone." *Abstracts, Weed Sci. Soc. Amer.,* 27, 175.

48. Sandmann, G., and P. Böger. 1986. "Interference of dimethazone with formation of terpenoid compounds." *Z. Naturforsch.,* 41c, 729–32.

49. Duke, S. O., R. N. Paul, J. M. Becerril, and J. H. Schmidt. 1991."Clomazone causes accumulation of sesquiterpenoids in cotton (*Gossypium hirsutum L.).*" *Weed Sci.,* 39, 339–46.

50. Benveniste, P. 1987. "Rational design of inhibitors of sterol biosynthesis in plants." Soc. of the Chem. Industries Symp., London, U.K..

50a. Lützow, M., P. Beyer, and H. Kleinig. 1990. "The herbicide command does not inhibit the prenyl diphosphate-forming enzymes in plastids." *Z. Naturforsch.,* 45c, 856–58.

51. Bartels, P. G., and A. Hyde. 1970. "Buoyant density studies of chloroplast and nuclear DNA from control and ATA-treated wheat seedlings, *Triticum vulgare*." *Plant Physiol.,* 46, 825–30.

52. Feierabend, J., T. Winkelhüsener, P. Kemmerich, and U. Schulz. 1982. "Mechanism of bleaching in leaves treated with chlorosis inducing herbicides." *Z. Naturforsch.,* 37c, 898–907.

53. Ferguson, I. B., and S. J. Dunning. 1986. "Effect of 3-amino-1,2,4 triazole, a catalase inhibitor, on peroxide content of suspension-cultured pear fruit cells." *Plant Sci.,* 43, 7–11.

54. Smith, I. K., A. C. Kendall, A. J. Keys, and J. C. Turner. 1984. "Increased levels of gluthatione in a catalase-deficient mutant of barley (*Hordeum vulgare* L.)." *Plant Sci. Letters,* 37, 29–33.

55. Smith, I. K. 1985. "Stimulation of glutathione synthesis in photorespiring plants by catalase inhibitors." *Plant Physiol.,* 79, 1044–47.

56. Bartels, P. G., and F. T. Wolf. 1965. "The effect of amitrole upon nucleic acid and protein metabolism of wheat seedlings." *Physiol. Plant.,* 18, 805–13.

57. Brown, J. C., and M. C. Carter. 1968. "Influence of amitrole upon protein metabolism in bean plants." *Weed Sci.,* 16, 222–26.

58. Burt, G. W., and T. J. Muzik. 1970. "Effect of 3-amino-1,2,4-triazol on the alcohol soluble nitrogen compounds in *Cirsium arvense*." *Physiol. Plant.*, 23, 498–504.

59. Davies, M. E. 1971. "Regulation of histidine biosynthesis in cultured plant cells: evidence from studies on amitrol toxicity." *Phytochemistry*, 10, 783–88.

60. Vivekandanan, M., and A. Gnanam. 1975. "Studies on the mode of action of aminotriazole in the induction of chlorosis." *Plant Physiol.*, 55, 526–31.

61. Suzuki, Y., H. Emoto, and H. Mitsuda. 1978. "Intracellular accumulation of S-adenosylmethionine in a high riboflavinogenic *Eremothecium ashbyii* grown in the presence of 3-amino-1,2,4-triazole." *Biochim. Biophys. Acta*, 541, 115–18.

62. Dörfling, P., W. Dummler, and D. Mucke. 1970. "Das Auftreten von Koproporphyrin in Kulturen von *Poteriochromonas stipitata* nach Inkubation mit 3-Amino-1,2,4-Triazol (Amitrol)." *Experientia*, 26, 728.

63. Hilton, J. L. 1969. "Inhibition of growth and metabolism by 3-amino-1,2,4-triazole (Amitrole)." *J. Agric. Food Chem.*, 17, 182–98.

64. Konotsune, T., K. Kawakubo, and T. Yanai. 1979. "Synthesis and herbicidal activity of 4-acyl-pyrazole derivatives." In *Advances in Pesticide Science, Part 2*, H. Geissbühler, ed. New York: Pergamon, pp. 94–98.

65. Yamaoka, K., M. Nakagawa, and M. Ishida. 1987. "Hydrolysis of the rice herbicide pyrazolate in aqueous solutions." *J. Pestic. Sci.*, 12, 209–12.

66. Ando, M., K. Yamaoka, Y. Shigematsu, M. Nakagawa, and M. Ishida. 1988. "Metabolism of DTP, the herbicidal entity of pyrazolate, in rice plants." *J. Pestic. Sci.*, 13, 579–85.

67. Sandmann, G., H. Reok, and P. Böger. 1984. "Herbicidal mode of action on chlorophyll formation." *J. Agric. Food Chem.*, 32, 868–72.

68. Kawakubo, K., M. Shindo, and T. Konotsune. 1979. "A mechanism of chlorosis caused by 1,3-dimethyl-4-(2,4-dichlorobenzoyl)-5-hydroxypyrazole, a herbicidal compound." *Plant Physiol.*, 64, 774–79.

69. Sarma, R., A. E. Flood, and R. L. Wain. 1977. "Studies on plant growth regulating substances. XLVII. Ultrastructural studies on emergent leaves of *Avena sativa* treated with 3,5-(diiodo and dibromo)-4-hydroxyphenyl methylsulphone." *Ann. Appl. Bot.*, 86, 429–32.

70. Dodge, A. D., and G. B. Lawes. 1972. "Some effects of the herbicide haloxydine on the structure and development of chloroplasts." *Ann. Bot.*, 36, 315–23.

71. Geronimo, J., and J. W. Herr. 1970. "Ultrastructural changes of tobacco chloroplasts induced by pyrichlor." *Weed Sci.*, 18, 48–53.

72. Matsunaka, S. 1969. "Acceptor of light energy in photoactivation of diphenylether herbicides." *J. Agric. Food Chem.*, 17, 171–75.

73. Teraoka, T., G. Sandmann, P. Böger, and K. Wakabayashi. 1987. "Effect of cyclic imide herbicides on pigment formation in plants." *J. Pestic. Sci.*, 12, 499–504.

74. Frosch, S., M. Jabben, R. Bergfeld, H. Kleinig, and H. Mohr. 1979. "Inhibition of carotenoid biosynthesis by the herbicide SAN 9789 and its consequences for the action of phytochrome on plastogenesis." *Planta*, 145, 497–505.

75. Mayfield, S. P., and W. C. Taylor. 1987. "Chloroplast photooxidation inhibits the expression of a set of nuclear genes." *Mol. Gen. Genet.*, 208, 309–14.

76. Bartels, P. G., K. Matsuda, A. Siegel, and T. E. Weier. 1967. "Chloroplastic ribosome formation: inhibition by 3-amino-1,2,4-triazole." *Plant Physiol.*, 42, 736–41.

77. Axelsson, L., C. Dahlin, and H. Ryberg. 1982. "The function of carotenoids during chloroplast development. V. Correlation between carotenoid content, ultrastructure and chlorophyll b to chlorophyll a ratio." *Physiol. Plant.*, 55, 111–16.
78. Bartels, P. G., and A. Hyde. 1970. "Chloroplast development in metflurazon treated wheat seedlings." *Plant Physiol.*, 45, 807–10.
79. Blume, D. E., and J. W. McClure. 1980. "Developmental effects of Sandoz 6706 on activities of enzymes of phenolic and general metabolism in barley shoots grown in the dark or under low or high intensity light." *Plant Physiol.*, 65, 238–44.
80. Deane-Drummond, C. E., and C. B. Johnson. 1980. "Absence of nitrate reductase activity in SAN 9789 bleached leaves of barley seedlings (*Hordeum vulgare* cv. Midas)." *Plant Cell Envir.*, 3, 303–08.
81. Duke, S. O., K. C. Vaughn, and S. H. Duke. 1982. "Effects of norflurazon (SAN 9789) on light increased extractable nitrate reductase activity in soybean (*Glycine max* (L.) Merr.) seedlings." *Plant Cell Envir.*, 5, 155–62.
82. Duke, S. H., and S. O. Duke, 1984. "Light control of extractable nitrate reductase activity in higher plants." *Physiol. Plant.*, 62, 485–93.
83. Kakefuda, G., S. H. Duke, and S. O. Duke. 1983. "Differential light induction of nitrate reductases in greening and photobleached soybean seedlings." *Plant Physiol.*, 73, 56–60.
84. Feierabend, J., and B. Schubert. 1978. "Comparative investigation of the action of several chlorosis-inducing herbicides on the biogenesis of chloroplasts and leaf microbodies." *Plant Physiol.*, 61, 1017–22.
85. Feierabend, J., and T. Winkelhüsener. 1982. "Nature of photooxidative events in leaves treated with chlorosis-inducing herbicides." *Plant Physiol.*, 70, 1277–82.
86. Lay, M. M., J. M. Henstrand, S. R. Lawrence, and T. H. Cromartie. 1985. "Studies on the mode of action of the herbicide fluorochloridone." In Proc. Brit. Crop Prot. Conf.—Weeds, Vol. 1. Croydon, U.K.: BCPC Publ., pp. 170–86.
87. Ensminger, M., F. D. Hess, and J. T. Bahr. 1985. "Nitro free radical formation of diphenylether herbicides is not necessary for their toxic action." *Pestic. Biochem. Physiol.*, 23, 163–70.
88. Wakabayashi, K., K. Matsuya, T. Teraoka, G. Sandmann, and P. Böger. 1986. "Effect of cyclic imide herbicides on chlorophyll formation in higher plants." *J. Pestic. Sci.*, 11, 635–40.
89. Wakabayashi, K., G. Sandmann, H. Ohta, and P. Böger. 1988. "Peroxidizing herbicides: comparison of dark and light effects." *J. Pestic. Sci.*, 13, 461–71.
90. Ohta, H., S. Suzuki, H. Watanabe, T. Jikihara, K. Matsuya, and K. Wakabayashi. 1976. "Structure activity relationship of cyclic imide herbicides." *Agr. Biol. Chem.*, 40, 745–51.
91. Matringe, M., J. L. Dufour, J. Lherminier, and R. Scalla. 1987. "Characterization of the mode of action of the experimental herbicide LS-82-556 (S)-N-(methylbenzyl)carbamoyl-5-propionyl-2,6-lutidine." *Pestic. Biochem. Physiol.*, 27, 267–74.
92. Matringe, M., and R. Scalla. 1987. "Induction of tetrapyrrole accumulation by diphenylether-type herbicides." In Proc. Brit. Crop Prot. Conf.—Weeds, Vol. 3. Thornton Heath, U.K.: BCPC Publ., pp. 981–88.
93. Duke, S. O., J. Lydon, and R. N. Paul. 1989. "Oxadiazon activity is similar to that of p-nitro-diphenyl ether herbicides." *Weed Sci.*, 37, 152–60.

94. Anon. 1988. "A new look at porphyrin nomenclature." *TIBS.*, 13, 458.
95. Matringe, M., J.-M. Camadro, P. Labbe, and R. Scalla. 1989. "Protoporphyrinogen oxidase as a molecular target for diphenyl ether herbicides." *Biochem. J.*, 260, 231–35.
96. Matringe, M., J.-M. Camadro, P. Labbe, and R. Scalla. 1989. "Protoporphyrinogen oxidase inhibition by three peroxidizing herbicides: oxadiazon, LS 82-556 and M&B 39279." *FEBS Lett.*, 245, 35–38.
97. Kouji, H., T. Masuda, and S. Matsunaka. 1988. "Action mechanism of diphenyl ether herbicides: light-dependent O_2 consumption in diphenylether-treated tobacco cell homogenate." *J. Pestic. Sci.*, 13, 495–99.
98. Matringe, M., and R. Scalla. 1988. "Effects of acifluorfen-methyl on cucumber cotyledons: porphyrin accumulation." *Pestic. Biochem. Physiol.*, 32, 164–72.
99. Matringe, M., and R. Scalla. 1988. "Studies on the mode of action of acifluorfen-methyl in nonchlorophyllous soybean cells. Accumulation of tetrapyrroles." *Plant Physiol.*, 86, 619–22.
100. Lydon, J., and S. O. Duke. 1988. "Porphyrin synthesis is required for photobleaching activity of the p-nitrosubstituted diphenyl ether herbicides." *Pestic. Biochem. Physiol.*, 31, 74–83.
101. Witkowski, D. A., and B. P. Halling. 1988. "Accumulation of photodynamic tetrapyrroles induced by acifluorfen-methyl." *Plant Physiol.*, 87, 632–37.
102. Witkowski, D. A., and B. P. Halling. 1989. "Inhibition of plant protoporphyrinogen oxidase by the herbicide acifluorfen-methyl." *Plant Physiol.*, 90, 1239–42.
103. Tietjen, K. Personal communication. 1990.
104. Devlin, R., S. J. Karczmarczyk, and I. I. Zbiec. 1983. "Influence of norflurazon on the activation of substituted diphenylether herbicides by light." *Weed Sci.*, 31, 109–12.
105. Gaba, V., N. Cohen, Y. Shaaltiel, A. Ben-Amotz, and J. Gressel. 1988. "Light-requiring acifluorfen action in the absence of bulk photosynthetic pigments." *Pestic. Biochem. Physiol.*, 31, 1–12.
106. Ensminger, M. P., and F. D. Hess. 1985. "Action spectrum of the activity of acifluorfen-methyl, a diphenyl ether herbicide, in *Chlamydomonas eugametos*." *Plant Physiol.*, 77, 503–05.
107. van Stone, D. E., and E. Stobbe. 1979. "Light requirement of the diphenylether herbicide oxyfluorfen." *Weed Sci.*, 27, 88–91.
108. Kunert, K. J., C. Homringhausen, H. Böhme, and P. Böger. 1985. "Oxyfluorfen and lipid peroxidation: protein damage as a phytotoxic consequence." *Weed Sci.*, 33, 766–70.
109. Gillham, D. J., and A. D. Dodge. 1987. "Studies into the action of the diphenyl ether herbicides acifluorfen and oxyfluorfen. Part I. Activation by light and oxygen in leaf tissue." *Pestic. Sci.*, 19, 19–24.
110. Ensminger, M. E., and F. D. Hess. 1985. "Photosynthesis involvement in the mechanism of action of diphenyl ether herbicides." *Plant Physiol.*, 78, 46–50.
111. Bowyer, J. R., B. J. Hallahan, S. A. Lee, and P. Camilleri. 1987. "The role of photosynthetic electron transport in the mode of action of nitrodiphenylether herbicides." In Proc. Brit. Prot. Conf.—Weeds, Vol. 3. Thornton Heath, U.K.: BCPC Publ., pp. 989–96.

112. Duke, S. O., K. C. Vaughn, and R. L. Meeusen. 1984. "Mitochondrial involvement in the mode of action of acifluorfen." *Pestic. Biochem. Physiol.,* 21, 368–76.

113. Duke, S. O., and W. H. Kenyon. 1986. "Photosynthesis is not involved in the mechanism of action of acifluorfen in cucumber (*Cucumis sativus* L.)." *Plant Physiol.,* 81, 882–88.

114. Lambert, R., P. M. H. Kroneck, and P. Böger. 1984. "Radical formation and peroxidative activity of phytotoxic diphenyl ethers." *Z. Naturforsch.,* 39c, 486–91.

115. Orr, G. L., C. M. Elliott, and M. E. Hogan. 1983. "Determination of redox behavior in vitro of nitrodiphenylether herbicides using cyclic voltammetry." *J. Agric. Food Chem.,* 31, 1192–95.

116. Orr, G. L., and C. M. Elliott. 1983. "Activity in vivo and redox states in vitro of chlorodiphenylether herbicide analogs." *Plant Physiol.,* 73, 939–44.

117. Orr, G. L., C. M. Elliott, and M. E. Hogan. 1984. "Electrochemical characterization and redox reaction scheme of the diphenylether herbicide nitrofen." *Pestic. Biochem. Physiol.,* 21, 242–47.

118. Orr, G. L., and M. E. Hogan. 1985. "UV-A photooxidation of β-carotene in Triton X-100 micelles by nitrodiphenylether herbicides." *J. Agric. Food Chem.,* 33, 968–72.

119. Elich, T. D., and J. C. Lagarias. 1988. "4-Amino-5-hexynoic acid—a potent inhibitor of tetrapyrrole biosynthesis in plants." *Plant Physiol.,* 88, 747–51.

120. Larson, R. A. 1988. "The antioxidants of higher plants." *Phytochemistry,* 27, 969–78.

121. Halling, B. P., and G. R. Peters. 1984. "Influence of chloroplast development on the activation of the diphenylether herbicide acifluorfen-methyl." *Plant Physiol.,* 84, 1114–20.

122. Corriveau, J. L., and S. I. Beale. 1986. "Influence of gabaculine on growth, chlorophyll synthesis, and δ-aminolevulinic acid synthase in *Euglena gracilis*." *Plant Sci.,* 45, 9–17.

123. May, T. B., J. A. Guikema, R. L. Henry, M. K. Schuler, and P. P. Wong. 1987. "Gabaculine inhibition of chlorophyll biosynthesis and nodulation in *Phaseolus lunatus* L." *Plant Physiol.,* 84, 1309–13.

124. Kenyon, W. H., S. O. Duke, and K. C. Vaughn. 1985. "Sequence of effects of acifluorfen on physiological and ultrastructural parameters in cucumber cotyledon discs." *Pestic. Biochem. Physiol.,* 24, 240–50.

125. Derrick, P. M., A. H. Cobb, and K. E. Pallett. 1988. "Ultrastructural effects of the diphenylether herbicide acifluorfen and the experimental herbicide M&B 39279." *Pestic. Biochem. Physiol.,* 32, 153–63.

126. Kenyon, W. H., and S. O. Duke. 1985. "Effects of acifluorfen on endogenous antioxidants and protective enzymes in cucumber (*Cucumis sativus* L.) cotyledons." *Plant Physiol.,* 79, 862–66.

127. Rebeiz, C. A., A. Montazer-Zouhoor, H. J. Hopen, and S. M. Wu. 1984. "Photodynamic herbicides: 1. Concept and phenomenology." *Enzyme Microb. Technol.,* 6, 390–401.

128. Rebeiz, C. A., A. Montazer-Zouhoor, J. M., Mayasich, B. C. Tripathy, S.-M. Wu, and C. C. Rebeiz. 1988. "Photodynamic herbicides. Recent developments and molecular basis of selectivity." *CRC Critical Reviews in Plant Sciences,* 6, 385–436.

129. Moreland, D. E., W. J. Blackmon, H. G. Todd, and F. S. Farmer. 1970. "Effects of diphenyl ether herbicides on reactions of mitochondria and chloroplasts." *Weed Sci.*, 18, 636–42.

130. Bugg, M. W., J. Whitmarsch, C. E. Rieck, and W. S. Cohen. 1980. "Inhibition of photosynthetic electron transport by diphenyl ether herbicides." *Plant Physiol.*, 65, 47–50.

131. Draber, W., H. J. Knops, and A. Trebst. 1981. "Mode of inhibition of photosynthetic electron transport by substituted diphenyl ethers." *Z. Naturforsch.*, 36c, 848–52.

132. Bowyer, J. R., B. J. Smith, P. Camilleri, and S. A. Lee. 1987. "Mode of action studies on nitrodiphenyl ether herbicides." *Plant Physiol.*, 83, 613–20.

133. Hoagland, R. E., D. J. Hunter, and M. L. Salin. 1986. "Acifluorfen and oxidized ubiquinone in soybean mitochondria." *Plant Cell Physiol.*, 27, 11–15.

134. Lambert, R., G. Sandmann, and P. Böger. 1983. "Correlation between structure and phytotoxic activities of nitrodiphenyl ethers." *Pestic. Biochem. Physiol.*, 19, 309–20.

135. Huchzermeyer, B. 1982. "Energy transfer inhibition induced by nitrofen." *Z. Naturforsch.*, 37c, 787–92.

136. Ridley, S. M. 1983. "Interaction of chloroplasts with inhibitors." *Plant Physiol.*, 72, 461–68.

137. Böger, P. 1984. "Multiple modes of action of diphenyl ethers." *Z. Naturforsch.*, 39c, 468–75.

138. Cosio, E. G., G. Weissenböck, and J. C. McClure. 1985. "Acifluorfen induced isoflavonoids and enzymes of their biosynthesis in mature soybean leaves." *Plant Physiol.*, 78, 14–19.

139. Kömives, T., and J. E. Casida. 1982. "Diphenyl ether herbicides: effects of acifluorfen on phenylpropanoid biosynthesis and phenylalanine ammonia-lyase activity in spinach." *Pestic. Biochem. Physiol.*, 18, 191–96.

140. Kömives, T., and J. E. Casida. 1983. "Acifluorfen increases the leaf content of phytoalexins and stress metabolites in several crops." *J. Agric. Food Chem.*, 31, 751–55.

141. Farrington, J. A., M. Ebert, E. J. Land, and K. Fletcher. 1973. "Bipyridylium quaternary salts and related compounds. V. Pulse radiolysis studies of the reaction of paraquat radical with oxygen. Implications for the mode of action of bipyridylium herbicides." *Biochim. Biophys. Acta*, 314, 372–81.

142. Bowyer, J. R., P. Camilleri, and M. T. Clark. 1986. "Mode of action studies and herbicidal properties of some PS I electron acceptors." In Int. Union of Pure and Applied Chem., 6th Pesticide Chem. Congr. Abstr. 3B-06..

143. Camilleri, P., J. R. Bowyer, and P. H. McNeil. 1987. "The effects of photosystem I electron acceptors on leaf discs." *Z. Naturforsch.*, 42c, 829–33.

144. Chia, L. S., J. E. Thompson, and E. B. Dumbroff. 1981. "Simulation of the effects of leaf senescence on membranes by treatment with paraquat." *Plant Physiol.*, 67, 415–20.

145. Chia, L. S., D. G. McRae, and J. E. Thompson. 1982. "Light-dependence of paraquat-induced membrane deterioration in bean plants. Evidence for the involvement of superoxide." *Physiol. Plant.*, 56, 492–99.

146. Birchem, R., W. G. Henk, and C. L. Brown. 1979. "Ultrastructure of paraquat-treated pine-cells (*Pinus elliottii* Engelm.) in suspension culture." *Ann. Bot.,* 43, 683–91.
147. Harris, N., and A. D. Dodge. 1972. "The effect of paraquat on flax cotyledon leaves: physiological and biochemical changes." *Planta,* 104, 210–19.
148. Law, M. Y., S. A. Charles, and B. Halliwell. 1983. "Glutathione and ascorbic acid in spinach (*Spinacea oleracea*) chloroplasts. The effect of hydrogen peroxide and of paraquat." *Biochem. J.,* 210, 899–903.
149. Boehler-Kohler, B. A., G. Läpple, V. Hellmann, and P. Böger. 1982. "Paraquat induced production of hydrocarbon gases." *Pestic. Sci.,* 13, 323–29.
150. van Rensen, J. J. S. 1975. "Lipid peroxidation and chlorophyll destruction caused by diquat during photosynthesis in *Scenedesmus.*" *Physiol. Plant.,* 33, 42–46.
151. Vaughn, K. C., and S. O. Duke. 1983. "In situ localization of the sites of paraquat action." *Plant Cell Envir.,* 6, 13–20.
152. Harris, N., and A. D. Dodge. 1972. "The effect of paraquat on flax cotyledon leaves: changes in fine structure." *Planta,* 104, 201–09.
153. Summers, L. A. 1980. *The Bipyridinium Herbicides.* London: Academic, 449 pp.
154. Harvey, B. M. R., J. Muldoon, and D. B. Harper. 1978. "Mechanism of paraquat tolerance in perennial ryegrass. I. Uptake, metabolism, and translocation of paraquat." *Plant Cell Envir.,* 1, 203–09.
155. Harper, D. B., and B. M. R. Harvey. 1978. "Mechanism of paraquat tolerance in perennial ryegrass. II. Role of superoxide dismutase, catalase, and peroxidase." *Plant Cell Envir.,* 1, 211–15.
156. Faulkner, J. S., and B. M. R. Harvey. 1981. "Paraquat tolerant *Lolium perenne* L.: effects of paraquat on germinating seedlings." *Weed Res.,* 21, 29–36.
157. Faulkner, J. S., C. B. Lambe, and B. M. R. Harvey. 1980. "Towards an understanding of paraquat tolerance in *Lolium perenne.*" In Proc. Brit. Crop Prot. Conf.—Weeds, Vol. 2. Nottingham, U.K.: Boots Company, pp. 445–52.
158. Vaughn, K. C., and E. P. Fuerst. 1985. "Structural and physiological studies of paraquat resistant *Conyza.*" *Pestic. Biochem. Physiol.,* 24, 86–94.
159. Fuerst, E. P., H. Y. Nakatani, A. D. Dodge, D. Penner, and C. J. Arntzen. 1985. "Paraquat resistance in *Conyza.*" *Plant Physiol.,* 77, 984–89.
160. Shaaltiel, Y., and J. Gressel. 1986. "Multienzyme radical detoxifying system correlated with paraquat resistance in *Conyza bonariensis.*" *Pestic. Biochem. Physiol.,* 26, 22–28.
161. Shaaltiel, Y., A. Glazer, P. F. Bocion, and J. Gressel. 1988. "Cross tolerance to herbicidal and environmental oxidants of plant biotypes tolerant to paraquat, sulfur dioxide, and ozone." *Pestic. Biochem. Physiol.,* 31, 13–23.
162. Tanaka, Y., H. Chisaka, and H. Saka. 1986. "Movement of paraquat in resistant and susceptible biotypes of *E. philadelphicus* and *E. canadensis.*" *Physiol. Plant.,* 66, 605–08.
163. Vaughn, K. C., M. A. Vaughan, and P. Camilleri. 1989. "Lack of cross-resistance of paraquat-resistant hairy fleabane (*Conyza bonariensis*) to other toxic oxygen generators indicates enzymatic protection is not the resistance mechanism." *Weed Sci.,* 37, 5–11.

164. Pölös, E., J. Mikulas, Z. Szigeti, B. Matkovics, Do Quy Hai, A. Parducz, and E. Lehoczki. 1988. "Paraquat and atrazine co-resistance in *Conyza canadensis* (L.) Cronq." *Pestic. Biochem. Physiol.,* 30, 142–54.

165. Powles, S. B. 1986. "Appearance of a biotype of the weed, *Hordeum glaucum* Steud., resistant to the herbicide paraquat." *Weed Res.,* 26, 167–72.

166. Powles, S. B., and G. Cornic. 1987. "Mechanism of paraquat resistance in *Hordeum glaucum*. I. Studies with isolated organelles and enzymes." *Aust. J. Plant Physiol.,* 14, 81–89.

167. Hughes, K. W., D. Negrotto, M. E. Daub, and R. L. Meeusen. 1984. "Free-radical stress response in paraquat-sensitive and resistant tobacco plants." *Envir. Exp. Bot.,* 24, 151–57.

168. Tanaka, K., I. Furusawa, N. Kondo, and K. Tanaka. 1988. "SO_2 tolerance of tobacco plants regenerated from paraquat-tolerant callus." *Plant Cell Physiol.,* 29, 743–46.

169. Carroll, E. W., O. J. Schwarz, and L. G. Hickok. 1988. "Biochemical studies of paraquat-tolerant mutants of the fern *Ceratopteris richardii*." *Plant Physiol.,* 87, 651–54.

170. Lewinsohn, E., and J. Gressel. 1984. "Benzyl-viologen mediated counteraction of diquat and paraquat phytotoxicities." *Plant Physiol.,* 76, 125–30.

171. Rabinowitch, H. D., and I. Fridovich. 1985. "Growth of *Chlorella sorokiniana* in the presence of sulfite elevates cell content of superoxide dismutase and imparts resistance towards paraquat." *Planta,* 164, 524–28.

172. Hassan, H. M., and I. Fridovich. 1978. "Superoxide radical and the oxygen enhancement of the toxicity of paraquat in *Escherichia coli*." *J. Biol. Chem.,* 253, 8143–48.

Chapter 9

Oxygen Toxicity and Herbicidal Action

9.1 PATHWAYS OF EXCITATION ENERGY DISSIPATION IN HERBICIDALLY DISTURBED PHOTOSYNTHESIS

In the presence of light, the photosynthetic chlorophyll pigments Chl, P680, and P700 (for details see Section 7.3) are excited by the absorption of photons ($h \cdot \nu$, see Figure 9.1) and thereby form higher energy intermediates (Chl*, P680*, P700*). Chlorophyll occurs normally in its singlet state and, after absorption of a photon, forms the very short-lived (10^{-8} to 10^{-6} s) excited singlet state, which may be converted to the more stable triplet state (10^{-4} to 10^{-3} s) [1]:

$$^1\text{Chl} \longrightarrow {}^1\text{Chl}^* \longrightarrow {}^3\text{Chl}^*$$

In the normal active photosynthetic electron transport system, singlet chlorophyll transfers one electron to plastoquinone (PQ), giving rise (stepwise) to the semiquinone and the plastohydroquinone. The electron is replaced immediately by one from the water-splitting system. When the reduction of plastoquinone is inhibited by a herbicide, the excitation energy cannot be disposed of in the usual way. The emission of heat and fluorescence ($h \cdot \nu_{\text{fluo}}$) rises to a maximum (see Section 7.3), and chlorophyll accumulates in the more stable triplet state. The accessory pigment β-carotene (Car) can quench some of the excited triplet chlorophyll [1] and re-emit the absorbed energy in a nonradiative way within a few microseconds:

$$^{3}Chl^{*} + {}^{1}Car \longrightarrow {}^{1}Chl + {}^{3}Car^{*}$$
$$^{3}Car^{*} \longrightarrow {}^{1}Car + \text{heat}$$

The reaction of β-carotene with excited triplet chlorophyll is so efficient that the half life of $^{3}Chl^{*}$, which is 10^{-4} s in benzene, is reduced to 10^{-8} s in the thylakoid membrane, where chlorophyll is surrounded by the accessory pigments (Figure 9.1).

The pathways that serve to quench excess excitation energy in the photosynthetic pigment system are quite efficient, and are adequate under normal environmental conditions. In herbicidally inhibited leaves in normal daylight, however, the energy quenching by β-carotene is overloaded, and the excess triplet chlorophyll states can react with ground state triplet oxygen to form the highly reactive singlet oxygen [2, 3]:

$$^{3}Chl^{*} + {}^{3}O_2 \longrightarrow {}^{1}Chl + {}^{1}O_2^{*}$$

Singlet oxygen is very reactive (see Section 9.3) and can induce pigment bleaching and lipid peroxidation. Some singlet oxygen can be produced under conditions of normal environmental fluctuations, and the chloroplast is therefore capable of removing singlet oxygen by reaction with α-tocopherol or β-carotene [1];

$$^{1}Car + {}^{1}O_2^{*} \longrightarrow {}^{3}Car^{*} + {}^{3}O_2$$

The excited triplet carotene produced in this reaction decays within a few microseconds, as described above. However, this pathway is not sufficiently

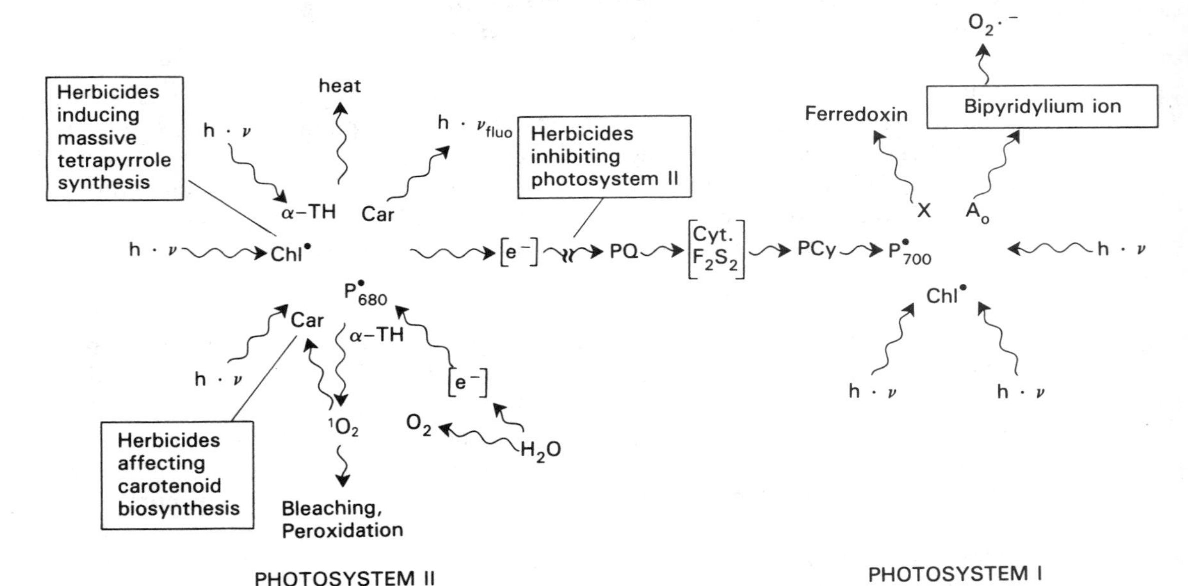

Figure 9.1 Light excitation of the photosynthetic pigment systems and excitation energy usage in the active and inhibited chloroplast. Excitation energy and electron movement, protective systems, and herbicide interference mechanisms are incorporated into this figure.

effective to remove the large amount of singlet oxygen produced during herbicidal inhibition of photosynthetic electron transport. Herbicidal damage can therefore develop as a consequence of singlet oxygen action (Section 7.6 and 9.3).

Herbicides may also interfere with normal photosynthesis by mechanisms other than electron transport inhibition (Figure 9.1). One important mechanism is the inhibition of carotenoid biosynthesis (Sections 8.1, 8.2, 8.3). Photosynthetic pigment systems devoid of carotenoids are not stable in air in the presence of light, as can easily be concluded from the foregoing and following discussions. Mutant plants free of carotenoids are bleached, except in very weak light, and become essentially white. Moreover, the absence of carotenoids leads to inhibition of chloroplast development through an intricate control system that correlates nuclear and chloroplast genome expression (see Section 8.3) [4]. Carotenoid-free mutants of photosynthetic bacteria and algae grow normally under an O_2-free atmosphere in the light, but bleach and die when exposed to an oxygenous atmosphere [1].

A third herbicidal mechanism of toxic oxygen production is provided by the electron drain induced by bipyridylium herbicides (Section 8.5). These redox herbicides react with molecular oxygen to produce the superoxide anion $O_2^{\cdot -}$. This type of reaction is also called a Mehler reaction. It forms the basis for "pseudocyclic electron transport" because $O_2^{\cdot -}$ is transformed back to O_2 by a series of reactions described in Section 9.4. Since this electron transport includes both photosystems and contains energy conservation sites, ATP is produced with no other net metabolic change. Pseudocyclic electron transport in vivo is catalyzed by ferredoxin (Figure 7.2) and may play a role in dissipating energy (and thus, preventing singlet oxygen formation) under adverse environmental conditions [5]. The superoxide anion ($O_2^{\cdot -}$) can also start a chain of oxidations and peroxidations leading to herbicidal action (Sections 9.2, 9.3).

A fourth mechanism involving oxygen activation is the production of excessive amounts of tetrapyrrole compounds in the presence of nitrodiphenylether (and similarly acting) herbicides. These tetrapyrroles act as photosensitizers in inducing photooxidative tissue damage (Section 8.4).

9.2 THE CHEMISTRY AND BIOCHEMISTRY OF OXYGEN ACTIVATION AND OXYGEN REDUCTION

Oxygen is an essential molecule for all aerobic organisms. It is very reactive, and is responsible for the conservation (in the form of ATP) of much of the energy obtained through eventual reaction with hydrogen donors in respiration. At the same time, its reactivity, and in particular the reactivity of the intermediate reduction products superoxide anion ($O_2^{\cdot -}$) and hydroxyl radical ($\cdot OH$), forms the basis for a number of damaging oxidation and peroxidation reactions [3, 6–8].

Ground-state oxygen is an unusual molecule because it contains two unpaired electrons with parallel spin. This triplet state oxygen ($^3O_2$) can be converted into the singlet state ($^1O_2$) by the absorption of excitation energy (e.g., from excited chlorophyll, see Section 9.1) and spin reversal of one electron. This can be shown diagrammatically as follows [3]:

$$^3O_2 \text{ or } \uparrow O_2 \uparrow \longrightarrow {}^1O_2 \text{ or } \uparrow O_2 \downarrow$$

Two singlet oxygen states may be formed, but one of these is energetically favored ($O_2{}^1\Delta g$ with 92 kJ above ground state). Singlet oxygen is much more reactive than ground-state triplet oxygen (Section 9.3) because the spin restriction is now alleviated. Singlet oxygen can break normal chemical bonds with antiparallel spins by forming two new bonds with antiparallel spins.

Oxygen is also known to be a lipophilic molecule and to accumulate preferentially in the lipid membrane interior. The oxygen concentration in the lipid membrane phase exceeds the concentration in the surrounding aqueous phase by a factor of 7–8. This lipophilic partitioning, supported by the liberation of oxygen in the thylakoid interior during photosynthesis, leads to a fairly high oxygen concentration in the thylakoid membrane.

As mentioned in Section 9.1, oxygen can accept electrons from the photosynthetic electron transport system via ferredoxin (naturally) or via bipyridylium herbicides (artificially) and form the superoxide anion $O_2^{\cdot -}$ (Figure 9.2). The superoxide anion can behave as an oxidant or, by interacting with a second superoxide anion, as a reductant. In effect, one $O_2^{\cdot -}$ molecule reduces a second $O_2^{\cdot -}$ molecule by donating an electron, yielding one O_2 and one molecule of hydrogen peroxide (H_2O_2). This spontaneous reaction is greatly increased by the

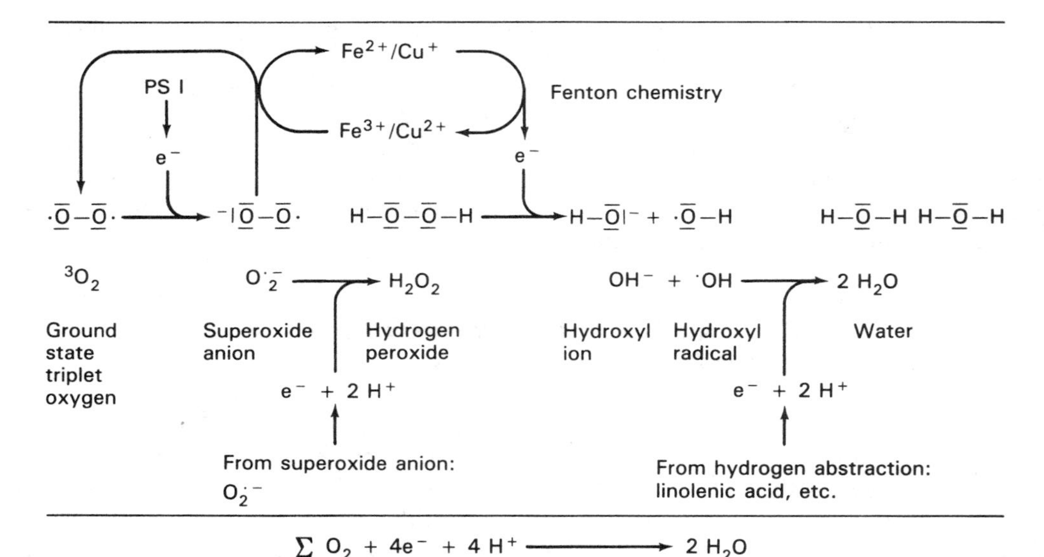

Figure 9.2 Four-step reduction of oxygen to water by successive electron donations. The suggested individual reduction mechanisms and the participating molecules, ions, and radicals are presented in a sequence from left to right.

activity of superoxide dismutase (Section 9.4), which holds the level of $O_2^{\cdot -}$ in the chloroplast below 0.01 μM. Hydrogen peroxide and the superoxide anion are themselves not very harmful to cells, but in the presence of traces of iron or copper ions they can give rise to the very reactive and toxic hydroxyl radical ($\cdot OH$). Hydrogen peroxide inhibits the Calvin cycle at concentrations in excess of 10 μM through the inactivation of seduheptulose-bis-phosphatase.

The hydroxyl radical is produced by the reduction of hydrogen peroxide with reduced iron or copper salts, called the Fenton reaction (Figure 9.2). Because of the potentially very toxic products of this reaction, it is doubtful that Fe^{2+} or Cu^+ are available in the normal healthy chloroplast. These metals are sequestered by special binding proteins and enzymes, and also occur largely in the oxidized state. However, in the event of excess production of superoxide anions (e.g., in the presence of bipyridylium herbicides) and after some original and/or initiating damage has occurred, Fe^{2+} and Cu^+ become available and catalyze the reduction of H_2O_2 to OH^- and $\cdot OH$ [9]. The hydroxyl radical reacts at a diffusion-controlled rate with unsaturated membrane lipids and extracts a hydrogen atom, thereby forming water. The newly produced unsaturated lipid radical starts a peroxidizing chain reaction (Section 9.3).

To sum up over the entire four-step reduction sequence, one oxygen molecule is sequentially reduced by four electrons and forms two water molecules. Of these four electrons, three come from photosystem I (one directly, one via $O_2^{\cdot -}$, and one via $O_2^{\cdot -}$ and Fe^{2+}/Cu^+) and only one comes from an unsaturated fatty acid. It is this fourth electron that initiates lipid peroxidation after the application of bipyridylium herbicides, and the hydroxyl radical appears to fulfill the role of the highly reactive intermediate.

9.3 THE PEROXIDATION OF LIPIDS

Thylakoid membranes contain about 90 percent unsaturated fatty acids in their membrane lipids, mainly linolenic acid (C18:3) and some linoleic acid (C18:2). These function to maintain the fluid character of the otherwise very heterogeneous thylakoid membranes. It has already been mentioned that oxygen concentrates into the lipid membrane phase by a factor of 7–8. Any singlet oxygen produced by herbicide action in the thylakoid membrane cannot, therefore, be expected to leave the lipid membrane phase rapidly. Moreover, the lifetime of $^1O_2$, which is between 25 and 100 μs in nonpolar solvents, appears to be long enough to allow diffusion in the membrane and also between aqueous and nonpolar phases [2]. Singlet oxygen will therefore diffuse in the membrane and, because of its electrophilic character, react with electron-rich functionalities. Such functionalities are present in linoleic acid and linolenic acid, and the reaction with singlet oxygen leads to the formation of peroxy-acids (Figure 9.3) [3].

The second oxygen species that initiates peroxidations in unsaturated lipids is the hydroxyl ion (Figure 9.3). The peroxy radical that is originally produced

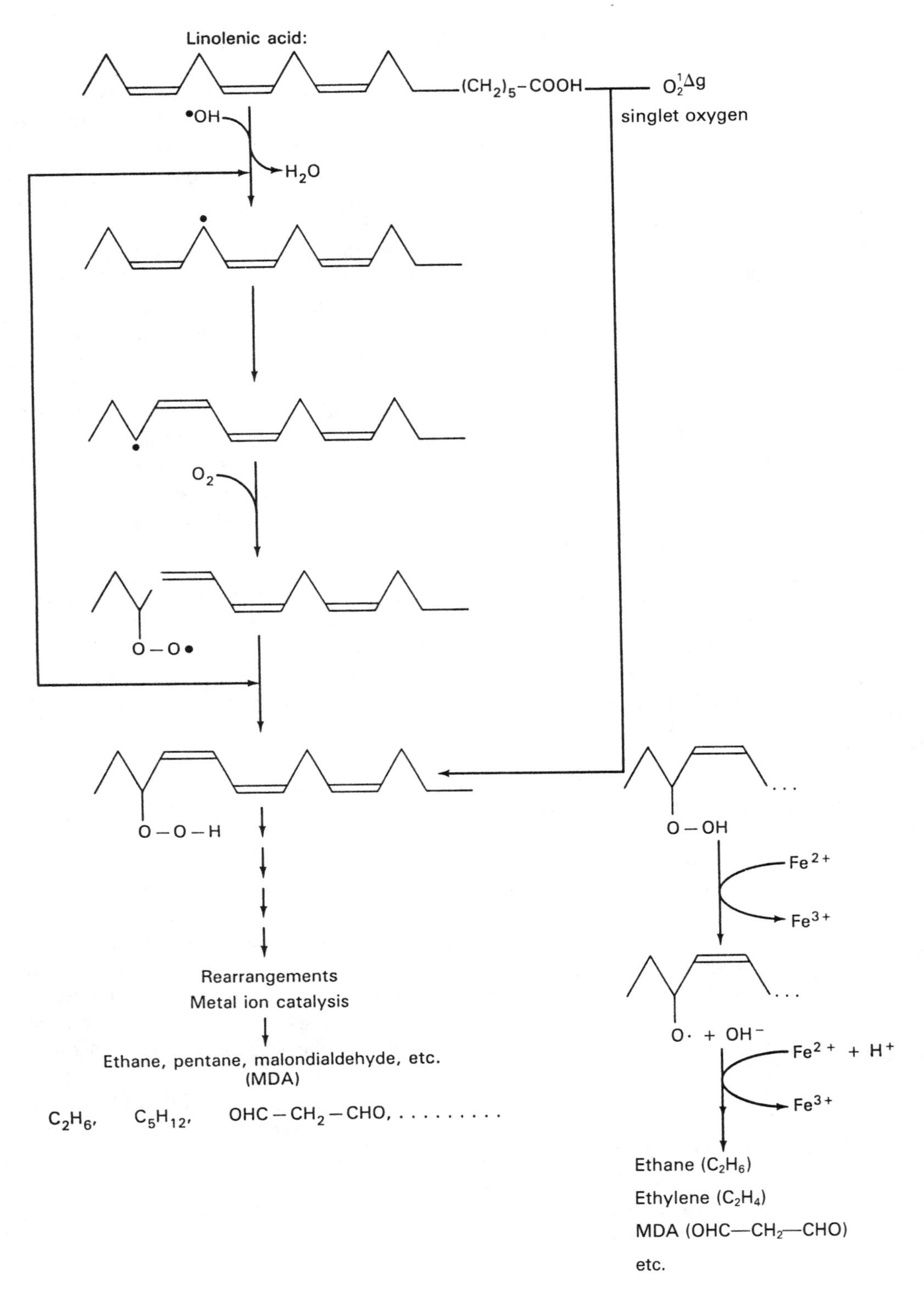

Figure 9.3 Peroxidation of polyunsaturated fatty acids by singlet oxygen and by the hydroxyl radical (·OH).

extracts a second hydrogen from another unsaturated fatty acid, thereby perpetuating the lipid peroxidation chain reaction. The fatty acid peroxide can subsequently be reduced to the fatty acid oxide radical, again by reduced transition metal ions (Fe^{2+} or Cu^{+}), and further through cleavage of the carbon chain, to an aldehyde and the remaining short alkane chain. In this way linolenic acid yields ethane, and linoleic acid yields pentane, both of which are found during the process of lipid peroxidation. In algae some quite different hydrocarbons are produced, reflecting the unusual unsaturated lipids that occur in these organisms [9]. Generally, the length of the liberated alkane/hydrocarbon chain equals the number of C-atoms beyond the position of the fatty acid double bond most distant from the carboxyl group (the ω-position). The length of the liberated alkanes is therefore ω-1. Malondialdehyde is also formed in large amounts during the peroxidation of unsaturated lipids, and can be measured easily (Sections 7.6, 8.4). Other molecules that can be found in natural or artificial stress situations leading to peroxidations include ethanol, acetaldehyde, and ethylene [10]. Ethylene is formed almost exclusively as a secondarily induced stress response; the synthesis of this "stress-induced" ethylene can be blocked by the inhibitor of enzymatic ethylene synthesis AVG (aminoethoxyvinylglycine), indicating that it is not formed directly from fatty acid catabolism.

The chemistry of lipid peroxidation described above and in Figure 9.3 is certainly a gross oversimplification of the many complex reactions, peroxidations, and interactions that take place in the thylakoid membrane. Singlet oxygen can also react with other molecules carrying electron-rich functionalities, such as the amino acids histidine, methionine, and tryptophan [3]. Peroxidation is not restricted to lipids, therefore, but extends to proteins, nucleic acids (most notably guanine), and pigments. Other classes of compounds may quench as well as react with singlet oxygen, examples being the phenol α-tocopherol (vitamin E) and certain amines. The highly reactive hydroxyl radical can interact in a similar manner (in this case by hydrogen extraction) with many other molecules besides unsaturated hydrocarbon chains. The radical molecules thereby produced react further and start oxidizing and peroxidizing chain reactions. Pigments are generally bleached by these oxidations/peroxidations, an effect that is very easily detected after herbicide application (Section 9.5).

Excess singlet oxygen formation is not only induced by herbicides that inhibit photosystem II electron transport. Photodynamic dyes such as rose bengal [11], eosin [12], and hypericin [13] have been demonstrated to lead to singlet oxygen generation, pigment bleaching, and damage in plant tissues. In addition, some secondary plant metabolites have been demonstrated to produce singlet oxygen after light absorption [14]. It has been speculated that this singlet oxygen formation plays a role in the plant's defense against pathogens [2]. On the other hand, many plant metabolites have antioxidant activity and may function as quenchers of naturally produced singlet oxygen. The artifical singlet oxygen quencher DABCO (Section 7.6) decreases the herbicidal activity of photosystem II inhibitors in this manner.

9.4 PROTECTIVE SYSTEMS

The important role of endogenous protective systems in the avoidance of damage due to the generation of activated oxygen species has already been mentioned several times. The protective systems that are present in the cell counteract most of the highly reactive or intermediary molecular species, namely singlet oxygen, superoxide anion, hydrogen peroxide, triplet chlorophyll, and lipid peroxide radicals. In the case of the hydroxyl radical the strategy seems to be mainly the avoidance of its generation (Section 9.2). However, the quenching of lipid peroxy radicals by α-tocopherol also stops chain reactions initiated by hydroxyl radicals.

The series of reactions that together eliminate superoxide anions and hydrogen peroxide is listed in Scheme 9.1 [8, 15, 16]. Superoxide anions ($O_2^{\cdot -}$) and hydrogen peroxide (H_2O_2) are produced during normal environmental fluctuations of photosynthesis, for example after transfer from low light to high light, during greening of etiolated tissue, or in water stress situations with limiting CO_2 supply. The spontaneous rate of superoxide dismutation (second-order rate constant of 10^5 $M^{-1}\,s^{-1}$ at pH 7) [6] is considerably enhanced by superoxide dismutase (a) which is present in the chloroplast stroma and thylakoids, and keeps the concentration of $O_2^{\cdot -}$ below 0.01 μM.

Superoxide can also be reduced directly by ascorbate (b), which is present in the chloroplast stroma at a concentration of 5 to 15 mM. The mechanism of this reaction is more complicated than indicated by equation (b); it is a stepwise reduction involving mono-dehydroascorbate as an intermediary product. The product of reactions (a) and (b), hydrogen peroxide, can be destroyed by catalase (f). However, catalase is not a chloroplast enzyme, but is present with high activity in peroxisomes. Since peroxisomes are often found to be closely associated with chloroplasts, and since H_2O_2 can diffuse freely and penetrate membranes, peroxisomal catalase might under certain conditions contribute significantly to the destruction of H_2O_2. Catalase is known to be strongly inhibited by elevated concentrations (0.1-1 mM) of the herbicide aminotriazole (Section 8.1.2).

$$\text{(a)}\ O_2^{\cdot -} + O_2^{\cdot -} + 2\,H^+ \longrightarrow O_2 + H_2O_2$$

$$\text{(b)}\ 2O_2^{\cdot -} + \text{asc.} + 2\,H^+ \longrightarrow 2\,H_2O_2 + \text{DHasc.}$$

$$\text{(c)}\ 2\,\text{DHasc.} + 4\,H_2O \longleftarrow 2\,H_2O_2 + 2\,\text{asc.}$$

$$\text{(d)}\ \text{DHasc.} + 2\,\text{GSH} \longrightarrow \text{asc.} + \text{GSSG}$$

$$\text{(e)}\ NADP^+ + 2\,\text{GSH} \longleftarrow NADPH + H^+ + \text{GSSG}$$

$$\text{(f)}\ O_2 + 2\,H_2O \longleftarrow 2\,H_2O_2$$

Scheme 9.1 Enzymatic and spontaneous reactions involved in the detoxification of the activated oxygen species superoxide anion ($O_2^{\cdot -}$) and hydrogen peroxide (H_2O_2). (a) superoxide dismutase, (b) spontaneous, (c) ascorbate peroxidase, (d) dehydroascorbate reductase, (e) glutathione reductase, (f) catalase.

The main reaction involved in dissipating H_2O_2 in the chloroplast is catalyzed by ascorbate peroxidase which produces dehydroascorbate and water (c). Ascorbate is regenerated by reactions (d) and (e) through the sequential activity of dehydroascorbate reductase and glutathione reductase (Scheme 9.1). Summing up, the reducing equivalents for the dissipation of the potentially harmful oxygen species $O_2^{\cdot -}$ and H_2O_2 eventually come from photosynthetic electron transport via NADPH. The inhibition of this electron flow by herbicides therefore not only increases the production of the activated oxygen species but also indirectly inhibits their removal by reductive enzymatic reactions. As a consequence, ascorbate and glutathione contents decrease rapidly in herbicide-treated tissues (Chapters 7 and 8).

A normal plant can adapt to a certain extent to hyperoxygenic conditions. Glutathione reductase is increased in plants grown in a 75 percent O_2 atmosphere [17]. Normally, approximately 60 percent of the cellular glutathione is present in the chloroplasts, and about 70 percent of this glutathione is present in the reduced state (GSH), as compared to 97–98 percent in the reduced state in the cytoplasm [18]. As plants grow older, their capacities for reductive defense reactions, and their concentrations of ascorbate and glutathione, decrease. Peroxidized lipids therefore accumulate in older, and especially in senescing, plant tissues. Membranes lose their fluidity and semipermeable behavior, and the tissue progressively deteriorates [19, 20]. Peroxidative herbicide damage and normal aging are therefore based on similar mechanisms, but of course they occur on very different time scales.

The quenching of excess activated chlorophyll states by carotenoids has already been mentioned (Section 9.1). In the thylakoid membrane, chlorophyll is closely associated with a large amount of accessory pigments (carotenes, xanthophylls) and with other protective molecules such as α-tocopherol (Figure 9.1). β-Carotene, α-tocopherol, and ascorbate also serve to quench singlet oxygen, which is constantly produced in the pigment system by energy spillover. In addition, α-tocopherol reacts with lipid peroxide radicals (instead of another unsaturated lipid) by donating an electron, thereby terminating the peroxidative chain reaction (Section 9.3). The α-tocopherol radical produced in this reaction can be reduced back to α-tocopherol by ascorbate. α-Tocopherol also reacts quickly with the hydroxyl radical.

In conclusion, plants contain an impressive arsenal of defense reactions in order to cope with the special situation of oxygen production by pigment systems in the light. In spite of chlorophyll being an extremely light- and oxygen-sensitive pigment, its close association with pigment quenchers and scavengers of activated oxygen species allows for a high level of stability and an optimum photosynthetic rate even in excessive radiation or under strong environmental fluctuations. Only in the presence of herbicides exerting one of the different mechanisms of interaction with the photosynthetic system are these protective systems overloaded, with the eventual consequences of herbicidal action. These different mechanisms of interaction include:

- generation of singlet oxygen by inhibition of photosynthetic electron transport (Section 7.6);
- generation of superoxide anions by diverting electrons from photosystem I to oxygen (Section 8.5);
- inhibition of carotenoid biosynthesis through interference with desaturase enzymes or earlier steps in the carotenoid biosynthetic pathway, resulting in unstable conditions in the thylakoid membrane (Section 8.1);
- induction of massive tetrapyrrole biosynthesis, leading to excessive photosensitized oxidations (Section 8.4); and
- other types of pigment bleaching (Sections 8.2, 8.3, 9.5).

9.5 DIFFERENT TYPES OF CHLOROSIS INDUCTION

Chlorosis is principally a visual symptom and, since it can easily be observed and measured, it is often reported after herbicide application. However, the visual symptoms of chlorosis tell us little about the underlying mechanism(s) by which it is produced. Additional information can be obtained when the conditions necessary for the appearance of chlorosis are known. For example, one might ask the following questions: Does chlorosis occur in low light, or only in high light? Is only newly developed tissue (formed after herbicide application) chlorotic or is established green tissue also bleached by the herbicidal action? Is the tissue evenly bleached or is the chlorosis restricted to certain types of tissue? Is the tissue entirely white or is the color yellowish? Answers to these questions can help to classify the type of metabolic interaction. Additional measurements, for example of ethane evolution, malondialdehyde production (Section 9.3), or phytoene accumulation (Section 8.1.1), can contribute greatly to identifying the primary site of action of a herbicide [21, 22].

When applied in the dark or in high light, inhibitors of carotenoid biosynthesis produce new tissue that is essentially white; in low light, however, they give rise to tissue that is light green (Sections 8.1, 8.2, 8.3). This treated tissue that is green in low light turns white in high light, with the simultaneous production of photooxidation products such as ethane and malondialdehyde. However, untreated green tissue is not visibly affected when treated with inhibitors of carotenoid biosynthesis.

Inhibitors of photosynthetic electron transport can bleach leaves to plain white, depending on the irradiance level, herbicide distribution in the tissue, and photosynthetic activity. The chlorotic effects induced by lipophilic compounds (e.g., simazine) are, because of the restricted movement of such herbicides, often confined to the veinal areas. In contrast, high photosynthetic activities in the interveinal areas can lead to enhanced chlorotic effects in these areas if the herbicide is more evenly distributed. Thus, more hydrophilic compounds (e.g., metribuzin), which are more evenly distributed in the leaf, cause more uniform chlorosis, from yellow or light green to white.

The chlorotic effects described for photosystem II inhibitors occur slowly after root uptake. However, foliar application of these compounds causes more rapid action that leaves little time for pigment bleaching. Under these conditions membrane lipid peroxidation occurs rapidly, resulting in necrosis and desiccation. The same is true for other herbicides interfering with the photosynthetic pigment system: slow action preferentially produces chlorotic effects, whereas rapid action leads to necrosis and desiccation.

Chlorotic tissue can also be formed after the application of herbicides with completely different mechanisms of action. Slow action leading to some type of tissue starving seems to be a prerequisite for these chlorotic effects. Diclofop-methyl causes chlorosis in wild oat [23] and glyphosate causes chlorosis in maize seedlings [24]. The inhibitors of acetyl-CoA carboxylase, alloxydim and sethoxydim (Section 11.2), indirectly inhibit chloroplast pigment biosynthesis and cause leaf chlorosis [25–27]. Of course, decaying plant tissue, by whatever reason, will be bleached to a certain extent through the liberation of hydrolytic enzymes, deteriorating repair systems, and increased pigment and membrane peroxidations. Only a carefully controlled kinetic experiment, studying the sequence of inhibitions and toxic effects after herbicide application, can help to solve the question of the primarily inhibited or affected metabolic pathway(s).

REFERENCES

1. Siefermann-Harms, D. 1987. "The light-harvesting and protective functions of carotenoids in photosynthetic membranes." *Physiol. Plant.*, 69, 561–68.
2. Knox, J. P., and A. D. Dodge, 1985. "Singlet oxygen in plants." *Phytochemistry*, 24, 889–96.
3. Halliwell, B. 1984. "Toxic oxygen species and herbicide action." In *Biochemical and Physiological Mechanisms of Herbicide Action*, S. O. Duke, ed. Southern Section, American Society of Plant Physiology. pp. 31–44.
4. Mayfield, S. P., and W. C. Taylor. 1987. "Chloroplast photooxidation inhibits the expression of a set of nuclear genes." *Mol. Gen. Genet.*, 208, 309–14.
5. Robinson, J. M. 1988. "Does O_2 photoreduction occur within chloroplasts in vivo?" In *Oxygen Interactions in the Chloroplast*. Proc. 1987 Annu. Symp. Southern Section, American Society of Plant Physiology. pp. 657–98. (See also *Physiol. Plant*, 72, 666–80.)
6. Fee, J. A. 1982. "Is superoxide important in oxygen poisoning?" *TIBS*, 7, 84–86.
7. Foyer, C. H., and D. O. Hall. 1980. "Oxygen metabolism in the active chloroplast." *TIBS*, 5, 188–91.
8. Elstner, E. F. 1982. "Oxygen activation and oxygen toxicity." *Annu. Rev. Plant Physiol.*, 33, 73–96.
9. Aust, S. D., and B. A. Svingen. 1982. "The role of iron in enzymatic lipid peroxidation." In *Free Radicals in Biology*, Vol. V, W. A. Wagner, ed. New York: Academic, pp. 1–28.

10. Kimmerer, T. W., and T. T. Kozlowski. 1982. "Ethylene, ethane, acetaldehyde, and ethanol production by plants under stress." *Plant Physiol.*, 69, 840–47.

11. Knox, J. P., and A. D. Dodge. 1984. "Photodynamic damage to plant leaf tissue by rose bengal." *Plant Sci. Lett.*, 37, 3–7.

12. Knox, J. P., and A. D. Dodge. 1985. "The photodynamic action of eosin, a singlet-oxygen generator." *Planta*, 164, 22–29.

13. Knox, J. P., and A. D. Dodge. 1985. "Isolation and activity of the photodynamic pigment hypericin." *Plant Cell Envir.*, 8, 19–25.

14. Towers, G. H. N., and J. T. Arnason. 1988. "Photodynamic herbicides." *Weed Technol.* 2, 545–49.

15. Badger, M. R. 1985. "Photosynthetic oxygen exchange." *Annu. Rev. Plant Physiol.*, 36, 27–53.

16. Gillham, D. J., and A. D. Dodge. 1985. "Chloroplast protection in greening leaves." *Physiol. Plant.*, 65, 393–96.

17. Foster, J. G., and J. L. Hess. 1982. "Oxygen effects on maize leaf superoxide dismutase and glutathione reductase." *Phytochemistry*, 21, 1527–32.

18. Smith, I. K., A. C. Kendall, A. J. Keys, J. S. Turner, and P. J. Lea. 1985. "The regulation of the biosynthesis of glutathione in leaves of barley (*Hordeum vulgare*)." *Plant Sci.*, 41, 11–17.

19. Mayak, S., R. L. Legge, and J. E. Thompson. 1983. "Superoxide radical production by microsomal membranes from senescing carnation flowers: an effect on membrane fluidity." *Phytochemistry*, 22, 1375–80.

20. Pauls, K. P., and J. E. Thompson. 1984. "Evidence for the accumulation of peroxidized lipids in membranes of senescing cotyledons." *Plant Physiol.*, 75, 1152–57.

21. Sandmann, G., and P. Böger. 1982. "Mode of action of herbicidal bleaching." In *Biochemical Responses Induced by Herbicides*, D. E. Moreland, J. B. St. John, F. D. Hess, eds. ACS Symposium Series 181, Amer. Chem. Soc. Washington, D.C., pp. 111–30.

22. Sandmann, G., and P. Böger. 1987. "Herbicides affecting plant pigments." In Proc. Brit. Crop Prot. Conf.—Weeds, Vol. 1. Thornton Heath, U.K.: BCPC Publ., pp. 139–48.

23. Hoerauf, R. A., and R. H. Shimabukuro. 1979. "The response of resistant and susceptible plants to diclofop-methyl." *Weed Res.*, 19, 293–99.

24. Croft, S. M., C. J. Arntzen, L. Vanderhoef, and C. S. Zettinger. 1974. "Inhibition of chloroplast ribosome formation by *N,N*-bis(phosphonomethyl)glycine." *Biochim. Biophys. Acta*, 335, 211–17.

25. Asare-Boamah, N. K., and R. A. Fletcher. 1983. "Physiological and cytological effects of BAS 9052 OH on corn (*Zea mays*) seedlings." *Weed Sci.*, 31, 49–55.

26. Ikai, K., T. Uezono, H. Otha, K. Hamada, and F. Tanaka. 1982. "Effect of alloxydim on growth, chlorophyll content and anthocyanin accumulation of crabgrass (*Digitaria adscendens* Henr.)." *Weed Res.*, (Japan) 27, 121–25.

27. Lichtenthaler, H. K., K. Kobek, and K. Ishii. 1987. "Inhibition by sethoxydim of pigment accumulation and fatty acid biosynthesis in chloroplasts of *Avena* seedlings." *Z. Naturforsch.*, 42c, 1275–79.

Chapter 10

Microtubule Disruptors

10.1 INTRODUCTION

The plant cytoskeleton is composed of two primary structural components—microtubules and microfilaments. These proteinaceous structural elements form a dynamic, three-dimensional network that provides cellular form and functions in cell division, growth, and morphogenesis. Obviously, any compound that would interfere with the proper functioning of either of these structures in a plant cell could be herbicidal. Many herbicides and phytotoxins interfere with cell division, growth, and/or morphogenesis. However, in most cases the effects on these processes are indirect, mediated through effects at molecular sites of action outside the plant cytoskeleton. Usually, when a compound does not directly affect microtubules, it stops cell division during interphase and prevents entry into mitosis. Compounds that directly affect microtubules generally cause abnormal, arrested mitotic figures, resulting in an increased proportion of cells found in a mitotic configuration.

Considerable literature exists on the direct effects of pharmaceutical compounds, fungicides, and nematicides on microtubules of fungal and animal cells. Recently, proof of direct effects of herbicides on plant microtubules has begun to accumulate. Several recent reviews have summarized some of these findings [1, 2]. No herbicides are known to act by direct interaction with microfilaments. Only those compounds with proven or strongly suspected primary interactions with microtubules, microtubule-organizing centers, and microtubule-associated proteins will be discussed in this chapter.

10.2 MICROTUBULE COMPOSITION AND FUNCTION

Microtubules have been the subject of intensive study and exhaustive reviews (e.g., [3–5]) and books (e.g., [6]). The overwhelming majority of these studies have been with animal sources of material. Plant and animal microtubules share many features; however, there are also many dissimilarities. For instance, plant microtubules appear to be more cross-bridged than those of animal cells [7], and their susceptibility to various chemicals generally differs from that of animal tubulin [8, 9]. This dissimilarity fulfills a prerequisite for the design of herbicides that act at the tubulin level of plant cells without affecting mammalian tubulin. Nevertheless, much about microtubules in plants has been assumed or extrapolated from animal studies.

All microtubules are composed of polymeric assemblies of protein heterodimers of α- and β-tubulin (Figure 10.1). The two subunits are of similar molecular

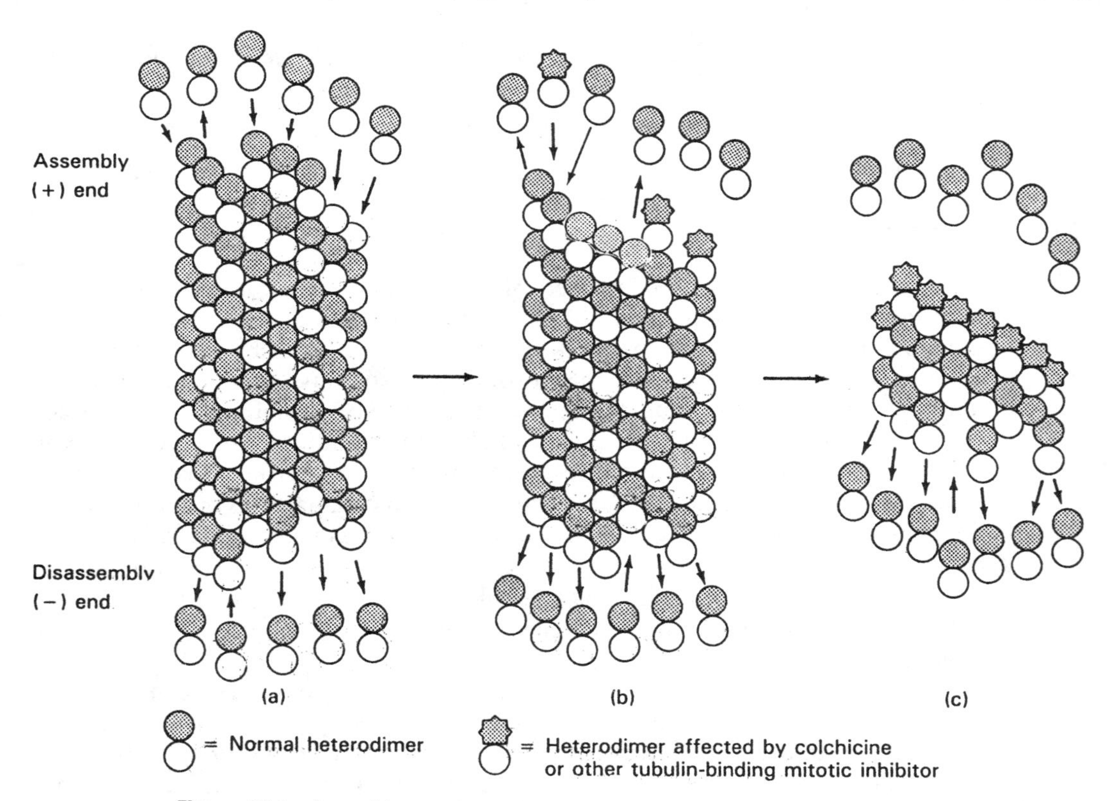

Figure 10.1 Assembly and disassembly of tubulin heterodimers into microtubules. Heterodimers from the free pool normally preferentially assemble into the assembly (+) end of the microtubule (A). When the heterodimer is altered by compounds such as colchicine or certain dinitroaniline herbicides, assembly at this end is prevented; however, disassembly is not prevented (B and C).

mass and the dimer is about 110 kD. Within the same species, there is about 50 percent homology between α- and β-tubulin, suggesting a common evolutionary origin. Antibodies to animal tubulins generally cross-react with plant tubulins, indicating homology in the antigenic site. Although antibodies to rose β-tubulin cross-react with brain tissue β-tubulin, there is no similar cross reaction between the α-tubulins [10]. More than one gene for each type of tubulin monomer exists within the genome, and there is not complete homology between them (e.g., [11]). Different pools of tubulin are apparently available for the synthesis of different types of microtubules within a cell. Furthermore, specific posttranslational modifications of tubulins occur (e.g., glycosylation, phosphorylation, methylation, acetylation). Isotypes of tubulins may be due to either genetic or posttranslational differences. Multiple tubulin isotypes have been found in a variety of higher plant species (e.g., [12–14]). It is not certain whether particular microtubule isotypes are homopolymers of particular tubulins all modified in the same way. If not, the variety of tubulins and posttranslational modifications could result in microtubule isotypes with an almost infinite variety of properties. Heteropolymeric microtubules in vivo are suggested by the finding that, in vitro, all of the tubulin isotypes of carrot will copolymerize [12], and even plant and brain tissue tubulins will copolymerize [15]. However, particular types of microtubules are almost certainly composed of specific tubulin isotypes. For example, the flagellar tubulin of *Chlamydomonas reinhardtii* does not contain the unmodified tubulin gene product but is composed of a specific tubulin isotype produced by posttranslational processing [16].

Microtubule heterogeneity in higher plants is assumed from a large amount of indirect evidence. For example, differential stability to various fixatives for microscopy, to colchicine, and to cold suggests different microtubule types [3]. The different isotypes of tubulin may provide the basis for microtubule heterogeneity. Heterogeneity could also be due to differences in microtubule-associated proteins (MAPs).

Microtubules are highly flexible cylinders, with walls of helical chains of tubulin dimers. This structure has been confirmed by both in vivo and in vitro electron micrographs (Figure 10.2). In the case of cortical microtubules, the cylinder is composed of 13 helical tubulin filaments. Microtubule dimensions are 25 nm in diameter with variable length up to more than several μm.

Tubulin in a living cell exists in a dynamic cycle between free tubulin and microtubules. The proportion of tubulin partitioned into microtubules can vary considerably, reaching as high as 90 percent. Microtubules spontaneously self-assemble when tubulin is in the proper environment. Starting from free tubulin, microtubule assembly is a two-step process, starting with condensation or nucleation and followed by polymerization. Assembly apparently occurs primarily at one end (+ or A) of the microtubule, while disassembly progresses predominantly at the other end (− or D; Figure 10.1). Thus, microtubules are polar structures of constant length only when assembly and disassembly are in equilibrium. Tubulin that is added to the assembly end of the microtubule will eventually disassociate

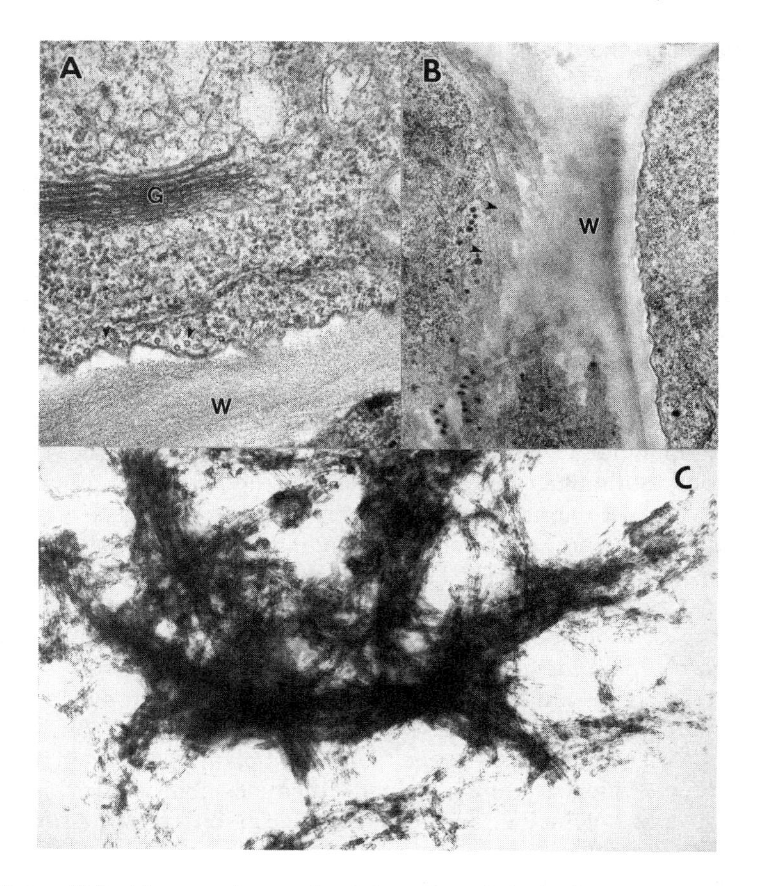

Figure 10.2 Electron micrographs of plant microtubules in vivo (arrows) in cross (A) and longitudinal (B) sections. G = Golgi apparatus, W = cell wall. (C) In vitro polymerized microtubules from tubulin of goosegrass (*Eleusine indica*). This preparation was negatively stained with aqueous uranyl acetate on a formvar-coated grid. (Courtesy of K. C. Vaughn).

from the other end, a process termed "treadmilling." Polarity is an essential feature for microtubule function. Any factor that favors either assembly or disassembly will cause the microtubule to lengthen or shorten, respectively. Many factors, including low tubulin concentration, high Ca^{2+}, and improper pH or temperature, can increase the rate of disassembly. Rapid changes in certain conditions can cause catastrophic depolymerization of the entire microtubule in a nonpolar fashion.

Tubulin heterodimers apparently assemble as heterodimer-guanosine triphosphate (GTP) and disassemble as heterodimer-GDP (Figure 10.3). The form of guanyl nucleotide in the heterodimer that binds to the microtubule is not known. The pool of heterodimer-GTP is essentially the pool of tubulin capable of polymerizing. Thus, anything that would affect GTP supply may strongly influence formation of microtubules.

Free tubulin heterodimers inhibit the synthesis of tubulin at a posttranscriptional level (Figure 10.3). Therefore, anything that causes depolymerization or inhibits polymerization of microtubules will inhibit tubulin synthesis [17]. Conversely, increased polymerization and/or decreased depolymerization will stimulate tubulin synthesis. In *Chlamydomonas reinhardtii,* removal of the flagellum by mild acid treatment results in greatly increased tubulin synthesis and polymerization as the flagellum is regenerated [18]. Regrowth begins after a lag of ca. 15 min., and all four tubulin genes (two α and two β) are activated [19]. This system has been useful in the study of herbicide effects on tubulin [20–22].

In certain regions of the cytoplasm termed microtubule-organizing centers (MTOC), the disassembly end is apparently "capped," preventing disassembly and assembly at this end. MTOCs are sites of microtubule initiation, sometimes

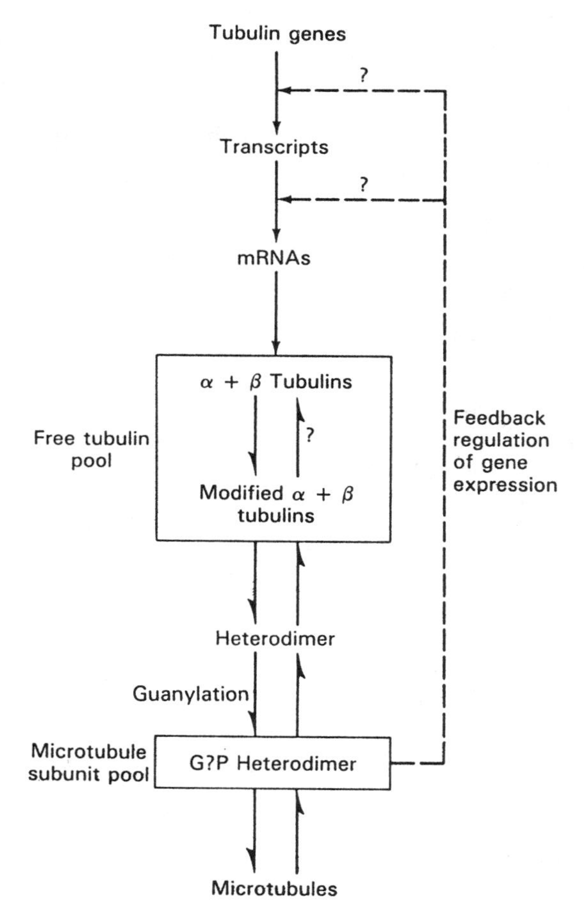

Figure 10.3 Model for tubulin pool movement into heterodimers and microtubules.

called nucleation. In animals and lower plants, the centriole or a centriole-like structure acts as a MTOC in organizing microtubules for cell division. The basal bodies of flagella which are composed of microtubule bundles also prevent disassembly at the "slow" end of the microtubule. MTOCs, basal bodies, and other structures offer an explanation for the finding that the polarity of microtubule arrays such as found in mitotic spindles or flagella is always uniform. Nuclear envelopes, organelle envelopes, small vesicles associated with cell wall deposition, and kinetichores have MTOC properties. The phragmoplast is a MTOC involved in cell plate formation.

Microtubules can be copolymerized with a variety of MAPs. Little is known of these proteins in plants; however, they have been shown to have profound influences on the polymerization and functioning of animal microtubules. For instance, some MAPs lower the critical concentration of tubulin heterodimers required for self-assembly to begin. There is apparently more evolutionary divergence between animal and plant MAPs than between their tubulins because there is little immunochemical cross reactivity between these proteins. Lack of such antibodies has hampered the study of MAPs in plants. Plant MAPs have only recently been directly confirmed by isolating nontubulin proteins from carrot cells which bind to and promote polymerization of tubulin [23]. These proteins attach to microtubules at regular intervals and cause bundling of microtubules in vitro, much like that observed in vivo. By immunofluorescence microscopy, a 76,000 M_r MAP was found to associate with cortical microtubules in vivo.

Microtubules are involved in a wide array of plant cell processes that are outlined in Scheme 10.1. In addition to these activities, microtubules provide the

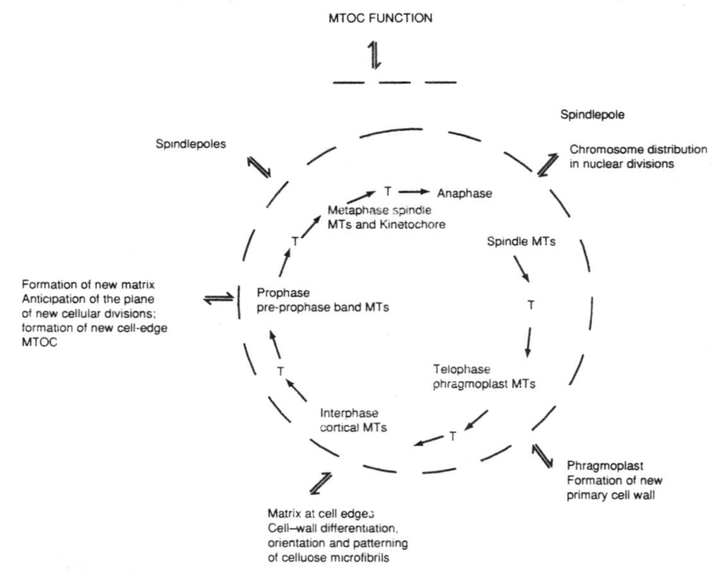

Scheme 10.1 Microtubular functions during the cell cycle of a higher plant. MT = microtubule, T = tubulin, MTOC = microtubule-organizing center.

cytoplasm with internal structure and, to some extent, shape—hence, cytoskeleton. Our understanding of the arrangement and functioning of microtubules in cells has been enhanced by immunocytochemical methods that allow observation of all tubulin structures in living cells [5] (Figure 10.4).

In higher plants, microtubule function in cell division (mitosis and cytokinesis) is perhaps most dramatic and obvious. The mitotic spindle that organizes and separates the chromosomes is composed of microtubules. During prophase, microtubule assembly at kinetichores is intense, forming a network between the poles of the two new cells that will form. During metaphase the microtubules form a spindle that orients the chromosomes in the central plane of the mitotic cell. During late prophase, the poles of the mitotic spindle begin to become apparent. At the poles, membranes are more concentrated than in most other cytoplasmic

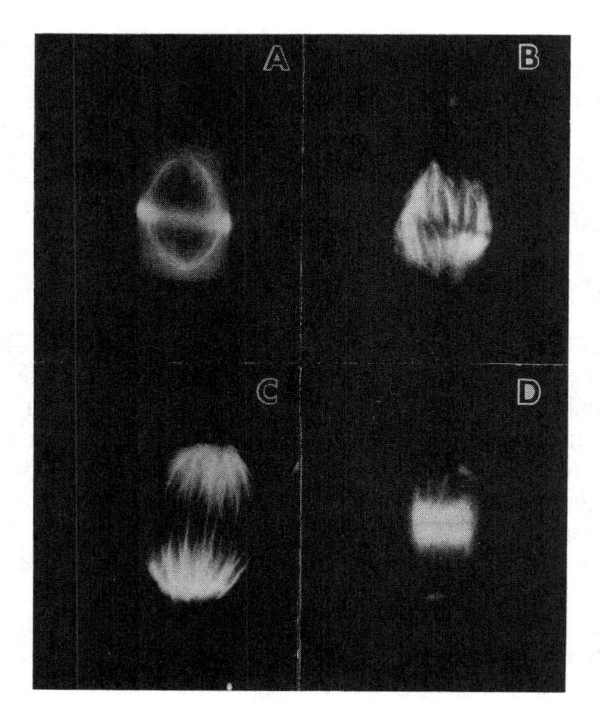

Figure 10.4 Immunocytochemical visualization of microtubules of onion root cells in the different stages of mitosis. A. Preprophase band microtubules predict the plane of future cytokinesis. Spindle microtubules emanate from the periphery of the preprophase nucleus. B. Chromosomes (in negative relief) are aligned at the metaphase plate with their kinetochore microtubules appearing as cone-shaped bundles with apices oriented toward the spindle poles. C. Chromosomes separate at anaphase and spindle microtubules extend between the poles at late anaphase. D. Phragmoplast microtubules form in the region between the two daughter nuclei. (Courtesy of K. C. Vaughn).

regions. The chromosomes separate and move along the microtubules to opposite ends of the cell during anaphase.

At telophase, microtubules play a role in cytokinesis and formation of the cell wall that separates the two new cells by the phragmoplast. Although cytokinesis is part of cell division, the phragmoplast-associated microtubules are separate and distinct from spindle microtubules. The microtubules of the phragmoplast are oriented with their + end toward the developing cell plate. The MTOCs associated with the mitotic spindle and phragmoplast microtubules have not been characterized.

Cortical microtubule arrays are found just inside the plasmalemma, along cell walls (Figure 10.2B). Immunofluorescence microscopy reveals a complex netlike array of cortical microtubules, defining a cytoskeleton. Although some investigators have questioned the role of microtubules in cell wall formation [3], several lines of evidence strongly suggest a linkage. Conditions that disrupt microtubules stop or disorganize cell wall growth (e.g., [24]). Furthermore, a biotype of goosegrass (*Eleucine indica*) with a mutant β-tubulin has occasional disorganized and irregular cell walls [2, 25]. Whether there is a direct linkage between the enzymes of cell wall components and these microtubule arrays is not known.

10.3 MICROTUBULE DISRUPTORS

10.3.1 Gross Effects and Methods

Numerous commercial herbicides slow or inhibit mitosis; however, few herbicides have been demonstrated to affect tubulin or microtubules directly. Simply slowing or blocking metabolism of a meristem will not cause an increase in the proportion of metaphase cells, nor will it cause the production of malformed cell walls. Evidence of cortical microtubule dysfunction, such as deranged cell walls, cellular swelling, and lack of secondary cell wall thickening, provides further evidence that a compound might interact with tubulin. Cessation of root growth, accompanied by swelling of root tips, has generally been the first clue to the interference of herbicides and other chemicals with microtubules. To avoid confusion with secondary effects, studies should be done with concentrations of a compound that reduce growth by 50 to 90 percent and evaluation should be made shortly after inhibition of growth can first be observed (usually less than 24 h after treatment) [1]. Direct interference of a compound with microtubules can be inferred from several observations. The simplest and most classical approach is to determine the effect of the compound on mitosis of meristematic tissues, with emphasis on chromosomal patterns during mitosis. The effect of several compounds that affect microtubules or mitotic figures is illustrated in Figure 10.5. The first antimitotic agent to draw the attention of cytologists was colchicine, a plant alkaloid that dramatically increases the number of mitotic figures in plant meristems by slowing and arresting mitosis, leaving many cells in an abnormal prometaphase at any particular time. In addition

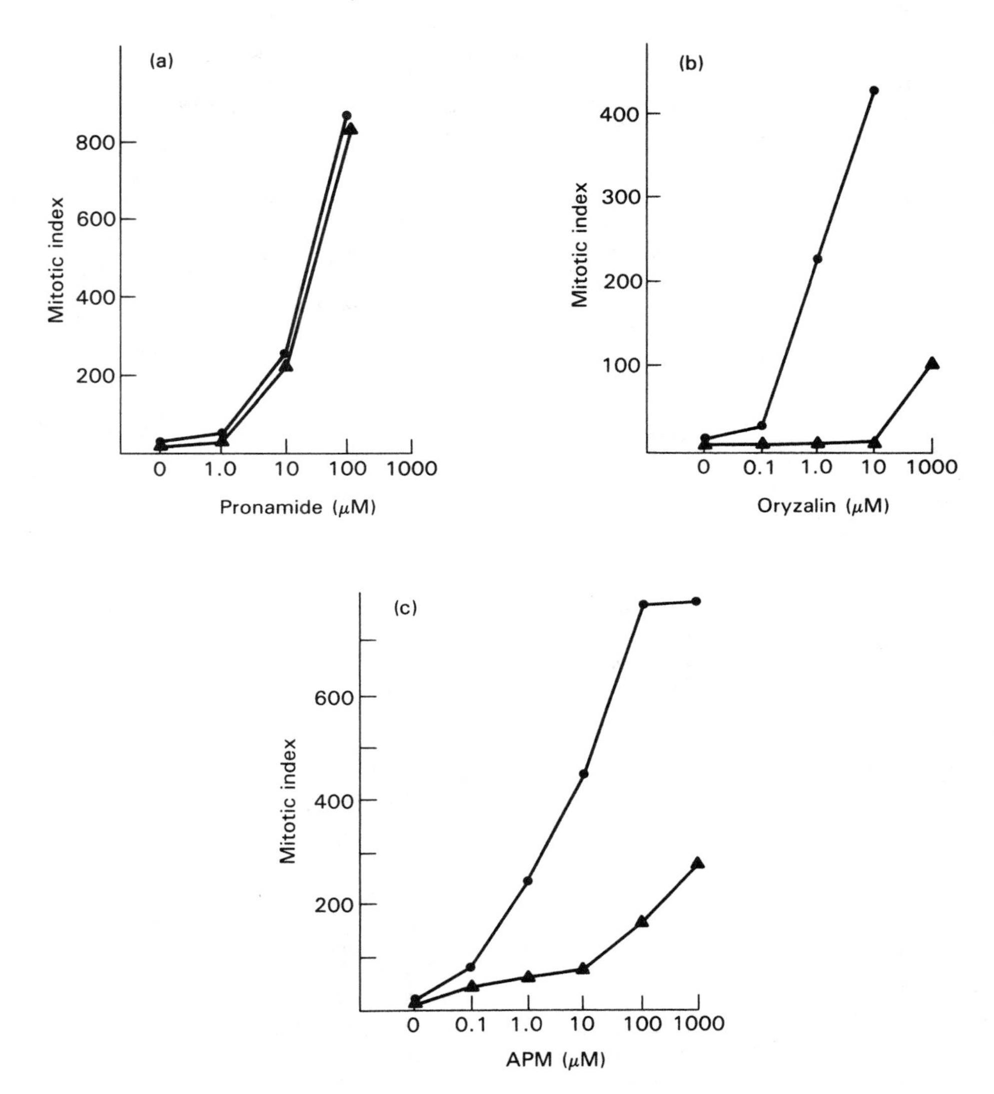

Figure 10.5 Mitotic indices (number of cells with arrested mitotic figures per 1,000 cells) in dinitroaniline-resistant (triangles) and susceptible (circles) biotypes of goosegrass as affected by various concentrations of three mitosis-inhibiting herbicides [24].

to colchicine, many other natural (see Chapter 18) and synthetic compounds have similar effects [26, 27]. Table 10.1 provides a survey of herbicides that cause such effects. In virtually all of these studies, the effects have been equated with the effects of colchicine. Most of these compounds have been linked to direct effects

TABLE 10.1 COMMERCIAL HERBICIDES THAT CAUSE ARRESTED PROMETAPHASE OR DISORDERED CELL WALLS IN HIGHER PLANTS. THE SUBHEADINGS WITHIN THE TABLE REFER TO MAJOR EFFECTS OBSERVED.

Herbicide	Genera affected	Reference
Interference with mitotic spindles		
Amiprophos-methyl	*Daucus*	28
	Solanum	28
	Happlopappus	28
Asulam	*Allium*	29
	Apium	30
Barban	*Vicia*	31
	Triticum	32
Butamifos	*Allium*	33
Carbetamide	*Allium*	34
Chlorpropham	*Vicia*	31
	Allium	35
	Hordeum	35
	Glycine	36
DCPA	*Setaria*	37
Nitralin	*Zea*	38, 39
Oryzalin	*Zea*	38
Pendimethalin	*Allium*	40
Pronamide	*Avena*	41
	Cucumis	41
Propham	*Pisum*	42, 43
	Secale	44
	Allium	45
	Avena	45, 46
	Haemanthus	47, 48
	Vicia	49
	Gossypium	49
Trifluralin	*Haemanthus*	50
	Gossypium	51
	Allium	52, 53
	Nicotiana	54
Dysfunction or lack of cortical microtubules		
Amiprophos-methyl	*Triticum*	55
Barban	*Triticum*	32
Chlorthal-dimethyl	*Avena*	56
Oryzalin	*Zea*	57
Pronamide	*Avena*	41
	Cucumis	41
Propham	*Allium*	58
	Haemanthus	59
Trifluralin	*Glycine*	60
	Allium	61, 62
	Triticum	61
	Gossypium	62

on tubulin. Herbicides that stop mitosis more indirectly (e.g., by interference with metabolism) generally cause a decrease in mitotic figures because they block synthesis of proteins, nucleic acids, or other requisites for mitosis (e.g., [63]). Hess [1] has reviewed these indirect effects of herbicides on cell division. Secondary effects are considered further in Chapter 16.

Ultrastructural and immunocytochemical observations of the effect of a compound on microtubule arrays and individual microtubule structure also provide evidence of direct involvement with tubulin. Still, unless the effect is rapid and dramatic, these methods cannot differentiate between direct and indirect effects of a compound on tubulin or microtubules. For example, glyphosate has been shown to cause root tip swelling, ultrastructural indications of microtubule loss, and reductions in tubulin levels [64]. However, glyphosate is known to exert these effects indirectly through its effects on amino acid synthesis and, ultimately, on protein synthesis and, perhaps, on Ca^{2+} concentration (see Chapter 13). These effects are much slower than those normally observed with compounds known to interact directly with microtubules.

Several plant species have been used extensively in microtubule studies. Blood lily (*Haemanthus katherinae* Baker) endosperm cells are especially good for this type of work because individual mitotic cells can be isolated easily and they have large mitotic spindles that are ideal for light microscopy studies (e.g., [47, 48]). The flagella of *Chlamydomonas reinhardtii* have been useful for the study of both tubulin synthesis and its polymerization into microtubules. Its two flagella can be removed by mild acid treatment and then tubulin synthesis and flagellum formation can be observed after a short (12–15 min) lag [20–22]. Cortical and spindle microtubules of several filamentous algae (e.g., *Spirogyra* and *Oedogonium*) are ideal for observation by immunofluorescence. Onion root tip cells have been used extensively to study mitotic inhibitors because of the ease in obtaining good root tip squashes for observation of mitotic cells.

The most direct approach to studying interference of herbicides with microtubules is to determine the effects of the compound on polymerization of tubulin in vitro. Until recently, this approach was limited to studies with animal tubulin; however, advances in methodology now allow similar studies with plant tubulin. Still, such in vitro methods may overlook effects on microtubules through effects on MAPs, MTOCs, Ca^{2+} concentration, and other factors that directly affect microtubules. Also, in vitro polymerization of microtubules may not always occur exactly as it does in vivo because of differences in tubulin isoforms and other factors.

10.3.2 Noncommercial Inhibitors of Microtubule Function

Much more is known about the mechanism of action of a number of pharmaceutical, fungicidal, and other compounds that interact with tubulin than about commercial herbicides with similar mechanisms of action. Several of the former compounds that have effects on plant tubulin will be discussed in this section. The most studied of these is colchicine, an alkaloid derived from autumn crocus (*Colchicum autumnale*) that has been used as an anticancer drug. It has been shown to bind directly to animal tubulin heterodimers in vitro with a stoichiometry of 1 : 1 [65]. Initial binding is fast and reversible; however, this is followed by slow, tight binding. The heterodimer-colchicine complex binds to the + end of the microtubule and prevents subsequent polymerization [66], or channels polymerization toward a nonmicrotubule product. Thus, the presence of the bound heterodimer-colchicine complex inhibits the addition of more tubulin (either liganded to colchicine or not). Whether this is because colchicine itself sterically hinders access to a heterodimer attachment site or because it causes a conformational change in tubulin is not known. Under some conditions, heterodimer attachment sites on heterodimer-colchicine complexes are available; however, the polymer formed is not a microtubule. Binding of the heterodimer-colchicine complex to microtubules is reversible and the binding constant is essentially the same as for the unliganded heterodimer [67]. Thus, the length and number of microtubules in colchicine-treated cells decrease.

Structure-activity studies with colchicine analogues such as podophyllotoxin, tropolone methyl ether, *N*-acetylmescaline, MTC [methoxy-5-(2,3,4-trimethoxyphenyl)-2,4,6-cycloheptatriene-1-one), and MTPC [2-methoxy-5-(3-(3,4,-5-trimethoxyphenyl)propionylamino)-2,4,6-cycloheptatriene-1-one] have shown that both the tropolone ring and trimethoxyphenyl ring bind to tubulin [68]. Colchicine is about three orders of magnitude less effective in inhibiting mitosis of plants than of animals and the binding affinity is also much lower (e.g., [69]). The autumn crocus, which produces colchicine, is 100 to 1,000 times more resistant to colchicine than are other plant species. The mechanism for this resistance is unknown. Microscopic studies indicate that colchicine affects different microtubule types to varying degrees (e.g., [70]). Whether this is due to different chemical characteristics of the tubulin composing the microtubules or to other factors is not known. In plant cells, colchicine disrupts cell division and cell wall formation, the two most easily observed processes dependent on microtubules. A disproportionate number of meristematic cells are found in arrested prometaphase, sometimes called C-metaphase or C-mitosis (the "C" meaning colchicine). This aberrant mitotic state has sometimes erroneously been equated with metaphase. Unlike normal chromosomes at prometaphase, no microtubules are observed to be associated with chromosomes of colchicine-treated cells at the ultrastructural level. After metaphase is arrested, the chromatids separate and nuclear membranes reform around the chromosomes, resulting in polyploid cells with irregular, lobed nuclei. At the gross anatomical level, colchicine causes characteristic cessation of longitudinal growth, and swelling of root tips. No direct effects on synthesis of cell wall components [71, 72] or protein secretion [71] have been found.

Several other mitotic disruptors of plant origin have been demonstrated to affect tubulin directly [3, 4, 33, 73]. These include podophyllotoxin, vinblastine, vincristine, maytansine, and taxol (see Chapter 18). As mentioned above, podophyllotoxin is a colchicine analogue and functions in essentially the same manner, although less efficiently, as colchicine. The *Vinca* alkaloids, vinblastine and vincristine (both products of *Vinca rosea*), and maytansine inhibit microtubule polymerization by binding to tubulin [74]. Based on virtually identical binding affinities, the *Vinca* alkaloids apparently bind to both α- and β-tubulin at a site with a high degree of homology between the two subunits. Vinblastine and vincristine do not stop polymerization; however, they induce tubulin to aggregate in vitro into double helical crystals consisting of two microtubule protofilament spirals about 18 to 20 nm in diameter [75]. These crystals have not been observed in vivo in treated plant cells. In plants, these compounds depress mitosis by slowing entry into mitosis and disrupting normal spindle development and function. This is manifested in multipolar divisions and colchicine-like prometaphases [24, 76–78]. These alkaloids also stabilize the binding of colchicine to plant tubulin [79]. Maytansine can competitively inhibit the binding of *Vinca* alkaloids to tubulin, prevents *Vinca* alkaloid-induced crystal formation, and is a potent inhibitor of tubulin polymerization [74]. Compared to other plant-derived mitotic inhibitors, the maytansinoids are highly effective in disrupting mitosis in both plants and animals [80]. In plants, maytansine causes colchicine-like effects in mitotic cells.

The action of taxol, a diterpene from western yew (*Taxus brevifolia* L.), is quite different from that of most other mitotic inhibitors. It strongly promotes polymerization of tubulin into microtubules in both plant and animal cells [80–82]. Because of this, it has been used to promote in vitro polymerization of tubulin (e.g., [83]). As with colchicine-like mitotic inhibitors, arrested prometaphases are abundant in taxol-treated meristematic tissues. The hyperstability of the microtubules prevents movement of the chromosomes. In vitro, taxol treatment results in more, but shorter microtubules, apparently due to increased initiation (nucleation) of tubule polymerization [84]. The initiation of tubulin polymerization is more rapid and the critical concentration of tubulin required for polymerization is lower in the presence of taxol. Most evidence indicates that taxol binds to the microtubule rather than directly to tubulin subunits. It strongly retards or prevents depolymerization of already-formed microtubules and reduces treadmilling of microtubules. Colchicine does not bind at the taxol binding site; however, it apparently prevents binding of taxol by elimination of microtubules. The unique mechanism of action of taxol has made it extremely useful in probing the mechanism of action of and resistance to herbicides that interact with tubulin [2].

10.3.3 Dinitroaniline Herbicides

The dinitroaniline herbicides, such as nitralin, trifluralin, and oryzalin (Figure 10.6), are preemergence herbicides, used primarily for grass weed control in dicotyledonous crops. In plant cells, these herbicides cause mitotic and ultrastructural effects similar to those caused by colchicine, but are much more effective than colchicine. However, they have no effect on mitosis of most animal cells (e.g., [20]). They stop growth of root tips and cause root tip swelling that is indistinguish-

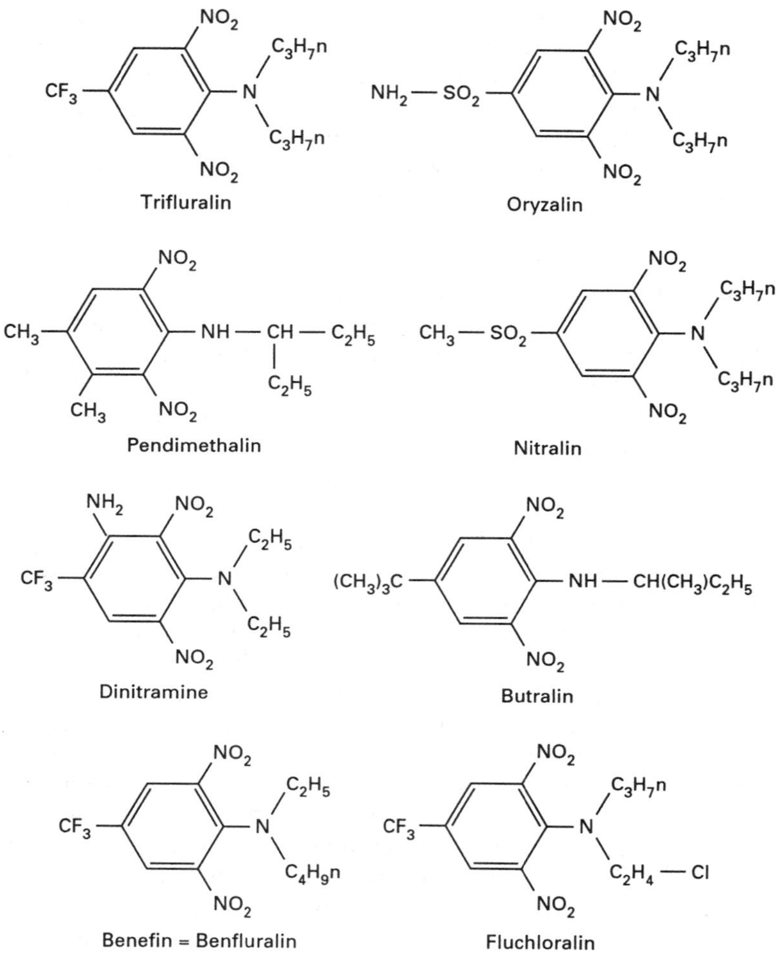

Trifluralin

Oryzalin

Pendimethalin

Nitralin

Dinitramine

Butralin

Benefin = Benfluralin

Fluchloralin

Figure 10.6 Structures of some dinitroaniline herbicides.

able from that induced by colchicine [38, 62, 85]. Furthermore, these herbicides also inhibit IAA-induced cell elongation of coleoptile segments [38]. Elongation of root tips is stopped before swelling occurs [57]. Swelling is the result of a loss of cortical microtubule-mediated longitudinal cell growth. Cell growth becomes isodiametric with no microtubule control of orientation. Trifluralin is generally effective on plant cells at concentrations near 1 μM (e.g., [61]).

Early mode of action work with the dinitroaniline herbicides has been considered in detail in other reviews [86–88], but it took many years to link these effects conclusively to direct effects on tubulin. Trifluralin and oryzalin treatment results in the loss of all microtubules from several types of plant cells, including meristems [51, 89, 90] (Figure 10.7), endosperm [50], and *Oocystis* cells [91]. Furthermore, it shortens flagella of *Chlamydomonas* and prevents regrowth of amputated flagella [21, 22]. No corresponding effects at the low concentrations that cause effects on plant tubulin have been found in animal cells [20]. However, at concentrations of 0.2 to 1.0 μM, trifluralin slows regeneration of the oral band of the ciliate protozoan *Stentor coeruleus,* a process known to be dependent on microtubules and sensitive to other compounds that interact with tubulin [92].

Immunofluorescence studies, using tubulin antibodies, reveal rapid loss of microtubules in oryzalin-treated root tip cells of an alga (*Mougeota* sp.) [93] and higher plants [94]. Nongrass monocots and dicots require much longer for these effects to occur; however, the effects are qualitatively the same in all species. Within any particular species there is a wide range of sensitivity of different microtubule types to the herbicide. In a similar study, the microtubules of the protonema of the moss *Funaria hygrometrica* were found to be differentially sensitive to oryzalin [95]. The cytoplasmic microtubules responsible for organelle movement were affected, and nonpolar growth was caused by low concentrations (< 1 μM). Mitotic spindles and phragmoplast microtubules were only affected by

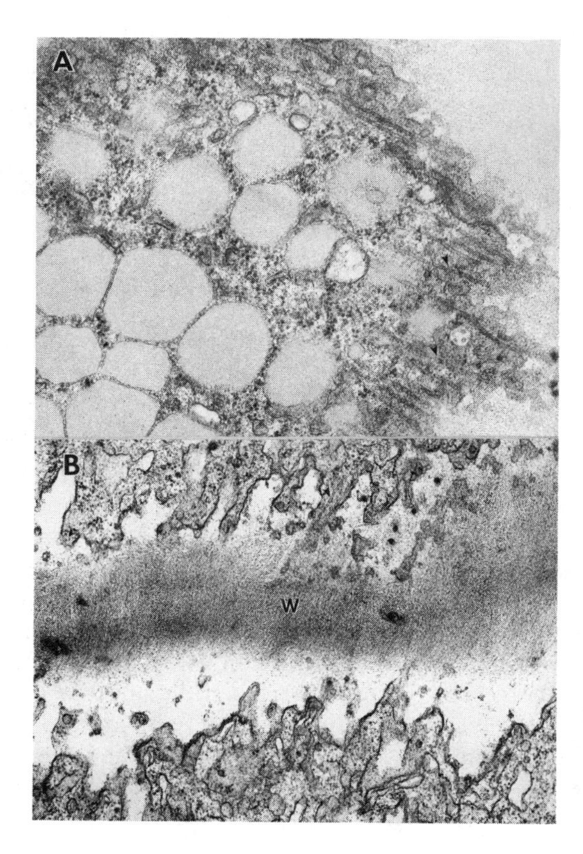

Figure 10.7 Effects of oryzalin on cortical microtubules of onion root tip cells. A = untreated, B = treated with 1 μM oryzalin for 24 h. Only a few microtubules (arrows) are noted in the treated root tip cells, although they are numerous in the control (Courtesy of K. C. Vaughn)

higher concentrations. The loss of polar growth was proposed to be due to loss of a feedback system dependent on microtubular connection of the nucleus and growing cell tip. This may be analogous to disorientation of longitudinal growth in plant meristems by dinitroaniline herbicides. Similar effects of dinitroaniline herbicides on microtubules have been observed in the green alga *Acetabularia acetabulum* [96].

Trifluralin is particularly difficult to use for in vitro studies because of its low aqueous solubility and its high affinity for glass [97]. It is also slightly volatile and subject to photodecomposition. Oryzalin and other dinitroanilines have more desirable properties than trifluralin in both categories. Both trifluralin and oryzalin have been reported to inhibit mammalian brain tubulin polymerization and promote brain tubulin depolymerization [91], although this finding has been disputed [22, 86, 98, 99]. Tubulin of the parasitic protozoan *Leishmania mexicana amazonensis* binds trifluralin, and trifluralin selectively inhibits proliferation of this intestinal parasite [99]. Consequently, trifluralin has been suggested as a chemotherapeutic agent for this disease. Trifluralin and orzyalin bind to tubulin isolated from the central tubules of *Chlamydomonas* flagella [20, 98], indicating that they might prevent tubulin polymerization in vivo. The molar binding ratio of oryzalin to tubulin was 0.98, and thermodynamic results indicate that binding is spontaneous [98]. Tubulin isolated from rose cell cultures is bound by oryzalin, and it prevents taxol-induced polymerization of rose tubulin [100] in a similar manner to colchicine-inhibited, taxol-induced rose tubulin polymerization [83]. Little or no binding of oryzalin to microtubules, denatured tubulin, or nontubulin proteins has been detected, indicating that it binds only to tubulin monomers [1]. Reversal of trifluralin-induced growth disruption by sulfhydryl-reducing agents [61, 101] indicates that the dinitroaniline binding site may contain sulfhydryl groups.

Three higher plants are extremely resistant to dinitroaniline herbicides: carrot (*Daucus carota* L.) [102], a goosegrass (*Eleusine indica* (L.) Gaertn.) biotype [2], and a green foxtail (*Setaria viridis* L.) biotype [103]. The dinitroaniline-resistant (R) goosegrass biotype has become common throughout cotton-growing regions of southeast North America where dinitroaniline herbicides have been used continuously for more than two decades. The R biotype is apparently as biologically fit as the sensitive (S) biotype [104]. Based on effects on mitotic indices, the R biotype is 1,000 to 10,000-fold more resistant to trifluralin than the S biotype [25] (Figure 10.5). It has similar levels of resistance to other dinitroaniline herbicides; however, it is fully sensitive to the plant alkaloid tubulin poisons colchicine, vinblastine, and podophyllotoxin [24] (Table 10.2). Tubulin isolated from the R biotype polymerizes into apparently normal microtubules in the presence of 10 μM oryzalin, whereas S biotype tubulin will not polymerize in this medium [2]. Tubulin from both biotypes polymerizes in the absence of oryzalin. These results indicate differences in tubulin between the two biotypes.

On one-dimensional polyacrylamide gels, only one α- and one β-tubulin isoform are apparent by immunoblotting in the S biotype [2]. The α-tubulins of the two biotypes appear identical. However, two β-tubulin isoforms, one with differ-

TABLE 10.2 RESPONSE OF DINITROANILINE-RESISTANT AND -SENSITIVE BIOTYPES OF GOOSEGRASS TO VARIOUS GROUPS OF MITOTIC INHIBITORS BASED ON EFFECTS ON MITOTIC INDICES AND ULTRASTRUCTURAL EFFECTS ON MITOSIS.

Group 1
(R and S biotypes equally affected—no cross-resistance)
Colchicine, podophyllotoxin, vinblastine, pronamide, terbutol, DCPA, oncodazole
Group 2
(R more affected than S—negative cross-resistance)
Griseofulvin, propham, chlorpropham
Group 3
(S more affected than R—cross-resistance)
All dinitroanlines and amiprophosmethyl

ent electrophoretic mobility than that of the S biotype, are observed in R biotypes from various locations in the southeastern United States. The only exception is an intermediate goosegrass biotype (I) that is found in only a narrow geographic area of South Carolina. It is only 10 to 50 times less susceptible to dinitroanilines than the S biotype, and apparently does not have an altered β-tubulin, or at least does not have the same alteration as the R biotype [2, 105]. The mechanism of its tolerance has not been established. Whether the alterations in β-tubulin in the R biotype are due to differences in the tubulin gene or to differences in posttranslational processing has not yet been conclusively determined. However, unpublished Southern blot analysis indicates that different β-tubulin genes are present in the R biotype.

The new β-tubulin isotype present in the R biotype may not bind dinitroaniline herbicides differently than that of the S biotype, but may simply lend hyperstability to microtubules [2]. In animal cells resistant to microtubule disruptors such as colchicine, microtubules have been found to be hyperstable [106] and, as a result, more sensitive to taxol, a compound that disrupts microtubule function by hyperstabilization (see section 10.3.2). Similarly, the R biotype is 100 times more sensitive to taxol than the S biotype [24]. The I biotype is not taxol hypersensitive [105]. These results can be explained by the model of stability/instability of Cabral et al. [107] (Figure 10.8) in which the level of sensitivity of microtubules to a compound that affects stability of the microtubule is described as a function of the preexisting stability. Thus, an alteration in tubulin that affects stability can profoundly influence the effect of such a compound. In fact, the S biotype treated with taxol has the occasional cell wall abnormalities of the untreated R biotype [2], indicating that enhanced stabilization of the tubulin (either genetically based or by chemical treatment) yields the same result. Because the taxol-treated S biotype is phenotypically similar to the R biotype, one might also expect it to be resistant to dinitroaniline herbicides. This is indeed the case [2]. Thus, the mechanism of resistance to dinitroaniline herbicides appears to be a

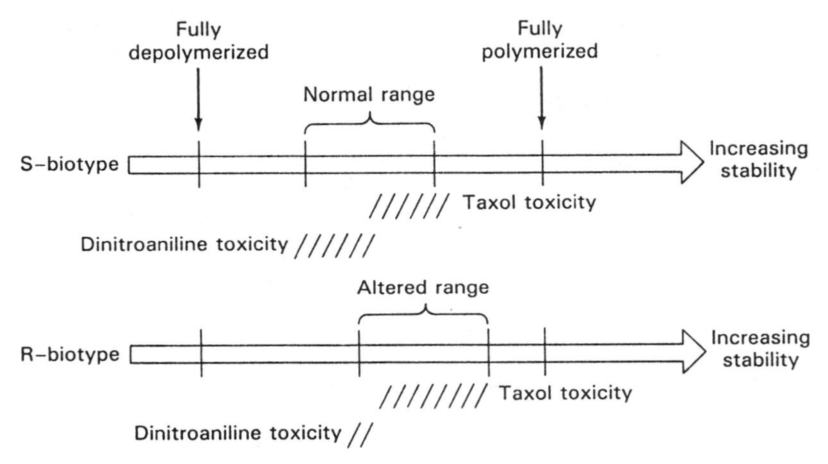

Figure 10.8 Model of microtubule stability/instability adapted from [107].

β-tubulin alteration that renders microtubules more stable. Cross-resistance to herbicides structurally unrelated to dinitroanilines that also affect tubulin (discussed below) also supports this hypothesis.

The resistance of carrot to dinitroaniline herbicides [102] is not as well understood. Mitosis is not disrupted in root tips of carrot at 10 μM concentrations of a wide range of dinitroaniline herbicides, and at 1 μM oryzalin. No ultrastructural effects of dinitroaniline herbicides can be seen on any microtubule type. There is no cross-resistance to colchicine or most other mitotic disruptors except for amiprophosmethyl and hexanitrodiphenylamine (see below). Carrot has three α- and four β-tubulin isotypes (106), but the effect of dinitroaniline herbicides on the polymerization of these has not been determined. Unpublished work, however, has shown that isolated protoplasts of carrot are not resistant to dinitroaniline herbicides, indicating that resistance of tissues may be due to limited uptake of the herbicide (unpublished data, K. C. Vaughn).

Dinitroaniline herbicide effects on intracellular compartmentalization of Ca^{2+} have been proposed by Hertel et al. [108, 109] to be indirectly responsible for depolymerization of microtubules. Normally, Ca^{2+} levels are kept low enough for polymerization of microtubules by mitochondrial uptake of the ion. Trifluralin, at about 1 μM and oryzalin, at about 25 μM, have been shown to reduce mitochondrial uptake of Ca^{2+} by about 50 percent [109]. Calcium accumulation in cytoplasmic vesicles is stimulated by trifluralin and inhibited by oryzalin [108]. These results do not agree well with what is known of the herbicidal action of these or other herbicides that affect microtubules. First, the concentrations required for the mitochondrial effects are higher than those required for in vivo disruption of microtubules. Second, the different effects of the two herbicides on Ca^{2+} uptake by cytoplasmic vesicles do not agree with the theory that microtubule depolymerization is a result of effects on Ca^{2+} transport. Third, mitochondrial respiratory inhibitors should have profound effects on microtubules if the Ca^{2+} uptake theory were correct; however, they do not have such effects. Fourth, Ca^{2+} distribution in both the S and R biotypes of goosegrass are similarly affected by dinitroaniline herbicides as determined by cytochemical methods [2].

Dinitroaniline herbicides also have effects on photosynthesis, oxidative phosphorylation, and membrane function. These effects are dealt with in Chapter 15.

10.3.4 Phosphoric Amides

As with dinitroaniline herbicides, the phosphoric amides such as isophos, amiprophosmethyl, and butamifos (Figure 10.9) cause swollen root tips, aberrant mitotic figures, and malformed cell walls (Table 10.1). Of these compounds, amiprophosmethyl (APM) is the most studied. At the light microscope level, the effects of APM and oryzalin on mitotic spindle organization are indistinguishable [110]. This compound causes loss of microtubules as seen at the ultrastructural [91, 111] and immunofluorescence light microscope [112–115] levels. It inhibits flagellar regeneration of *Chlamydomonas* [22, 116], and inhibition of postmitotic nuclear migration (a microtubule-dependent process) in *Micrasterias denticulata* [117, 118] and *Acetabularia mediterranea* [119]. Suspension cultures of potato, carrot, *Happlopappus gracilis,* and tobacco cells treated with APM rapidly undergo metaphase

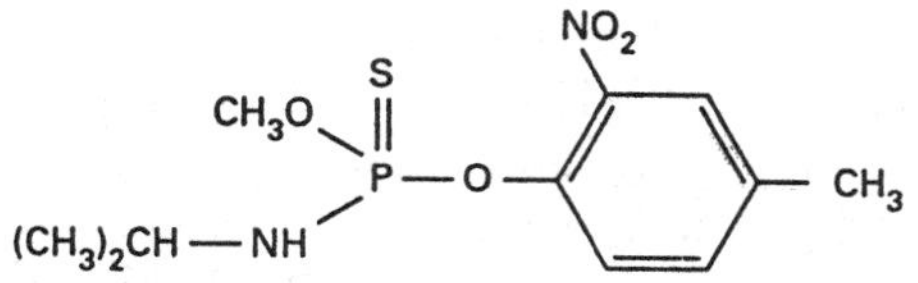

Amiprophosmethyl

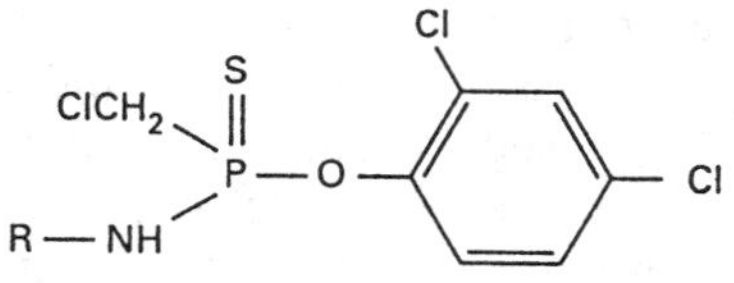

R = Sec. butyl = Isophos 1

R = Isopropyl = Isophos 2

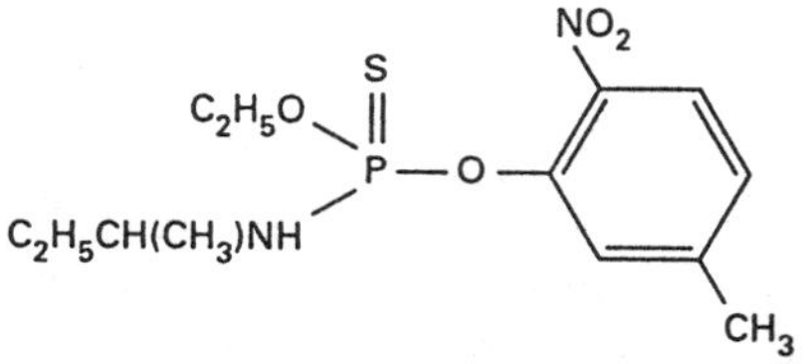

Butamifos

Figure 10.9 Structures of some phosphoric amide herbicides.

arrest and produce micronuclei containing two or more chromosomes each [28, 120]. After removal of the herbicide from cultures of potato, alfalfa, tobacco, lettuce, carrot, and *Happlopappus,* mitosis and growth return to normal [28, 114]. APM is more effective on "nonstabilized" cortical microtubules in rapidly dividing cells than on "stabilized" microtubules in elongating cells [114]. In tobacco, the frequency of lobed micronuclei in APM-treated cells increases and mitoses with double chromosome numbers appear in the next cell division [120]. Plant geneticists have taken advantage of the micronuclei produced in APM-treated plant cells to isolate and sort different chromosomes by flow cytometry [121]. Disruption of microtubules by APM in oat coleoptile or mesocotyl tissue results in loss of cellulose microfibril orientation [122]. In the filamentous green alga *Chamaedoris orientalis,* APM disrupts cortical microtubules and this causes abnormal microfibril orientation changes [123]. However, there is no change in the pattern of microfibril orientation in the protonema tip of the fern *Adiantum* associated with APM disruption of microtubules [115]. Similarly, the deposition pattern of microfibrils of cortical cells of *Nicotiana tabacum* is not altered by treatment with butamifos [124].

APM inhibits in vitro polymerization of plant tubulin into microtubules [9]. This direct effect is more likely to be the mode of action of APM than the indirect effects on Ca^{2+} compartmentalization described by Hertel et al., which require higher concentrations of APM than are required for effects on tubulin [109, 110]. Furthermore, it has similar effects on Ca^{2+} compartmentalization in both animal and plant mitochondria, although it has little or no effect on polymerization or depolymerization of mammalian brain [91] and frog heart [109] tubulin in vivo. One effect that has not been reported for dinitroanilines is that APM inhibition of flagellar regeneration is accompanied by cessation of tubulin synthesis [22].

Chlamydomonas mutants that are resistant to APM are strongly cross-resistant to butamiphos, another phosphoric amide, and to orzyalin [125]. These mutants are only weakly resistant (ca. 10-fold) to these herbicides. Two Mendelian genes are associated with resistance. Mutants homozygous for one of the genes are more resistant than those homozygous for the other, indicating different gene products. Dinitroaniline-resistant goosegrass and carrot are also resistant to APM [24, 102] (Table 10.2), indicating a common mechanism or site of action with dinitroaniline herbicides. Taxol prevents some of the microtubule depolymerization caused by butamifos in the protonema of *Physcomitrella patens,* resulting in partial reversal of some of the abnormal morphogenic effects of the herbicide [126]. This is similar to the effects of taxol on dinitroaniline-microtubule interactions (see Section 10.3.3).

10.3.5 Carbamates

In a review of *N*-phenyl carbamates, Tissut et al. [127] described three major effects of this herbicide group: inhibition of mitosis; inhibition of PSII electron transport; and uncoupling of oxidative phosphorylation. The last two effects are dealt with in Chapters 7 and 15, respectively. *N*-phenyl carbamates of the general structures in Figure 10.10 are shown in their order of decreasing efficacy as mitotic inhibitors. The *N*-phenyl carbamates such as propham, chlorpropham, asulam, and barban (Figure 10.11) have been known for many years to affect mitosis of plant cells (e.g., [128]). These herbicides are especially active on grasses. They do not cause the loss of microtubules, but specifically cause loss of spindle microtubule function; consequently, they have been categorized as spindle function inhibitors [1].

Mitosis is stopped or altered by these carbamates by interference with the mitotic spindle of higher plants (Table 10.1) and algae [111, 118, 129–131]. At the light microscope level, the *N*-phenyl carbamates cause lagging chromosomes, multipolar anaphases, chromosome stickiness, anaphase bridges, chromosome fragmentation, and multinucleate cells. The effects of asulam and barban are not as well characterized as those of propham and chlorpropham. Propham and chlorpropham induce multipolar spindles as well as other cell division abnormalities in higher plants [47, 48, 132–136], and also in some animals [137]. Unlike compounds known to interact directly with tubulin, propham and chlorpropham decrease the mitotic index of affected root tips [136]. Neither propham [91, 138] nor chlorpropham [91] binds to brain tubulin or affects its in vitro assembly/disassembly or its assembly into MTOCs in *Polytomella* [111]. In *Euglena gracilis*, propham causes partial inhibition of flagellar regeneration, as well as arrest of cell nuclei in the G2 phase and many structural aberrations of the nucleolus, chromosomes, pellicle,

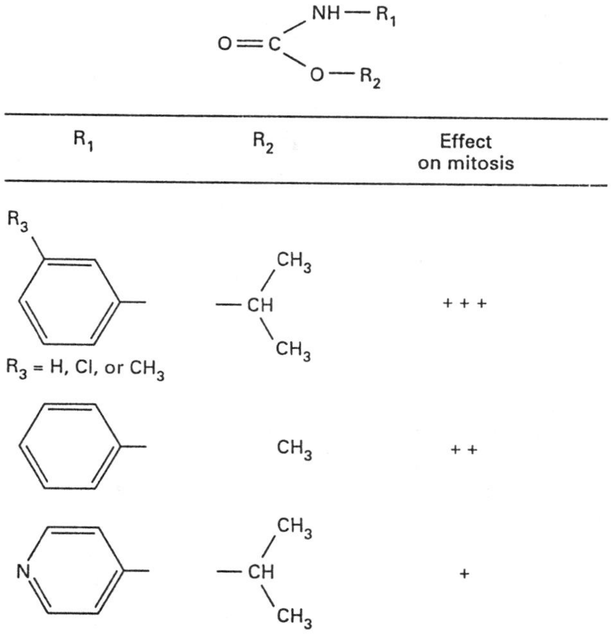

O=C(NH—R_1)(O—R_2)

R_1	R_2	Effect on mitosis
R_3-phenyl; R_3 = H, Cl, or CH_3	—CH(CH_3)(CH_3)	+ + +
phenyl	CH_3	+ +
N (pyridyl)	—CH(CH_3)(CH_3)	+

Figure 10.10 Mitotic inhibitor structure-activity relationships of *N*-phenyl carbamates as determined by Tissut et al. (127). + = $>100\ \mu M$, ++ = $100\ \mu M$, and +++ = $< 100 \mu M$ required to inhibit mitosis.

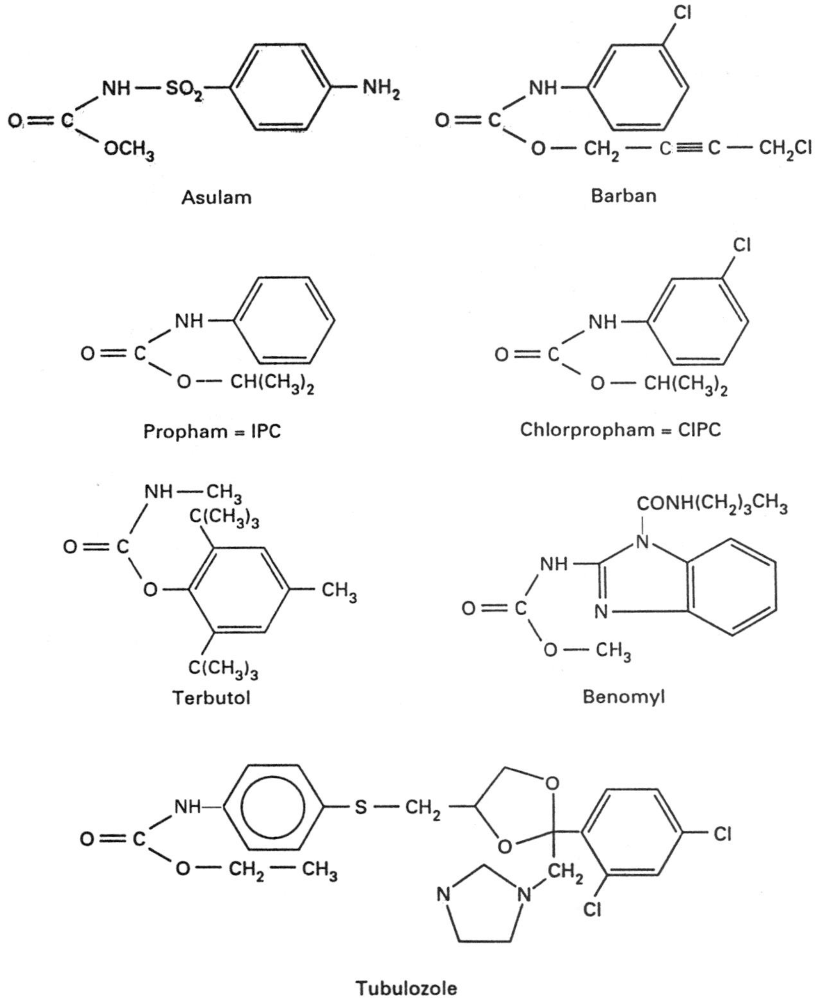

Asulam

Barban

Propham = IPC

Chlorpropham = CIPC

Terbutol

Benomyl

Tubulozole

Figure 10.11 Structures of some herbicidal, fungicidal, and nematicidal carbamates that interfere with microtubules.

mitochondria, chloroplasts, and dictyosomes [139]. Most of these effects are thought to be due to interference with MTOCs. Ultrastructural studies show that although the microtubules of the spindle are present, they are oriented in all directions in *N*-phenyl carbamate-treated tissues [47], rather than in a normal (parallel) fashion. Although spindle organization is disrupted in chlorpropham-treated onion cells, resulting in tripolar or multipolar spindles giving rise to multiple nuclei [134], there are no effects on cortical microtubules, as determined by immunofluorescence microscopy. However, phragmoplast orientation and resulting cell walls are disrupted. Asulam also causes phragmoplast dysfunction [135]. Gunning and Hardham's [4] analysis of the data available through 1981 and a more recent analysis by Hidalgo et al. [136] has led them to suggest that the *N*-phenyl carbamates might act only at the MTOC level.

The only evidence that suggests that these herbicides might bind to plant tubulin is that the methyl ester of propham has been used on an affinity column for the purification of both plant and animal tubulin [140]. However, the movement of tubulin through the column was only retarded, indicating that the separation of tubulin by this column may not have been due to the propham. There is no cross-resistance with dinitroaniline-resistant carrot [102], and dinitroaniline-resistant goosegrass biotypes are hypersensitive to both propham and chlorpropham [24] (Table 10.2).

Although terbutol is often grouped with other carbamate herbicides, its structure is considerably different (Figure 10.11). It is chemically related to the methyl carbamates, physostigmine (a naturally occurring, cholinergic alkaloid) and to the insecticide carbaryl. It is also closely related to the carotenoid synthesis inhibitor, dichlormate (see Figure 8.2). Terbutol was a highly selective herbicide for crabgrass (*Digitaria sanguinalis*) control in turf, but is no longer commercially available. Relatively little study of its mode of action has been conducted. Lehnen et al. [141] found terbutol to cause mitotic abnormalities quite unlike those caused by the *N*-phenyl carbamates. So-called "star anaphase" arrangements of microtubules are found in terbutol-treated onion root tips. Immunofluorescence microscopy with antitubulin and electron microscopy reveal that clusters of microtubules radiate from abnormally diffuse MTOCs of poles of the dividing cell. Phragmoplast MTOCs are also abnormal, resulting in incomplete or abnormally oriented cell walls. These abnormalities have not been observed with other herbicides, indicating that terbutol may have a unique mechanism of action. Dinitroaniline-resistant goosegrass [24] (Table 10.2) and carrot [102] are not cross-resistant to terbutol.

Related carbamates with antimitotic activities, the benzimidazole-2-yl carbamates, are used as fungicides and antihelminthics. Examples of the benzimidazole carbamates include parbenazole, nocodazole (= oncodazole), tubulozole, benomyl (Figure 10.11), and MBC (carbendazim). These compounds inhibit cell division of fungi, and inhibition of growth of fungi can in some cases be correlated with efficacy in inhibiting tubulin polymerization [142, 143]. The benzimidazole-2-yl carbamates apparently inhibit mitosis of fungi by binding to tubulin. Tubulozole-C, parbendazole, and nocodazole inhibit mammalian brain tubulin polymerization in vivo and in vitro [144–146]; however, MBC has little or no effect on this process with brain tubulin [91, 143]. The antihelmintic benzimidazoles (with no carbamate moiety) parbendazole, mebendazole, fenbendazole, thiabendazole, and oxfendazole all inhibit the in vitro polymerization of nematode tubulin more than brain tubulin [147]. MBC has no effect on nuclear migration in *Micrasterias* [118]. However, it can alter cell wall morphology in *Oocystis* [91]. MBC and benomyl weakly inhibit phragmoplast function and spindle function in onion root tip cells [148]. The effects described are much like those caused by *N*-phenyl carbamates. Nocodazole inhibits nuclear migration in *Micrasterias* [118]. The benzimidazole-2-yl carbamates bind to the colchicine-binding site of tubulin in fungi, and apparently act by an entirely different mechanism of action than do the *N*-phenyl carbamates in higher plants. Mutant fungi with strong resistance to these compounds are produced with relative ease and have been crucial in the identification of tubulin genes in a wide variety of fungi [149]. Two mechanisms of resistance have been identified in *Aspergillis nidulans*—altered binding affinity to tubulin and increased microtubule stability (hyperstability). Hyperstability and decreased affinity for benzimidazoles are both due to altered β-tubulins. Since different tubulins are involved in vegetative growth and conidiation, two mutations must occur for complete resistance of the fungus.

Both *N*-phenyl and benzimidazole carbamates cause the release of Ca^{2+} from plant mitochondria in vivo [109]. The effect could be a secondary effect of raising cAMP levels, since MBC inhibits cAMP phosphodiesterase in *Aspergillus* [150]. However, depolymerization of microtubules due to high Ca^{2+} concentrations cannot be the mode of action of these herbicides, because they do not cause the loss of microtubules.

10.3.6 Other Herbicides Affecting Microtubules

DCPA (chlorthal-dimethyl) (Figure 10.12) is structurally unrelated to other mitotic disruptors. It stops mitosis in root tips of corn [151], oat [46, 152], and foxtail millet (*Setaria italica*) [153] during prophase, similar to other compounds that disrupt mitosis by interference with microtubules. Holmsen and Hess [46] found similarities between DCPA, colchicine, and propham in an analysis of mitotic stages in oat root meristems, resulting from treatment with each of the compounds. At the light microscope and ultrastructural level, Vaughan and Vaughn [154] found several types of cytological effects to be caused by DCPA in wheat root tips. In epidermal cells, cell walls were incomplete and/or curved, as well as being oriented out of their normal axial planes. Phragmoplast microtuble arrays in this tissue were oriented in many directions, rather than in one, as in untreated tissues. In other

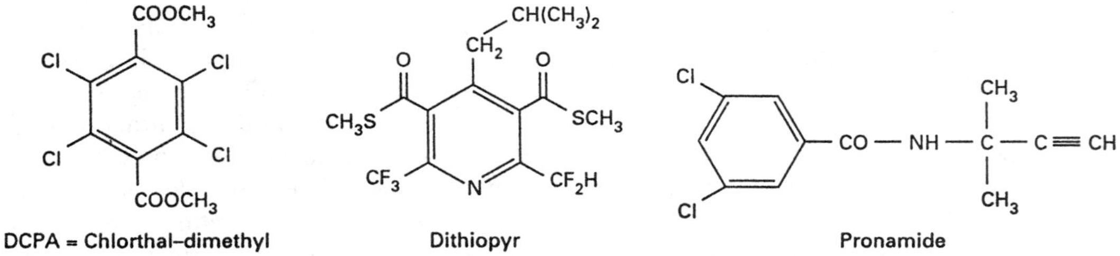

DCPA = Chlorthal–dimethyl

Dithiopyr

Pronamide

Figure 10.12 Structures of some miscellaneous herbicides that interfere with microtubules.

non-meristematic layers of the root tip, cell walls do not form between daughter nuclei of the dividing cells, creating binucleate and, eventually, multinucleate cells. This agrees with previous observations of unusually high numbers of binucleate and multinucleate cells in DCPA-treated corn [154] and foxtail millet [128] roots. Multinucleate cells have not been observed as a major effect for other herbicides. In meristematic cells, prometaphase figures with only a few dispersed microtubules and multipolar mitoses are the predominant abnormalities—effects that are similar to those of *N*-phenyl carbamates. This indicates a greater effect of DCPA in wheat on phragmoplast microtubules than on spindle microtubules. Unpublished results of Holmsen (see [1]) with oat roots gave similar results. Dinitroaniline-resistant goosegrass is not cross-resistant to DCPA [24] (Table 10.2).

Pronamide (Figure 10.12) also disrupts mitosis in a manner similar to other microtubule disruptors. It causes root tip swelling [41, 155] and rapid effects on mitosis that are similar to those caused by colchicine or dinitroanilines [41]. At the ultrastructural level, there is loss of both cortical and spindle microtubules in pronamide-treated corn and wheat root tips [89]. Further ultrastructural and immunofluorescent studies in onion root tips have revealed that pronamide-treated cells have greatly shortened microtubules located only in the kinetichore region [156]. Abnormal secondary xylem walls are associated with normal, but markedly fewer, cortical microtubules. Fluorescence microscopy of cultured tobacco cells treated with pronamide reveals that both spindle and cortical microtubules are rapidly disrupted [157]. This effect is readily reversible when pronamide is removed from the growth medium. Pronamide at 10 μM completely inhibits polymerization of tobacco tubulin in vitro [157]. However, it has no effect on brain tubulin [89, 91, 157]. Tobacco tubulin can be affinity-purified with a pronamide-analogue-linked Sepharose chromatography column [157]. No cross-resistance occurs with dinitroaniline-resistant carrot [103] or dinitroaniline-resistant biotypes of goosegrass [24] (Table 10.2).

Dithiopyr (formerly MON-7200) (Figure 10.12) stops cell division in a similar manner to dinitroanilines [158, 159]. It causes loss of both spindle and cortical microtubules by its effects on tubulin polymerization. However, it was found to bind to a nontubulin protein of M_r 68,000, composed of two M_r 38,000 subunits (possibly MAPs—see below). No evidence of binding to tubulin was found.

10.3.7 Interactions with Microtubule-Associated Proteins

No herbicides are known to affect plant MAPs. This is hardly surprising since plant MAPs have only recently been conclusively isolated and identified [23]. MAPs may be good targets for herbicide action for compounds with little mammalian toxicity because plant and animal MAPs are likely to be more evolutionarily divergent than tubulin. Already, one compound, estramustine phosphate {estradiol-3,[*N*-bis-(2-chloroethyl) carbamate] 17β-phosphate} has been demonstrated to inhibit brain microtubule assembly predominantly by binding to MAPs [160]. Taxol has been proposed to exert part or all of its influence on microtubules by altering the composition of MAPs with microtubules or the cross linking of the microtubules [161]. The protein with which dithiopyr (discussed in Section 10.3.6) interacts may be a MAP.

10.3.8 Microfilaments and Disrupting Agents

Microfilaments are polymeric strands of the proteins actin and myosin. They are involved in protoplasmic streaming, contractile movements, cell division, and in providing protoplasmic shape. They are often found in bundles of variable dimensions. Microfilaments are not easily observed with electron microscopy; however, they can be detected with immunofluorescence or fluorescence-labeled phalloidin, a compound that binds to actin. Like microtubules, microfilaments have polarity. Only two agents have been identified that interfere with plant microfilaments—cytochalasins and phalloidin [3]. The phalloidins prevent depolymerization of microfilaments and the cytochalasins enhance depolymerization. As may be expected, each inhibits the effect of the other. Both agents prevent protoplasmic streaming and cause abnormal cell division planes. No commercial herbicides are known to have a direct effect on microfilaments, although little research has been conducted to discover such an interaction. The microtubule-depolymerizing herbicide butamifos has no effect on microfilament arrays in moss protonema, despite complete loss of microtubules [125]. Conversely, cytochalasin disrupts microfilaments in these cells without having any effect on microtubules.

10.4 SUMMARY

There are numerous herbicides that act primarily by causing dysfunction of plant microtubules. The dinitroanilines and phosphoric amides apparently achieve this effect primarily through binding to tubulin and preventing polymerization. The *N*-phenyl carbamates have a different mechanism, perhaps by interacting with MTOCs. The mechanisms of action of mitotic disruptors such as DCPA, terbutol, and pronamide are not as well studied or understood.

The plant cytoskeleton has components other than microtubules (e.g., MTOCs, microfibrils, MAPs). Interference with any component of the plant cytoskeleton should result in cessation of growth and other phytotoxic symptoms. We have no good evidence that any existing herbicide acts directly on any of these other components. This is probably partly due to our relative ignorance of these components, especially in plants. These other cytoskeleton constituents could be sites of action of current and future herbicides.

Cell division and microtubular function are processes common to all eucaryotic organisms. For toxicological reasons, such fundamental processes may seem like poor choices as targets for herbicides. However, the proteins involved in these processes are apparently different enough between fungi, mammals, and green plants for a great deal of selectivity in affecting microtubular function. Improved knowledge of the proteins involved in higher plant microtubular function may allow for targeting these proteins as specific molecular sites of herbicide action.

REFERENCES

1. Hess, F. D. 1987. "Herbicide effects on the cell cycle of meristematic plant cells." *Rev. Weed Sci.,* 3, 183–203.
2. Vaughn, K. C., and M. A. Vaughan. 1990. "Structural and biochemical characterization of dinitroaniline resistant *Eleusine*." *Amer. Chem. Soc. Symp. Ser.,* 421, 364–75.
3. Seagull, R. W. 1989. "The plant cytoskeleton." *CRC Crit. Rev. Plant Sci.,* 8, 131–67.
4. Gunning, B. E. S., and A. R. Hardham. 1982. "Microtubules." *Annu. Rev. Plant Physiol.,* 33, 651–98.
5. Lloyd, C. W. 1987. "The plant cytoskeleton: the impact of fluorescence microscopy." *Annu. Rev. Plant Physiol.,* 38, 119–39.
6. Soifer, D., ed. 1986. *Dynamic Aspects of Microtubule Biology*. Ann. New York Acad. Sci., Vol. 466.
7. Hardham, A. R., and B. E. S., Gunning. 1978. "Structure of cortical microtubule arrays in plant cells." *J. Cell Biol.,* 77, 14–34.
8. Van't Hof, J., G. B. Wilson, and A. Colon. 1960. "Studies on the control of mitotic activity: The use of colchicine in the tagging of a synchronous population of cells in the meristem of *Pisum sativum*." *Chromosoma,* 11, 313–21.
9. Morejohn, L. C. and D. E. Fosket. 1984. "Inhibition of plant microtubule polymerization in vitro by the phosphoric amide herbicide amiprophos-methyl." *Science,* 224, 874–76.
10. Morejohn, L. C., T. E. Bureau, L. P. Tocchi, and D. E. Fosket. 1984. "Tubulins from different higher plant species are immunologically nonidentical and bind colchicine differently." *Proc. Natl. Acad. Sci.,* USA, 81, 1440–44.
11. Silflow, C. D., and J. Youngblum. 1986. "*Chlamydomonas reinhardtii* tubulin gene structure." *Ann. N.Y. Acad. Sci.,* 466, 18–30.
12. Dawson, P. J., and C. W. Lloyd. 1985. "Identification of multiple tubulins in taxol microtubules purified from carrot suspension cells." *EMBO J.,* 4, 2451–55.
13. Hussey, P. J., and K. Gull. 1985. "Multiple isotypes of α and β tubulin in the plant *Phaseolus vulgaris*." *FEBS Lett.,* 181, 113–118.
14. Mizuno, K., J. Perkin, F. Sek, and B. Gunning. 1985. "Some biochemical properties of higher plant tubulins." *Cell Biol. Intl. Rep.,* 9, 5–12.
15. Yadav, N. S., and P. Filner. 1983. "Tubulin from cultured tobacco cells. Isolation and identification based on similarities to brain tubulin." *Planta,* 157, 46–52.
16. Brunke, K. J., P. S. Collis, and D. P. Weeks. 1982. "Post-translational modification of tubulin dependent on organelle assembly." *Nature,* 297, 516–18.

17. Cleveland, D. W., M. A. Lopata, P. Sherline, and M. W. Kirschner. 1981. "Unpolymerized tubulin modulates the level of tubulin mRNAs." *Cell,* 25, 537–46.

18. Weeks, D. P., and P. S. Collis. 1976. "Induction of microtubule protein synthesis in *Chlamydomonas reinhardtii* during flagellar regeneration." *Cell,* 9, 15–27.

19. Brunke, K. J., E. E. Young, B. E. Buchbinder, and D. P. Weeks. 1982. "Coordinate regulation of the four tubulin genes of *Chlamydomonas reinhardtii.*" *Nucleic Acid Res.,* 10, 1295–1310.

20. Hess, F. D., and D. E. Bayer. 1977. "Binding of the herbicide trifluralin to *Chlamydomonas* flagellar tubulin." *J. Cell Sci.,* 24, 351–60.

21. Hess, F. D. 1979. "The influence of the herbicide trifluralin on flagellar regeneration in *Chlamydomonas.*" *Exp. Cell Res.,* 119, 99–106.

22. Quader, H., and P. Filner. 1980. "The action of antimitotic herbicides on flagellar regeneration of *Chlamydomonas reinhardtii:* a comparison with the action of colchicine." *Eur. J. Cell Biol.,* 21, 301–04.

23. Cyr, R. J., and B. A. Palevitz. 1989. "Microtubule-binding proteins from carrot I. Initial characterization and microtubule bundling." *Planta,* 177, 245–60.

24. Vaughn, K. C., M. D. Marks, and D. P. Weeks. 1987. "A dinitroaniline-resistant mutant of *Eleusine indica* exhibits cross-resistance and supersensitivity to antimicrotubule herbicides and drugs." *Plant Physiol.,* 83, 956–64.

25. Vaughn, K. C. 1986. "Cytological studies of dinitroaniline-resistant *Eleusine.*" *Pestic. Biochem. Physiol.,* 26, 66–74.

26. Deysson, G. 1975. "Microtubules and antimitotic substances." In *Microtubules and Microtubule Inhibitors,* M. Borgers and M. de Brabander, eds., Amsterdam: North-Holland, pp. 427–51.

27. Wilson, L., J. R. Bamburg, S. B. Mizel, L. M. Cirsham, and K. M. Creswell. 1974. "Interaction of drugs with microtubule proteins." *Fed. Proc.,* 33, 158–166.

28. Ramulu, K. S., H. A. Verhoeven, P. Dijkhuis, and L. J. W. Gilissen. 1988. "Chromosome behavior and formation of micronuclei after treatment of cell suspension cultures with amiprophos-methyl in various plant species." *Plant Sci.,* 56, 227–39.

29. Sterett, R. B., and T. A. Fretz. 1975. "Asulam-induced mitotic irregularities in onion root-tips." *Hort. Sci.,* 10, 161–62.

30. Watts, M. J., and H. A. Collin. 1979. "The effect of asulam on the growth of tissue cultures of celery." *Weed Res.,* 19, 33–37.

31. Mann, J. D., and W. B. Storey. 1966. "Rapid action of carbamate herbicides upon plant cell nuclei." *Cytologia,* 31, 203–07.

32. Burström, H. G. 1968. "Root growth activity of barban in relation to auxin and other growth factors." *Physiol. Plant.,* 21, 1137–55.

33. Sumida, S., and M. Ueda. 1976. "Effect of O-ethyl O-(3-methyl-6-nitrophenyl) N-*sec*-butylphosphorothioamidate (S-2846), an experimental herbicide, on mitosis in *Allium cepa.*" *Plant Cell Physiol.,* 17, 1351–54.

34. Badr, E. A. 1983. "Mitodepressive and chromotoxic activities of two herbicides in *Allium cepa.*" *Cytologia,* 49, 451–57.

35. Nasta, A., and E. Günther. 1973. "Mitoseanomalien bei *Allium cepa* und *Hordeum vulgare* nach Einwirkung eines Carbamatherbizids." *Biol. Zentralbl.,* 92, 27–36.

36. Davis, D. G., R. A. Hoerauf, K. E. Dusbabek, and D. K. Dougall. 1977. "Isopropyl *m*-chlorocarbanilate and its hydroxylated metabolites: their effects on cell suspensions and cell division in soybean and carrot." *Physiol. Plant.*, 40, 15–20.

37. Chang, C. T., and D. Smith. 1972. "Effect of DCPA on ultrastructure of foxtail millet cells." *Weed Sci.*, 20, 220–25.

38. Upadhyaya, M. K., and L. D. Nooden. 1977. "Mode of dinitroaniline herbicide action. I. Analysis of the colchicine-like effects of dinitroaniline herbicides." *Plant Cell Physiol.*, 18, 1319–30.

39. Gentner, W. A., and L. G. Burk. 1966. "Gross morphological and cytological effects of nitralin on corn roots." *Weed Sci.*, 16, 259–60.

40. Beuret, E. 1980. "Influence de la pendiméthaline sur les méristèmes radiculaires de *Allium cepa* L." *Weed Res.*, 20, 83–86.

41. Carlson, W. C., E. M. Lignowski, and H. J. Hopen. 1975. "The mode of action of pronamide." *Weed Sci.*, 23, 155–61.

42. Rost, T. L., and D. E. Bayer. 1976. "Cell cycle population kinetics of pea root tip meristems treated with propham." *Weed Sci.*, 24, 81–87.

43. Rost, T. L., and S. L. Morrison, 1984. "The comparative cell cycle and metabolic effects of chemical treatments on root tip meristems II. Propham, chlorpropham, and 2,4-dinitrophenol." *Cytologia*, 49, 61–72.

44. Doxey, D. 1949. "The effects of isopropyl phenyl carbamate on mitosis in rye (*Secale cereale*) and onion (*Allium cepa*)." *Ann. Bot.*, 13, 329–36.

45. Ennis, W. B. 1948. "Some cytological effects of O-isopropyl N-phenyl carbamate upon *Avena*." *Amer. J. Bot.*, 35, 15–21.

46. Holmsen, J. D., and F. D. Hess. 1985. "Comparison of the disruption of mitosis and cell plate formation in oat roots by DCPA, colchicine and propham." *J. Exp. Bot.*, 36, 1504–13.

47. Jackson, W. T. 1969. "Regulation of mitosis. II. Interaction of isopropyl *N*-phenylcarbamate and melatonin." *J. Cell. Sci.*, 5, 745–55.

48. Hepler, P. K., and W. T. Jackson. 1969. "Isopropyl *N*-phenylcarbamate affects spindle microtubule orientation in dividing endosperm cells of *Haemanthus katherinae* Baker." *J. Cell. Sci.*, 5, 727–43.

49. Amer, S. M., and O. R. Farah. 1974. "Cytological effects of pesticides. VII. Mitotic effects of isopropyl-*N*-phenyl carbamate and "Duphar." *Cytologia*, 40, 21–29.

50. Jackson, W. T., and D. A. Stetler. 1973. "Regulation of mitosis. IV. An *in vitro* and ultrastructural study of effects of trifluralin." *Can. J. Bot.*, 51, 1513–18.

51. Hess, F. D., and D. Bayer. 1974. "The effect of trifluralin on the ultrastructure of dividing cells of the root meristem of cotton (*Gossypium hirsutum* L. 'Acala 4-42')." *J. Cell. Sci.*, 15, 429–41.

52. Delcourt, A., and G. Deysson. 1976. "Effects of trifluralin on root meristems of *Allium sativum* L." *Cytologia*, 41, 75–84.

53. Lignowski, E. M., and E. G. Scott. 1972. "Effect of trifluralin on mitosis." *Weed Sci.*, 20, 267–270.

54. Young, L. W., and N. D. Camper. 1979. "Trifluralin effects on tobacco callus tissue: Mitosis and selected metabolic effects." *Pestic. Biochem. Physiol.*, 12, 117–23.

55. Draber, W., and C. Fedtke. 1979. "Herbicide interaction with plant biochemical systems." In *Adv. Pestic. Sci.,* Part 3., H. Geissbühler, ed., Oxford: Pergamon, pp. 475–86.
56. Shaybany, B., and J. L. Anderson. 1981. "Effect of chlorthal dimethyl on oat and foxtail seedlings anatomy." *Weed Res.,* 21, 164–68.
57. Upadhyaya, M. K., and L. D. Nooden. 1978. "Relationship between the induction of swelling and the inhibition of elongation caused by oryzalin and colchicine in corn roots." *Plant Cell Physiol.,* 19, 133–38.
58. Palevitz, B. R., and P. K. Hepler. 1976. "Cellulose microfibril orientation and cell shaping in developing guard cells of *Allium*. The role of microtubules and ion accumulation." *Planta,* 132, 71–93.
59. Sanger, J. M. 1971. "Fine structure study of pollen development in *Haemanthus katherinae* Baker. II. Microtubules and elongation of the generative cells." *J. Cell Sci.,* 8, 303.
60. Kust, C. A., and B. E. Struckmeyer. 1971. "Effects of trifluralin on growth, nodulation and anatomy of soybeans." *Weed Sci.,* 19, 147–52.
61. Lignowski, E. M., and E. G. Scott. 1971. "Trifluralin and root growth." *Plant Cell Physiol.,* 12, 701–08.
62. Bayer, D. E., C. L. Foy, T. E. Mallory, and E. G. Cutter. 1967. "Morphological and histological effects of trifluralin on root development." *Amer. J. Bot.,* 54, 945–52.
63. DiTomaso, J. M., T. L. Rost, and F. M. Ashton. 1988. "The comparative cell cycle and metabolic effects of the herbicide napropamide on root tip meristems." *Pestic. Biochem. Physiol.,* 31, 166–74.
64. Vaughn, K. C., and S. O. Duke. 1986. "Ultrastructural effects of glyphosate on *Glycine max* seedlings." *Pestic. Biochem. Physiol.,* 26, 56–65.
65. Luduena, R. F. 1979. "Biochemistry of tubulin." In *Microtubules,* K. Rober and J. S. Hyams, eds. London: Academic, pp. 65–116.
66. Margolis, R. L., and L. Wilson. 1977. "Addition of colchicine-tubulin complex to microtubule ends: the mechanism of substoichiometric colchicine poisoning." *Proc. Natl. Acad. Sci. USA,* 74, 3466–470.
67. Lambeir, A., and Y. Engelborghs. 1980. "A quantitative analysis of tubulin-colchicine binding to microtubules." *Eur. J. Biochem.,* 21, 619–24.
68. Andreu, J. M., and S. N. Timasheff. 1986. "Tubulin-colchicine interactions and polymerization of the complex." *Ann. N.Y. Acad. Sci.,* 466, 676–86.
69. Morejohn, L. C., T. E. Bureau, L. P. Tocchi, and D. E. Fosket. 1987. "Resistance of *Rosa* microtubule polymerization to colchicine results from a low-affinity interaction of colchicine and tubulin." *Planta,* 170, 230–41.
70. Hardham, A. R., and B. E. S. Gunning. 1980. "Some effects of colchicine on microtubules and cell division in roots of *Azolla pinnata*." *Protoplasma,* 102, 31–51.
71. Chrispeels, M. J. 1972. "Failure of colchicine or cytochalasin to inhibit protein secretion by plant cells." *Planta,* 108, 283–87.
72. Robinson, D. G., I. Grimm, and H. Sachs. 1976. "Colchicine and microfibril orientation." *Protoplasma,* 89, 375–80.
73. Vaughn, K. C., and M. A. Vaughan. 1988. "Mitotic disrupters from higher plants. Effects on plant cells." *Amer. Chem. Soc. Symp. Ser.,* 380, 273–93.

74. Luduena, R. F., W. H. Anderson, V. Prasad, M. A. Jordan, K. C., Ferrigni, M. C. Roach, P. M. Horowitz, D. B. Murphy, and A. Fellous. 1986. "Interactions of vinblastine and maytansine with tubulin." *Ann. N.Y. Acad. Sci.*, 466, 718–32.

75. Maranz, R., and M. Shelanski. 1970. "Structure of microtubular crystals induced by vincristine *in vitro*." *J. Cell Biol.*, 44, 234–38.

76. Hillman, G., and A. Ruthman. 1982. "Effect of mitotic inhibitors on the ultrastructure of root meristem cells." *Planta*, 155, 124–32.

77. Segawa, M., and K. Kondo. 1978. "Effects of vinblastine on meristematic cells of *Allium cepa*, I." *Experientia*, 34, 996–99.

78. Vaughn, K. C., and M. A. Vaughan. 1987. "Ultrastructural and cytological effects of vinblastine and vincristine on *Catharanthus roseus*." *Amer. J. Bot.*, 74, 627–28.

79. Okamura, S. 1980. "Binding of colchicine to a soluble fraction of carrot cells in suspension culture." *Planta*, 149, 350–54.

80. Vaughan, M. A., and K. C. Vaughn, 1988. "Mitotic disrupters from higher plants and their potential uses as herbicides." *Weed Technol.*, 2, 533–39.

81. Manfredi, J. J., and S. B. Horwitz. 1984. "Taxol: An antimitotic agent with a new mechanism of action." *Pharmac. Ther.*, 25, 83–125.

82. Schiff, P. B., T. Fant, and S. B. Horwitz. 1979. "Promotion of microtubule assembly *in vitro* by taxol." *Nature* (London), 277, 665–67.

83. Morejohn, L. C., and D. E. Fosket. 1984. "Taxol-induced rose microtubule polymerization *in vitro* and its inhibition by colchicine." *J. Cell Biol.*, 99, 141–47.

84. Horwitz, S. B., L. Lothstein, J. J. Manfredi, W. Mellado, J. Parness, S. N. Roy, P. B. Schiff, L. Sorbara, and R. Zeheb. 1986. "Taxol: Mechanisms of action and resistance." *Proc. N.Y. Acad. Sci.*, 466, 733–44.

85. Hacskaylo, V., and V. A. Amato. 1968. "Effect of trifluralin on roots of corn and cotton." *Weed Sci.*, 16, 513–15.

86. Parka, S. J., and O. F. Soper. 1977. "The physiology and mode of action of the dinitroaniline herbicides." *Weed Sci.*, 25, 79–87.

87. Probst, G. W., T. Golab, and W. L. Wright. 1975. "Dinitroanilines." In *Herbicides: Chemistry, Degradation and Mode of Action*, P. C. Kearney and D. D. Kaufman, eds. New York: Marcel Dekker, pp. 453–95.

88. Appleby, A. P., and B. E. Valverde. 1989. "Behavior of dinitroaniline herbicides in plants." *Weed Technol.*, 3, 198–206.

89. Bartels, P. G., and J. L. Hilton. 1973. "Comparison of trifluralin, oryzalin, pronamide, propham and colchicine treatments on microtubules." *Pestic. Biochem. Physiol.*, 3, 462–72.

90. Jackson, W. T., and D. A. Stetler. 1973. "Regulation of mitosis. IV. An *in vitro* and ultrastructural study of effects of trifluralin." *Can. J. Bot.*, 51, 1513–18.

91. Robinson, D. G., and W. Herzog. 1977. "Structure, synthesis and orientation of microfibrils. III. A survey of the action of microtubule inhibitors on microtubules and microfibril orientation in *Oocystis solitaria*." *Cytobiologie*, 15, 462–74.

92. Banerjee, S., J. K. Kelleher, and L. Margulis. 1975. "The herbicide trifluralin is active against microtubule-based oral morphogenesis in *Stentor coeruleus*." *Cytobios*, 12, 171–78.

93. Galway, M. E., and A. R. Harham. 1989. "Oryzalin-induced microtubule disassembly and recovery in regenerating of the alga *Mougeotia*." *J. Plant Physiol.*, 135, 337–45.

94. Cleary, A. L., and A. R. Hardham. 1988. "Depolymerization of microtubule arrays in root tip cells by oryzalin and their recovery with modified nucleation patterns." *Can. J. Bot.*, 66, 2353–66.

95. Wacker, I., H. Quader, and E. Schnepf. 1988. "Influence of the herbicide oryzalin on cytoskeleton and growth of *Funaria hygrometrica* protonemata." *Protoplasma*, 142, 55–67.

96. Menzel, D. 1988. "Perturbation of cytoskeletal assemblies in cyst domain morphogenesis in the green alga *Acetabularia*." *Eur. J. Cell Biol.*, 46, 217–26.

97. Strachan, S. D., and F. D. Hess. 1982. "Dinitroaniline herbicides adsorb to glass." *J. Agric. Food Chem.*, 30, 389–91.

98. Strachan, S. D., and F. D. Hess. 1983. "The biochemical mechanism of action of the dinitroaniline herbicide oryzalin." *Pestic. Biochem. Physiol.*, 20, 141–50.

99. Chan, M. M.-Y., and D. Fong. 1990. "Inhibition of leishmanias but not host macrophages by the antitubulin herbicide trifluralin." *Science*, 249, 924–26.

100. Morejohn, L. C., T. E. Bureau, J. Mole-Bajer, A. S. Bajer, and D. E. Fosket. 1987. "Oryzalin, a dinitroaniline herbicide, binds to plant tubulin and inhibits microtubule polymerization *in vitro*." *Planta*, 172, 252–64.

101. Shahied, S. I., and J. Giddens. 1970. "Effect of cysteine on the action of trifluralin on lateral roots of cotton and corn." *Agron. J.*, 62, 306–07.

102. Vaughan, M. A., and K. C. Vaughn. 1988. "Carrot microtubules are dinitroaniline resistant. I. Cytological and cross-resistance studies." *Weed Res.*, 28, 73–83.

103. Morrison, I. N., B. G. Todd, and K. M. Nawalsky. 1989. "Confirmation of trifluralin-resistant green foxtail (*Setaria viridis*) in Manitoba." *Weed Technol.*, 3, 544–51.

104. Murphy, T. R., B. J. Gossett, and J. E. Toler. 1986. "Growth and development of dinitroaniline-susceptible and -resistant goosegrass (*Eleusine indica*) biotypes under noncompetitive conditions." *Weed Sci.*, 34, 704–10.

105. Vaughn, K. C., M. A. Vaughan, and B. J. Gossett. 1990. "A biotype of goosegrass (*Eleusine indica*) with an intermediate level of dinitroaniline herbicide resistance." *Weed Technol.*, 4, 157–62.

106. Hussey, P. J., J. A. Traas, K. Gull, and C. W. Lloyd. 1987. "Isolation of cytoskeletons from synchronized plant cells: the interphase microtubule array utilizes multiple tubulin types." *J. Cell Sci.*, 88, 225–30.

107. Cabral, F. R., R. C. Brady, and M. J. Schibler. 1986. "A mechanism of cellular resistance to drugs that interfere with microtubule assembly." *Ann. N.Y. Acad. Sci.*, 466, 745–56.

108. Hertel, C., and D. Marmé. 1983. "Herbicides and fungicides inhibit Ca^{2+} uptake by plant mitochondria: A possible mechanism of action." *Pestic. Biochem. Physiol.*, 19, 282–90.

109. Hertel, C., H. Quader, D. G. Robinson, and D. Marmé. 1980. "Antimicrotubular herbicides and fungicides affect Ca^{2+} transport in plant mitochondria." *Planta*, 149, 336–40.

110. Bajer, A. S., and J. Molè-Bajer. 1986. "Drugs with colchicine-like effects that specifi-

cally disassemble plant but not animal microtubules." *Ann. N.Y. Acad. Sci.*, 466, 767–84.

111. Stearns, M. E., and D. L. Brown. 1981. "Microtubule organizing centers (MTOCs) of the alga *Polytomella* exert spatial control over microtubule initiation *in vivo* and *in vitro*." *J. Ultrastruct. Res.*, 77, 366–78.
112. Falconer, M. M., G. Donaldson, and R. W. Seagull. 1988. "MTOCs in higher plant cells. An immunofluorescent study of microtubule assembly sites following depolymerization by APM." *Protoplasma*, 144, 46–55.
113. Hogetsu, T. 1987. "Re-formation and ordering of wall microtubules in *Spirogyra* cells." *Plant Cell Physiol.*, 28, 875–83.
114. Falconer, M. M., and R. W. Seagull. 1987. "Amiprophos-methyl (APM): a rapid, reversible, anti-microtubule agent for plant cell cultures." *Protoplasma*, 136, 118–24.
115. Murata, T., and M. Wada. 1989. "Effects of colchicine and amiprophos-methyl on microfibril arrangement and cell shape in *Aniantum* protonemal cells." *Protoplasma*, 151, 81–87.
116. Collis, P. S., and D. P. Weeks. 1978. "Selective inhibition of tubulin synthesis by amiprophos-methyl during flagellar regeneration in *Chlamydomonas reinhardtii*." *Science*, 202, 440–42.
117. Kiermayer, O., and C. Fedtke. 1977. "Strong anti-microtubule action of amiprophos-methyl (APM) in *Micrasterias*." *Protoplasma*, 92, 163–66.
118. Meindl, U. 1983. "Cytoskeletal control of nuclear migration and anchoring in developing cells of *Micrasterias denticulata* and the change caused by the anti-microtubular herbicide amiprophos-methyl (APM)." *Protoplasma*, 118, 75–90.
119. Koop, H.-U., and O. Kiermayer. 1980. "Protoplasmic streaming in the giant unicellular green alga *Acetabularia mediterranea*. II. Differential sensitivity of movement systems to substances acting on microfilaments and microtubuli." *Protoplasma*, 102, 295–306.
120. Ramulu, K. S., H. A. Verhoeven, and P. Dijkhuis. 1988. "Mitotic dynamics of micronuclei induced by amiprophos-methyl and prospects form chromosome-mediated gene transfer in plants. *Theor. Appl. Genet.*, 75, 575–84.
121. De Laat, A. M. M., H. A. Verhoeven, K. S. Ramulu, and P. Dijkhuis. 1987. "Efficient induction by amiprophos-methyl and flow cytometric sorting of micronuclei in *Nicotiana plambaginifolia*." *Planta*, 1972, 473–478.
122. Iwata, K., and T. Hogetsu. 1989. "Orientation of wall microfibrils in *Avena* coleoptiles and mesocotyls and in *Pisum* epicotyls." *Plant Cell Physiol.*, 30, 749–57.
123. Mizuta, S., U. Kurogi, K. Okuda, and R. M. Brown. 1989. "Microfibrillar structure, cortical microtubule arrangement and the effect of amiprophos-methyl on microfibril orientation in the thallus cells of the filamentous green alga, *Chamaedoris orientalis*." *Ann. Bot.*, 64, 383–94.
124. Wilms, F. H. A., A. M. C. Wolters-Arts, and J. Derksen. 1990. "Orientation of cellulose microfibrils in cortical cells of tobacco explants." *Planta*, 182, 1–8.
125. James, S. W., L. P. W. Ranum, C. D. Silflow, and P. A. Lefebvre. 1988. "Mutants resistant to anti-microtubule herbicides map to a locus on the *uni* linkage group in *Chlamydomonas reinhardtii*." *Genetics*, 118, 141–47.

126. Doonan, J. H., D. J. Cove, and C. W. Lloyd. 1988. "Microtubules and microfilaments in tip growth: evidence that microtubules impose polarity on protonemal growth in *Physcomitrella patens*." *J. Cell Sci.*, 89, 533–40.

127. Tissut, M., F. Nurit, P. Ravanel, S. Mona, N. Benevides, and D. Macherel. 1986. "Herbicidal modes of action depending on substitution in a phenylcarbamate series." *Physiol. Vég.*, 24, 523–35.

128. Templeton, W. G., and W. A. Sexton. 1945. "Effect of some arylcarbamic esters and related compounds upon cereals and other plant species." *Nature*, 156, 630.

129. Brown, D. L., and G. B. Bouck. 1974. "Microtubule biogenesis and cell shape in *Ochromonas*. III. Effects of the herbicidal mitotic inhibitor isopropyl *N*-phenylcarbamate on shape and flagellum regeneration." *J. Cell Biol.*, 61, 514–36.

130. Marchant, H. J., and E. R. Hines. 1979. "The role of microtubules and cell-wall deposition in elongation of regenerating protoplasts of *Mougeotia*." *Planta*, 146, 41–48.

131. Coss, R. A., and J. D. Pickett-Heaps. 1974. "The effects of isopropyl *N*-phenyl carbamate on the green alga *Oedogonium cardiacum* I. Cell division." *J. Cell Biol.*, 63, 84–98.

132. Brower, D. L., and P. K. Hepler. 1976. "Microtubules and secondary wall deposition in xylem: the effects of isopropyl *N*-phenylcarbamate." *Protoplasma*, 87, 91–111.

133. Palevitz, B. A., and P. K. Hepler. 1974. "The control of the plane of division during stomatal differentiation in *Allium*. II. Drug studies." *Chromosoma*, 46, 327–41.

134. Clayton, L., and C. W. Lloyd. 1984. "The relationship between the division plate and spindle geometry in *Allium* cells treated with CIPC and griseofulvin: an anti-tubulin study." *Eur. J. Cell Biol.*, 34, 248–53.

135. Rao, B. V., B. G. S. Rao, and C. B. S. R. Sharma. 1988. "Cytological effects of herbicides and insecticides on *Allium cepa* root meristems." *Cytologia*, 53, 255–61.

136. Hidalgo, A., J. A. Gonzalez-Reyes, P. Navas, and G. Garcia-Herdugo. 1989. "Abnormal mitosis and growth inhibition in *Allium cepa* roots induced by propham and chloropropham." *Cytobios*, 57, 7–14.

137. Magistrini, M., and D. Szöllösi. 1980. "Effects of cold and of isopropyl-*N*-phenylcarbamate on the second meiotic spindle of mouse oocytes." *Eur. J. Cell Biol.*, 22, 699–707.

138. Coss, R. A., R. A., Bloodgood, D. L. Brower, J. D. Pickett-Heaps, and J. R. McIntosh. 1975. "Studies on the mechanism of action of isopropyl *N*-phenyl carbamate." *Exp. Cell Res.*, 92, 394–98.

139. Vannini, G. L., D. Mares, and G. Dall'Olio. 1982. "Structural alterations in *Euglena gracilis* exposed to the herbicide isopropyl-*N*-phenylcarbamate (IPC)." *Protoplasma*, 111, 189–94.

140. Mizuno, K., M. Loyama, and H. Shibaoka. 1981. "Isolation of plant tubulin from azuki bean epicotyls by ethyl *N*-phenylcarbamate-sepharose affinity chromatography." *J. Biochem.*, 89, 329–32.

141. Lehnen, L. P., M. A. Vaughan, and K. C. Vaughn. 1990. "Terbutol affects spindle microtubule organizing centres." *J. Exp. Bot.*, 41, 537–46.

142. Davidse, L. C., and W. Flach. 1977. "Differential binding of methyl benzimidazol-2-yl

carbamate to fungal tubulin as a mechanism of resistance to this antimitotic agent in mutant strains of *Aspergillus nidulans*." *J. Cell Biol.*, 72, 174–93.

143. Quinlan, R. A., A. Roobol, C. I. Pogson, and K. Gull. 1981. "A correlation between *in vivo* effects of the microtubule inhibitor, methyl benzimidazol-2-yl-carbamate (MBC) on nuclear division and the cell cycle in *Saccharomyces cerevisiae*." *J. Cell. Sci.*, 46, 341–52.
144. De Brabander, M., G. Geuens, R. Nuydens, R. Willebrords, M. Moeremans, R. Van Ginckel, W. Distelmans, C. Dragonetti, and M. Mareel. 1986. "Tubulozole: A new sterioselective microtubule inhibitor." *Ann. N.Y. Acad. Sci.*, 466, 757–66.
145. Hoebeke, J., G. Van Nijen, and M. De Brabander. 1976. "Interaction of nocodazole (R 17934), a new antitumoral drug, with rat brain tubulin." *Biochem. Biophys. Res. Commun.*, 69, 319–24.
146. Havercroft, J. C., R. A. Quinlan, and K. Gull. 1981. "Binding of parbendazole to tubulin and its influence on microtubules in tissue culture cells as revealed by immunofluorescence microscopy." *J. Cell Sci.*, 49, 195–204.
147. Dawson, P. J., W. E. Gutteridge, and K. Gull. 1984. "A comparison of the interaction of antihelminthic benzimidazoles with tubulin isolated from mammalian tissue and the parasitic nematode *Ascaridia galli*." *Biochem. Pharmacol.*, 33, 1069–74.
148. Richmond, D. V., and A. Phillips. 1975. "The effect of benomyl and carbendazim on mitosis in hyphae of *Botrytis cinera* Pers. ex. Fr. and roots of *Allium cepa* L." *Pestic. Biochem. Physiol.*, 5, 67–79.
149. Davidse, L. C. 1987. "Advances in understanding fungicidal modes of action and resistance." In *Pesticide Science and Biotechnology*, R. Greenhalgh and T. R. Roberts, eds. Oxford: Blackwell, pp. 169–76.
150. Künkel, W., and W. Römer. 1980. "Das cyclische Nucleotidsystem von *Aspergillus nidulans* unter dem Einfluss von Methylbenzimidazol-2-ylcarbamat (MBC)." *Z. Allg. Mikrobiol.*, 20, 195–207.
151. Bingham, S. W. 1968. "Effect of DCPA on anatomy and cytology of roots." *Weed Sci.*, 16, 449–52.
152. Holmsen, J. D., and F. D. Hess. 1984. "Growth inhibition and disruption of mitosis by DCPA in oat (*Avena sativa*) roots." *Weed Sci.*, 32, 732–38.
153. Chang, C. T., and D. Smith. 1972. "Effect of DCPA on ultrastructure of foxtail millet cells." *Weed Sci.*, 20, 220–25.
154. Vaughan, M. A., and K. C. Vaughn, 1990. DCPA causes cell plate disruption in wheat roots." *Ann. Bot.*, 65, 379–88.
155. Peterson, R. L., and L. W. Smith. 1971. "Effects of *N*-(1,1-dimethylpropynyl-3,5-dichlorobenzamide on the anatomy of *Agropyrons repens* (L.) Beauv." *Weed Res.*, 11, 84–87.
156. Vaughan, M. A., and K. C. Vaughn. 1987. "Pronamide disrupts mitosis in a unique manner." *Pestic. Biochem. Physiol.*, 28, 182–93.
157. Akashi, T., K. Izumi, E. Nagano, M. Enomoto, K. Mizuno, and H. Shibaoka. 1988. "Effects of propyzamide on tobacco cell microtubules in vivo and in vitro." *Plant Cell Physiol.*, 29, 1053–62.
158. Molin, W. T., B. L. Armbruster, C. A. Porter, and M. W. Bugg. 1988. "Inhibition of microtubule polymerization by MON 7200." *Weed Sci. Soc. Amer. Abstr.*, 28, 69.

159. Molin, W. T., T. C. Lee, and M. W. Bugg. 1988. "Purification of a protein which binds MON 7200." *Weed Sci. Soc. Amer. Abstr.*, 28, 69.
160. Wallin, M., J. Deinum, and B. Friden. 1985. "Interaction of estramustine phosphate with microtubule-associated proteins." *FEBS Lett.*, 179, 289–93.
161. Black, M. M., and K. Peng. 1986. "*In vivo* taxol treatment alters the solubility properties of microtubule-associated proteins (MAPs) of cultured neurons." *Ann. N.Y. Acad. Sci.*, 466, 426–28.

Chapter 11

Herbicide Effects on Lipid Synthesis

11.1 INTRODUCTION

Plant lipid synthesis as discussed in this chapter refers mainly to the pathways of fatty acid synthesis and elongation. Other lipid classes, for example, sterols, sterol esters, and the large family of terpenoids, are of relatively minor importance in the formation of plant membranes and wax layers, and are not considered here. Inhibition of terpenoid desaturation, which is similar to fatty acid desaturation (enzymes F and G in Scheme 11.1), is an important mechanism of chlorosis induction and is discussed in detail in Section 8.1.1.

Scheme 11.1 represents a current view of the pathway(s) of fatty acid synthesis in higher plants. This pathway has been studied mainly in green tissues or in chloroplasts. Quantitatively, the most important location for fatty acid synthesis in plant tissues is the active or developing chloroplast. However, there are other subcellular and enzymatic systems that are less important quantitatively, and are not as well characterized. Lipid and fatty acid compositions are different in chloroplast and cytoplasmic membranes: whereas MGDG (monogalactosyldiacylglyceride) represents the major chloroplast lipid (26–46%), PC (phosphatidylcholine) is the major membrane lipid in cytoplasmic membranes (7–24%). Up to 65 percent of the total leaf fatty acids in the different lipid fractions are of the C18 : 3 (linolenic acid) type; that is, they are highly unsaturated.

The herbicidal inhibition sites shown in Scheme 11.1 represent several classes of herbicides; these have predominantly, or exclusively, graminicidal ac-

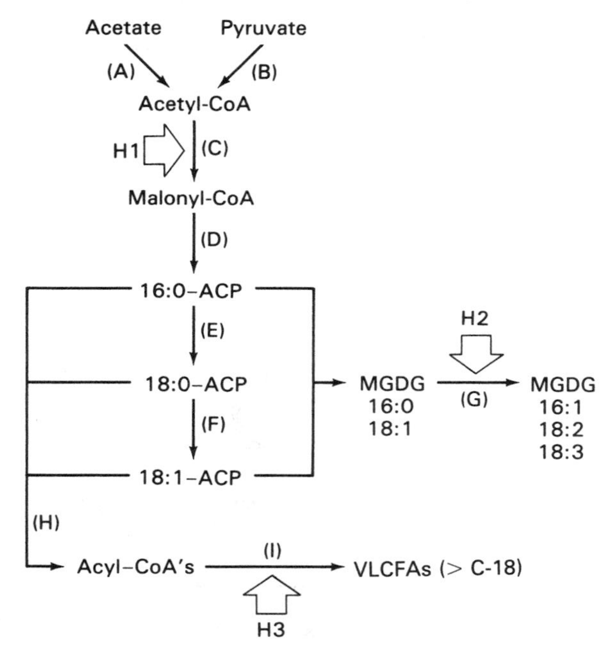

H1: aryl-propanoic acids and similar structures; cyclohexanediones

H2: pyridazinones

H3: thiocarbamates, halogenated acids (?)

Scheme 11.1 Major synthesis routes of saturated, unsaturated, and very long chain length fatty acids (VLCFAs) in leaves [1]. H1, H2, and H3 denote herbicidal inhibition sites. The enzymes are (A) acetate thiokinase, (B) pyruvate dehydrogenase, (C) acetyl-CoA carboxylase, (D) type II fatty acid synthetase complex, (E) palmitoyl-ACP elongase, (F) and (G) different desaturases, (H) an acyl-CoA transfer system (plastid→cytoplasm), (I) type III fatty acid synthetase complex and specific elongases. ACP = acyl carrier protein; MGDG = monogalactosyldiacylglyceride.

tion [2, 3]. The structures of a selection of important compounds from each group are shown in Figure 11.1. The structural group of "aryl-propanoic acids" and their esters has also been named phenoxypropionic acids, phenoxy-phenoxypropionic acids, polycyclic alkanoic acids, aryloxy-phenoxypropionic acids, and so on, and their respective esters. They are also often called "fops," as contrasted by the "dims," the second group of herbicides with the same target enzyme (Figure 11.1, Section 11.2). These and similar names refer to the structures of most of the herbicides in this group, but do not comprise all of their structural aspects. The term used here, aryl-propanoic acids, is suggested as a compromise. In Figure 11.1 only the structures of the herbicidally active free acids are shown. The commercial herbicides are different esters of these acids (e.g., methyl-, ethyl-, isopropyl-,

Common name	Structure
Aryl-propanoic acids	
Diclofop	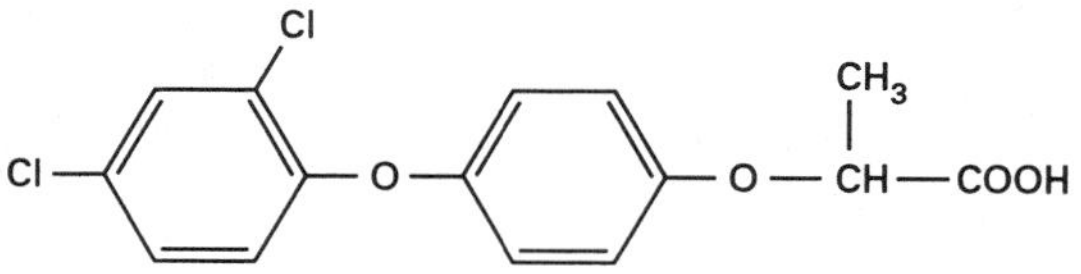
Fenoxaprop	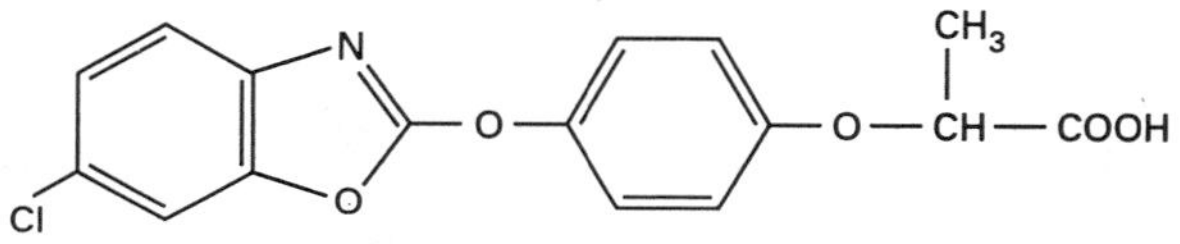
Fenthiaprop	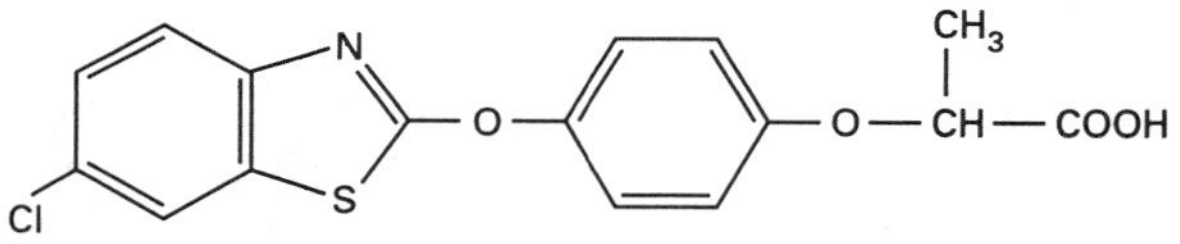
Fluazifop	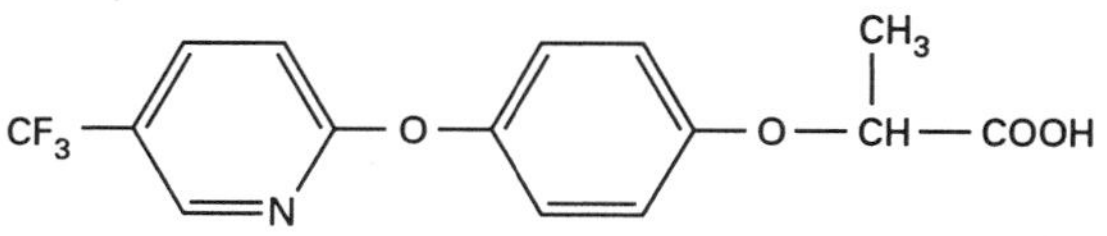
Haloxyfop	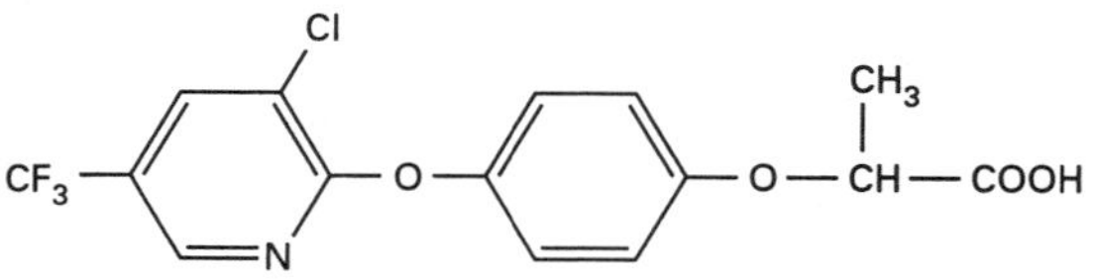
Cyclohexanediones	
Alloxydim	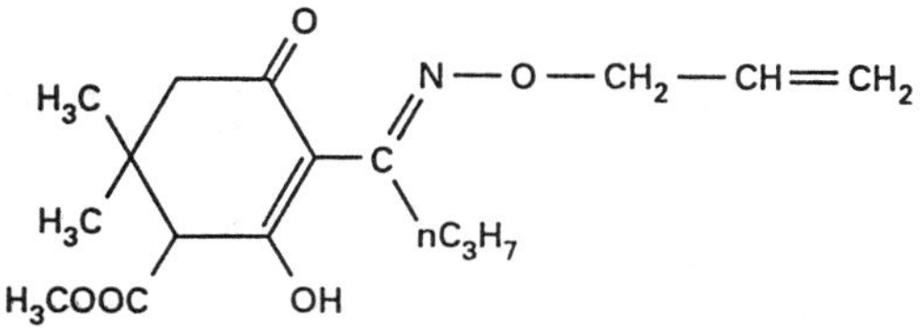

Figure 11.1 Herbicidal structures that have been reported to inhibit the synthesis of lipids.

Common name	Structure
Clethodim	H_3C; $CH-CH_2$; H_5C_2-S; O; OH; $N-O-CH_2-CH=CHCl$; C; C_2H_5
Sethoxydim	H_3C; $CH-CH_2$; H_5C_2-S; O; OH; $N-O-C_2H_5$; C; nC_3H_7
Thiocarbamates	
CDEC	$N(C_2H_5)_2$; $S=C$; $S-CH_2-C=CH_2$; Cl
Diallate	$N[CH(CH_3)_2]_2$; $O=C$; $S-CH_2-C=CH$; Cl Cl
EPTC	$N(CH_2-CH_2-CH_3)_2$; $O=C$; $S-C_2H_5$
Triallate	$N[CH(CH_3)_2]_2$; $O=C$; $S-CH_2-C=C-Cl$; Cl Cl

Figure 11.1 (Continued).

Common name	Structure
Chloroacetamides	
Alachlor	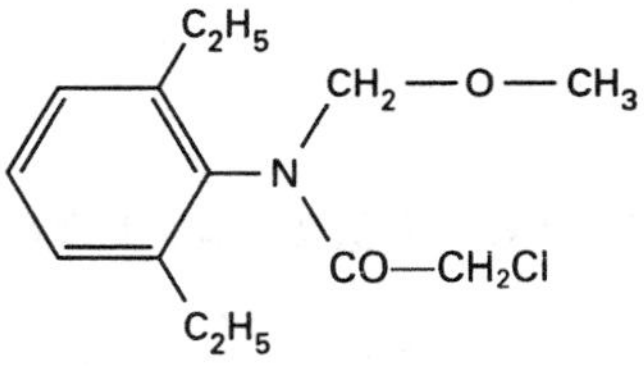 C_2H_5; $CH_2—O—CH_3$; N; $CO—CH_2Cl$; C_2H_5
Metolachlor	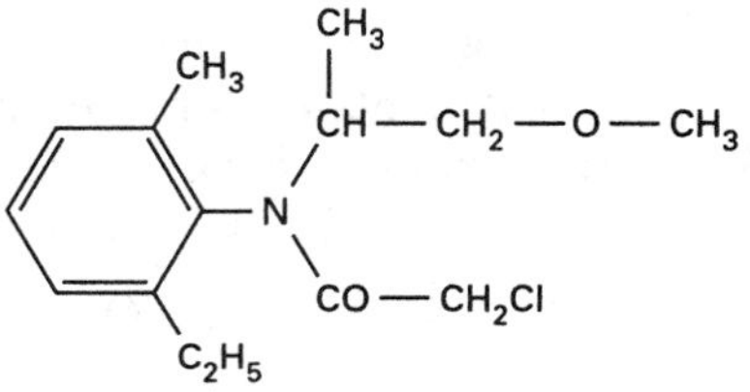 CH_3; CH_3; $CH—CH_2—O—CH_3$; N; $CO—CH_2Cl$; C_2H_5
Miscellaneous	
Ethofumesate	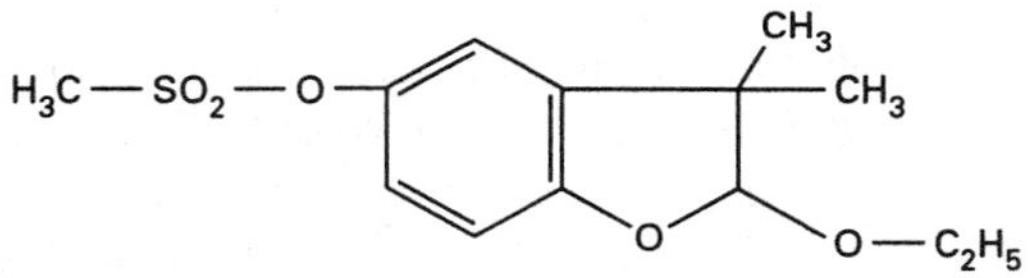 $H_3C—SO_2—O$; CH_3; CH_3; O; $O—C_2H_5$
Dalapon	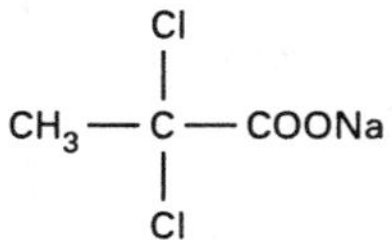 Cl; $CH_3—C—COONa$; Cl
TCA	$Cl_3C—COONa$

Figure 11.1 (Continued).

ethoxyethyl-esters) that are used to facilitate absorption into plant tissue. A non-herbicidal inhibitor of the type II fatty acid synthetase complex (enzyme D, Scheme 11.1) is the microbial metabolite cerulenin [(2S),(3R)-2,3-epoxy-4-oxo-7,10-dodeca-dienoyl-amide]. This compound induces chlorotic effects and growth inhibition in intact plants and algae at comparatively high doses [4, 5]. The antibiotic thiolactomycin [(4S)-(2E,5E)-2,4,6-trimethyl-3-OH-2,5,7-octatriene-4-thiolide] is also a specific inhibitor of the fatty acid synthetase complex [6]. Another interference site, also mentioned above, is collectively represented as "different desaturases" in Scheme 11.1 (F and G). The herbicides that interfere with "different desaturases" (H2 in Scheme 11.1) are listed and discussed in Section 8.1.1; the herbicides that form groups H1 and H3 are discussed in this chapter.

11.2 INHIBITION OF ACETYL-COA CARBOXYLASE

Acetyl-CoA carboxylase (ACCase) is the site of inhibition of the aryl-propanoic acids and the cyclohexanedione herbicides (Figure 11.1). ACCase is a complex enzyme that contains three functional sites: a biotin carboxyl carrier site, an ATP-dependent biotin carboxylase, and a carboxyltransferase (acetyl-CoA transcarboxylase) [1]. The enzymatic reaction is as follows:

$$\text{Acetyl-CoA} + HCO_3^- + \text{ATP} \longrightarrow \text{Malonyl-CoA} + \text{ADP} + P_i$$

In plants, ACCase is formed by one multifunctional polypeptide chain. The enzyme is located primarily in the chloroplasts, and its activity is strongly enhanced by light. Because the apparent inhibition constant in the presence of cyclohexanedione herbicides is very sensitive to the level of acetyl-CoA, it has been suggested that these herbicides inhibit at the acetyl-CoA → malonyl-CoA transferase site [7]. In a computer modeling study of a series of 3-acyl tetrahydro-2*H*-pyran-2,4-diones, which act similarly to the cyclohexanediones and which have a very similar molecular volume, it was found that the molecular geometry of the inhibitor molecule is similar to that of the anticipated acetyl-CoA-carboxybiotin transition state [8]. More specifically, the proposed transient six-membered ring in the transition state model

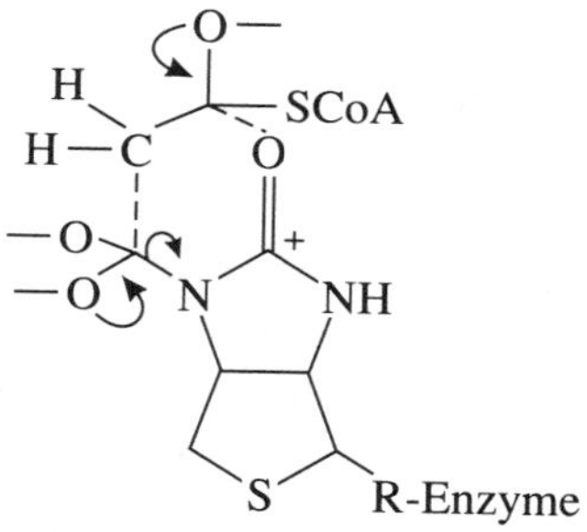

superimposes directly on the six-membered ring of the herbicide molecule. The two carbonyl groups of the cyclohexanedione molecule are equivalent to the carboxyl group being transferred and the carbonyl group in acetyl-CoA. These herbicides would therefore be transition state inhibitors, similar, perhaps, to glyphosate (Section 13.2).

The first definitive evidence that aryl-propanoic acid herbicides are potent lipid synthesis inhibitors was the inhibition of ^{14}C-acetate incorporation into the

TABLE 11.1 INHIBITION OF ^{14}C-ACETATE INCORPORATION INTO THE FREE FATTY ACIDS (FFAs) OF ISOLATED CORN CHLOROPLASTS BY DIFFERENT ARYL-PROPANOIC ACIDS AND THEIR RESPECTIVE ESTERS [9]. Lipids were extracted after 1 h of treatment.

	% Inhibition of incorporation	
	Herbicide concentration	
Inhibitor	0.1 μM	1 μM
Diclofop-methyl	8	24
Diclofop	45	83
Fenthiaprop-ethyl	48	92
Fenthiaprop	92	96
Fenoxaprop-ethyl	8	43
Fenoxaprop	60	89
D-Diclofop	54	92
L-Diclofop	18	35

FFAs (free fatty acids) in corn leaves and in isolated corn chloroplasts by several of these compounds (Table 11.1). It is clear from these and other data that the free acids are the herbicidally active structures (Tables 11.1, 11.2). Moreover, all the known aryl-propanoic acid herbicides contain an optically active C-atom close to the carboxyl group. Of the two possible stereoisomers, only the R-enantiomer is inhibitory, while the S-enantiomer has very little or no activity as a herbicide [11–15], or as an ACCase inhibitor [16, 17]. In early studies, the incorporation of

TABLE 11.2 INHIBITION OF ^{14}C-ACETATE INCORPORATION INTO THE FREE FATTY ACIDS (FFAs) IN ISOLATED OAT CHLOROPLASTS BY DIFFERENT ARYL-PROPANOIC ACIDS AND THEIR RESPECTIVE ESTERS [10].

	I_{50}-concentration (μM)	
Inhibitor	Acid	Ester
Diclofop	0.1	10
5-OH-diclofop	2	—
Fenoxaprop	0.1	10
Fluazifop	3	100
Haloxyfop	0.3	50

^{14}C-propionate into FFAs was also found to be sensitive to these herbicides; this was obviously due to decarboxylation of propionate to acetate, followed by its incorporation into fatty acids, in untreated tissue. The incorporation of malonyl-CoA into FFAs is not sensitive to these herbicides [10].

In addition to the aryl-propanoic acids and their herbicidal esters, the cyclohexanediones are also very potent inhibitors of ACCase (Table 11.3) [18, 19]. Both groups of herbicides are selective graminicides; this selectivity is also found in ^{14}C-acetate incorporation experiments at the chloroplast level [9, 20, 21] and in tests of ACCase from species of different sensitivities (Table 11.3). However, although ACCase from dicotyledonous plants is insensitive to these herbicides, and that from graminaceous plants usually is sensitive [17, 18], the ACCase from different grasses is not always equally sensitive. ACCase from corn is more sensitive than that from wheat or barley (Table 11.3). ACCase from sethoxydim-resistant *Festuca rubra* plants has been found to be insensitive to sethoxydim (I_{50} >1 mM), while ACCase from sensitive *Festuca arundinacea* plants is sensitive (I_{50} 6.9 μM) [22]. The enzyme from both *Festuca* species is sensitive to haloxyfop, although not to the same extent (I_{50} 118 and 5.8 μM for ACCase from *F. rubra* and *F. arundinacea*, respectively). This difference in enzyme sensitivity is an unusual and interesting mechanism of herbicide selectivity, fundamentally different from the more common mechanism of tolerance by herbicide detoxification (Section 6.1). Sethoxydim is very rapidly transformed into about nine metabolites by photo-and thermal transformations. Within 24 h, 98 percent of this herbicide is degraded in tolerant as well as in sensitive species [23, 24]. Nevertheless, it is clear that enough untransformed herbicide molecules reach the chloroplast to inhibit ACCase in sensitive species.

In addition, not all tissues of one plant species are equally sensitive to cyclohexanediones. Sethoxydim rapidly inhibits ^{14}C-acetate incorporation into lipids in corn root tips (0–2 mm), whereas no inhibition is found in the more proliferative root regions (10–15 mm) [25]. It can be concluded, therefore, that the target ACCase appears to be present in rapidly dividing cells and in active chloro-

TABLE 11.3 INHIBITION CONSTANTS (±SD) FOR THE INHIBITION OF THE CHLOROPLAST ACCase (ACETYL-COA CARBOXYLASE) FROM DIFFERENT PLANT SOURCES BY CYCLOHEXANEDIONES [7].

	Plant source				
	Barley	Wheat	Corn	Spinach	Mung bean
		Inhibition constant			
Inhibitor	I_{50} (μM)	K_i			
Clethodim	0.12 ± 0.02	0.14 ± 0.03	0.02 ± 0.003	1240 ± 280	53 ± 15
Sethoxydim	0.94 ± 0.14	0.96 ± 0.19	0.47 ± 0.11	2160 ± 500	1880 ± 620
Alloxydim	4.88 ± 0.44	1.95 ± 0.34	0.88 ± 0.17	—	—

plasts. Accordingly, visible symptoms of herbicidal activity are most rapidly and most strongly observed in meristematic regions and, on an ultrastructural level, in the chloroplast [26, 27]. Secondary effects of herbicide action at the biochemical level include chlorosis (inhibition of chlorophyll and carotenoid biosynthesis) [21, 28], altered fatty acid chain length distribution in the chloroplast (decrease of C-18 acids and increase of ≤ C-16 acids) [2, 17], and inhibition of DNA synthesis and mitosis at the cellular level [29]. The changed C-16 : C-18 ratio can be interpreted as a consequence of the altered acetyl-CoA : malonyl-CoA ratio after ACCase inhibition. Rapid ultrastructural damage at the chloroplast level presumably follows membrane perturbation after the loss of necessary structural elements. Loss of membrane function, and membrane semipermeability in particular, leads to an "intracellular mix" where catabolic enzymes and their substrates, which are separated into different compartments in the cell, now come into contact. In particular, γ-aminobutyric acid, the decarboxylation product of glutamic acid, accumulates in damaged tissue and can be taken as a biochemical indicator of the degree of damage [30, 31]. In corn roots, meristematic growth is arrested after 4 h of treatment with 0.1 μM sethoxydim [29]. Similar growth inhibition has been observed in alloxydim-treated oat leaf meristems, leading to cell enlargement, nuclei aggregation, and necrosis [32].

An independent confirmation of ACCase as the site of action of these herbicides comes from resistant corn tissue cultures that were selected on 10 μM sethoxydim [33]. The cells exhibit sevenfold resistance to sethoxydim and twofold resistance to haloxyfop. An increased rate of fatty acid synthesis (fivefold), which can be traced back to increased ACCase activity (normally rate limiting), has been found in the herbicide-resistant cells. This increased target enzyme expression/activity could well explain the observed resistance, in accordance with similar target enzyme increases that have been found for glyphosate and glufosinate (Chapter 13).

11.3 AUXIN INTERACTIONS OF THE ARYL-PROPANOIC ACIDS

The aryl-propanoic acids are known for another aspect of their physiological effects on plant tissues, one that does not appear to be related to their inhibition of ACCase: they have been repeatedly described as antagonists of auxin activities and auxin effects. The availability of a free carboxyl group in the active herbicide molecule gives additional support to this type of interference [34–36]. Similar physiological effects and interferences have not been described for the cyclohexanedione herbicides and seem, therefore, to be specific to the aryl-propanoic acids. The reported effects include: (a) antagonism of diclofop-methyl and flamprop-methyl with many (but not all) auxin herbicides in the field [37–43] and in some physiological systems [36] (the herbicidal/inhibitory activity of the aryl-propanoic acid compound is reduced in the presence of the auxin), (b) reverse antagonism where diclofop-methyl antagonizes the action of the auxin herbicide

2,4-D [36], and (c) antagonism of the aryl-propanoic acid herbicide chlorfenprop-methyl with indolylacetic acid (IAA) and naphthyl-1-acetic acid in several auxin-dependent systems, with only the herbicidal L(-)-enantiomer being active [44]. In other studies, however, no evidence of antagonism at the coleoptile level has been found [45].

The antagonism between these two classes of herbicides is probably due to an increased rate of aryl-propanoic herbicide detoxification by conjugation reactions [37–39]. Interestingly, most of the herbicidally active auxins induce increased herbicide conjugation (Section 14.3), but exceptions exist both ways in that active auxins may not be inducers (e.g., fluroxypyr) and auxin-inactive compounds can be strong inducers (e.g., triiodobenzoic acid = TIBA). The reason for the antagonism may therefore lie in the similarity between auxin and aryl-propanoic acid molecules: auxin-induced auxin conjugation systems may have a sufficiently low substrate specificity to enable them to detoxify the aryl-propanoic acid herbicides rapidly. The antagonism between these two groups of herbicides is dealt with in detail in Section 17.2.

The mechanism of antagonism described above may not be valid in all instances. In rapid tests, for example with coleoptiles or membrane systems, much higher concentrations (10–100 μM) are commonly applied [46, 47]. Membrane permeability and interference effects observed in these conditions may not be relevant for the antagonism observed in the field or with other whole-plant systems.

11.4 INHIBITION OF ELONGATION OF VERY LONG CHAIN FATTY ACIDS

The thiocarbamates, the halogenated acids (dalapon and TCA), and the herbicide ethofumesate (Figure 11.1) have all been shown to inhibit the synthesis of very long chain fatty acids (VLCFAs) (Scheme 11.1) [48–52]. At field-related, low (micromolar) concentrations, a decrease in the chain lengths of epicuticular waxes is a major effect of these herbicides. Both EPTC and ethofumesate cause a reduction in total leaf wax deposition on *Brassica oleracea* leaves (Table 11.4). The n-nonocosane

TABLE 11.4 DECREASE OF EPICUTICULAR WAXES ON THE SURFACE OF LEAVES OF *Brassica oleracea* AFTER TREATMENT WITH EPTC OR ETHOFUMESATE [50].

Treatment	Rate (kg/ha)	Leaf wax (μg/cm^2)	n-Nonocosane (%)	n-Nonocosane-15-one (%)	Long chain wax esters (%)
Control	—	153.5	100	100	100
EPTC	0.84	82.4	21.8	37.4	95.8
EPTC	3.36	70.4	9.6	18.2	122.1
Ethofumesate	0.84	63.1	4.7	12.3	156
Ethofumesate	2.24	75.1	2.4	7.2	173.1

(C-29) and n-nonocosane-15-one fractions decrease significantly, whereas the long-chain wax esters are unchanged or increase, presumably because of increased use of the accumulating shorter chain-length FFAs (C-16 to C-22). In scanning electron micrographs of the leaf surfaces, decreased amounts of epicuticular waxes (or no epicuticular wax at all) were observed in these experiments. In other instances the total amount of wax deposited on the leaf surface remained unchanged, but the structure of the wax was drastically altered. Under these conditions, increased transpiration of water from the leaf surface can be responsible for increased uptake of xylem-mobile herbicides (e.g., atrazine or phenmedipham; see Section 17.2) from the soil [48, 53, 54].

At higher herbicide concentrations (ca. 100 μM) the chain lengths of the FFAs become increasingly shorter (<C-20), and other reactions, such as fatty acid desaturation, are inhibited (possibly indirectly) [52]. However, many of the experimental plant materials or tissues commonly used for these experiments are not as sensitive as the target plants in the field, and might therefore require higher concentrations for inhibition. Different *Brassica* sp. leaves and aged potato discs may not be representative of the germinating grass species that are the major targets for these herbicides in the field. However, some important findings and leads for further research have been found in these and other tissues. For example, the inhibition of incorporation of ^{14}C-stearic acid into certain surface lipids of pea leaves by different thiocarbamate herbicides is shown in Table 11.5; similar results were obtained when ^{14}C-acetate was supplied as a precursor. Here again, fatty acid chain elongation enzymes are inhibited in the presence of the thiocarbamate herbicides. Within the different leaf surface lipid fractions, VLCFAs (C-26 to C-31) decrease, whereas the C-22 chain-length fatty acids increase. This shift in chain lengths can also be clearly seen in Table 11.6, taken from an experiment with corn leaves.

The biosynthesis of VLCFAs has not been studied in much detail [1]. A microsomal enzyme system is known to synthesize VLCFAs with malonyl-CoA as the C-2 donor. Specific elongases seem to be present for the different elongation

TABLE 11.5 EFFECT OF DIFFERENT THIOCARBAMATES ON THE INCORPORATION OF ^{14}C-STEARIC ACID INTO THE SURFACE LIPID COMPONENTS OF PEA LEAVES [55].

		Incorporation % of control			
Herbicide	Conc. (μM)	Hydrocarbon	Secondary alcohol	Wax ester	Primary alcohol
Diallate	3	72	91	94	156
	12	25	41	95	70
EPTC	3	57	74	96	100
	12	21	32	99	71
CDEC	3	43	42	120	131
	12	18	30	101	125

TABLE 11.6 VLCFA (VERY LONG CHAIN FATTY ACID) COMPOSITION IN THE EPICUTICULAR LEAF WAX OBTAINED FROM CONTROL AND 10^{-4} M EPTC-TREATED CORN LEAVES [56].

Treatment	Wax fraction	Carbon chain length distribution (%)								
		24	25	26	27	28	29	30	31	32
Control	Ester +	8.2	*	4.4	*	1.9	*	3.0	2.5	80.0
EPTC	aldehyde	16.6	*	7.6	*	4.6	*	9.6	2.0	59.6
Control	Hydro-	8.6	5.2	3.2	14.4	2.4	18.1	2.1	46.0	*
EPTC	carbon	14.1	8.9	6.4	13.7	4.9	30.1	4.3	17.7	*

* = less than 0.3%.

steps leading to C-20, C-22, and C-24 acids in potato. However, the mechanism of herbicide action here is not known, nor is it clear whether the inhibition is caused directly or indirectly. In addition, it is not clear how, or whether at all, the inhibition of VLCFA synthesis is related to herbicidal action. In view of these unanswered questions, many suggestions have been advanced for the possible explanation of various aspects of the mechanism of action of thiocarbamates. One intriguing finding is the inhibition of kaurene and gibberellic acid synthesis at the oxygenation step by both thiocarbamate and chloroacetamide herbicides [57–59] (see Section 15.4). These herbicide groups have other features in common, such as detoxification by glutathione conjugation or safening in corn and some other crops by the same safeners (Section 17.4). This may point to similarities with respect to the site of action, but a detailed understanding of this is still lacking. Alkylation of coenzyme A and of other metabolic intermediates containing sulfhydryl groups has been suggested as one possible mechanism of action [60]. Another similarity that has been pointed out is the inhibition of monooxygenases (mixed function oxidase type enzymes) by these different structures (Section 17.4). More information is obviously needed for a better understanding of the herbicidal action of these compounds.

The situation is particularly complex in the case of the thiocarbamate herbicides; some of the major factors regulating the activity of EPTC are illustrated in Scheme 11.2. EPTC, like other thiocarbamates, is known to be oxygenated to the sulfoxide and further to the sulfone. EPTC-sulfoxide is somewhat phytotoxic, particularly to dicotyledonous species, as is EPTC-sulfone, in this case to corn. However, both oxygenation products are very rapidly detoxified in plant tissue, one by glutathione conjugation and the other by a carbamoylation reaction. The phytotoxic potential of these metabolites is therefore not sufficient to explain the herbicidal action, which appears to be caused by the thiocarbamate parent molecule. Glutathione antagonizes the herbicidal activity because of more rapid detoxification, whereas a monooxygenase inhibitor (e.g., PBO) increases the herbicidal activity through inhibition of the detoxification pathway. Further metabolism of

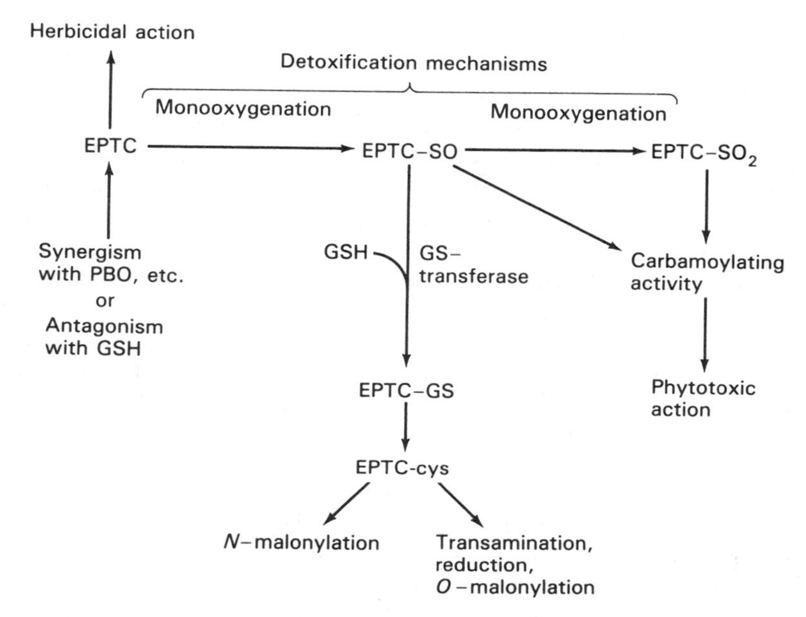

Scheme 11.2 Metabolic conversion of thiocarbamate herbicides in plants, exemplified by the behavior of EPTC. SO = sulfoxide, SO_2 = sulfone, GSH = glutathione, cys = cysteine, PBO = piperonylbutoxide [61, 62].

the EPTC-glutathione conjugate follows the pattern of metabolism known for a number of glutathione conjugates [63] (Section 6.3).

11.5 OTHER MECHANISMS OF LIPID SYNTHESIS INHIBITION

The microbial inhibitor of 3-hydroxymethylglutaryl-CoA (HMG-CoA) reductase, mevinolin (1,2,6,7,8,8a-hexahydro-β-δ-dihydroxy-2,6-dimethyl-8-(2-methyl-1-oxobutoxy)-1-naphthalene-heptanoic acid], does not have sufficient phytotoxic potential to be used as a herbicide. However, mevinolin is a very efficient inhibitor of isoprenoid synthesis, in which HMG-CoA reductase is an early, feedback-regulated enzyme in the pathway leading to mevalonic acid [64–66]. This inhibition therefore affects the synthesis of all terpenoid compounds (see Section 8.1), including sterols.

The α-chloroacetamides, and the α-chloroacetanilides in particular (Figure 11.1), constitute another group of herbicides that have been shown to affect lipid synthesis in some studies. The chloroacetamides have already been mentioned in Section 11.4 with reference to some similarities with thiocarbamate herbicides. These similarities have been stressed in a recent summary [60]. Unlike the thiocarbamates, chloroacetamides are not known to inhibit VLCFA synthesis. However, there have been some reports of the inhibition of lipid synthesis from ^{14}C-acetate as a precursor [67, 68]. This inhibition has been observed in the 10–100 μM concentration range, and is therefore of doubtful value for herbicidal mechanism of action considerations. In addition, the inhibition of lipid synthesis by chloroacetamides has not been confirmed or reproduced in other studies [69]. One interesting similarity between thiocarbamates and chloroacetamides, also mentioned above, is the inhibition of oxygenation steps in the synthesis of kaurene and gibberellic acid [58] (compare with Section 15.4). The inhibition of anthocyanin and lignin biosynthesis by different chloroacetanilides at 20 μM might also involve oxygenation steps [70, 71] (see also Sections 15.5 and 16.3). However, thiocarbamates do not have a similar effect on the synthesis of these phenolic compounds. Thus, there are some interesting similarities between the mechanisms of action of thiocarbamates and chloroacetamides/chloroacetanilides, but the basis of these similarities is not yet understood. It might be that the interference with the same safener structures (Section 17.4) holds a key to the understanding of the modes of action of these herbicides *and* their safeners at the same time.

Besides their doubtful effects on lipid synthesis, chloroacetamides have been recognized as inhibitors of cell division and cell growth [72]. In this respect, their physiological mechanism of action is definitely different from that of the thiocarbamates. In a study with the unicellular green alga *Chlamydomonas reinhardtii* and several other diagnostic test systems, chloroacetamides and thiocarbamates have also been found to act differently on a physiological basis [73]. Algal growth is sensitive only to the chloroacetamides, and other test systems also show differences between the two groups of herbicides. The herbicidal mode of action of chloroacetamides seems, therefore, to lie in (an) enzymatic reaction(s) that are of fundamental importance in plant cell growth and division. In contrast, the enzymatic site(s) responsible for the herbicidal action of the thiocarbamates seem to be specific to higher plants, in particular to the monocotyledonous family. VLCFA synthesis comprises one pathway which, if inhibited, could lead to depletion of compounds with particular importance in higher plants, but other pathways may also be inhibited, primarily and secondarily, in the course of the herbicidal action of the thiocarbamates.

REFERENCES

1. Harwood, J. L. 1988. "Fatty acid metabolism." *Annu. Rev. Plant Physiol. Plant Mol. Biol.,* 39, 101–38.
2. Harwood, J. L., K. A. Walker, and D. Abulnaja. 1987. "Herbicides affecting lipid metabolism." In Proc. Brit. Crop Prot. Conf.—Weeds, Vol. 1. Thornton Heath, U.K.: BCPC Publ., pp. 159–69.
3. Harwood, J. L. 1988. "The site of action of some graminaceous herbicides is identified as acetyl-CoA carboxylase." *TIBS,* 13, 330–31.
4. Lehoczki, E., and T. Farker. 1984. "Lipid composition of cerulenin-treated *Chlorella pyrenoidosa* in relation to herbicide resistance." *Plant Sci. Lett.,* 36, 125–30.

5. Harwood, J. L., and N. J. Russell. 1984. *Lipids in Plants and Microbes*. London: Allen & Unwin, pp. 75–83.
6. Nishida, I., A. Kawaguchi, and M. Yamada. 1984. "Selective inhibition of type II fatty acid synthetase by the antibiotic thiolactomycin." *Plant Cell Physiol.*, 25, 265–68.
7. Rendina, A. R., and J. M. Felts. 1988. "Cyclohexanedione herbicides are selective and potent inhibitors of acetyl-CoA carboxylase from grasses." *Plant Physiol.*, 86, 983–86.
8. Winkler, D. A., A. J. Liepa, J. E. Anderson-McKay, and N. K. Hart. 1989. "A molecular graphics study of factors influencing herbicidal activity of 3-acyltetrahydro-2*H*-pyran-2,4-diones." *Pestic. Sci.*, 27, 45–63.
9. Hoppe, H. H., and H. Zacher. 1985. "Inhibition of fatty acid biosynthesis in isolated bean and maize chloroplasts by herbicidal phenoxyphenoxypropionic acid derivatives and structurally related compounds." *Pestic. Biochem. Physiol.*, 24, 298–305.
10. Kobek, K., M. Focke, and H. K. Lichtenthaler. 1987. "Fatty acid biosynthesis and acetyl-CoA carboxylase is a target of diclofop, fenoxaprop and other aryloxy-phenoxy-propionic acid herbicides." *Z. Naturforsch.*, 43c, 47–54.
11. Uchiyama, M., N. Washio, T. Ikai, H. Igarashi, and K. Suzuki. 1986. "Stereospecific responses to (R)-(+)- and (S)-(−)-quizalofop-ethyl in tissues of several plants." *J. Pestic. Sci.*, 11, 459–67.
12. Wink, O., and U. Luley. 1988. "Enantioselective transformation of the herbicides diclofop-methyl and fenoxaprop-ethyl in soil." *Pestic. Sci.*, 22, 31–40.
13. Nestler, J. H., and H. Bieringer. 1980. "Synthese und herbizide Wirksamkeit der D- und L-Enantiomeren des 2-[4-(2,4-dichlorphenoxy)phenoxy]propionsäure-methylesters." *Z. Naturforsch.*, 34, 366–71.
14. Schmidt, T., C. Fedtke, and R. R. Schmidt. 1976. "Synthesis and herbicidal activity of d(+)- and l(−)-[methyl-2-chloro-3-(4-chlorophenyl)-propionate.]" *Z. Naturforsch.*, 31c, 252–54.
15. Scott, R. M., A. J. Sampson, and D. Jordan. 1976. "WL-43425—a chiral form of flamprop-isopropyl." In Proc. Brit. Crop Prot. Conf—Weeds, Vol. 2. Nottingham, U.K. BCPC: The Boots Comp., pp. 723–30.
16. Secor, J., and C. Czeke. 1988. "Inhibition of acetyl-CoA carboxylase activity by haloxyfop and tralkoxydim." *Plant Physiol.*, 86, 10–12.
17. Walker, K. A., S. M. Ridley, T. Lewis, and J. L. Harwood. 1988. "Fluazifop, a grass selective herbicide which inhibits acetyl-CoA carboxylase in sensitive plant species." *Biochem. J.*, 254, 307–10.
18. Burton, J. D., J. W. Gronwald, D. A. Somers, J. A. Connelly, B. G. Gengenbach, and D. L. Wyse. 1987. "Inhibition of acetyl-CoA carboxylase by the herbicides sethoxydim and haloxyfop." *Biochem. Biophys. Res. Comm.*, 148, 1039–44.
19. Focke, M., and H. K. Lichtenthaler. 1987. "Inhibition of acetyl-CoA carboxylase of barley chloroplasts by cycloxydim and sethoxydim." *Z. Naturforsch.*, 42c, 1361–63.
20. Kobek, K., M. Focke, H. K. Lichtenthaler, G. Retzlaff, and B. Würzer. 1988. "Inhibition of fatty acid biosynthesis in isolated chloroplasts by cycloxydim and other cyclohexane-1,3-diones." *Physiol. Plant.*, 72, 492–98.
21. Lichtenthaler, H. K., and K. Kobek. 1987. "Inhibition by sethoxydim of pigment accumulation and fatty acid biosynthesis in chloroplasts of *Avena* seedlings." *Z. Naturforsch.*, 42c, 1275–79.
22. Stoltenberg, D. E., J. W. Gronwald, D. L. Wyse, J. D. Burton, D. A. Somers, and

B. G. Gengenback. 1989. "Effect of sethoxydim and haloxyfop on acetyl-CoA carboxylase activity in *Festuca* species." *Weed Sci.*, 37, 512–16.

23. Campbell, J. R., and D. Penner. 1985. "Sethoxydim metabolism in monocotyledonous and dicotyledonous plants." *Weed Sci.*, 33, 771–73.
24. Campbell, J. R., and D. Penner. 1987. "Retention, absorption, translocation, and distribution of sethoxydim in monocotyledonous and dicotyledonous plants." *Weed Res.*, 27, 179–86.
25. Hosaka, H., and M. Takagi. 1987. "Biochemical effects of sethoxydim in excised root tips of corn (*Zea mays*)". *Weed Sci.*, 35, 612–18.
26. Brezeanu, A. G., D. G. Davis, and R. H. Shimabukuro. 1976. "Ultrastructural effects and translocation of methyl 2-[4-(2,4-dichloro-phenoxy)phenoxy]propanoate in wheat (*Triticum aestivum*) and wild oat (*Avena fatua*)." *Can. J. Bot.*, 54, 2038–48.
27. Chandrasena, J.P.N.R., and G. R. Sagar. 1987. "Effect of fluazifop-butyl on the chlorophyll content, fluorescence and chloroplast ultrastructure of *Elymus repens* (L.) Gould. leaves." *Weed Res.*, 27, 103–12.
28. Asare-Boamah, N. K., and R. A. Fletcher. 1983. "Physiological and cytological effects of BAS 9052 OH on corn (*Zea mays*) seedlings." *Weed Sci.*, 31, 49–55.
29. Hosaka, H., and M. Tagaki. 1987. "Physiological responses to sethoxydim in tissues of corn (*Zea mays*) and pea (*Pisum sativum*)." *Weed Sci.*, 35, 604–11.
30. Wallace, W., J. Secor, and L. E. Schrader. 1984. "Rapid accumulation of aminobutyric acid and alanine in soybean leaves in response to an abrupt transfer to lower temperature, darkness, or mechanical manipulation." *Plant Physiol.*, 75, 170–75.
31. Fedtke, C. 1972. "Mechanism of action of the selective herbicide chlorfenprop-methyl." *Weed Res.*, 12, 325–36.
32. Ishikawa, H., S. Okunuki, T. Kawana, and Y. Hirino. 1980. "Histological investigation of the herbicidal effects of alloxydim-sodium in oat." *J. Pestic. Sci.*, 5, 547–51.
33. Parker, W. B., D. A. Somers, D. L. Wyse, J. W. Gronwald, B. G. Gengenbach, and J. D. Burton. 1988. "Selection and characterization of corn cell lines tolerant to sethoxydim." Abstr., Weed Sci. Soc. Amer. Meet., 28, 64.
34. Fletcher, R. A., and D. M. Drexler. 1980. "Interactions of diclofop-methyl and 2,4-D in cultivated oats (*Avena sativa*)." *Weed Sci.*, 28, 363–66.
35. Olson, W. A., and J. D. Nalewaja. 1981. "Antagonistic effects of MCPA on wild oat (*Avena fatua*) control with diclofop." *Weed Sci.*, 29, 566–71.
36. Shimabukuro, R. H., W. C. Walsh, and R. A. Hoerauf. 1986. "Reciprocal antagonism between the herbicides, diclofop-methyl and 2,4-D, in corn and soybean tissue culture." *Plant Physiol.*, 80, 612–17.
37. Taylor, H. F., and M. P. C. Loader. 1984. "Research on the control of wild oats and broad-leaved weeds by herbicide mixtures." *Outl. Agric.*, 13, 58–68.
38. Taylor, H. F., and M. P. C. Loader. 1984. "Metabolism of diclofop-methyl with reference to its interaction with other compounds." *Pestic. Sci.*, 15, 527–28.
39. Taylor, H. F., and M. P. C. Loader, and S. J. Norris. 1983. "Compatible and antagonistic mixtures of diclofop-methyl and flamprop-methyl with herbicides used to control broad-leaved weeds." *Weed Res.*, 24, 185–90.
40. Qureshi, F. A., and W. H. Vanden Born. 1979. "Interaction of diclofop-methyl and MCPA on wild oats." *Weed Sci.*, 27, 202–05.

41. O'Sullivan, P. A., and W. H. Vanden Born. 1980. "Interaction between benzoylprop-ethyl, flamprop-methyl or flamprop-isopropyl and herbicides used for broadleaved weed control." *Weed Res.*, 20, 53–57.

42. Todd, B. G., and E. H. Stobbe. 1980. "The basis for the antagonistic effect of 2,4-D on diclofop-methyl toxicity to wild oat (*Avena fatua*)." *Weed Sci.*, 28, 371–77.

43. Hill, B. D., B. G. Todd, and E. H. Stobbe. 1980. "Effect of 2,4-D on the hydrolysis of diclofop-methyl in wild oat." *Weed Sci.*, 28, 725–29.

44. Andrev, G. K., and N. Amrhein. 1976. "Mechanism of action of the herbicide 2-chloro-3-(4-chlorophenyl)propionate and its methyl ester: interaction with cell responses mediated by auxin." *Physiol. Plant.*, 37, 175–82.

45. Olson, W. A., J. D. Nalewaja, G. L. Schroeder, and M. E. Duysen. 1981. "Diclofop and MCPA influence on coleoptile growth." *Weed Sci.*, 29, 597–600.

46. Wright, J. P., and R. H. Shimabukuro. 1987. "Effects of diclofop and diclofop-methyl on the membrane potentials of wheat and oat coleoptiles." *Plant Physiol.*, 85, 188–93.

47. Fitzsimons, P. J., and P. R. Miller. 1987. "Auxin induced H^+-efflux: herbicide activity and antagonism." Proc. Brit. Crop Prot. Conf.—Weeds, Vol. 1. Thornton Heath, U.K.: BCPC Publ., pp. 179–86.

48. Siebert, R., and H. Köcher. 1972. "Retention und Penetration von Phenmedipham in Abhängigkeit von der Zeit und der zusätzlichen Verwendung einiger anderer Herbizide." *Z. Pfl.Krankh. Pfl.Schutz*, 79, 463–70.

49. Templier, J., C. Largeau, and E. Casadevall. 1987. "Effect of various inhibitors on biosynthesis of non-isoprenoid hydrocarbons in *Botryococcus braunii*." *Phytochemistry*, 26, 377–83.

50. Leavitt, J. R., D. N. Duncan, D. Penner, and W. F. Meggitt. 1978. "Inhibition of epicuticular wax deposition on cabbage by ethofumesate." *Plant Physiol.*, 61, 1034–36.

51. Barta, I., T. Kömives, and F. Dutka. 1983. "Radioanalytical study of lipid biosynthesis in EPTC-treated corn plants." *Radiochem. Radioanalyt. Lett.*, 58, 357–62.

52. Bolton, P., and J. L. Harwood. 1976. "Effect of thiocarbamate herbicides on fatty acid synthesis by potato." *Phytochemistry*, 15, 1507–09.

53. Rivera, C. M., and D. Penner. 1979. "Effect of herbicides on plant cell membrane lipids." *Residue Rev.*, 70, 45–76.

54. Leavitt, J. R. C., and D. Penner. 1979. "Prevention of EPTC-induced epicuticular wax aggregation on corn (*Zea mays*) with R-25788." *Weed Sci.*, 27, 47–50.

55. Kolattukudy, P. E., and L. Brown. 1974. "Inhibition of cuticular lipid biosynthesis in *Pisum sativum* by thiocarbamates." *Plant Physiol.*, 53, 903–06.

56. Barta, I. C., T. Kömives, and F. Dutka. 1985. "Effects of EPTC and its antidotes on epicuticular wax of corn." In Fat Science, Proc. 16th ISF Congr., J. Hollo, ed. Budapest, Hungary: Akademiai Kiado, pp. 277–83.

57. Wilkinson, R. E., and D. Ashley. 1979. "EPTC induced modification of gibberellin biosynthesis." *Weed Sci.*, 27, 270–74.

58. Wilkinson, R. E. 1981. "Metolachlor [2-chloro-*N*-(2-ethyl-6-methylphenyl)-*N*-(2-methoxy-1-methylethyl)acetamide] inhibition of gibberellin precursor biosynthesis." *Pestic. Biochem. Physiol.*, 16, 199–205.

59. Wilkinson, R. E. 1982. "Alachlor influence on sorghum growth and gibberellin precursor synthesis." *Pestic. Biochem. Physiol.*, 17, 177–84.

60. Fuerst, E. P. 1987. "Understanding the mode of action of the chloroacetamide and thiocarbamate herbicides." *Weed Technol.,* 1, 270–77.

61. Adams, D. O., and R. Y. Lee. 1983. "Uptake and metabolism of S-ethyl-*N,N*-dipropylthiocarbamate (EPTC) and EPTC-sulfoxide by suspension cells of *Zea mays* (Black Mexican Sweet)." *Plant Physiol. Suppl.,* 72, 174–76.

62. Dutka, F., and T. Kömives. 1983. "On the mode of action of EPTC and its antidotes." In *Human Welfare and the Environment,* Vol. 3, IUPAC Pesticide Chemistry, J. Mijamomto, P. C. Kearny, eds. Oxford: Pergamon, pp. 213–18.

63. Lamoureux, G. L., and D. G. Rusness. 1987. "EPTC metabolism in corn, cotton, and soybean: identification of a novel metabolite derived from the metabolism of a glutathione conjugate." *J. Agric. Food Chem.,* 35, 1–7.

64. Bach, T. J. 1985. "Selected natural and synthetic enzyme inhibitors of sterol biosynthesis as molecular probes for in vitro studies concerning the regulation of plant growth." *Plant Sci.,* 39, 183–87.

65. Ceccarelli, N., and R. Lorenzi. 1984. "Growth inhibition by competitive inhibitors of 3-hydroxymethylglutarylcoenzyme A reductase in *Helianthus tuberosus* tissue explants." *Plant Sci. Lett.,* 34, 269–76.

66. Schindler, S., T. J. Bach, and H. K. Lichtenthaler. 1985. "Differential inhibition by mevinolin of prenyllipid accumulation in radish seedlings." *Z. Naturforsch.,* 40c, 208–14.

67. Chang, S.-S., F. M. Ashton, and D. E. Bayer. 1985. "Butachlor influence on selected metabolic processes of plant cells and tissues." *J. Plant Growth Regul.,* 4, 1–9.

68. Weisshaar, H., and P. Böger. 1987. "Primary effects of chloroacetamides." *Pestic. Biochem. Physiol.,* 28, 286–93.

69. Mellis, J. M., P. Pillai, D. E. Davis, and B. Truelove. 1982. "Metolachlor and alachlor effects on membrane permeability and lipid synthesis." *Weed Sci.,* 30, 399–404.

70. Molin, W. T., E. J. Anderson, and C. A. Porter. 1986. "Effects of alachlor on anthocyanin and lignin synthesis in etiolated sorghum (*Sorghum bicolor* (L.) Moench) mesocotyls." *Pestic. Biochem. Physiol.,* 25, 105–11.

71. Molin, W. T., C. A. Porter, J. P. Chupp, and K. Naylor. 1990. "Differential inhibition of anthocyanin synthesis in etiolated sorghum (*Sorghum bicolor* (L.) Moench) mesocotyls by rotameric 2-halo-*N*-methyl-*N*-phenylacetamides." *Pestic. Biochem. Physiol.,* 36, 277–80.

72. Deal, L. M., and F. D. Hess. 1980. "An analysis of the growth inhibitory characteristics of alachlor and metolachlor." *Weed Sci.,* 28, 168–75.

73. Fedtke, C. 1987. "Physiological activity spectra of existing graminicides and the new herbicide 2-(2-benzothiazolyl-oxy)-*N*-phenylacetamide (mefenacet)." *Weed Res.,* 27, 221–28.

Chapter 12

Nucleic Acid and Protein Synthesis Inhibitors

12.1 INTRODUCTION

No commercially available herbicides are known to affect nucleic acid or protein synthesis directly, although reports of apparent indirect effects of herbicides on these processes are common (e.g., [1–7]). Furthermore, no definitive proof of a significant direct effect has been produced in any of the many reports of effects of commercial herbicides on these fundamental processes. Most studies of the effects of herbicides on nucleic acid or protein synthesis have only measured effects on protein or nucleic acid content or in vivo incorporation of radiolabeled precursors into these pools. Generally, no attempt is made to separate organelle from nuclear nucleic acid or protein synthesis, or to assay effects on in vitro synthetic systems or specific enzymes. Furthermore, effects of the herbicides on uptake of radiolabeled precursors are often not taken into account. In experiments of short duration, effects on uptake of radiolabeled amino acids can often account for most or all of the effect of the herbicide on protein synthesis (e.g., [8]).

In the few cases in which in vitro effects have been searched for to explain in vivo effects, no significant effect has been found (e.g., [9]). All herbicides will eventually affect these processes (see Chapter 16), so an effect can always be measured at some point after exposure to a herbicide. As expected, herbicides that disrupt amino acid metabolism (see Chapter 13) will also affect protein synthesis. However, the effect is usually no greater than the effects of herbicides with entirely different mechanisms of action (e.g., [10]). Acetolactate synthase inhibitors sometimes have rapid effects on nucleic acid synthesis [11], but no direct effect has been discovered.

12.2 EXAMPLES OF INDIRECT EFFECTS OR UNKNOWN MECHANISMS

The mechanisms of secondary effects are often difficult to deduce, based on knowledge of the primary effect. In studies in which the relative effectiveness of a diverse group of herbicides on protein and RNA synthesis has been determined, there are no obvious relationships with the primary mechanisms of action (Table 12.1). Generally, protein synthesis is inhibited more than RNA synthesis; however, in a few cases, RNA synthesis is affected more than protein synthesis. Any compound that inhibits respiration (e.g., dinoseb) will obviously have a rapid and profound effect on these energy-dependent processes. Other than this obvious cause/effect relationship, the mechanisms of inhibition of these processes by other herbicides are generally unclear. Furthermore, reported effects of a herbicide on

TABLE 12.1 INHIBITION OF RNA AND PROTEIN SYNTHESIS IN SOYBEAN HYPOCOTYLS (AFTER 6 H) AND RNA SYNTHESIS IN CORN MESOCOTYLS (AFTER 8 H) TREATED WITH 0.6 mM CONCENTRATIONS (EXCEPT WHERE ANNOTATED 0.2 mM) OF VARIOUS HERBICIDES [2].

Herbicide	Soybean RNA	Soybean Protein	Corn RNA
		(% inhibition)	
Dinoseb	80	98	91
Ioxynil (0.2 mM)	78	97	79
Propanil	64	90	78
Chlorproham	72	89	81
Pyriclor	56	88	77
2,4,5-T	44	67	41
Diuron	62	42	37
Fenac	13	70	25
Karsil	27	46	19
EPTC	42	24	2
Propachlor	14	44	41
Dichlobenil	25	33	21
CDEC (0.2 mM)	12	33	13
Atrazine (0.2 mM)	27	14	−48
Dicamba	14	24	−1
Trifluralin (0.2 mM)	4	21	2
Picloram	11	−1	−27
MH	2	0	43

protein or nucleic acid synthesis can vary considerably, depending on the concentration used, plant species, plant tissue or organ assayed, and time frame and environmental conditions of the assay.

For example, culture age of *Sorghum nigrum* cells greatly influences the effect of amitrole and vernolate on incorporation of ^{14}C-leucine into protein [10]. In this study, herbicides were generally stronger inhibitors than nonherbicides and, among herbicides, photosynthetic inhibitors were generally, but not always, weaker inhibitors than other herbicides. With this system, thiocarbamates were found to be especially active protein synthesis inhibitors. However, among thiocarbamates, there was a poor correlation between the herbicidal efficacy and the effect on protein synthesis.

Early work with chlorsulfuron, an inhibitor of branched-chain amino acid (leucine, valine, and isoleucine) synthesis (Chapter 13), demonstrated rapid (< 1 h) effects on incorporation of nucleotides into DNA [11]. Protein and RNA syntheses are affected to a much lesser degree. Later work demonstrated that this herbicide has no effect on DNA synthesis in isolated plant nuclei, or on DNA polymerase or thymidine kinase [12], and that its effects cannot be reversed by supplemental nucleotide precursors. Imidazolinones, another class of branched chain amino acid synthesis inhibitors, also strongly inhibit DNA synthesis [13, 14]; however, in treatments in which symptoms develop more slowly, DNA synthesis is inhibited only after growth has stopped [15]. Supplemental branched-chain amino acids can reverse the effect of chlorsulfuron [16] and imidazolinones [14] on DNA synthesis. Isoleucine-deficient Chinese hamster ovary (CHO) cells grown in isoleucine-deficient media are arrested in the G_1 stage of cell division [17], as are pea root tip cells treated with chlorsulfuron [18]. Isoleucine may be important in initiation of DNA synthesis in plants, as has been suggested in the case of CHO cells [17].

Difenzoquat (Figure 12.1) inhibits ^{14}C-thymidine incorporation into apical meristems of wheat shoots by 50 percent within 15 min [19]. This effect is not the result of inhibited thymidine uptake, and it precedes inhibited cell division and growth. Although difenzoquat can act like paraquat [20, 21], this is apparently not its primary mechanism of action. Higher doses of difenzoquat are required for the paraquat-like effect than are required for inhibition of mitosis and, for the paraquat-like effect, a difenzoquat-tolerant wheat variety is more susceptible than a difenzoquat-susceptible variety [19, 21]. Unlike paraquat, the effects of difenzoquat on

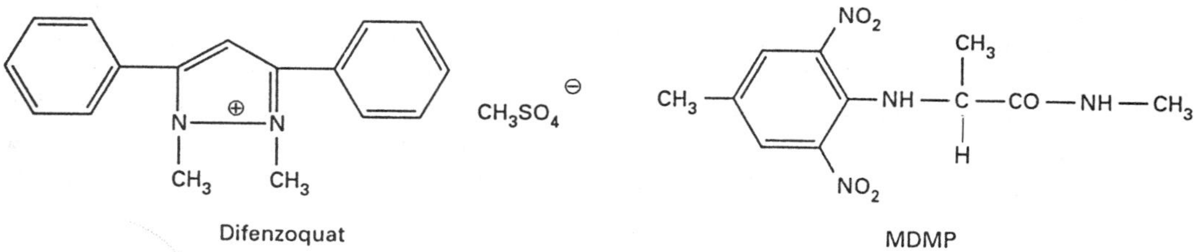

Difenzoquat: N—N (⊕), CH$_3$, CH$_3$; $CH_3SO_4^{\ominus}$

MDMP: NO$_2$, CH$_3$, CH$_3$—, —NH—C—CO—NH—CH$_3$, H, NO$_2$

Figure 12.1 Structures of difenzoquat and MDMP.

target weed species are quite slow, requiring a week or more for plant mortality. Difenzoquat can interfere directly with plasmalemma function [22]; however, it is not clear whether this is its primary site of action. No in vitro studies of its effects on DNA synthesis have been conducted.

Herbicides that inhibit the synthesis of carotenoids eventually cause loss of plastid ribosomes by photobleaching, resulting in cessation of plastid protein synthesis [23]. This effect has been useful in determining whether certain plastid proteins are nuclear- or plastid-coded [24], although this method is complicated by the fact that plastid products can affect expression of nuclear genes encoding proteins destined for the plastid [25, 26]. Expression of genes coding for extraplastidic enzymes with functions related to intact plastids (e.g., nitrate reductase and peroxisomal enzymes) appears to be most affected by photobleaching of plastids.

Any herbicide that inhibits cell division without affecting DNA synthesis can cause increases in the amount of DNA per cell as the cell becomes polyploid. For example, DCPA (chlorthal in some publications, see Chapter 10) causes the nuclear DNA content of meristematic cells of dodder (*Cuscuta lupuliformis*) to increase fourfold [27]. Other herbicides dramatically increase protein and nucleic acid content of plant tissues. For instance, naproanilide and 2,4-D cause tenfold increases in RNA synthesis in roots of *Cyperus difformis* L. [28]. Fenoxaprop (Chapter 11) causes a twofold increase in the DNA content of corn shoots [29]. Herbicides with auxin activity (2,4-D, 2,4,5-T, MCPA, and picloram) can have strong effects on the distribution of polyploidy in cells of treated plants [30]. Extensive literature exists on phenoxyalkanoic acid herbicide-induced increases in proteins and nucleic acids in plants (summarized in [31]). Aminotriazole has been shown to stimulate RNA synthesis by 50 percent in corn mesocotyls, but to have no effect on protein synthesis [2]. No direct effects of any of these herbicides on these processes have been found and the importance of these effects in the mechanism of action of these herbicides is not clear.

Only one specific effect of a herbicide on mitochondrial protein synthesis has been reported: aminotriazole (see Chapters 8 and 13) has been shown to inhibit mitochondrial protein synthesis in *Neurospora crassa* [32].

12.3 COMPOUNDS THAT DIRECTLY AFFECT PROTEIN OR NUCLEIC ACID SYNTHESIS

MDMP [2-(4-methyl-2,6-dinitroanilino)-*N*-methyl propionamide] (Figure 12.1), an unmarketed herbicide, directly affects protein synthesis in higher plants by interference with 60 S ribosome subunit interaction with the 40 S ribosomal subunit-RNA-Met-tRNA complex [33], resulting in the inhibition of initiation of peptide synthesis, but not of peptide elongation. The compound has no effect on organellar protein synthesis [34].

Dinoseb, ioxynil, and pyrichlor, three structurally unrelated herbicides, inhibit in vitro RNA polymerase activity by about 20 percent at concentrations of

1.5, 0.6, and 1.5 mM, respectively [2]. Effects at such high concentrations are unlikely to have any significance in the mode of action of these compounds.

Several fungal toxins are known to have specific effects on chloroplast nucleic acid synthesis or protein synthesis (see Chapter 18). Since these processes are similar in plastids and prokaryotic pathogens, these phytotoxins are also often good antibiotics. Tagetitoxin, a bacterial (*Pseudomonas syringae* pv. *tagetis*) phytotoxin, inhibits RNA synthesis in plastids by inhibiting plastid RNA polymerase [35]; however, it also inhibits eukaryote RNA polymerase III, the polymerase that transcribes genes for 5S rRNA, tRNAs, 7SK and 7SL RNA, U6 snRNA, and other small RNAs [36]. Fungal toxins such as albicidin [37], erythromcyin [38], actinomycin [39], and mitomycin [40], which inhibit prokaryote DNA or protein synthesis, often also inhibit the same processes in plastids, causing arrested chloroplast development and bleaching. The bacterial DNA gyrase inhibitors nalidixic acid and novobiocin inhibit chloroplast DNA synthesis [40]. These compounds could form the basis for structure-activity endeavors to develop herbicides with similar mechanisms of action.

12.4 SUMMARY

Direct effects of commercial herbicides on protein or nucleic acid synthesis have not yet been discovered, probably because neither of these sites are primary sites of action of any commercial herbicides. Nuclear-directed protein and nucleic acid synthesis are universal processes of eucaryotic organisms and are sufficiently evolutionarily conserved that they are unlikely to be satisfactory herbicide target sites because of potential toxicological problems. These objections may not be true for the same processes in the plastid. In fact, microbial toxins with rather specific effects on nucleic acid synthesis of plastids are known and have potential for exploitation as herbicides.

REFERENCES

1. Mann, J. D., L. S. Jordan, and B. E. Day. 1965. "A survey of herbicides for their effect upon protein synthesis." *Plant Physiol.,* 40, 840–43.
2. Moreland, D. E., S. S. Malhotra, R. D. Gruenhagen, and E. H. Shokraii. 1969. "Effects of herbicides on RNA and protein synthesis." *Weed Sci.,* 17, 556–63.
3. Ashton, F. M., O. T. de Villiers, R. K. Glenn, and W. B. Duke. 1977. "Localization of metabolic sites of action of herbicides." *Pestic. Biochem. Physiol.,* 7, 122–41.
4. Yohshida, Y., K. Nakamura, and A. Hiura. 1983. "Contraction of chromosomes and depression of RNA synthesis by isopropyl-N-(3-chlorophenyl) carbamate (CIPC) in *Vicia faba* root tip cells." *Cytologia,* 48, 707–17.
5. Zama, P., and K. K. Hatzios. 1987. "Interactions between the herbicide metolachlor and the safener CGA-92194 at the levels of uptake and macromolecular synthesis in sorghum leaf protoplasts." *Pestic. Biochem. Physiol.,* 27, 86–96.

6. DiTomaso, J. M., T. L. Rost, and F. M. Ashton. 1988. "The comparative cell cycle and metabolic effects of the herbicide napropamide on root tip meristems." *Pestic. Biochem. Physiol.*, 31, 166–74.
7. Bellinder, R. R., K. K. Hatzios, and H. P. Wilson. 1985. "Mode of action investigations with the herbicides HOE-39866 and SC-0224." *Weed Sci.*, 33, 779–85.
8. Camper, N. D., and K. L. Ellers. 1989. "Protein synthesis in trifluralin-treated corn root tips." *J. Plant Growth Regul.*, 8, 205–14.
9. Ochoa-Alejo, N., and O. J. Crocomo. 1987. "Influence to ametryn on chromatin activity and on RNA synthesis in a non-chlorophyllaceous sugarcane cell suspension." *J. Plant Physiol.*, 126, 355–63.
10. Egli, M. A., D. Low, K. R. White, and J. A. Howard. 1985. "Effects of herbicides and herbicide analogs on [^{14}C]leucine incorporation by suspension-cultured *Solanum nigrum* cells." *Pestic. Biochem. Physiol.*, 24, 112–18.
11. Ray, T. B. 1982. "The mode of action of chlorsulfuron: a new herbicide for cereals." *Pestic. Biochem. Physiol.*, 17, 10–17.
12. Ray, T. B. 1982. "The mode of action of chlorsulfuron: the lack of direct inhibition of plant DNA synthesis." *Pestic. Biochem. Physiol.*, 18, 262–66.
13. Shaner, D. L., and M. L. Reider. 1986. "Physiological responses of corn (*Zea mays*) to AC 243,997 in combination with valine, leucine, and isoleucine." *Pestic. Biochem. Physiol.*, 25, 248–57.
14. Pillmoor, J. B., and J. C. Caseley. 1987. "The biochemical and physiological effects and mode of action of AC 222, 293 against *Alopecurus myosuroides* Huds. and *Avena fatua* L." *Pestic. Biochem. Physiol.*, 27, 340–49.
15. Shaner, D. L., M. Stidham, M. Muhitch, M. L. Reider, P. Robson, and P. Anderson. 1985. "Mode of action of the imidazolinones." Brit. Crop Protect. Conf. 1985, 147–54.
16. Rost, T. L., and T. Reynolds. 1985. "Reversal of chlorsulfuron-induced inhibition of mitotic entry by isoleucine and valine." *Plant Physiol.*, 77, 481–82.
17. Toby, R. A. 1973. "Production and characterization of mammalian cells reversibly arrested in G1 by growth in isoleucine-deficient medium." *Methods Cell Biol.*, 6, 67–112.
18. Rost, T. L. 1984. "The comparative cell cycle and metabolic effects of chemical treatments on root tip meristems. III. Chlorsulfuron." *J. Plant Growth Regul.*, 3, 51–63.
19. Pallett, K. C., and J. C. Caseley. 1980. "Differential inhibition of DNA synthesis in difenzoquat tolerant and susceptible United Kingdom spring wheat cultivars." *Pestic. Biochem. Physiol.*, 14, 144–52.
20. Halling, B. P., and R. Behrens. 1983. "Effects of difenzoquat on photoreactions and respiration in wheat (*Triticum aestivum*) and wild oat (*Avena fatua*)." *Weed Sci.*, 31, 693–99.
21. Pallett, K. E. 1982. "The contact activity of difenzoquat in two United Kingdom spring wheat cultivars." *Weed Res.*, 22, 329–35.
22. Cohen, A. S., and I. N. Morrison. 1982. "Differential inhibition of potassium ion absorption by difenzoquat in wild oat and cereals." *Pestic. Biochem. Physiol.*, 18, 174–79.

23. Bartels, P. G., and A. Hyde. 1970. "Chloroplast development in 4-chloro-5-(dimethylamino)-2-(α,α,α-trifluoro-*m*-tolyl)-3 (2H)-pyridazinone (Sandoz 6706)-treated wheat seedlings. A pigment, ultrastructural, and ultracentrifugal study." *Plant Physiol.*, 45, 807–10.

24. Feierabend, J., and B. Schubert. 1978. "Comparative investigation of the action of several chlorosis-inducing herbicides on the biogenesis of chloroplasts and leaf microbodies." *Plant Physiol.*, 61, 1017–22.

25. Oelmüller, R., I. Levitan, R. Bergfeld, V. K. Rajasekhar, and H. Mohr. 1986. "Expression of nuclear genes as affected by treatments acting on plastids." *Planta*, 168, 482–92.

26. Oelmüller, R. 1989. "Photooxidative destruction of chloroplasts and its effect on nuclear gene expression and extraplastidic enzyme levels." *Photochem. Photobiol.*, 49, 229–39.

27. Agaev, R., F. M. Lecocq, A. Fer, and J. N. Hallet. 1988. "Étude de l'action d'un herbicide, le chlorthal, sur la prolifération cellulaire et la croissance de la tige du *Cuscuta lupuliformis*." *Can. J. Bot.*, 66, 328–38.

28. Kobayashi, K., H. Hyakutake, and K. Ishizuka. 1981. "Selective action of naproanilide on growth and RNA synthesis between smallflower umbrellaplant and rice." *Weed Res.* (Japan), 26, 30–36.

29. Snipes, C. E., J. E. Street, and D. S. Luthe. 1987. "Physiological influence of fenoxaprop on corn (*Zea mays*)." *Pestic. Biochem. Physiol.*, 28, 333–40.

30. Nagl, W. 1988. "Genome changes induced by auxin-herbicides in seedlings and calli of *Zea mays* L." *Environ. Exper. Bot.*, 28, 197–206.

31. Cherry, J. H. 1976. "Action on nucleic acid and protein metabolism." In *Herbicides, Vol. I*, L. J. Audus, ed. New York: Academic, pp. 525–46.

32. Kumar, C. C., and G. Padmanaban. 1980. "3-amino-1,2,4-triazole is an inhibitor of protein synthesis on mitoribosomes in *Neurospora crassa*." *Biochim. Biophys. Acta*, 607, 339–49.

33. Baxter, R., V. Knell, H. J. Somerville, H. M. Swain, and D. P. Weeks. 1973. "Effect of MDMP on protein synthesis in wheat and bacteria. *Nature* (London) *New Biol.*, 243, 139–42.

34. Ellis, R. J. 1975. "Inhibition of chloroplast protein synthesis by lincomycin and 2-(4-methyl-2,6-dinitroanilino)-*N*-methyl propionamide." *Phytochemistry*, 14, 89–93.

35. Mathews, D. E., and R. D. Durbin. 1990. "Tagetitoxin inhibits RNA synthesis directed by RNA polymerases from chloroplasts and *Escherichia coli*." *J. Biol. Chem.*, 265, 493–98.

36. Steinberg, T. H., D. E. Mathews, R. D. Durbin, and R. R. Burgess. 1990. "Tagetitoxin: A new inhibitor of eukaryotic transcription by RNA polymerase III." *J. Biol. Chem.*, 265, 499–505.

37. Birch, R. G., and S. S. Patil. 1987. "Evidence that an albicidin-like phytotoxin induces chlorosis in sugarcane leaf scald disease by blocking plastid DNA replication." *Physiol. Mol. Plant Pathol.*, 30, 207–14.

38. Tassi, F., F. M. Restivo, C. Ferrari, and P. P. Puglisi. 1983. "Erythromycin as a tool for discriminating *in vivo* between mitochondrial and chloroplastic protein synthesis in *Nicotiana sylvestris*." *Plant Sci. Lett.*, 29, 215–25.

39. Ebringer, L. 1972. ''Are plastids derived from prokaryotic microorganisms? Action of antibiotics on chloroplasts of *Euglena gracilis*.'' *J. Gen. Microbiol.*, 71, 35–52.

40. Mills, W. R., M. Reeves, D. L. Fowler, and S. F. Capo. 1989. ''DNA synthesis in chloroplasts III. The DNA gyrase inhibitors nalidixic acid and novobiocidin inhibit both thymidine incorporation into DNA and photosynthetic oxygen evolution by isolated chloroplasts.'' *J. Exp. Bot.*, 40, 425–29.

Chapter 13

Inhibition of Amino Acid Biosynthesis

13.1 INTRODUCTION

Plants synthesize all of their essential amino acids. Consequently, there are numerous potential sites of herbicide action in this highly complex and normally well-coordinated array of enzymes and processes. Animals have little of this biochemical capacity and, therefore, many of the enzymes of amino acid synthesis of plants are potentially toxicologically safe sites of herbicide action. These sites have only relatively recently been utilized for herbicides. Nevertheless, some of the most economically important, efficacious, and environmentally safe herbicides belong to this category of herbicides, and herbicide discovery efforts are being focused on unexploited enzymatic sites of amino acid synthesis.

Virtually all of the enzymes of amino acid synthesis in eukaryotic green plants are nuclear-coded, cytoplasmically synthesized as inactive pre-enzymes, transported across the plastid envelope, and activated to become functional in amino acid synthesis. In green tissues, the activity of the metabolic pathways to which these enzymes belong is generally highly dependent on photosynthetic activity and, thus, the metabolic damage caused by a herbicide affecting these sites of action in green tissues may be expected to be greater in the light than in darkness.

Inhibition of specific sites of amino acid synthesis eventually results in depleted metabolic pools of certain amino acids, resulting in slowing or cessation of all processes dependent on amino acids. However, total free amino acid pools are often misleading because of the inability to distinguish between metabolically active pools and the amino acids stored in the vacuole. The metabolic stress caused by inhibition of synthesis of a particular amino acid or amino acid family can result in increased protein turnover, resulting in increases in free pools of amino acids—even in increases in total (metabolic plus vacuolar) free pools of the amino acids whose synthesis is blocked.

Discovery that a herbicide's site of action is involved in amino acid synthesis has resulted from observations that exogenous amino acids fed to microorganisms or plant cell cultures can reverse the growth-inhibiting effects of the herbicides. Inhibited protein synthesis is an obvious secondary effect of inhibited amino acid synthesis, although other secondary effects may occur more quickly and be more drastic, depending on the site of action. The effects are generally slow in mature tissues and more rapid in meristematic tissues. Perhaps one reason that inhibitors of amino acid synthesis were developed much later than other herbicides is their relatively slow action, which may have been missed in early herbicide discovery screens designed to observe only rapid effects. Also, some of the amino acid synthesis inhibitors have more complex chemistry that has only recently been tested for herbicide activity.

Blockage of a pathway can disrupt plant metabolism by means other than reduction of its product (see Section 1.2). A product of the pathway often regulates the activity and/or level of an early enzyme of the pathway. Thus, blockage can cause deregulation of the pathway, resulting in uncontrolled flow of metabolites that are important to other pathways into the herbicidally blocked pathway. This exacerbates the accumulation of amino acid precursors or amino acid precursor products that normally may be found in only small concentrations in plant tissues, but may be phytotoxic at higher-than-normal concentrations.

Herbicides that inhibit amino acid synthesis have been the subject of reviews [1, 2]. Only three molecular sites of action in amino acid synthesis have been conclusively demonstrated to be the primary sites of action of commercially available herbicides. These three sites will be discussed in detail, and several other enzymatic sites of amino acid synthesis known to be affected by herbicides or other phytotoxins will be described briefly.

13.2 INHIBITION OF AROMATIC AMINO ACID SYNTHESIS

The three essential aromatic amino acids, phenylalanine, tyrosine, and tryptophan, are products of the shikimate pathway (Figure 13.1). Furthermore, many aromatic secondary plant products (e.g., lignins, alkaloids, flavonoids, benzoic acids) that are important in plant growth and development and in interactions with other organisms are also products of this pathway. The shikimate pathway is not found in animals but is important in the metabolism of plants, fungi, and bacteria. Chapter 16 provides more detail on the effects of herbicides on shikimate pathway-derived secondary products. As much as 20 percent of the carbon fixed by photosynthetic

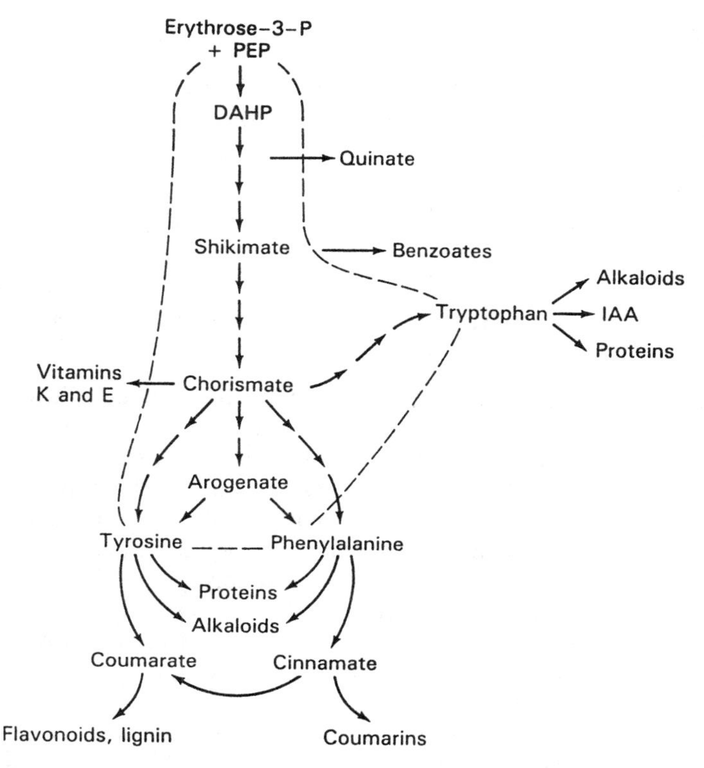

Figure 13.1 The shikimate pathway (enclosed by dashed line) and its relationship to other metabolic pathways.

plants flows through this highly regulated pathway [3]. All of the enzymes of the pathway have been found in the plastid; however, cytosolic forms of some of the enzymes apparently exist. The metabolic role of the enzymes of the shikimate pathway found in the cytosol is not understood, although evidence for dual (cytosolic and plastidic) shikimate pathways exists [4]. The plastidic enzymes are highly regulated, whereas those of the cytosol are under less control. Synthesis of aromatic amino acids in green tissues is strongly enhanced by light [5, 6].

Glyphosate (Figure 13.2) is the only compound developed as a herbicide that blocks an enzymatic step of the shikimate pathway. Its mechanism of action has been previously reviewed from various perspectives [1, 7, 8] and its general properties, toxicology, environmental effects, and herbicidal use have been the topics of a book [9] and reviews [7, 10, 11]. Glyphosate has very low mammalian toxicity [7, 11]. In fact, the acute toxicity of the surfactant with which it is formulated is three times higher than that of glyphosate [12].

At physiological pH levels, glyphosate is a zwitterion, existing predominantly as a divalent anion that can be strongly complexed with certain divalent

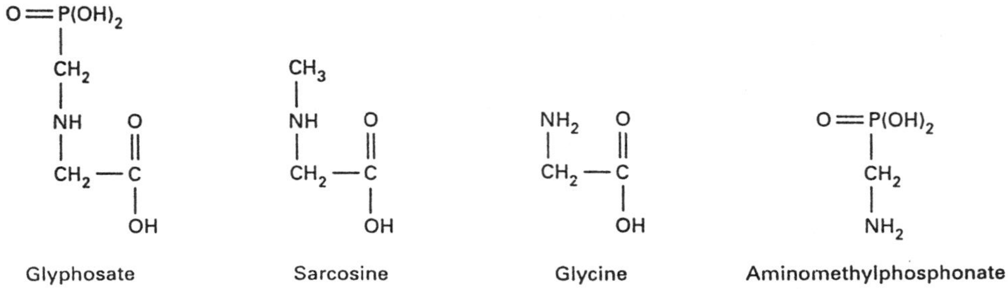

Glyphosate Sarcosine Glycine Aminomethylphosphonate

Figure 13.2 Structures of glyphosate and its metabolites.

metal cations. It is normally sold as the isopropylamine salt. Other salts, such as various sulfonium salts, have been under development for herbicide use [5]. There is no evidence that the cation with which the glyphosate anion is coupled contributes significantly to the activity of glyphosate. On the contrary, the presence of certain divalent cations in the application solution often causes a reduction of glyphosate activity. Thus, the glyphosate anion is the active form of the molecule. Its physicochemical properties make it highly phloem-mobile, and it is readily translocated with photosynthate from foliar sites of application to distant metabolic sinks (see Chapter 5). This, coupled with its relatively slow activity, make it an ideal herbicide for control of perennial weeds with subterranean regenerative organs such as rhizomes or tubers.

There is good evidence that this nonselective herbicide has only one significant site of action. It inhibits 5-enolpyruvylshikimate-3-phosphate synthase (EPSPS), the enzyme that condenses shikimate-3-phosphate (S3P) and phosphoenolpyruvate (PEP) to yield EPSP and inorganic phosphate [7] (Figure 13.3). Inhibition is competitive with respect to PEP with a K_i of 0.08 to 0.50 μM and uncompetitive with respect to S3P with a K_i of 10 μM [1, 13]. EPSPS is a monomer with a molecular mass of 44,000 to 51,000 M_r in plants and bacteria, and is a component of an *arom* complex in fungi [1, 13]. The inhibition of EPSPS by glyphosate is complex, but accumulated data indicate that glyphosate binds the region of EPSPS that binds the phosphate moiety of PEP [1]. The EPSPS of mutant glyphosate-resistant petunia cell lines have little change in their K_m values for EPSP or S3P; however, the K_m for PEP of the glyphosate-resistant EPSPS is more than 40-fold higher than that of the wild type [1]. Furthermore, production of a thiocyano-EPSPS does not alter the K_m of S3P, whereas its I_{50} for glyphosate and K_m for PEP are both increased [14]. Glyphosate binds the EPSPS-S3P complex 115-fold tighter and 20-fold slower than PEP, and the dissociation rate is 2,300-fold slower than that of PEP [15]. Whether glyphosate is a transition state or ground state analogue of PEP is not clear; however, recent data indicate that glyphosate binds the active site of the enzyme with the S3P to mimic the tetrahedral intermediate of the enzyme reaction [16]. Chemical modification investigations have demonstrated that Lys, Arg, Cys, Glu, and His residues are components of the binding

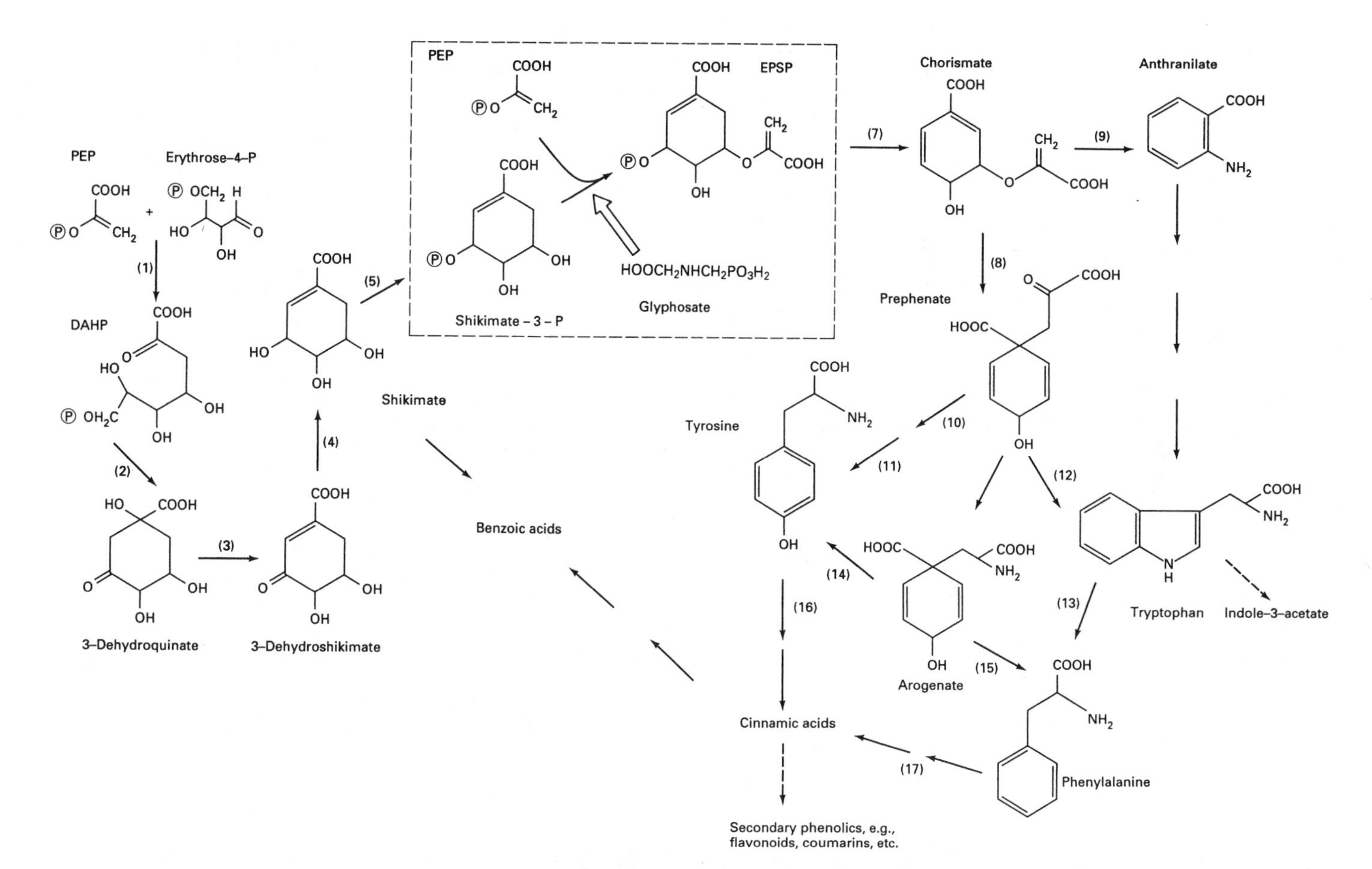

Figure 13.3 Enzymatic sites of the shikimate pathway. Enzymes listed by number: (1) 3-deoxy-*D*-arabino-heptulosonate-7-phosphate (DAHP) synthase; (2) 3-dehydroquinate synthase; (3) 3-dehydroquinate dehydratase; (4) shikimate dehydrogenase; (5) shikimate kinase; (6) 5-enolpyruvylshikimic acid-3-phosphate (EPSP) synthase; (7) chorismate synthase; (8) chorismate mutase; (9) anthranilate synthase; (10) prephenate dehydrogenase; (11) tyrosine aminotransferase; (12) prephenate dehydratase; (13) phenylalanine aminotransferase; (14) arogenate dehydrogenase (15) arogenate dehydratase (16) tyrosine ammonia-lyase; (17) phenylalanine ammonia-lyase. The dotted box highlights the primary site of action of glyphosate.

site(s) of the enzyme [1, 14, 17, 18], and that an Arg residue (either Arg^{28} and/or Arg^{131}), the Glu^{418} residue, and the Cys^{408} residue are components of the glyphosate-binding site [14, 17, 18]. By site-directed mutagenesis, modification of the Lys^{23} to Ala or Glu have been shown to inactivate the petunia EPSPS [19]. Replacement of the Lys^{23} with Arg produces an enzyme with normal K_m values for its substrates, but with an I_{50} for glyphosate that is five-fold higher than the unmodified enzyme. A single amino acid change of Pro to Ser in the 101 position of EPSPS from *Salmonella typhimurium* leads to resistance to glyphosate [20].

Inhibition of EPSPS leads to deregulation of the shikimate pathway due to increased activity of 3-deoxy-D-*arabino*-heptulosonate-7-phosphate synthase (DAHPS) [21–23], the enzyme that catalyzes the condensation of erythrose-4-phosphate with PEP. Increased DAHPS activity of the Mn^{2+}-dependent, plastidic form of DAHPS is apparently due to lowered levels of arogenate, a potent inhibitor of this enzyme [24], and to apparent derepression or increased stabilization of the enzyme [23] (Figure 13.1). The other DAHPS form, the Co^{2+}-requiring form, is inhibited by caffeic acid, an aromatic amino acid derivative that is reduced in concentration in plant tissues treated with glyphosate [25]. Blockage and deregulation of the pathway leads to accumulation of very high levels of shikimate [26] and, in some cases, shikimate-derived benzoic acids [27]. Shikimate and shikimate-3-phosphate can account for up to 16 percent of the dry weight of sink tissues in which glyphosate accumulates [28]. The amount of shikimate that is phosphorylated by the deregulated pathway is an energy drain from other metabolic pathways. In fact, discovery of the induction of high levels of shikimate by glyphosate was the clue that led Amrhein's laboratory to discover the enzymatic site of action of the herbicide [29, 30]. Leaves of henbane (*Hyoscyamus niger* L.) accumulate up to 2 percent of their dry weight as shikimate when treated with glyphosate [31]. Radiolabeled shikimate can be economically produced by allowing incorporation of $^{14}CO_2$ after treatment of plant tissues with glyphosate. Shikimate-3-P does not accumulate, probably because it is hydrolyzed by endogenous phosphatases [32].

Uncontrolled flow of carbon into the shikimate pathway removes building blocks from other metabolic pathways [33], possibly disrupting other aspects of plant metabolism (see Section 5.5.2 for one example of this). However, only 4 percent of the $^{14}CO_2$ that would have normally gone into starch in glyphosate-treated sugar beet leaves was detected as shikimate [34]. The deregulation of the pathway is complicated by the fact that glyphosate also inhibits the activity of the cytosolic Co^{2+}-dependent DAHPS [35], but less effectively than EPSPS. This inhibition may be due to glyphosate complexing with Co^{2+} rather than to direct inhibition of the enzyme. Kishore and Shah [1] have speculated that the glyphosate-Co^{2+} complex may have a higher affinity for the enzyme than either the metal ion or the herbicide alone.

In green plants, EPSPS is a nuclear-coded protein with a transit peptide that allows movement into the plastid. In *Petunia hybrida,* for example, EPSPS is synthesized with a 72 amino acid transit peptide attached to the 444 amino acid mature enzyme [36]. The molecular mass of the transit peptide for *Corydalis sempervirens* is about 8,400 M_r [37]. The nucleotide sequences of several genes coding for EPSPS have been determined and there is a high level of conservation of amino acid sequences in the mature enzyme. The transit peptides, however, are highly divergent [1, 38]. EPSPS is the only nuclear-coded, plastid enzyme known to have activity in the cytoplasmic precursor state (pEPSPS) before it is transported into the plastid [39]. Thus, herbicide binding and inhibition can occur in the cytoplasm (Figure 13.4). Binding of glyphosate to pEPSPS has been shown to inhibit plastid uptake and processing of pEPSPS [40]. Import is completely inhibited at glyphosate concentrations of about 10 μM. The effect is apparently a direct consequence of glyphosate binding to the pEPSPS-S3P complex, because glyphosate is ineffective in the absence of S3P, and glyphosate has no effect on the transport and processing of other nuclear-coded proteins. Apparently, the structural stability of the pEPSPS-S3P complex precludes the required conformational flexibility for transport across the plastid envelope.

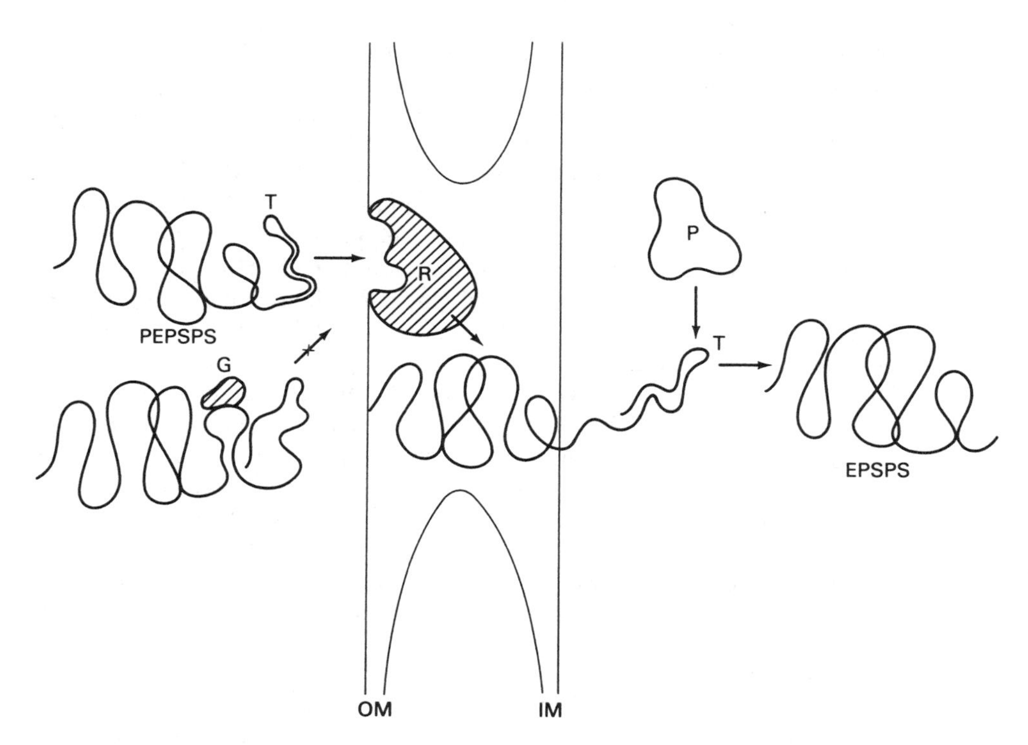

Figure 13.4 Model for pEPSPS import into plastids. R, receptor for imported proteins; P, stromal protease; OM, plastid outer envelope membrane; IM, plastid inner envelope membrane; G, glyphosate; T, transit peptide. See text for details. Note that glyphosate binding to pEPSPS inhibits its import into the plastid by altering conformation of the transit peptide.

Glyphosate is an ideal herbicide for development of herbicide-resistant crops because, for reasons given below, herbicide-resistant weeds are unlikely to arise rapidly in such crops. Also, it is a very toxicologically and environmentally safe herbicide [7]. Three strategies for developing a glyphosate-resistant crop plant are possible: (a) overproduction of EPSPS so that the herbicide is diluted out, (b) introduction or induction of a glyphosate-resistant EPSPS in the plant, or (c) introduction of an enzyme that catabolizes glyphosate. The first two of these options have been the focus of extensive efforts.

Selection of glyphosate-tolerant plant cell lines has generally resulted in selection for mutants that overproduce EPSPS. Elevated EPSPS activity has been found in plant cell cultures of tobacco [41, 42], carrot [43], tomato [44], *Corydalis sempervirens* [45], and petunia [46] selected by culturing surviving plant cells on media of increasing glyphosate concentrations. A glyphosate-tolerant tobacco cell line isolated by Dyer et al. [41] also overproduced DAHPS. Plants regenerated from these cultures generally retain some degree of glyphosate tolerance [36, 41–45]. In tobacco cell cultures, there is an excellent correlation between level of tolerance to glyphosate and specific activity of EPSPS from the cultures [42]. The elevated EPSPS activity has been found to be caused by overproduction of the enzyme in the three cases that have been examined at this level [36, 46]. In petunia and tobacco, overproduction is due to gene amplification (i.e., increased copies of the EPSPS gene) [46]. This was determined by analysis of the genome of the tolerant line using genomic blot analysis with EPSPS cDNA. In *C. sempervirens,* overproduction was not due to gene amplification [47]. In this case, a ten-fold increase in EPSPS mRNA was found in a glyphosate-tolerant *C. sempervirens* cell culture that had a 30-fold increase in extractable EPSPS activity. However, Southern and dot blot analysis of the glyphosate-tolerant and -susceptible cell lines revealed that gene amplification was not involved in the overexpression of EPSPS in this cell selection. The increased levels of EPSPS mRNA could be due to increased stability and/or a more active promotor region regulating transcription of the EPSPS genes in this cell line. The overproduced EPSPS was found (by immunocytochemical methods at the transmission electron microscope level) to reside entirely in the plastid [48]. However, the finding that most of the EPSPS in the overproducing cell lines is present as pEPSPS [37] is difficult to reconcile with this result, unless the primary antigenic site for the antibody used is present on EPSPS but not on pEPSPS. Also, this study only compared immunogold staining density in plastids versus cytoplasm, without consideration of the relative volumes of the plastids versus cytoplasm.

The overproduction strategy has been deliberately implemented by genetically engineering amplification of the gene encoding EPSPS. Initially, *E. coli* was made eightfold more tolerant of glyphosate by introduction of a gene encoding EPSPS (the *aro*A gene) on a high-copy plasmid [49]. Later, petunia callus transformed with EPSPS coupled with the highly active cauliflower mosaic virus promoter produced about 20-fold more EPSPS than untransformed plants [36]. The transformed callus was highly tolerant to glyphosate. Petunia plants regenerated from these calli were tolerant of treatments of commercially formulated glyphosate that were about four-fold the minimum lethal dose of the herbicide in untransformed callus. Similar results have been obtained with *Arabidopsis thaliana* [50] and flax (*Linum usitatissimum*) [51].

Unfortunately, overproduction of the site of action approach is not a good strategy for a highly phloem-mobile, metabolically stable herbicide like glyphosate. The concentrations of the herbicide in metabolic sinks such as meristematic regions can overwhelm the overexpressed enzyme, even though other parts of the plant are unaffected by the herbicide. For this reason, these plants mature later than untreated plants in the presence of glyphosate [1]. Other phytotoxicity problems could be expected to be found in flowers and fruits, for the same reason. However, the greater amounts of EPSPS in flowers than in leaves [38] in some species could counteract the higher accumulations of glyphosate in these tissues.

The second strategy for producing glyphosate-resistant plants, the introduction of glyphosate-resistant EPSPS, avoids these problems. Glyphosate-resistant EPSPS has been found in *Salmonella typhimurium* [20], *Aerobacter aerogenes* [52], and *E. coli* [53]. All of these mutant EPSPSs have higher K_m values for PEP than the corresponding wild-type enzymes, and the I_{50} values for glyphosate have ranged from 9- to almost 8,000-fold higher for the mutant enzyme than the wild types. Transgenic tobacco plants have been regenerated from cells transformed with the gene for glyphosate-resistant EPSPS from *E. coli* (with a cauliflower mosaic virus promoter) alone or fused with the cDNA for the transit peptide of petunia EPSPS [54]. In these plants the enzyme was found to be entirely cytoplasmic without the transit peptide, and to be highly expressed in the chloroplast with the transit peptide. As might be expected, tobacco plants transformed with the latter construct are more resistant to glyphosate than those without the capability of transporting resistant EPSPS into the plastid [55]. A chimeric gene of glyphosate-resistant *S. typhimurium* EPSPS (mutant *aro*A gene) and a promoter sequence has been used to transform tobacco [56], tomato [57], and *Populus* [58, 59] cells. The plants generated from these transformed cells are only moderately tolerant of glyphosate, perhaps because, without a transit peptide, the enzyme is apparently entirely cytoplasmic, rather than plastidic as it should be for proper function in a higher plant cell [60]. The transit peptide of the small subunit of RuBP carboxylase has been linked to the glyphosate-resistant *S. typhimurium* EPSPS; however, it does not affect the import of the procaryotic version of EPSPS into tobacco chloroplasts [61]. The transit peptide plus 24 amino acids of the small subunit of RuBP carboxylase fused to EPSPS is transported into the chloroplast both in vivo and in vitro. Plants regenerated from such transformed cells are more resistant to glyphosate than earlier regenerants containing only cytoplasmic expression of the *aro*A gene [62].

The gene for EPSPS is highly regulated in a tissue-specific manner [36, 63]. Histochemical staining of petunia tissues transformed with the promotor region of EPSPS of petunia fused with the β-glucuronidase (GUS) reporter gene reveals intense staining in the upper and lower leaf epidermis, vascular tissue, and the mesoderm of flower petals—all known as sites of highly active secondary metabolism. Fusion of the promoter region of EPSPS from higher plants with a microbial gene for resistant EPSPS could provide the highly regulated level of gene expression required for normal plant development.

Glyphosate-resistant higher plant EPSPS has been generated by site-directed mutagenesis [1]. The K_m for PEP is about 40-fold higher and the K_i for glyphosate almost four orders of magnitude higher in the EPSPS of glyphosate-resistant than of susceptible petunia. Because the K_m for PEP is increased dramatically, one would expect that this enzyme would have to be overexpressed in order for the plant to have normal shikimate metabolism. This is the only reported resistant EPSPS from a higher plant.

The third strategy for producing glyphosate-resistant plants, introducing genes for enzymes that metabolize glyphosate, could overcome the problem of accumulation of glyphosate in harvested portions of the crop. The advantages of this strategy are several. First, neither the herbicide nor its residues will accumulate in edible metabolic sinks (fruits, tubers, etc.). Second, the introduced enzyme does not have to be highly regulated in order to function properly. Finally, transformation with only one gene should be required. This strategy has not been attempted yet, although the genes for degradation are available in a variety of microorganisms.

Metabolic degradation of glyphosate by most higher plants either does not take place or is very slow [7]. However, in *Equisetum* sp., metabolism to aminomethylphosphonate (AMP, Figure 13.2) can be rapid [64]. Several soil microorganisms can readily degrade glyphosate, and some of them can use it as their sole source of phosphorus [65–69]. To use glyphosate as a sole phosphorus source, the C-P bond must be cleaved. In certain *Pseudomonas* sp. and *Arthrobacter* sp. strains that can utilize glyphosate as their sole phosphorus source, degradation begins by cleavage of the C-P bond to form sarcosine (*N*-methylglycine) (Figure 13.2) and P_i [66, 67]. In other *Pseudomonas* sp. strains and a *Flavobacterium* species, glyphosate is metabolized to AMP before the C-P bond is cleaved [68, 69]. In all of these cases, glyphosate is eventually completely metabolized to nonphytotoxic compounds (neither sarcosine nor AMP is phytotoxic [70]). Lack of progress in isolating an enzyme that breaks the C-P bond (C-P lyase) has prevented utilization of such an enzyme in genetic engineering of glyphosate-resistant crops. Only recently has a cell-free assay of a C-P lyase been successful [71]. This preparation from *Enterobacter aerogenes* releases phosphonate from glyphosate and other alkylphosphonic acids. Two different proteins in the preparations are required for C-P lyase activity [72].

No glyphosate-resistant weeds have occurred in the field, despite more than a decade of heavy use in some areas. Kishore and Shah (1) have speculated that this is largely due to two factors. The first is that the herbicide is inactive in most soils, because of complex formation with soil components and microbial degradation. Thus, selection pressure is short-lived. Second, since glyphosate-resistant EPSPS is relatively inefficient because of its low affinity for PEP, glyphosate-resistant EPSPS must have a higher level of expression than the wild-type gene in order for adequate EPSPS activity to be present in the plant for normal growth and development. Therefore, a minimum of two mutations is required for a viable resistant plant to occur. Furthermore, isozymes of EPSPS apparently exist in higher plants (e.g., [13]). If this is an indication of different genes encoding different forms of the enzyme, rather than different posttranslational forms of the enzyme, complete resistance may not occur through insensitivity of only one form of the plant's EPSPS to glyphosate.

Some weeds or weed biotypes are more tolerant of glyphosate than others. Differences in spray retention [73], absorption [74], translocation [75], and vegetative reproductive potential [76] have been invoked to explain differences in some cases. Specific activity of EPSPS among selections of birdsfoot trefoil (*Lotus corniculatus*) correlates well with tolerance to glyphosate [77]. Whether the enhanced EPSPS activity is due to gene amplification or to other factors is not yet known. In a selection system using the haploid phase of *Ceratopteris*, designed to increase the odds of obtaining viable mutants of herbicide-resistant green plants, glyphosate-tolerant (ten-fold) mutants grew only half as well as wild-type gametophytes [78]. However, the mechanism of tolerance was not determined.

Another proposed reason [7] why glyphosate-resistant weeds have not been found is that, in intact higher plants, a secondary site of action, unrelated to the shikimate pathway, may play a role in the mechanism of action of glyphosate at high herbicide levels. This possibility is becoming more unlikely with increasing evidence that secondary sites of action, if they exist, are of minimal importance in the mechanism of action of this herbicide.

No other significant primary sites of action of glyphosate, other than those in the shikimate pathway, have been identified. Because glyphosate is a strong competitive inhibitor of PEP with EPSPS, one might expect it to inhibit other PEP-dependent reactions; however, none has been discovered [1]. For example, 5 mM glyphosate reduces PEP carboxylase from maize by only 14 percent [79]. Although the evidence is strong that EPSPS is the predominant primary site of action, there is some indirect evidence that another site may play a role in the mechanism of action in higher plants. Theoretically, the phytotoxic effects of glyphosate should be ameliorated by supplying affected organisms with the proper concentrations of aromatic amino acids. In fact, Jaworski [80] first proposed inhibition of synthesis of aromatic amino acids as the mode of action of glyphosate, based on results of reversal studies on *Rhizobium japonicum* and *Lemna gibba* with exogenous aromatic amino acids. In intact, higher plants, however, reversal of the phytotoxic effects of glyphosate is rarely achieved by exposure to exogenous aromatic amino acids (e.g., [7, 8, 81, 82]). This could be due partly to insufficient transport of these amino acids to glyphosate-affected tissues or to the inability of aromatic amino acids to prevent deregulation of the shikimate pathway because glyphosate-reduced arogenate levels would be unaffected by aromatic amino acid pool sizes [22]. Furthermore, the high concentrations of shikimate, shikimate-3-phosphate, and benzoic acids could be phytotoxic by effects on proteins and enzymes remote from EPSPS. Shikimate is an effective growth inhibitor of the blue-green alga *Nostoc linckia* [83]; however, its phytotoxicity to higher plants has not been explored. Shikimate and shikimate-3-phosphate are Michael acceptors, capable of 1,4 addition to α,β-unsaturated carboxylate species in the plant [84].

Relatively high concentrations of glyphosate can inhibit photosynthesis in some systems. For example, 1 mM glyphosate causes ca. 50 percent inhibition of PS I electron transport and ca. 60 percent and 90 percent inhibition of PS II electron transport in alfalfa and clover chloroplasts, respectively [85]. At such high concentrations, the effects of pH can be important unless the assays are carefully buffered [86]. In more recent studies, no effect of glyphosate or its trimethylammonium salt at concentrations up to 10 mM could be found on photosynthetic electron transport [87]. Any effect on photosynthesis is unlikely to be of great significance in the mechanism of action of glyphosate, because the herbicide is unlikely to accumulate to such high concentrations in chloroplasts, except, perhaps, in metabolic sinks such as developing leaves.

The metal ion-complexing capacity of glyphosate at physiological pHs has been speculated to play a role in the mechanism of action of glyphosate [5], although no direct evidence of this has been produced. Metal ions strongly influence glyphosate absorption and translocation and vice versa; however, we do not know if glyphosate has a significant effect on intracellular distribution and availability of metal ions.

No significant effects of glyphosate analogues have been demonstrated on activity of EPSPS or accumulation of shikimate (e.g., [88]). However, several close analogues of glyphosate have strong herbicidal activity (e.g., [89, 90]). In some cases, the mode of action of these herbicidal analogues is known to have nothing to do with EPSPS (e.g., glufosinate, see Figures 13.2 and 13.10), whereas for others the mechanism of action is not known.

As mentioned above, no other herbicides that inhibit an enzyme of the shikimate pathway are used commercially. However, several other enzymes of the pathway have been the subject of studies to discover new herbicides. Nonisosteric phosphonate and isosteric homophosphonate analogues of erythrose-4-P have been studied as substrates for DAHPS, but their efficacy as herbicides is not known [1]. Fluoro-PEP is a very good inhibitor of DAHPS, but its effect on other PEP-requiring enzymes would probably preclude it from consideration as a herbicide because of toxicological considerations.

Dehydroquinate synthase (DHQS), the enzyme that converts DAHP to 3-dehyroquinate (Figure 13.3), is apparently a single protein in *E. coli* and higher plants [91, 92], although it is part of a multiprotein complex in fungi [93]. It is inhibited by phosphonate analogues of DAHP such as 3-deoxy-D-*arabino*-heptulosonic acid 7-phosphate (DAHAP) [91, 92]; this is a potent competitive inhibitor of DHQS from *E. coli* ($K_i = 1.1\ \mu M$) and pea ($K_i = 0.8\ \mu M$). DAHAP is herbicidal to *Setaria viridis* and somewhat less herbicidal to *Echinochloa crus-galli* and wild oat, with no activity against johnsongrass or pea, despite the in vitro activity on DHQS [92]. These compounds cause an increase in the dephosphorylated substrate of DHQS in the plant species to which they are herbicidal, although the increases are very small in comparison to the increases in shikimate found in the same plants treated with glyphosate [92]. Whether this disparity is due to more inefficient blockage of the shikimate pathway, instability of DAH in vivo, or other factors has not been determined.

Roisch and Lingens [94] reported glyphosate to be a weak inhibitor of DHQS. However, as with DAHPS, this may be due to metal ion chelation. No inhibitors of the next three enzymes in the shikimate pathway (3-dehydroquinate hydrolyase, shikimate dehydrogenase, and shikimate kinase) have been reported. The allylic isomer of EPSP is a potent inhibitor of chorismate synthase ($K_i = 8.7\ \mu M$), but its herbicidal activity has not been reported [1]. Inhibitors of chorismate mutase and anthranilate synthase have been discovered; however, they have no reported herbicidal activity [1]. One potent inhibitor (2-mercaptobenzimidazole) of *E. coli* tryptophan synthetase is highly phytotoxic to rice, while another inhibitor [4-(dimethylamino)pyridine] is not phytotoxic [95]. Whether this discrepancy is due to differential metabolism, differences between *E. coli* and higher plant tryptophan synthase, or to a second molecular target site has not been determined.

13.3 INHIBITION OF BRANCHED-CHAIN AMINO ACID SYNTHESIS

The amino acids valine, leucine, and isoleucine are products of the branched-chain amino acid pathway (Figure 13.5). Four of the enzymes of the pathway are common to synthesis of all three branched-chain amino acids. In the past decade, two commercialized herbicide chemical classes (the sulfonylureas and imidazolinones—Figure 13.6) have been found to act on acetolactate synthase (ALS; also called acetohydroxy acid synthase—AHAS), the first enzyme in this pathway. At least one other herbicide group (the 1,2,4-triazolo[1,5A]pyrimidines or sulfonanilides; Figure 13.6) that inhibits ALS is under development. Reviews have covered sulfonylureas [96–98], ALS inhibitors in general [1, 99, 100], triazolopyrimidines [101, 102], and the molecular biology of resistance to ALS-inhibiting sulfonylureas [103].

ALS condenses two pyruvate molecules to form 2-acetolactate, the precursor for leucine and valine, and CO_2. The enzyme can also perform the same reaction with pyruvate and 2-ketobutyrate to form CO_2 and 2-acetohydroxybutyrate, a precursor of isoleucine. ALS requires FAD, thiaminepyrophosphate (TPP), and Mg^{2+} or Mn^{2+} for activity, and the enzyme generates hydroxyethyl-TPP (HETPP) as an intermediate [1]. In the presence of FAD, ALS is predominantly in a tetrameric form [104]. This key enzyme in branched-chain amino acid synthesis is regulated by feedback inhibition of all three of its amino acid products [105, 106]. Two forms of ALS exist in some, if not all, higher plants [107–110]. Each of the forms has characteristic sensitivity to branched-chain amino acids and to ALS-inhibiting herbicides. The entire pathway can be found in the plastid and, as with

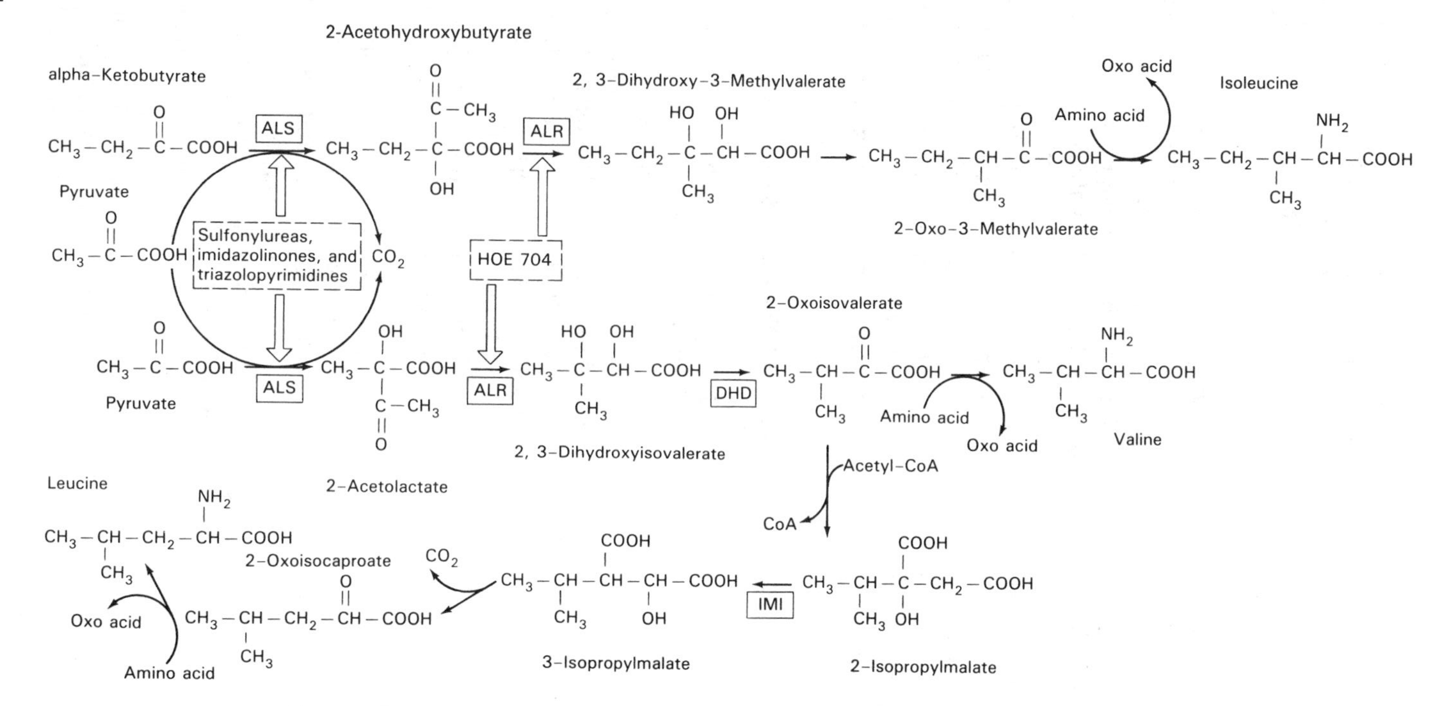

Figure 13.5 Branched-chain amino acid biosynthesis pathway. ALS, acetolactate synthase; ALR, acetolactate reductoisomerase; DHD, 2,3-dihydroxyacid dehydratase; IMI, isopropylmalate isomerase.

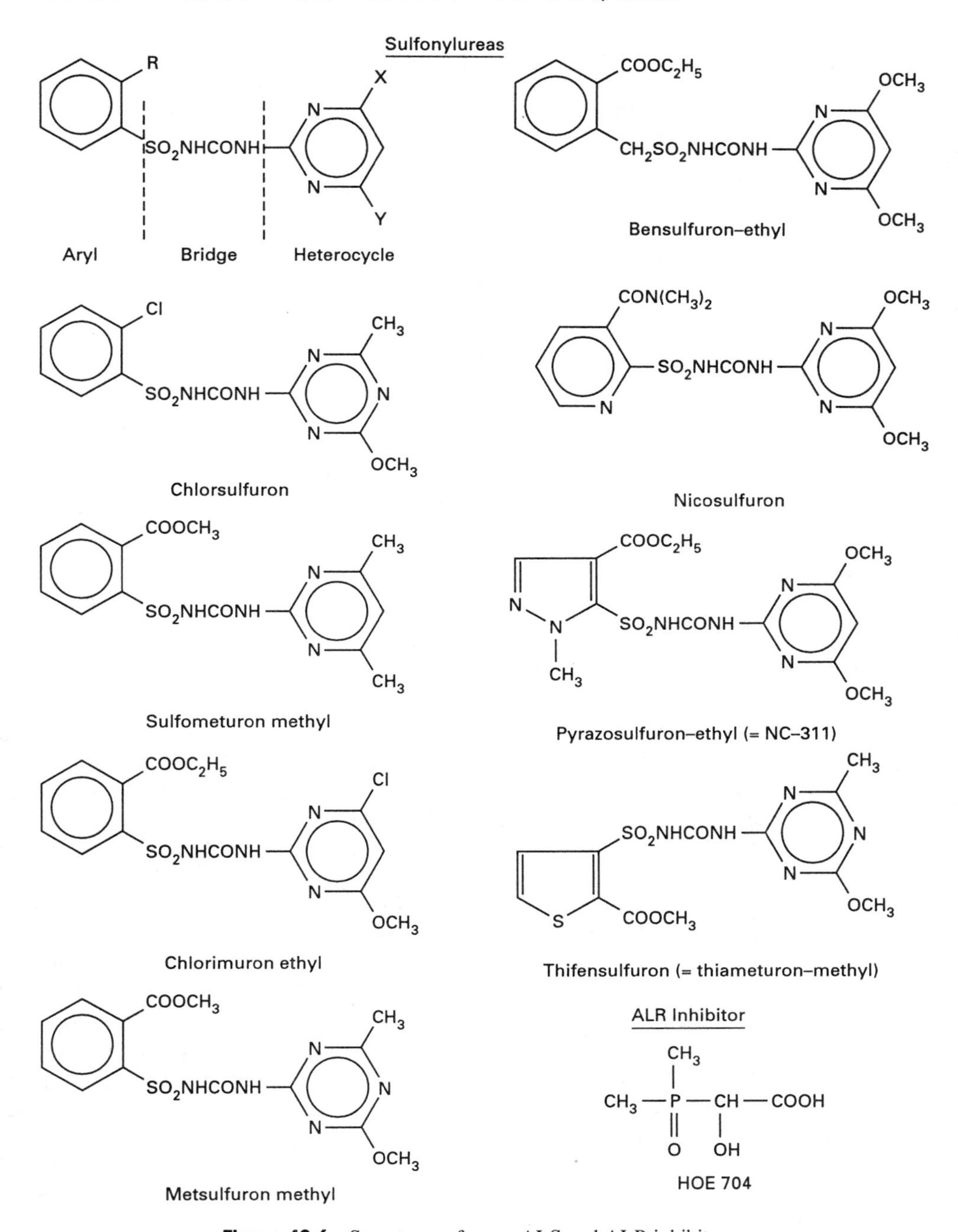

Figure 13.6 Structures of some ALS and ALR inhibitors

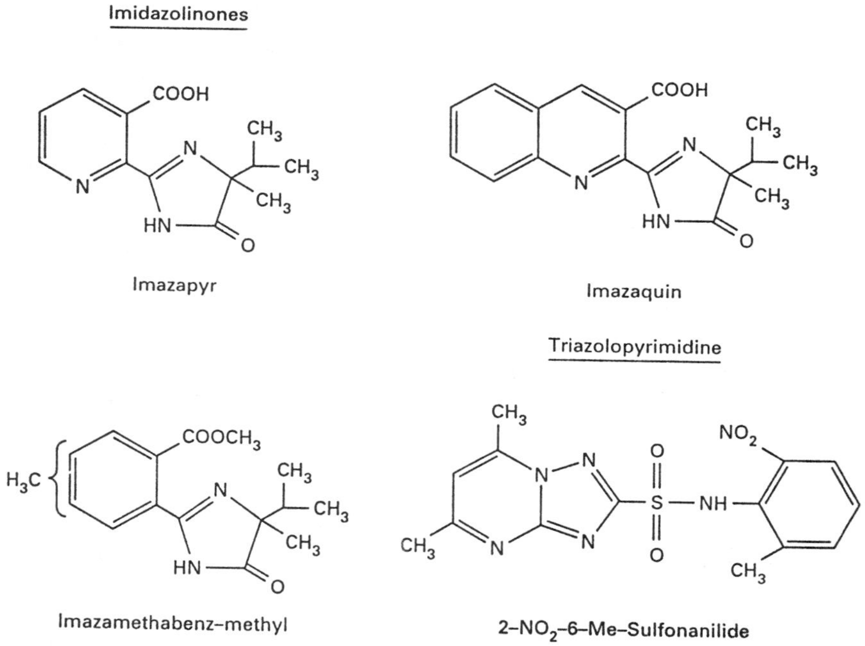

Imidazolinones

Imazapyr

Imazaquin

Triazolopyrimidine

Imazamethabenz-methyl

2-NO_2-6-Me-Sulfonanilide

Figure 13.6 (Continued).

the shikimate pathway, the enzymes are nuclear coded with transit peptides for proper movement into the plastid. Genes coding for ALS have been isolated and characterized from tobacco and *Arabidopsis* [111], with about 85 percent homology between the 667 and 670 amino acid protein products, respectively.

Pyruvate oxidase and ALS have a remarkable level of amino acid sequence homology, and it has been suggested that the quinone binding site of pyruvate oxidase and the herbicide binding site of ALS may have a common evolutionary origin [1, 112]. Ubiquinone homologues of the native pyruvate oxidase ubiquinone effectively inhibit ALS and will compete radiolabeled sulfometuron methyl off the ALS molecule [112]. The growing and diverse array of herbicidal ALS inhibitors and apparently overlapping sites of these compounds at a quinone-binding site is similar to the situation with the quinone (in this case plastoquinone) binding site of the D-1 protein of photosystem II (Chapter 7). This suggests that quinone-binding sites can be effectively bound by a relatively wide array of heterocyclic compounds. It should be pointed out that since the binding site of these herbicides has no functional role in catalysis, these herbicides would not likely have been discovered in a herbicide discovery process based on mechanistic considerations of ALS [113].

The three classes of inhibitors of ALS are structurally diverse (Figure 13.6) and each herbicide class has a unique set of interactions with the enzyme. All of these compounds can be considered growth inhibitors that cause plant death within a period ranging from several days to more than a week. Generally, the effects of these compounds are faster than those of glyphosate.

Sulfonylurea herbicides are composed of three moieties (Figure 13.6): an aryl group—usually a phenyl group substituted ortho to the sulfonylurea link; a heterocycle portion—often a symmetrical pyrimidine or triazine; and a sulfonylurea bridge that links the other two moieties. Small substitution differences in each part of the molecule can give rise to large shifts in biological activity and selectivity. These compounds are readily absorbed by both roots and leaves and are translocated in both phloem and xylem. They are highly active, with field use rates as low as 2 g/ha [98].

Much of the biochemical characterization of ALS interactions with sulfonylureas has been done with ALS from *Salmonella typhimurium* and sulfometuron methyl. Its binding with the bacterial ALS is slow (initial K_i = 660 nM), but eventually very tight (final K_i = 65 nM) [114]. It binds the ALS-FAD-TPP-Mg^{2+}-decarboxylated pyruvate complex at a site that overlaps the binding site of the second pyruvate [115]. The inhibition eventually becomes irreversible and the slow inhibition has been hypothesized to be due to slow inactivation of ALS rather than to slow-tight binding [99].

Plant ALS is much more difficult to work with than the bacterial ALS [106], and biochemical characterization of herbicide interactions with higher plant ALS is not complete. Higher plant ALSs are generally about an order of magnitude more sensitive to sulfonylureas than those of bacteria (Table 13.1). Inhibition of pea ALS by sulfonylureas is biphasic, with inhibition increasing with time [116, 117]. The ALS of tolerant crop species have all been found to be highly sensitive to sulfonylureas [98]; selectivity in these species is based on rapid detoxification of the herbicide (see Chapter 6).

Complete suppression of the growth-inhibiting effects of sulfonylurea herbicides by supplementing growth media with leucine, isoleucine, and valine [82, 106, 118] is good evidence that ALS is the only important site of action of this class of herbicides. Reversal of sulfonylurea-reduced growth of whole plant tissues with branched-chain amino acids is much more effective than reversal of glyphosate-stunted growth with aromatic amino acids [82]. There are some data that suggest that reversal of ALS inhibitor-inhibited cell growth with exogenous branched-chain amino acids is much more effective than reversal of cell division [119].

Accumulation of 2-ketobutyric acid clearly plays a role in the mode of action of ALS-inhibiting herbicides in some microorganisms. Isoleucine allosterically inhibits synthesis of threonine deaminase-generated 2-ketobutyric acid, and can reverse the growth-inhibiting effect of sulfometuron methyl on *S. typhimurium* by this mechanism (e.g., [120]). In a mutant with no isoleucine feedback regulation of threonine deaminase, isoleucine could not counteract the effect of the herbicide. In addition to the above evidence, mutants of *S. typhimurium* with hypersensitivity to sulfonylureas have been found that: (a) have lower 2-ketobutyrate utilization; (b) have various 2-ketobutyrate degrading pathways blocked; and (c) have in-

TABLE 13.1 RELATIVE EFFECTS OF SEVERAL ALS INHIBITORS ON ALS FROM A VARIETY OF SOURCES (DATA FROM [98, 102]).

Herbicide	ALS Source	I_{50}(nM)
(Sulfonylureas)		
Chlorsulfuron	Pea	21
	Wheat	19, 21
	Wild oats	16
	Wild mustard	11
	Tobacco	
	Sensitive	14
	Resistant	>8000
Sulfometuron methyl	Salmonella	65
	Yeast	120
	Wild oats	7
	Wheat	13
	Wild mustard	9
	Pea	15
Chlorimuron ethyl	Pea	6
	Soybean	8
	Morningglory	7
(Imidazolinones)		
Imazapyr	Pea	9,000
	Corn	12,000
	Bacteria	>100,000
Imazaquin	Pea	3,000
	Corn	3,400
	Bacteria	20,000
(Triazolopyrimidines)		
Triazolopyrimidine	Soybean	452
	Rice	124
	Barley	173
	Maize	200
	Lambsquarters	436

creased sensitivity of aspartate aminotransferase to 2-ketobutyrate [121]. In addition to inhibition of aspartate aminotransferase, 2-ketobutyrate can have other toxic effects. It combines with acyl-CoA to form propionyl-CoA, and high concentrations of propionyl-CoA may inhibit the normal functioning of the TCA cycle by mimicking the acyl-CoA derivatives of other 2-ketoacids that play a role in this pathway. There could also be inhibition of enzymes acting directly on 2-ketoacids. For example, nutritional studies with *S. typhimurium* mutants indicate that 2-ketobutyrate competes with 2-ketovalerate, an intermediate of pantothenate, L-leucine, and L-valine [121]. Coenzyme A synthesis is dependent on pantothenate synthesis and succinyl-CoA is dependent on CoA synthesis. Methionine and lysine (synthesis of both is dependent on succinyl-CoA) and pantothenate can restore sulfometuron methyl-inhibited growth of *S. typhimurium* [114]. Other metabolic disruptions caused by 2-ketobutyrate accumulation have been reviewed by LaRossa and van Dyk [121].

Whether 2-ketobutyrate plays a significant role in the mode of action of ALS-inhibiting herbicides in higher plants is still an unanswered question. There has been speculation that the reason that ALS is the target site of three highly effective herbicide classes in the branched-chain pathway is that it is the only site of many enzymatic sites of the pathway that will result in the accumulation of 2-ketobutyrate [97]. Chlorsulfuron causes accumulation of high levels (2.44% of the free amino acid pool) of α-amino-*n*-butyrate in *Lemna gibba,* apparently through amination of 2-ketobutyric acid [122]. In trichome cells of tobacco, chlorsulfuron causes increases in valeryl and butyryl esters of sucrose rather than other acyl acid esters that normally accumulate [123]. Both valeryl and butyryl components can be derived from 2-ketobutyrate. The relative contribution of isoleucine to the alleviation of growth retardation by ALS inhibitors (e.g., [116, 124]) in the presence of valine by feedback inhibition of the pathway is unknown.

HOE-704 inhibits the next enzyme in the branched-chain amino acid pathway, acetolactate reductoisomerase (ALR) (see below for more detailed discussion), resulting in accumulation of acetolactate and acetoin. Phytotoxic symptoms similar to those obtained with ALS inhibitors are induced by HOE-704 [125]. However, neither it nor *N*-isopropyl oxalylhydroxamate, a highly potent ALR inhibitor (K_i = 22 pM), are as herbicidal as sulfonylureas, which have much higher K_i values for ALS [125, 126]. This further indicates that 2-ketobutyrate may play a role in the mode of action of ALS-inhibiting herbicides.

Sulfonylurea-resistant forms of plant ALS have been produced by mutagenesis and cell culture selection methods with tobacco [116], flax [127], the cyanobacteria *Synechococcus* [128], *Chlorella emersonii* [119], and *Chlamydomonas reinhardtii* [129, 130], and by microspore mutagenesis methods with *Brassica napus* L. [131]. The selected tobacco cell lines are more than 100-fold less sensitive to chlorsulfuron than the wild type, although the ALS is 300 times less sensitive. This selection was for a single gene mutant in which only one of the ALS forms was resistant to the herbicide. The homozygous mutants are highly resistant, whereas heterozygous genotypes have an intermediate level of resistance [109]. Tobacco is resistant if either of the two ALS forms of the plant is resistant [108–110, 132]. In mutants homozygous for either ALS form, only about half of the extractable ALS is inhibited by sulfonylurea herbicides (Figure 13.7). The two nuclear genes conferring resistance in tobacco are unlinked [110] and both are semidominant [109]. The *SURB-Hra* gene isolated by a series of two genetic selections contains amino acid substitutions (Pro^{196} to Ala and Trp^{573} to Leu), whereas the *SURA-C3* has only one substitution (Pro^{196} to Gln) [132, 133].

Similar results have been obtained with whole plant mutagenesis and selection with *Arabidopsis* [134]. In this case, resistance is due to a single dominant nuclear gene. The gene encoding sulfonylurea-resistant ALS in these plants has been cloned and found to differ by only one base pair (resulting in conversion of Pro^{197} to Ser) from the susceptible plants [135]. When introduced into susceptible

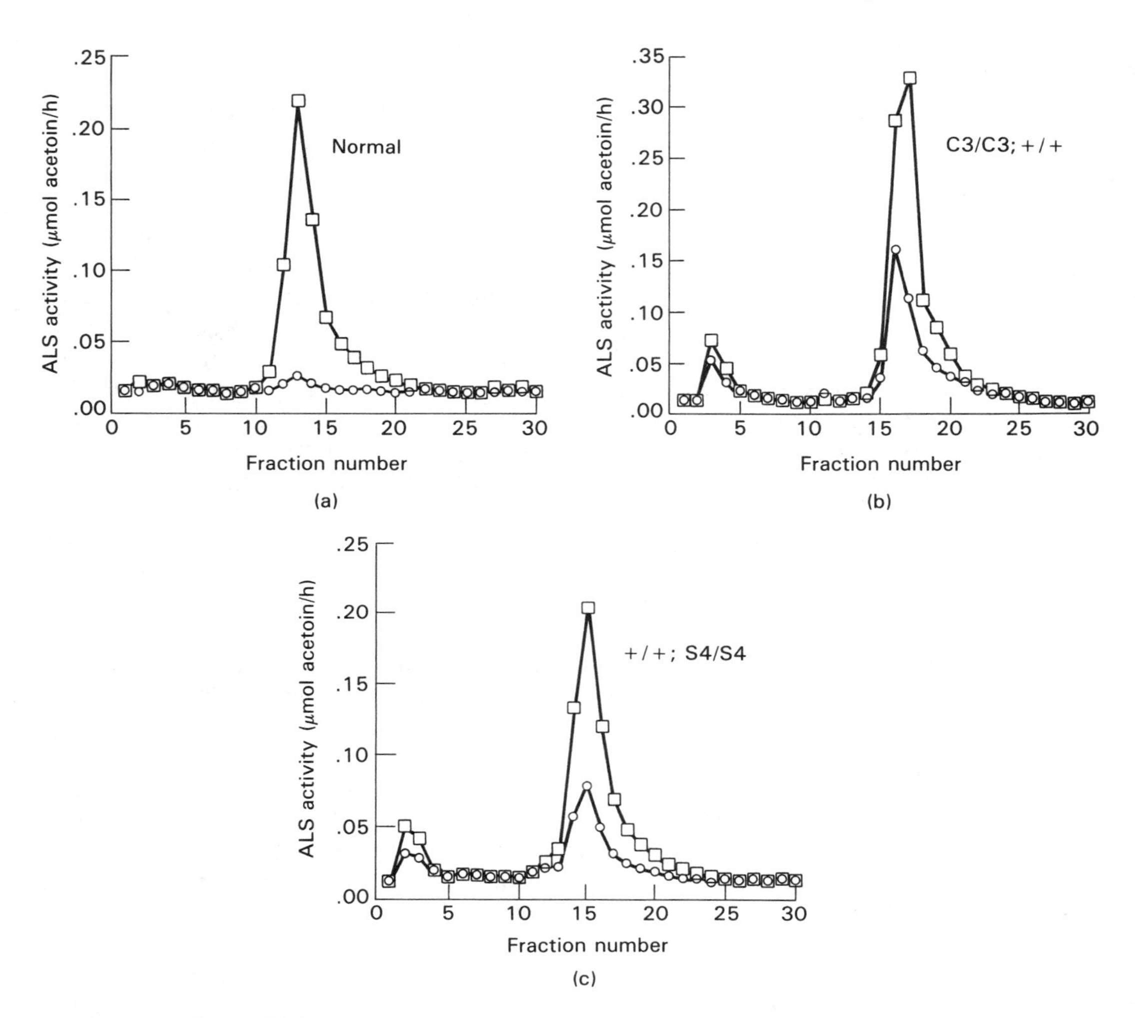

Figure 13.7 Fractionation by ion exchange chromatography of ALS activities in leaf extracts of wild type (normal)(+/+;+/+) and two homozygous mutant (*C3/C3*;+/+ and +/+;*S4/S4*) tobacco plants. Activity was assayed with (circles) or without (squares) the presence of 280 nM chlorsulfuron [110].

tobacco [135–137], flax [138], or canola [139], the gene for sulfonylurea-resistant ALS from *Arabidopsis* provides a high level of resistance to sulfonylurea herbicides at the whole plant level. Mutagenesis and selection at the whole plant level in soybean has led to tolerance to sulfonylurea herbicides that is apparently based on enhanced herbicide detoxification [140] and to a resistant line with ALS that is insensitive to sulfonylureas [141]. In the latter case, resistance is monogenic and semidominant.

ALS is generally inhibited slightly less effectively by the triazolopyrimidines than by sulfonylureas [101] (Table 13.1). As with sulfonylureas, branched-chain amino acids will almost completely alleviate growth retardation caused by these herbicides [101, 142]. Like the sulfonylureas, this herbicide class has slow, tight ALS binding kinetics. With respect to pyruvate and TPP, the inhibition kinetics appear to be a linear mixed type with K_i values around 50 nM. The triazolopyrimidine 2-NO_2-6-Me-sulfonanilide very effectively competes radiolabeled sulfometuron off the ALS molecule [112], indicating that they may share the same binding site on ALS.

The imidazolinones are much weaker ALS inhibitors in vitro than the sulfonylureas, with I_{50} values that are generally more than two orders of magnitude higher than those for sulfonylureas (Table 13.1). The K_i values for higher plant ALS range from about 1 to 10 μM [143]. Like the sulfonylureas, the imidazolinones have slow, tight binding to ALS. The initial K_i of imazaquin is 0.8 mM and its final K_i is 20 μM [1]; similarly, the initial K_i of imazapyr for corn ALS is 15 μM and the final K_i is 0.9 μM [144]. Inhibition has been reported to be uncompetitive with respect to pyruvate with corn ALS [143, 145] and noncompetitive with pea ALS [100], in contrast to the competitive inhibition of the sulfonylureas. When the enzyme is incubated with the herbicide in vitro for an extended time, the activity is not restored upon desalting, indicating that the herbicide does not readily dissociate from the enzyme [144]. Furthermore, extractable ALS is greatly decreased by treatment of plants with imidazolinones, indicating irreversible inhibition or destabilization in vivo [145]. Extractable ALS activity is not affected by sulfometuron treatment, and sulfometuron treatment can prevent loss of extractable activity caused by imidazolinone treatment.

Despite the great difference in sensitivity in ALS to imidazolinones and sulfonylureas, the inhibition of growth of cell cultures or intact plants by imidazolinones generally requires only about 10- to 20-fold higher concentrations than required for sulfonylureas. Thus, the in vitro ALS sensitivity does not entirely predict the in vivo herbicidal activity or vice versa. This could be due in part to different enzyme aggregation states in vivo than in vitro, because different aggregation states of the enzyme have different sensitivities to imidazolinones [104]. The growth-reducing activity of imidazolinones can be reversed easily by supplying exogenous branched-chain amino acids to cell cultures [124], to plant roots [146], or to intact plants [147, 148], indicating that inhibition of branched-chain amino acid synthesis is their sole mechanism of action.

Mutants tolerant of imidazolinones have been selected from corn tissue or cell cultures [149, 150] and by microspore mutagenesis and selection with canola [151]. Imidazolinone resistance in corn plants regenerated from these cultures is due to a resistant ALS and in some cases the regenerant is also resistant to sulfonylureas. Development of the corn germplasm into commercial lines is in progress. The molecular basis for the imidazolinone-resistant canola has not been reported; however, the genetics are similar to tobacco sulfonylurea resistance—two unlinked, semidominant genes for resistance that confer maximal resistance when a hybrid with both mutations is produced. No yield or quality penalty has been found with these canola mutants.

Examination of cross-resistance between sulfonylurea-, triazolopyrimidine-, and imidazolinone-resistant plants reveals no consistent pattern; that is, some sulfonylurea-resistant plants or plant cell cultures are cross-resistant to imidazolinones, some are not, and vice versa [101–104, 129, 142, 150, 152, 153]. *Chlamydomonas* lines resistant to imidazolinones have been reported that are even more sensitive to sulfonylureas than are the wild-type strain from which they were derived [129]. Conversely, a sulfonylurea-resistant culture of *Datura innoxia* has been produced that is more sensitive to imidazolinones than the wild type [152]. Imidazolinone-resistant cell lines of *D. innoxia* have been produced which are not resistant to chlorsulfuron [154]. Furthermore, the differences in K_i values of sulfonylureas and imidazolinones is about two orders of magnitude, whereas the efficiencies in competing radiolabeled sulfometuron from ALS differ by about three orders of magnitude [112]. In another study, all but 1 of 15 triazolopyrimidine-resistant mutant cell cultures of tobacco and soybean were also cross-resistant to both sulfonylureas and imidazolinones [142]. One culture was cross-resistant to sulfonylureas only. Affinities of ALS of triazolopyrimidine-resistant mutants of tobacco and cotton for pyruvate and TPP were unchanged, whereas feedback inhibition of the enzyme by valine and leucine was reduced in one cotton mutant [155]. Sulfonylurea-resistant *Stellaria media,* selected by repeated use of chlorsulfuron in the field, is highly cross-resistant to a triazolopyrimidine herbicide, but only marginally cross-resistant to an imidazolinone [156]. These data indicate that the molecular sites of interaction of the sulfonylureas and triazolopyrimidines have a high degree of overlap, whereas the binding domains of sulfonylureas and imidazolinones overlap only slightly.

The sulfonylurea and imidazolinone binding site is separate from the feedback regulation binding sites of valine and leucine. Mutations that confer sulfonylurea or imidazoline resistance do not directly influence valine or leucine effects on ALS activity [104] or vice versa [107].

There is some natural variation among higher plant species in both sensitivity of ALS to different herbicidal inhibitors and in the effects of these inhibitors on growth [139]. The range in sensitivity of ALS to both sulfonylureas and imidazolinones is about 20-fold for both chemical classes. There are no good correlations between the relative effects of different compounds, whether from the same chemical class or not, on ALS from different species. The range of growth responses to these compounds among the same species is much greater than 20-fold, apparently because of differences in factors other than ALS, such as metabolic degradation.

Mutations in the gene coding for ALS can increase the whole plant resistance to ALS inhibitors many fold. Weeds that have become highly resistant to ALS-inhibiting herbicides were first reported in 1987, after only 5 years of sulfonylurea herbicide use [157]. Only 106 g of sulfonylureas per hectare were used over this time period. The resistance has been demonstrated to be due to altered ALS, and cross-resistance to imidazolinones has been found in sulfonylurea-resistant biotypes of two weed species. Earlier predictions of rapid development of weed resistance to ALS inhibitors, based on the ease of producing resistant mutants under laboratory conditions, appear to be correct.

The apparent ease with which mutant plants that are highly resistant to these herbicides can be created, and the rapidly growing number of reports of weeds resistant to sulfonylureas and imidazolinones, suggest that the ALS gene(s) is highly plastic. The complex patterns of cross-resistance demonstrate that several resistant mutations are possible. To date, there are few data on the biological fitness of the different ALS mutants, but all indications are that many of them are not substantially less fit than wild types. Nevertheless, no weeds or crops have been reported with natural (that is, not selected for) tolerance to commercially available ALS inhibitors at the ALS level. All commercially available ALS inhibitor herbicides have been developed for crops that are tolerant due to rapid degradation (e.g., [158]). These facts suggest that there is a fitness price paid for ALS that is resistant to most inhibitors.

Cases of multiple resistances in which weeds sprayed with a herbicide with one mechanism of action become resistant to herbicides with other mechanisms of action to which they may not have been exposed have recently occurred. For example, in Australia annual ryegrass (*Lolium rigidum*) sprayed with diclofop-methyl has become resistant to both diclofop-methyl and chlorsulfuron [159]. Neither enzymatic site of action was altered in this case, suggesting that resistance is based on rapid metabolic degradation of both herbicides.

The second enzyme of branched-chain amino acid biosynthesis, acetolactate reductoisomerase (ALR, Figure 3.5), is inhibited by the experimental herbicide HOE-704 (2-methylphosphinoyl-2-hydroxyacetic acid; Figure 13.5) [125] and reaction-intermediates such as *N*-isopropyl oxalylhydroxamate [126]. HOE-704-inhibited *Lemna* growth can be reversed by addition of valine, leucine, and isoleucine to the media. Acetolactate and acetoin, the decarboxylation products of acetolactate, accumulate rapidly in treated *Lemna* and maize tissues. Acetolactate reductoisomerase has two substrates, 2-aceto-2-hydroxybutyrate (AHB) when synthesizing isoleucine, and acetolactate (AL) when synthesizing valine or leucine. The I_{50} values for the enzyme from carrot are 19.4 and 8.2 μM for AHB and AL, respectively. Little is known of this enzyme in plants, and the mechanism by which HOE-704 inhibits it is unknown. Only one enantiomer of HOE-704 may be active as a herbicide.

Isopropylmalate isomerase (IMI) catalyzes the isomerization of 2-isopropylmalate to 3-isopropylmalate (3-IPM) during leucine synthesis. It is inhibited competitively by 2-hydroxy-3-nitro-4-methylpentanoic acid, a nitro analogue of 3-IPM [1], with a K_i of 1 μM in its anionic form. Several effective inhibitors of 2,3-dihydroxyacid dehydratase (DHD), the enzyme that produces 2-oxoisovaleric acid in the valine synthesis pathway, have been discovered [160]. However, only one of these has significant herbicidal activity, and there is no proof that inhibition of 2,3-dihyroxyacid dehydratase is its mechanism of action.

13.4 INHIBITION OF GLUTAMINE SYNTHESIS

Glutamine synthetase (GS) is the initial enzyme in the pathway that assimilates inorganic nitrogen into organic compounds. It is a pivotal enzyme in nitrogen metabolism in that, in addition to assimilating ammonia produced by nitrite reductase, it recycles ammonia produced by other processes, including photorespiration and deamination reactions (Figure 13.8). A recent review discusses GS and its inhibitors in detail [162]. GS is a nuclear-coded enzyme that is found as different isoforms in the cytoplasm and plastid [163, 164]. The ratio of the two isoforms varies between cell types and species, but in green tissues the chloroplast isoform is always the predominant or only form found. The plastidic isoform is composed of eight subunits of four or six different types. Each of the eight subunits has a reaction center [165].

GS is inhibited by several compounds that share some structural similarities (Figure 13.9). These compounds are all variations of substituted glutamate and most of them are naturally occurring peptides. Three of these compounds, methionine sulfoximine (MSO) [166], phosphinothricin (PPT; herbicidal name—

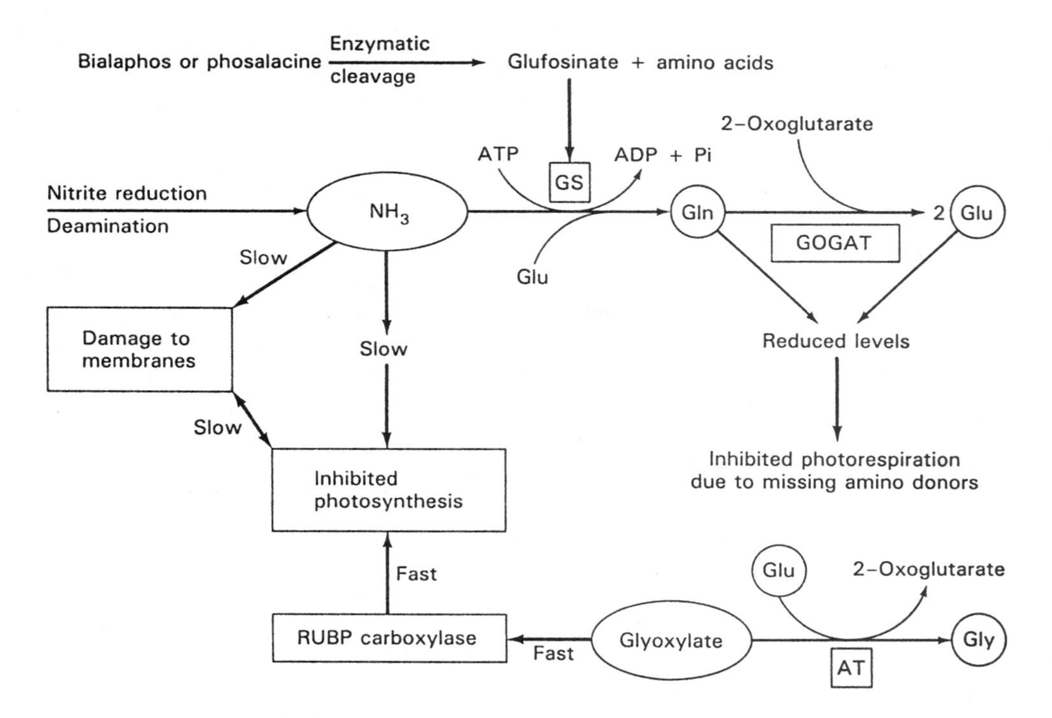

Figure 13.8 Effects of inhibition of glutamine synthetase (GS) inhibition on the physiology of a green plant cell. Bold arrows represent inhibition, ovals indicate increased levels, and circles indicate decreased levels. GOGAT = glutamate synthase, AT = aminotransferase. Adapted from [161].

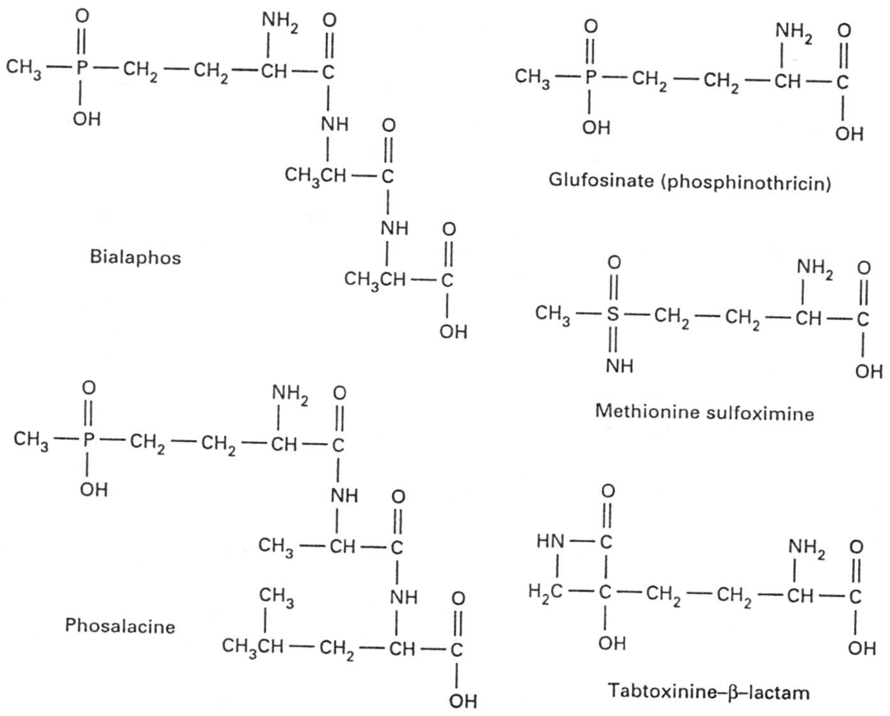

Bialaphos

Glufosinate (phosphinothricin)

Methionine sulfoximine

Phosalacine

Tabtoxinine–β–lactam

Figure 13.9 Structures of several glutamine synthetase inhibitors.

glufosinate), and tabtoxinine-β-lactam (tabtoxin) have been studied extensively. Chloroplastic GS from spinach is inhibited by MSO and glufosinate with K_i values of 100 and 6.1 μM, respectively [167]. The K_i for cytosolic GS from *Sorghum* sp. to phosphinothricin is 8 μM [168]. Several γ-oxygenated forms of glufosinate inhibit GS activity less effectively than does glufosinate [168]. Bialaphos is a tripeptide (phosphinothricyl-L-alanyl-L-alanine) produced by *Streptomyces hygroscopis* that is metabolized to glufosinate and alanine by plants [169]. Similarly, the tripeptide phosalacine (phosphinothricyl-L-alanyl-L-leucine) is also metabolized to glufosinate by plants [170]. Consequently, the mechanism of action of bialaphos is essentially the same as that of glufosinate. These four compounds are also discussed in Chapter 18, in the context of naturally occurring compounds as herbicides. Glufosinate and bialaphos are the only GS inhibitors that have been commercialized.

Tabtoxin, MSO, and glufosinate are irreversible inhibitors of GS in the presence of ATP [1]. This probably accounts for most of the rapid loss of extractable GS activity from glufosinate-treated plants [171]. The first step in the normal mechanism of GS activity is formation of the tetrahedral adduct, λ-glutamylphosphate, from ATP and glutamate. The inactivating activity of these three inhibitors is attributed to their close transition state analog properties when phosphorylated to the tetrahedral adduct of the substrates. This irreversible binding under physiological conditions was used by Höpfner et al. [165] to determine that chloroplast GS from mustard contains eight reaction centers per molecule; each GS molecule was labeled with eight ^{32}P-λ-glutamylphosphate adducts.

Only the L-isomer of the chemically synthesized glufosinate D- and L-isomer mixture is active as a GS inhibitor [172]. Structure-activity studies have yielded no significant improvement in herbicidal activity [173]. A synthesized version of the phosphorylated species of glufosinate that irreversibly binds GS when formed on the GS molecule was found to be a reversible inhibitor of GS; this is probably because conformational changes that occur during ATP binding, that are favorable for irreversible binding of glufosinate after it is phosphorylated, are unfavorable for prephosphorylated glufosinate [162].

Inhibition of GS activity by glufosinate leads to a rapid accumulation of high levels of ammonia under conditions in which NO_2^- is being photosynthetically reduced and/or conditions which support photorespiration [174–178]. Similarly, barley mutants lacking GS are detected by their inability to grow under conditions that favor photorespiration [179]. Under nonphotorespiratory conditions (high CO_2 and low O_2), glufosinate has little effect on photosynthesis [174, 175, 178]. Similarly, Turner et al. [180] showed that tabtoxin has no effect on ammonia accumulation and little phytotoxicity in plants grown under high CO_2.

Accumulation of ammonia caused by glufosinate is accompanied by cessation of photosynthesis, disruption of chloroplast structure, and vesiculation of the stroma. Although some have attributed the inhibition of photosynthesis in cells treated with GS inhibitors to effects of ammonia on photosynthesis, and on photophosphorylation in particular (e.g., [181, 182]), Sauer et al. [175] found glutamine depletion caused by glufosinate to be the primary cause of cessation of photosynthesis. Simultaneous treatment with glutamine and glufosinate resulted in reduced effects of glufosinate on photosynthesis, despite high ammonia levels (Figure 13.10) [178]. A similar effect of glutamine on MSO activity has also been reported [183]. In addition, exogenously supplied ammonia (up to 50 mmol per g fr wt) has comparatively little effect on photosynthesis [175, 183] or growth [184]. The cessation of photosynthesis caused by GS inhibitors has also been attributed to accumulation of glyoxylate (a RuBP carboxylase inhibitor) [183], inhibition of protein synthesis [173], and to depletion of Calvin cycle intermediates [175] (Figure 13.8). Cessation of the Calvin cycle results in channeling of some energy into photosynthetic reduction of molecular oxygen, nitrite, sulfur, and phosphorus. Photoinhibition will occur, ultimately resulting in membrane lipid peroxidation via triplet chlorophyll (see Chapter 9). Glutamate, proline, arginine, and a combination of the latter two amino acids have been shown to provide some amelioration of the growth-reducing effects of glufosinate on rice calli [185].

Glufosinate and other GS inhibitors are nonselective herbicides. Although there is considerable variation between species in sensitivity to glufosinate, the variation is not due to differences in sensitivity of their GS [186]. Plants tolerant or

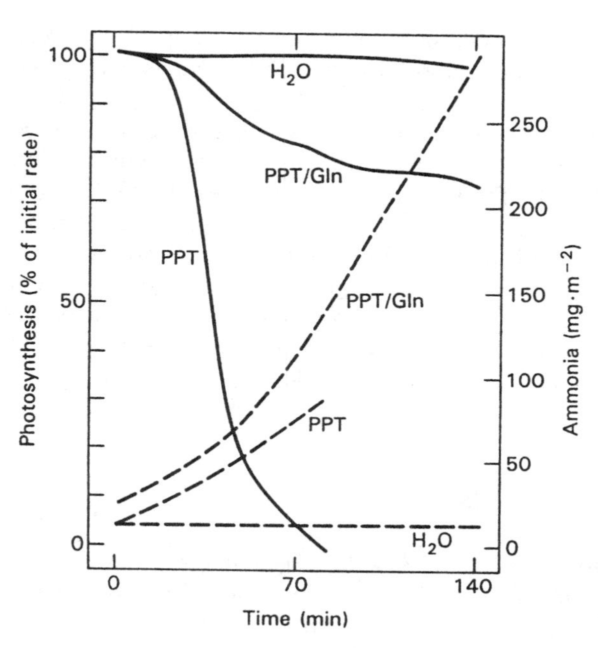

Figure 13.10 Photosynthetic activity (——) and ammonia accumulation (----) of mustard primary leaves treated with 1 mM glufosinate (PPT), 1 mM glufosinate plus 50 mM Gln (glutamine), or water [175].

resistant to glufosinate have been produced by two strategies. In one study, selection of cell lines of alfalfa on cultures of increasing concentration of glufosinate led to tolerant (20-fold) cell lines that owed their tolerance to gene amplification of a GS-encoding gene [187]; gene copy number correlated with the level of tolerance. Unfortunately, during selection, the ability to regenerate plants was lost, and cell fusion with highly regenerative lines did not restore the ability to regenerate intact plants [188]. In another study, selection for tolerance of intact oat plants to the pathogen that produces tabtoxin resulted in tolerant lines of oat with leaf-derived GS (both cytoplasmic and plastidic) with much lower sensitivity to tabtoxin than leaf GS from unselected oat plants [189]. This was the only mechanism found to account for insensitivity to the pathogen. The insensitive GS isoforms were sensitive to MSO, but their sensitivity to glufosinate was not reported. Further work is needed to determine whether genetic manipulation of GS quantity and/or quality can confer resistance to GS-inhibiting herbicides in crops without a metabolic penalty.

A highly successful strategy for producing glufosinate- or bialaphos-resistant crops has been based on the mechanism by which bialaphos-producing microorganisms protect themselves. The glufosinate resistance *bar* gene of *Streptomyces hygroscopicus* [190] and *pat* gene of *Streptomcyces viridochromogenes* [191] encode for PPT (glufosinate) acetyltransferase (PAT). There is considerable homology between the two genes. Soil microorganisms can also inactivate glufosinate by transamination and oxidative deamination [192]. PAT has high substrate specificity for glufosinate, and acetylated glufosinate is apparently herbicidally inactive and stable in vivo. Glufosinate-resistant rice callus has been produced by electroporation of plasmid DNA containing the *bar* gene with a promoter [185]. Bialaphos- and glufosinate-resistant corn cells have been generated by bombardment of suspension cell cultures with microprojectiles coated with plasmids containing the *bar* gene (193, 194). Transgenic, glufosinate-resistant tobacco, tomato, and potato plants have been produced by *Agrobacterium*-mediated transformation with the *bar* gene coupled with the cauliflower mosaic virus promoter [195], and similar resistant tobacco plants have been generated with the *pat* gene [196]. Transformed plants with the *bar* gene are completely resistant to field rates of glufosinate in the greenhouse [190, 194] and field [196]. Glufosinate-treated resistant plants produce normal amounts of viable seed [190]. They are resistant to both glufosinate and bialaphos.

13.5 INHIBITION OF HISTIDINE SYNTHESIS

The enzymology of histidine biosynthesis in higher plants is not well studied, with characterization of only one of the higher plant enzymes of this pathway, histinol dehydrogenase, having been accomplished so far [1]. Although amitrole is known to inhibit carotenoid biosynthesis (Chapter 8), there is evidence that it also inhibits histidine synthesis in higher plants. Inhibition of growth of yeast, *Neurospora*, and *Salmonella* by the herbicide amitrole can be reversed by including histidine in the media [197–199]. Histidine does not affect amitrole-inhibited root elongation in *Arabidopsis thaliana;* however, it does reverse the growth inhibition caused by the specific histidine synthesis inhibitor triazolealanine [200]. Imidazole glycerol phosphate (IGP) and imidazoleglycerol (IG) accumulate in amitrole-treated yeast and *Salmonella*. Isolated IGP dehydratase from *Salmonella* and yeast is competitively inhibited by amitrole. No other enzymes of the histidine synthesis pathway are affected by amitrole. IGP dehydratase inhibition by amitrole may be due to its structural similarity to IGP (Figure 13.11). Only one published paper has supported the view that IGP dehydratase from higher plants is affected by amitrole [201]. As with microorganisms, IG and IGP accumulate in the suspension culture of amitrole-treated rose cells; histidine added to the media stops this accumulation. Unpublished research by others has produced similar results [2].

Inhibition of histidine synthesis by amitrole is only part of the mechanism of action of this herbicide, because supplying histidine to higher plants treated with amitrole does not counteract the effects of the herbicide, and exogenous histidine strongly inhibits growth during early seedling development when histidine pools are high and unaffected or elevated by amitrole [202, 203]. Many other effects of amitrole on plants and microorganisms have been reported. Some are primary effects such as inhibition of carotenoid synthesis (Chapter 8) and catalase inhibition [204], and the mechanisms of other effects such as inhibition of purine synthesis [205] and interference with porphyrin metabolism [206] have not been deter-

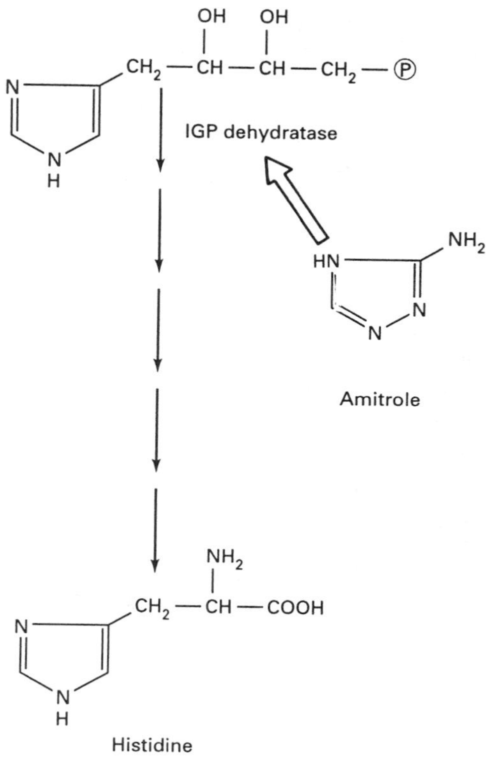

Figure 13.11 The step in the histidine synthesis pathway that is inhibited by amitrole.

mined. In *Arabidopsis,* amitrole inhibits growth at concentrations lower than those required to inhibit pigment synthesis [200]. Thus, although there is good evidence that amitrole inhibits histidine synthesis in higher plants, the effect is apparently masked by its effects on other sites of action. The primary molecular site of action of this herbicide remains a mystery.

13.6 INHIBITION OF SYNTHESIS OF OTHER AMINO ACIDS

Benzadox [(benzamidooxy)acetic acid] is a patented proherbicide that is metabolized to aminooxyacetic acid by plants [207]. Aminooxyacetate is a non-specific inhibitor of pyridoxyl phosphate-requiring enzymes, including many transaminases [208]. No other commercial herbicides are thought to have an enzyme of an amino acid synthesis pathway as their primary site of action. However, there are numerous unexploited enzymes of amino acid metabolism, and several inhibitor leads from which herbicides could be developed. For instance, phaseolotoxin, a phytotoxic tripeptide from *Pseudomonas syringae* that causes halo blight in beans, is a potent inhibitor of ornithine carbamoyltransferase [209, 210], one of the nine enzymes required for arginine synthesis. Exogenous arginine can counteract the phytotoxicity of this nonselective toxin [211]. Rhizobitoxin inhibits methionine synthesis by inhibition of β-cystathionase [212] and gostatin inhibits glutamate synthesis by inhibition of aspartate aminotransferase [213]. Root growth inhibition caused by an analog of aspartate, aminoethylcysteine, can be reversed by lysine [214], indicating that it inhibits an enzyme involved in lysine synthesis.

13.7 SUMMARY

Three enzymatic sites in amino acid synthesis pathways have been found to be the sites of action of several enormously successful herbicides. The interrelationships of these pathways and the general metabolism of the plant are illustrated in Figure 13.12. This schematic is an oversimplification that leaves out many intermediates and the feedback controls of many steps of the pathways. In each of the three established herbicidal sites of action, toxic precursors (in boxes) may be involved in the mode of action of the herbicide. The roles of shikimate and 2-ketobutyrate in the development of phytotoxic effects of glyphosate and ALS inhibitors, respectively, remain to be determined.

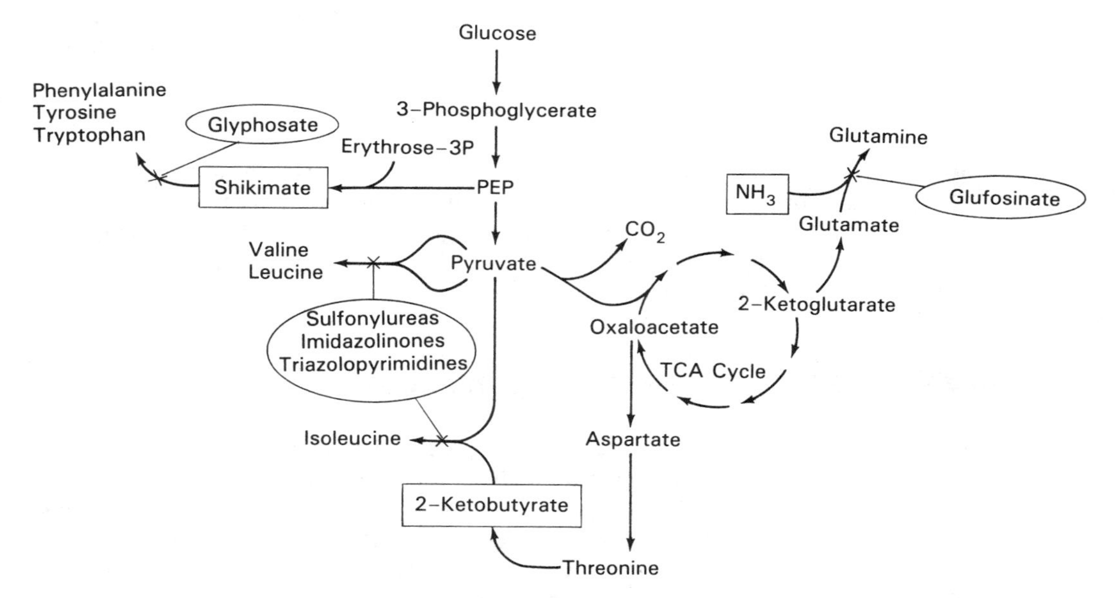

Figure 13.12 Known sites of herbicide action in amino acid biosynthesis and their relationship to general metabolism. Potentially phytotoxic precursors are in boxes and herbicides are in ovals.

The success of each of these herbicides has been due to different factors. However, in each of these three cases, only one route of synthesis of the amino acid product(s) is available to the plant. Enzymatic sites for which there are alternative pathways are not likely candidates as effective herbicide target sites. In each of these three successes, one of the precursors is a phytotoxin. Focusing a herbicide discovery effort only on the enzymes involved in synthesis of essential amino acids may seem to eliminate possible toxicological concerns. However, glufosinate is an example of a highly successful herbicide that inhibits an enzyme of amino acid synthesis that is shared by mammals, and there is no guarantee that inhibitors of enzymes involved in synthesis of essential amino acids will be toxicologically safe. Two sources of information exist that can provide leads for those interested in further exploiting enzymes of amino acid synthesis for herbicides. These are mutant plant amino acid auxotrophs and microbial phytotoxin inhibitors of enzymes involved in amino acid synthesis. This information can often clearly demonstrate the potential herbicidal effectiveness of a chemical that inhibits a particular enzyme.

The mechanisms of action of the EPSPS- and ALS-inhibiting herbicides were discovered after these compounds were found to have excellent herbicidal properties. In contrast, the mechanisms of action of phosphinothricin and bialaphos were known before their development as herbicides. Thus, none of the real commercial successes were discovered by biorational design or by consciously targeting a specific enzyme site. However, the success of these herbicides makes it quite clear that targeting enzymes of amino acid synthesis for biorational development of herbicides is a viable and promising strategy for herbicide discovery. There is every likelihood that more enzymes of amino acid biosynthesis will be target sites of new groups of herbicides in the future. These efforts should greatly increase our understanding of the biosynthesis of amino acids by plants, as well as provide useful new herbicides.

REFERENCES

1. Kishore, G. M., and D. M. Shah. 1988. "Amino acid biosynthesis inhibitors as herbicides." *Annu. Rev. Biochem.*, 57, 627–63.
2. Shaner, D. L. 1989. "Sites of action of herbicides in amino acid metabolism: primary and secondary physiological effects." *Rec. Adv. Phytochemistry*, 23, 227–61.
3. Conn, E. E., ed. 1986. *The Shikimate Pathway*, Rec. Adv. Phytochem., Vol. 20. New York: Plenum, 347 pp.
4. Morris, P. F., R.-L. Doong, and R. A. Jensen. 1989. "Evidence from *Solanum tuberosum* in support of the dual-pathway hypothesis of aromatic biosynthesis." *Plant Physiol.*, 89, 10–14.
5. Duke, S. O., and A. W. Naylor. 1976. "Light effects on phenylalanine ammonia-lyase substrate levels and turnover rate in maize seedlings." *Plant Sci. Lett.*, 6, 361–67.

6. Homeyer, U., and G. Schultz. 1988. "Activation by light of plastidic shikimate pathway in spinach." *Plant Physiol. Biochem.*, 26, 365–70.
7. Duke, S. O. 1988. "*Glyphosate*." In *Herbicides—Chemistry, Degradation, and Mode of Action,* P. C. Kearney and D. D. Kaufman, eds. New York: Dekker, pp. 1–70.
8. Cole, D. J. 1985. "Mode of action of glyphosate—a literature analysis." In *The Herbicide Glyphosate,* E. Grossbard and D. Atkinson, eds. London: Butterworths, pp. 48–74.
9. Grossbard, E., and D. Atkinson, eds. 1985. *The Herbicide Glyphosate*. London: Butterworths, 490 pp.
10. Carlisle, S. M., and J. T. Trevors. 1988. "Glyphosate in the environment." *Water, Air, Soil Pollution,* 39, 409–20.
11. Malik, J., G. Barry, and G. Kishore. 1989. "The herbicide glyphosate." *BioFactors,* 2, 17–25.
12. Sawada, Y., Y. Nagai, M. Ueyama, and I. Yamamoto. 1988. "Probable toxicity of surface-active agent in commercial herbicide containing glyphosate." *The Lancet,* 1(8580), 299.
13. Ream, J. E., H. C. Steinrücken, C. A. Porter, and J. A. Sikorski. 1988. "Purification and properties of 5-enolpyruvylshikimate-3-phosphate synthase from dark-grown seedlings of *Sorghum bicolor*." *Plant Physiol.*, 87, 232–38.
14. Padgette, S. R., Q. K. Huynh, S. Aykent, R. D. Sammons, J. A. Sikorski, and G. M. Kishore. 1988. "Identification of the reactive cysteines of *Escherichia coli* 5-enolpyruvylshikimate-3-phosphate synthase and their nonessentiality for enzymatic catalysis." *J. Biol. Chem.*, 263, 1798–1802.
15. Anderson, K. S., J. A. Sikorski, and K. A. Johnson. 1988. "Evaluation of 5-enolpyruvoylshikimate-3-phosphate synthase substrate and inhibitor binding by stopped-flow and equilibrium fluorescence measurements." *Biochemistry,* 27, 1604–10.
16. Anderson, K. S., J. A. Sikorski, and K. A. Johnson. 1988. "A tetrahedral intermediate in the EPSP synthase reaction observed by rapid quench kinetics." *Biochemistry,* 27, 7395–7406.
17. Padgette, S. R., C. E. Smith, Q. K. Huynh, and G. M. Kishore. 1988. "Arginine chemical modification of *Petunia hybrida* 5-enolpyruvylshikimate-3-phosphate synthase." *Arch. Biochem. Biophys.*, 266, 254–62.
18. Huynh, Q. K. 1988. "Evidence for a reactive λ-carboxyl group (glu-418) at the herbicide glyphosate binding site of 5-enolpyruvylshikimate-3-phosphate synthase from *Escherichia coli*." *J. Biol. Chem.*, 262, 11631–635.
19. Huynh, Q. K., S. C. Bauer, G. S. Bild, G. M. Kishore, M. Ganesh, and J. R. Borgmeyer. 1988. "Site-directed mutagenesis of *Petunia hybrida* 5-enolpyruvylshikimate-3-phosphate synthase: Lys-23 is essential for substrate binding." *J. Biol. Chem.*, 263, 11636–639.
20. Stalker, D. M., W. R. Hiatt, and L. Comai. 1985. "A single amino acid substitution in the enzyme 5-enolpyruvylshikimate-3-phosphate synthase confers resistance to the herbicide glyphosate." *J. Biol. Chem.*, 260, 4724–28.
21. Jenson, R. A. 1985. "The shikimate/arogenate pathway: link between carbohydrate metabolism and secondary metabolism." *Physiol. Plant.*, 66, 164–68.

22. Fischer, R. S., A. Berry, C. G. Gaines, and R. A. Jenson. 1986. "Comparative action of glyphosate as a trigger of energy drain in eubacteria." *J. Bacteriol.*, 168, 1147–54.

23. Pinto, J. E. B. P., W. E. Dyer, S. C. Weller, and K. M. Herrman. 1988. "Glyphosate induces 3-dexoxy-D-*arabino*-heptulosonate 7-phosphate synthase in potato (*Solanum tuberosum* L.) cells grown in suspension culture." *Plant Physiol.*, 87, 891–93.

24. Rubin, J. L., and R. A. Jensen. 1985. "Differentially regulated isozymes of 3-deoxy-D-*arabino*-heptulosonate-7-phosphate synthase from seedlings of *Vigna radiata* [L.] Wilczek." *Plant Physiol.*, 79, 711–18.

25. Ishikura, N., and Y. Takeshima. 1984. "Effects of glyphosate on caffeic acid metabolism in *Perrilla* cell suspension cultures." *Plant Cell. Physiol.*, 25, 185–89.

26. Holländer, H., and N. Amrhein. 1980. "The site of the inhibition of the shikimate pathway by glyphosate. I. Inhibition by glyphosate of phenylpropanoid synthesis in buckwheat (*Fagopyrum esculentum* Moench)." *Plant Physiol.*, 66, 823–29.

27. Lydon, J., and S. O. Duke. 1988. "Glyphosate induction of elevated levels of hydroxybenzoic acids in higher plants." *J. Agric. Food Chem.*, 36, 813–18.

28. Schulz, A., T. Munder, H. Holländer-Czytko, and N. Amrhein. 1990. "Glyphosate transport and early effects on shikimate metabolism and its compartmentation in sink leaves of tomato and spinach plants." *Z. Naturforsch.*, 45c, 529–34.

29. Amrhein, N., B. Deus, P. Gehrke, and H. C. Steinrücken. 1980. "The site of the inhibition of the shikimate pathway by glyphosate. II. Interference of glyphosate with chorismate formation *in vivo* and *in vitro*." *Plant Physiol.*, 66, 830–34.

30. Steinrücken, H. C., and N. Amrhein. 1980. "The herbicide glyphosate is a potent inhibitor of 5-enolpyruvylshikimic acid-3-phosphate synthase." *Biochem. Biophys. Res. Commun.*, 94, 1207–12.

31. Krüper, A., P. Gehrke, and N. Amrhein. 1990. "Facile and economical preparation of [^{14}C]-labeled shikimic acid." *J. Lab. Comp. Radiopharm.*, 28, 713–18.

32. Amrhein, N., D. Johanning, J. Schab, and A. Schulz. 1983. "Biochemical basis for glyphosate-tolerance in a bacterium and a plant tissue culture." *FEBS Lett.*, 157, 191–96.

33. Killmer, J., J. Widholm, and F. Slife. 1981. "Reversal of glyphosate inhibition of carrot cell culture growth by glycolytic intermediates and organic and amino acids." *Plant Physiol.*, 68, 1299–1302.

34. Geiger, D. R., S. W. Kapitan, and M. A. Tucci. 1986. "Glyphosate inhibits photosynthesis and allocation of carbon to starch in sugar beet leaves." *Plant Physiol.*, 82, 468–72.

35. Rubin, J. L., C. G. Gaines, and R. A. Jensen. 1982. "Enzymological basis for the herbicidal action of glyphosate." *Plant Physiol.*, 70, 833–39.

36. Shah, D. M., R. B. Horsch, H. J. Klee, G. M. Kishore, J. A. Winter, N. E. Tumer, C. M. Hironaka, P. R. Sanders, C. S. Gasser, S. Aykent, N. R. Siegel, S. G. Rogers, and R. T. Fraley. 1986. "Engineering herbicide tolerance in transgenic plants." *Science*, 233, 478–81.

37. Holländer-Czytko, H., and N. Amrhein. 1987. "5-Enolpyruvylshikimate-3-phosphate synthase, the target enzyme of the herbicide glyphosate, is synthesized as a precursor in a higher plant." *Plant Physiol.*, 83, 229–31.

38. Gasser, C. S., J. A. Winter, C. M. Hironaka, and D. M. Shah. 1988. "Structure, expression, and evolution of the 5-enolpyruvylshikimate-3-phosphate synthase genes of petunia and tomato." *J. Biol Chem.*, 263, 4280–89.

39. Della-Cioppa, G., C. Bauer, B. Klein, D. Shah, R. Fraley, and G. Kishore. 1986. "Translocation of the precursor of 5-enolpyruvylshikimate-3-phosphate synthase into chloroplasts of higher plants *in vitro*." *Proc. Natl. Acad. Sci. USA,* 83, 6873–77.

40. Della-Cioppa, G., and G. M. Kishore. 1988. "Import of a precursor protein into chloroplasts is inhibited by the herbicide glyphosate." *EMBO J.*, 7, 1299–1305.

41. Dyer, W. E., S. C. Weller, R. A. Bressan, and K. M. Herrman. 1988. "Glyphosate tolerance in tobacco (*Nicotiana tabacum* L.)." *Plant Physiol.*, 88, 661–66.

42. Goldsbrough, P. B., E. M. Hatch, B. Huang, W. G. Kosinski, W. E. Dyer, K. M. Herrman, and S. C. Weller. 1990. "Gene amplification in glyphosate tolerant tobacco cells." *Plant Sci.*, 72, 53–62.

43. Nafziger, E. D., J. M. Widholm, H. C. Steinrücken, and J. L. Killmer. 1984. "Selection and characterization of a carrot cell line tolerant to glyphosate." *Plant Physiol.*, 76, 571–74.

44. Smith, C. M., D. Pratt, and G. A. Thompson. 1986. "Increased 5-enolpyruvylshikimic acid-3-phosphate synthase activity in a glyphosate-tolerant variant strain of tomato cells." *Plant Cell Rep.*, 5, 298–301.

45. Smart, C. C., D. Johanning, G. Muller, and N. Amrhein. 1985. "Selective overproduction of 5-*enol*-pyruvylshikimic acid 3-phosphate synthase in a plant cell culture which tolerates high doses of the herbicide glyphosate." *J. Biol. Chem.*, 260, 16338–346.

46. Steinrücken, H. C., A. Schulz, N. Amrhein, C. A. Porter, and R. Fraley. 1986. "Overproduction of 5-enolpyruvylshikimate 3-phosphate synthase in a glyphosate-tolerant *Petunia hybrida* cell line." *Arch. Biochem. Biophys.*, 244, 169–78.

47. Holländer-Czytko, H., D. Johanning, H. E. Meyer, and N. Amrhein. 1988. "Molecular basis for the overproduction of 5-enolpyruvylshikimate 3-phosphate synthase in a glyphosate-tolerant cell suspension culture of *Corydalis sempervirens.*" *Plant Molec. Biol.*, 11, 215–20.

48. Smart, C. C., and N. Amrhein. 1987. "Ultrastructural localization by protein A-gold immunocytochemistry of 5-enolpyruvylshikimic acid 3-phosphate synthase in a plant cell culture which overproduces the enzyme." *Planta,* 170, 1–6.

49. Rogers, S. G., L. A. Brand, S. B. Holder, E. S. Sharps, and M. J. Bracklin. 1983. "Amplification of the *aro* gene from *Eschericia coli* results in tolerance to the herbicide glyphosate." *Appl. Environ. Microbiol.*, 46, 37–43.

50. Klee, H. J., Y. M. Muskopf, and C. S. Gasser. 1987. "Cloning of an *Arabidopsis thaliana* gene encoding 5-enolpyruvylshikimate-3-phosphate synthase: sequence analysis and manipulation to obtain glyphosate-tolerant plants." *Mol. Gen. Genet.*, 210, 437–42.

51. Jordan, M. C., and A. McHughen. 1988. "Glyphosate tolerant flax plants from *Agrobacterium* mediated gene transfer." *Plant Cell Rep.*, 7, 281–84.

52. Sost, D., A. Schulz, and N. Amrhein. 1984. "Characterization of a glyphosate-insensitive 5-enolpyruvylshikimic acid-3-phosphate synthase." *FEBS Lett.*, 173, 328–42.

53. Kishore, G. M., L. Brundage, K. Kolk, S. R. Padgette, D. Rochester, Q. K. Huynh, and G. Della-Cioppa. 1986. "Isolation, purification, and characterization of a glyphosate-tolerant mutant of *E. coli* EPSP synthase." *Fed. Proc. Amer. Soc. Expl. Biol.*, 45, 1506.

54. Della-Cioppa, G., S. C. Bauer, M. L. Taylor, D. E. Rochester, B. K. Klein, D. M. Shah, R. T. Fraley, and G. M. Kishore. 1987. "Targeting a herbicide-resistant enzyme from *Escherichia coli* to chloroplasts of higher plants." *Bio/Technology*, 5, 579–84.

55. Padgette, S. R., G. Della-Cioppa, D. M. Shah, R. T. Fraley, and G. M. Kishore. 1989. "Selective herbicide tolerance through protein engineering." In *Cell Culture and Somatic Cell Genetics of Plants*, Vol. 6, New York: Academic, pp. 441–76.

56. Comai, L., D. Facciotti, W. R. Hiatt, G. Thompson, R. E. Rose, and D. M. Stalker. 1985. "Expression in plants of a mutant *aro*A gene from *Salmonella typhimurium* confers tolerance to glyphosate." *Nature*, 327, 741–44.

57. Fillatti, J. J., J. Kiser, R. Rose, and L. Comai. 1987. "Efficient transfer of a glyphosate tolerance gene into tomato using a binary *Agrobacterium tumefaciens* vector." *Bio/Technology*, 5, 726–30.

58. Riemenschneider, D. E., B. E., Haissig, J. Sellmer, and J. J. Fillatti. 1988. "Expression of an herbicide tolerance gene in young plants of a transgenic hybrid poplar clone." In *Somatic Cell Genetics of Woody Plants*, M. R. Ahuja, ed. Boston: Kluwer Academic Publishers, pp. 73–80.

59. Fillatti, J. J., B. Haissig, B. McCown, L. Comai, and D. Riemenschneider. 1988. "Development of glyphosate-tolerant *Populus* plants through expression of a mutant *aroA* gene from *Salmonella typhimurium*." *In Genetic Manipulation of Woody Plants*, J. W. Hanover and D. E. Keathley, eds. New York: Plenum, pp. 243–49.

60. Thompson, G. A., W. R. Hiatt, D. Facciotti, D. M. Stalker, and L. Comai. 1987. "Expression in plants of a bacterial gene coding for glyphosate resistance." *Weed Sci.*, 35 (Suppl. 1), 19–23.

61. Comai, L., N. Larson-Kelly, J. Kiser, C. J. D. Mau, A. R. Pokalsky, C. K. Shewmaker, K. McBride, A. Jones, and D. M. Stalker. 1988. "Chloroplast transport of a RuBP carboxylase small subunit-EPSP synthase chimeric protein requires part of the mature small subunit in addition to the transit peptide." *J. Biol. Chem.*, 263, 5104–09.

62. Larson-Kelly, N., L. Comai, J. Kiser, C. Mau, A. R. Pokalsky, K. McBride, A. Jones, C. Shewmaker, and D. M. Stalker. 1988. "Chloroplast delivery of a bacterial EPSP synthase in transgenic plants and tolerance to glyphosate." *SAAS Bull.*, 1, 37–40.

63. Benfey, P. N., and N.-H. Chua. 1989. "Regulated genes in transgenic plants." *Science*, 244, 174–81.

64. Marshall, G., R. C. Kirkwood, and D. J. Martin. 1987. "Studies on the mode of action of asulam, aminotrizole and glyphosate in *Equisetum arvense* L (field horsetail). II: The metabolism of [^{14}C]asulam, [^{14}C]aminotriazole and [^{14}C]glyphosate." *Pestic. Sci.*, 18, 65–77.

65. Moore, J. K., H. D. Braymer, and A. D. Larson. 1983. "Isolation of a *Pseudomonas* sp. which utilizes the phosphate herbicide glyphosate." *Appl. Environ. Microbiol.*, 46, 316–20.

66. Kishore, G. M., and G. S. Jacob. 1987. "Degradation of glyphosate by *Pseudomonas* sp. PG2982 via a sarcosine intermediate." *J. Biol. Chem.*, 262, 12164–169.

67. Pipke, R. N., N. Amrhein, G. S. Jacob, J. Schaefer, and G. M. Kishore. 1987. "Metabolism of glyphosate by an *Arthrobacter* sp. GLP-1." *Eur. J. Biochem.*, 165, 267–73.

68. Balthazor, T. M., and L. E. Hallas. 1986. "Glyphosate-degrading microorganisms from industrial activated sludge." *Appl. Environ. Microbiol.*" 51, 432–34.

69. Jacob, G. S., J. R. Garbow, L. E. Hallas, N. M. Kimack, G. M. Kishore, and J. Schaefer. 1988. "Metabolism of glyphosate in *Pseudomonas* sp. strain LBr." *Appl. Environ. Microbiol.*, 54, 2953–58.

70. Hoagland, R. E. 1980. "Effects of glyphosate on metabolism of phenolic compounds. VI. Effects of glyphosine and glyphosate metabolites on phenylalanine ammonia-lyase activity, growth, and protein, chlorophyll, and anthocyanin levels in soybean (*Glycine max*) seedlings." *Weed Sci.*, 28, 393–400.

71. Murata, K., N. Higaki, and A. Kimura. 1989. "Carbon-phosphorus hydrolase: some properties of the enzyme in cell extracts of *Enterobacter aerogenes*." *Agric. Biol. Chem.*, 53, 1225–29.

72. Murata, K., N. Higaki, and A. Kimura. 1989. "Carbon-phosphorus hydrolase: functional association of two different proteins for the enzyme activity in *Enterobacter aerogenes*." *Agric. Biol. Chem.*, 53, 1419–20.

73. Gottrup, O., P. A. O'Sullivan, R. J. Schraa, and W. H. Vanden Born. 1976. "Uptake, translocation, metabolism and selectivity of glyphosate in Canada thistle and leafy spurge." *Weed Res.*, 16, 197–201.

74. Neal, J. C., W. A. Skroch, and T. J. Monaco. 1985. "Effects of plant growth stage on glyphosate absorption and transport in ligustrum (*Ligustrum japonicum*) and blue pacific juniper (*Juniperus conferta*)." *Weed Sci.*, 34, 115–21.

75. Waldecker, M. A., and D. L. Wyse. 1985. "Soil moisture effects on glyphosate absorption and translocation in common milkweed (*Asclepias syriaca*)." *Weed Sci.*, 33, 299–305.

76. Marquis, L. Y., R. D. Comes, and C. P. Yang. 1979. "Selectivity of glyphosate in creeping red fescue and reed canarygrass." *Weed Res.*, 19, 335–42.

77. Boerboom, C. M., D. L. Wyse, and D. A. Somers. 1990. "Mechanism of glyphosate tolerance in birdsfoot trefoil (*Lotus corniculatus*)." *Weed Sci.*, 38, 463–67.

78. Hickock, L. G. 1987. "Applications of *in vitro* selection systems: Whole-plant selection using the haploid phase of *Ceratopteris*." *Amer. Chem. Soc. Symp. Ser.*, 334, 53–65.

79. Podestá, F. E., D. H. Gozález, and C. S. Andrea. 1987. "Glyphosine inhibits maize leaf phosphoenolpyruvate carboxylase." *Plant Cell Physiol.*, 28, 375–78.

80. Jaworski, E. G. 1972. "Mode of action of *N*-phosphonomethyl-glycine: inhibition of aromatic amino acid biosynthesis." *J. Agric. Food Chem.*, 20, 1195–98.

81. Duke, S. O., and R. E. Hoagland. 1981. "Effects of glyphosate on metabolism of phenolic compounds. VII. Root-fed amino acids and glyphosate toxicity in soybean (*Glycine max*) seedlings." *Weed Sci.*, 29, 297–302.

82. Mugnier, J. 1988. "Behaviour of herbicides in dicotyledonous roots transformed by *Agrobacterium rhizogenes* I. Selectivity." *J. Exp. Bot.*, 39, 1045–56.

83. Kumar, D. 1989. "Biochemical characterization of a shikimic acid-resistant mutant of *Nostoc linckia*." *J. Basic Microbiol.*, 29, 353–59.

84. Plieninger, H., and K. Schneider. 1959. "Die Anlagerung von Ammoniak an Shikimisäure." *Ber. Dtsh. Chem. Ges.*, 92, 1587–93.

85. Munõz-Rueda, A., C. Gonzalez-Murua, J. M. Becerril, and M. F. Sánchez-Díaz. 1986. "Effects of glyphosate [*N*-(phophonomethyl)glycine] on photosynthetic pigments, stomatal response and photosynthetic electron transport in *Medicago sativa* and *Trifolium pratense*." *Physiol. Plant.*, 66, 63–68.

86. Richard, E. P., J. R. Goss, and C. S. Arntzen. 1979. "Glyphosate does not inhibit photosynthetic electron transport and phosphorylation of pea (*Pisum sativum*) chloroplasts." *Weed Sci.*, 27, 684–88.

87. Cooley, W. E., and C. L. Foy. 1989. "Effects of SC-0224 on photosynthetic electron transport." *Plant Growth Regul. Soc. Amer. Quart.*, 17, 31–38.

88. Bode, R., and D. Birnbaum. 1989. "Specificity of glyphosate action in *Candida maltosa*." *Biochem. Physiol. Pflanzen*, 184, 163–70.

89. Natchev, I. A. 1988. "Synthesis, enzyme-substrate interaction, and herbicidal activity of phosphoryl analogues of glycine." *Liebigs Ann. Chem.*, 1988, 861–67.

90. Hoagland, R. E. 1988. "Naturally occurring carbon-phosphorus compounds as herbicides." *Amer. Chem. Soc. Symp. Ser.*, 380, 182–210.

91. Duncan, K., M. R. Edwards, and J. R. Coggins. 1987. "The pentafunctional *arom* enzyme of *Saccharomyces cerevisiae* is a mosaic of monofunctional domains." *Biochem. J.*, 246, 375–86.

92. Myrvold, S., L. M. Reimer, D. L. Pompliano, and J. W. Frost. 1989. "Chemical inhibition of dehydroquinate synthase." *J. Amer. Chem. Soc.*, 111, 1861–66.

93. Pompliano, D. L., L. M. Reimer, S. Myrvold, and J. W. Frost. 1989. "Probing lethal metabolic perturbations in plants with chemical inhibition of dehydroquinate synthase." *J. Amer. Chem. Soc.*, 111, 1866–71.

94. Roisch, U., and F. Lingens. 1980. "The mechanism of action of the herbicide *N*-(phosphonomethyl)glycine: its effect on the growth and the enzymes of aromatic amino acid biosynthesis in *Eschericia coli*." *Hoppe Seyler's Z. Physiol. Chem.*, 361, 1049–58.

95. Shuto, A., M. Ohgai, and M. Eto. 1989. "Screening of tryptophan synthase inhibitors as leads of herbicide candidates." *J. Pestic. Sci.*, 14, 69–74.

96. Blair, A. M., and T. D. Martin. 1988. "A review of the activity, fate, and mode of action of sulfonylurea herbicides." *Pestic. Sci.*, 22, 195–219.

97. LaRossa, R. A., S. C. Falco, B. J. Mazur, K. J. Livak, J. V. Schloss, D. R. Smulski, T. K. Van Dyk, and N. S. Yadav. 1987. "Microbiological identification and characterization of an amino acid biosynthesis enzyme as the site of sulfonylurea herbicide action." *Amer. Chem. Soc. Symp. Ser.*, 334, 190–203.

98. Beyer, E. M. Jr., M. J. Duffy, J. V. Hay, and D. D. Schlueter. 1988. "Sulfonylureas." In *Herbicides—Chemistry, Degradation, and Mode of Action*, P. C. Kearney and D. D. Kaufman, eds. New York: Marcel Dekker, pp. 117–89.

99. Hawkes, T. R. 1989. "Studies of herbicides which inhibit branched chain amino acid biosynthesis." Brit. Crop Protect. Conf. Monograph Ser. 42, 131–38.

100. Hawkes, T. R., J. L. Howard, and S. E. Pontin. 1989. "Herbicides that inhibit the biosynthesis of branched chain amino acids." In *Herbicides and Plant Metabolism*, A. D. Dodge, ed. Cambridge: Cambridge University Press, pp. 113–36.

101. Subramanian, M. V., and B. C. Gerwick. 1989. "Inhibition of acetolactate synthase by triazolopyrimidines. A review of recent developments." *Amer. Chem. Soc. Symp. Ser.*, 389, 277–88.
102. Gerwick, B. C., M. V. Subramanian, V. I. Loney-Gallant, and D. P. Chandler. 1990. "Mechanism of action of 1,2,4-triazolo[1,5-*a*]pyrimidines." *Pestic. Sci.*, 29, 357–64.
103. Smith, J. E., C. J. Mauvais, S. Knowlton, and B. J. Mazur. 1988. "Molecular biology of resistance to sulfonylurea herbicides." *Amer. Chem. Soc. Symp. Ser.*, 379, 25–36.
104. Singh, B. K., K. E., Newhouse, M. A. Stidham, and D. L. Shaner. 1989. "Acetohydroxyacid synthase-imidazolinone interaction." Brit. Crop Protect. Conf. Monograph Ser. 42, 87–95.
105. Bryan, J. K. 1980. "Synthesis of the aspartate family and branched-chain amino acids." In *The Biochemistry of Plants, A Comprehensive Treatise. Vol. 5. Amino Acid and Derivatives,* B. F. Miflin, ed. New York: Academic, pp. 403–52.
106. Durner, J., and P. Böger. 1988. "Acetolactate synthase from barley (*Hordeum vulgare* L.): Purification and partial characterization." *Z. Naturforsch.*, 43c, 850–56.
107. Singh, B. K., M. A. Stidham, and D. L. Shaner. 1988. "Separation and characterization of two forms of acetohydroxy acid synthase from black Mexican sweet corn cells." *J. Chromatog.*, 444, 251–61.
108. Creason, G. L., and R. S. Chaleff. 1988. "A second mutation enhances resistance of a tobacco mutant to sulfonylurea herbicides." *Theor. Appl. Genet.*, 76, 177–82.
109. Chaleff, R. S., and C. J. Mauvais. 1984. "Acetolactate synthase is the site of action of two sulfonylurea herbicides in higher plants." *Science,* 224, 1443–45.
110. Chaleff, R. S., and N. F. Bascomb. 1987. "Genetic and biochemical evidence for multiple forms of acetolactate synthase in *Nicotiana tabacum.*" *Mol. Gen. Genet.*, 210, 33–38.
111. Mazur, B. J., C.-F. Chui, and J. K. Smith. 1987. "Isolation and characterization of plant genes coding for acetolactate synthase, the target enzyme for two classes of herbicides." *Plant Physiol.*, 85, 1110–17.
112. Schloss, J. V., L. M. Ciskanik, and D. E. Van Dyk. 1988. "Origin of the herbicide binding site of acetolactate synthase." *Nature,* 331, 360–62.
113. Schloss, J. V. 1990. "Acetolactate synthase, mechanism of action and its herbicide binding site." *Pestic. Sci.*, 29, 283–92.
114. LaRossa, R. A., and J. V. Schloss. 1984. "The sulfonylurea herbicide sulfometuron methyl is an extremely potent and selective inhibitor of acetolactate synthase in *Salmonella typhimurium.*" *J. Biol. Chem.*, 259, 8753–57.
115. LaRossa, R. A., and D. R. Smulski. 1984. "*ilv*B-encoded acetolactate synthase is resistant to the herbicide sulfometuron methyl." *J. Bacteriol.*, 160, 391–94.
116. Ray, T. B. 1985. "The site of action of the sulfonylureas." Proc. Brit. Crop Protect. Conf.—Weeds, 1, 131–38.
117. Chaleff, R. A., and T. B. Ray. 1984. "Herbicide-resistant mutants from tobacco cell cultures." *Science,* 223, 1148–52.
118. Ray, T. B. 1984. "Site of action of chlorsulfuron." *Plant Physiol.*, 75, 827–31.
119. Landstein, D., D. M. Chipman, S. Arad, and Z. Barak. 1990. "Acetohydroxy acid synthase activity in *Chlorella emersonii* under auto- and heterotrophic growth conditions." *Plant Physiol.*, 94, 614–20.

120. LaRossa, R. A., T. K. van Dyk, and D. R Smulski. 1987. "Toxic accumulation of α-ketobutyrate caused by inhibition of the branched-chain amino acid biosynthetic enzyme acetolactate synthase in *Salmonella typhimurium*." *J. Bacteriol.*, 169, 1372–78.

121. LaRossa, R. A., and T. van Dyk. 1987. "Metabolic mayhem caused by 2-ketoacid imbalances." *BioEssays*, 7, 125–30.

122. Rhodes, D., A. L. Hogan, L. Deal, G. C. Jamieson, and P. Haworth. 1987. "Amino acid metabolism of *Lemna minor* L. II. Responses to chlorsulfuron." *Plant Physiol.*, 84, 775–80.

123. Kandra, L., and G. J. Wagner. 1990. "Chlorsulfuron modifies biosynthesis of acyl acid substituents of sucrose esters secreted by tobacco trichomes." *Plant Physiol.*, 94, 906–12.

124. Anderson, P. C., and K. A. Hibberd. 1985. "Evidence for the interaction of an imidazolinone herbicide with leucine, valine, and isoleucine metabolism." *Weed Sci.*, 33, 479–83.

125. Schulz, A., P. Spönemann, H. Köcher, and R. Wengenmayer. 1988. "The herbicidally active experimental compound Hoe 704 is a potent inhibitor of the enzyme acetolactate reductoisomerase." *FEBS Lett.*, 238, 375–78.

126. Schloss, J. V., and A. Aulabaugh. 1990. "Acetolactate synthase and ketol-acid reductoisomerase: Targets for herbicides obtained by screening and *de novo* design." *Z. Naturforsch.*, 45c, 544–51.

127. Jordan, M. C., and A. McHughen. 1987. "Selection for chlorsulfuron resistance in flax (*Linum usitatissimum*) cell cultures." *J. Plant Physiol.*, 131, 333–38.

128. Friedburg, D., and J. Seiffers. 1988. "Sulfonlyurea-resistant mutants and natural tolerance of cyanobacteria." *Arch. Microbiol.*, 150, 278–81.

129. Winder, T., and M. H. Spalding. 1988. "Imazaquin and chlorsulfuron resistance and cross resistance in mutants of *Chlamydomonas reinhardtii*." *Mol. Gen. Genet.*, 213, 394–99.

130. Harnett, M. E., J. R. Newcomb, and R. C. Hodson. 1987. "Mutations in *Chlamydomonas reinhardtii* conferring resistance to the herbicide sulfometuron methyl." *Plant Physiol.*, 85, 898–901.

131. Swanson, E. B., M. P. Coumans, G. L. Brown, J. D. Patel, and W. D. Beversdorf. 1988. "The characterization of herbicide tolerant plants in *Brassica napus* L. after in vitro selection of microspores and protoplasts." *Plant Cell Rep.*, 7, 83–87.

132. Lee, K. Y., J. Townsend, J. Tepperman, M. Black, C. F. Chui, B. Mazur, P. Dunsmuir, and J. Bedbrook. 1988. "The molecular basis of sulfonylurea herbicide resistance in tobacco." *EMBO J.*, 7, 1241–48.

133. Mazur, B. J., and S. C. Falco. 1989. "The development of herbicide resistant crops." *Annu. Rev. Plant Physiol. Plant Mol. Biol.*, 40, 441–70.

134. Haughn, G. W., and C. R. Somerville. 1986. "Sulfonylurea-resistant mutants of *Arabidopsis thaliana*." *Mol. Gen. Genet.*, 204, 430–34.

135. Haughn, G. W., J. Smith, B. Mazur, and C. Somerville. 1988. "Transformation with a mutant *Arabidopsis* acetolactate synthase gene renders tobacco resistant to sulfonylurea herbicides." *Mol. Gen. Genet.*, 211, 266–71.

136. Olszewski, N. E., F. B. Martin, and F. M. Ausubel. 1988. "Specialized binary vector

for plant transformation: expression of the *Arabidopsis thaliana* AHAS gene in *Nicotiana tabacum*." *Nucl. Acids Res.*, 16, 10765–782.

137. Odell, J. T., P. G. Caimi, N. S. Yadav, and C. J. Mauvais. 1990. "Comparison of increased expression of wild-type and herbicide-resistant acetolactate synthase genes in transgenic plants, and indication of posttranscriptional limitation on enzyme activity." *Plant Physiol.*, 94, 1647–54.
138. McHughen, A. 1989. "Agrobacterium mediated transfer of chlorsulfuron resistance to commercial flax cultivars." *Plant Cell Rep.*, 8, 445–49.
139. Miki, B. L., H. Labbé, J. Hattori, T. Ouellet, J. Gabard, G. Sunohara, P. J. Charest, and V. N. Iyer. 1990. "Transformation of *Brassica napus* canola cultivars with *Arabidopsis thaliana* acetohydroxyacid synthase genes and analysis of herbicide resistance." *Theor. Appl. Genet.*, 80, 449–58.
140. Sebastian, S. A., and R. S. Chaleff. 1987. "Soybean mutants with increased tolerance for sulfonylurea herbicides." *Crop Sci.*, 27, 948–52.
141. Sebastian, S. A., G. M. Fader, J. F. Ulrich, D. R. Forney, and R. S. Chaleff. 1989. "Semidominant soybean mutation for resistance to sulfonylurea herbicides." *Crop Sci.*, 29, 1403–08.
142. Subramanian, M. V., and L. Pao. 1989. "Mechanism of action of 1,2,4-triazolo[1,5-a]pyrimidine sulfonamide herbicides." Brit. Pest Control Conf. Monograph Ser., 42, 97–100.
143. Shaner, D. L., P. C. Anderson, and M. A. Stidham. 1984. "Imidazolinones—Potent inhibitors of acetohydroxyacid synthase." *Plant Physiol.*, 76, 545–46.
144. Muhitch, M. J., D. L. Shaner, and M. A. Stidham. 1987. "Imidazolinones and acetohydroxyacid synthase from higher plants—Properties of the enzyme from maize suspension culture cells and evidence for the binding of imazapyr to acetohydroxyacid synthase *in vivo*." *Plant Physiol.*, 83, 451–56.
145. Shaner, D. L., B. K. Singh, and M. A. Stidham. 1990. "Interaction of imidazolinones with plant acetohydroxy acid synthase: evidence for in vivo binding and competition with sulfometuron methyl." *J. Agric. Food Chem.*, 38, 1279–82.
146. Rost, T. L., D. Gladish, J. Steffen, and J. Robbins. 1990. "Is there a relationship between branched chain amino acid pool size and cell cycle inhibition in roots treated with imidazolinone herbicides?" *J. Plant Growth Regul.*, 9, 227–32.
147. Shaner, D. L., and M. L. Reider. 1986. "Physiological responses of corn (*Zea mays*) to AC 243,997 in combination with valine, leucine, and isoleucine." *Pestic. Biochem. Physiol.*, 25, 248–57.
148. Pillmoor, J. B., and J. C. Caseley. 1987. "The biochemical and physiological effects and mode of action of AC 222,293 against *Alopecurus myosuroides* Huds. and *Avena fatua* L." *Pestic. Biochem. Physiol.*, 27, 340–49.
149. Shaner, D. L., and P. C. Anderson. 1985. "Mechanism of action of the imidazolinones and cell culture selection of tolerant maize." In *Biotechnology in Plant Science*, M. Zaitlin, P. R. Day, and A. Hollaender, eds. New York: Academic, pp. 287–300.
150. Anderson, P. C., and M. Georgeson. 1990. "Herbicide-tolerant mutants of corn." *Genome*, 31, 994–99.
151. Swanson, E. B., M. J. Herrgesell, M. Arnoldo, D. W. Sippell, and R. S. C. Wong.

1989. "Microspore mutagenesis and selection: canola plants with field tolerance to the imidazolinones." *Theor. Appl. Genet.*, 78, 525–30.

152. Saxena, P. K., and J. King. 1988. "Herbicide resistance in *Datura innoxia*, cross-resistance of sulfonylurea-resistant cell lines to imidazolinones." *Plant Physiol.*, 86, 863–67.

153. Gabard, J. M., P. J. Charest, V. N. Iyer, and B. L. Miki. 1989. "Cross-resistance to short residual sulfonylurea herbicides in transgenic tobacco plants." *Plant Physiol.*, 91, 574–89.

154. Saxena, P. K., and J. King. 1990. "Lack of cross-resistance of imidazolinone-resistant cell lines of *Datura innoxia* P. Mill. to chlorsulfuron." *Plant Physiol.*, 94, 1111–15.

155. Subramanian, M. V., H.-Y. Hung, J. M. Dias, V. W. Miner, J. H. Butler, and J. J. Jachetta. 1990. "Properties of mutant acetolactate synthases resistant to triazolopyrimidine sulfonanilide." *Plant Physiol.*, 94, 239–44.

156. Hall, L. M., and M. D. Devine. 1990. "Cross-resistance of a chlorsulfuron-resistant biotype of *Stellaria media* to a triazolopyrimidine herbicide." *Plant Physiol.*, 93, 962–66.

157. Primiani, M. M., L. L. Saari, and J. C. Cotterman. 1990. "Resistance of kochia (*Kochia scoparia*) to sulfonylurea and imidazolinone herbicides." *Weed Technol.*, 4, 169–72.

158. Harms, C. T., A. L. Montoya, L. S. Privalle, and R. W. Briggs. 1990. "Genetic and biochemical characterization of corn inbred lines tolerant to the sulfonyurea herbicide primisulfuron." *Theor. Appl. Genet.*, 80, 353–58.

159. Matthews, J. M., J. A. M. Holtum, D. R. Liljegren, B. Furness, and S. B. Powles. 1990. "Cross-resistance to herbicides in annual ryegrass (*Lolium rigidum*) I. Properties of the herbicide target enzymes acetyl-Coenzyme A carboxylase and acetolactate synthase." *Plant Physiol.*, 94, 1180–86.

160. Pirrung, M. C., H. H. Joon, and C. P. Holmes. 1989. "Purification and inhibition of spinach α,β-dihyroxyacid dehydratase." *J. Org. Chem.*, 54, 1543–48.

161. Ziegler, C., and A. Wild. 1989. "The effect of bialaphos on ammonium-assimilation and photosynthesis. II. Effect on photosynthesis and photorespiration." *Z. Naturforsch.*, 44c, 103–08.

162. Lea, P. J., and S. M. Ridley. 1989. "Glutamine synthetase and its inhibition." In *Herbicides and Herbicide Metabolism*, A. D. Dodge, ed. Cambridge: Cambridge University Press, pp. 137–70.

163. Botella, J. R., J. P. Verbelen, and V. Valpuesta. 1988. "Immunocytolocalization of glutamine synthetase in green leaves and cotyledons of *Lycopersicon esculentum*." *Plant Physiol.*, 88, 943–46.

164. McNally, S., B. Hirel, P. Gadal, A. F. Mann, and G. R. Stewart. 1983. "Glutamine synthetases of higher plants." *Plant Physiol.*, 72, 22–25.

165. Höpfner, M., G. Reifferscheid, and A. Wild. 1988. "Molecular composition of glutamine synthetase of *Sinapsis alba* L." *Z. Naturforsch.*, 43c, 194–98.

166. Rowe, W. B., R. A. Ronzio, and A. Meister. 1969. "Inhibition of glutamine synthetase by methionine sulfoximine. Studies on methionine phosphate." *Biochemistry*, 8, 2674–80.

167. Ericson, M. C. 1985. "Purification and properties of glutamine synthetase from spinach leaves." *Plant Physiol.*, 79, 923–27.

168. Walker, D. M., J. F. McDonald, J. E. Franz, and E. W. Logusch. 1990. "Design and synthesis of γ-oxygenated phosphinothricins as inhibitors of glutamine synthetase." *J. Chem. Soc. Perkin Trans.*, 1, 659–66.

169. Wild, A., and C. Ziegler. 1989. "The effect of bialaphos on ammonium-assimilation and photosynthesis. I. Effect on the enzymes of ammonium assimilation." *Z. Naturforsch.*, 44c, 97–102.

170. Omura, S., M. Murata, H. Hanaki, K. Hinotozawa, R. Oiwa, and H. Tanaka. 1984. "Phosalacine, a new herbicidal antibiotic containing phosphinothricin." *J. Antibiot.*, 37, 939–40.

171. Lacuesta, M., B. Gonzalez-Moro, C. Gonzalez-Murua, and A. Muñoz-Rueda. 1990. "Temporal study of the effect of phosphinothricin on the activity of glutamine synthetase, glutamine dehydrogenase and nitrate reductase in *Medicago sativa* L." *J. Plant Physiol.*, 136, 4109–14.

172. Manderscheid, R., and A. Wild. 1986. "Studies on the mechanism of inhibition by phosphinothricin of glutamine synthetase isolated from *Triticum aestivum* L." *J. Plant Physiol.*, 123, 135–42.

173. Maier, L., and P. J. Lea. 1983. "Synthesis and properties of phosphinothricin derivatives." *Phosphor. Sulf.*, 17, 1–19.

174. Wild, A., H. Sauer, and W. Rühle. 1987. "The effect of phosphinothricin (glufosinate) on photosynthesis I. Inhibition of photosynthesis and accumulation of ammonia." *Z. Naturforsch.*, 42c, 263–69.

175. Sauer, H., A. Wild, and W. Rühle. 1987. "The effect of phosphinothricin (glufosinate) on photosynthesis II. The causes of inhibition of photosynthesis." *Z. Naturforsch.*, 42c, 270–78.

176. Rhodes, D., L. Deal, P. Haworth, G. C. Jamieson, C. C. Reuter, and M. C. Ericson. 1986. "Amino acid metabolism of *Lemna minor* L. I. Responses to methionine sufloximine." *Plant Physiol.*, 82, 1057–62.

177. Turner, J. G. 1981. "Tabtoxin, produced by *Pseudomonas tabaci,* decreases *Nicotiana tabacum* glutamine synthetase *in vivo* and causes accumulation of ammonia." *Physiol. Plant Pathol.*, 19, 57–67.

178. Wendler, C., M. Barniske, and A. Wild. 1990. "Effect of phosphinothricin (glufosinate) on photosynthesis and photorespiration of C_3 and C_4 plants." *Photosyn. Res.*, 24, 55–61.

179. Wallsgrove, R. M., J. C. Turner, N. P. Hall, A. C. Kendall, and S. W. J. Bright. 1987. "Barley mutants lacking chloroplast glutamine synthetase—biochemical and genetic analysis." *Plant Physiol.*, 83, 155–58.

180. Turner, J. B., R. R. Taha, and J. Debbage. 1986. "Effects of tabtoxin on nitrogen metabolism." *Physiol. Plant.*, 67, 649–53.

181. Franz, T. A., D. M. Peterson, and R. D. Durbin. 1982. "Sources of ammonium in oat leaves treated with tabtoxin or methionine sulfoximine." *Plant Physiol.*, 69, 345–48.

182. Platt, S. G., and G. E. Anthon. 1981. "Ammonia accumulation and inhibition of photosynthesis in methionine sulfoximine treated spinach." *Plant Physiol.*, 67, 509–13.

183. Ikeda, M., W. L. Ogren, and R. H. Hageman. 1984. "Effect of methionine sulfoximine on photosynthetic carbon metabolism in wheat leaves." *Plant Cell Physiol.*, 25, 447–52.

184. Krieg, L. C., M. A. Walker, T. Senaratna, and B. D. McKersie. 1990. "Growth, ammonia accumulation and glutamine synthetase activity in alfalfa (*Medicago sativa* L.) shoots and cell cultures treated with phosphinothricin." *Plant Cell Rep.*, 9, 80–83.

185. Dekeyser, R., B. Claes, M. Marichal, M. Van Montagu, and A. Caplan. 1989. "Evaluation of selectable markers for rice transformation." *Plant Physiol.*, 90, 217–23.

186. Ridley, S. M., and S. F. McNally. 1985. "Effects of phosphinothricin on the isozymes of gluatamine synthetase isolated from plant species which exhibit varying degrees of susceptibility to the herbicide." *Plant Sci.*, 39, 31–36.

187. Donn, G., E. Tischer, J. A. Smith, and H. M. Goodman. 1984. "Herbicide-resistant alfalfa cells: an example of gene amplification in plants." *J. Molec. Appl. Genet.*, 2, 621–35.

188. Deak, M., G. Donn, A. Feher, and D. Dudits. 1988. "Dominant expression of a gene amplification-related herbicide resistance in *Medicago* cell hybrids." *Plant Cell Rep.*, 7, 158–61.

189. Knight, T. J., D. R. Bush, and P. J. Langston-Unkefer. 1988. "Oats tolerant of *Pseudomonas syringae* pv. *tabaci* contain tabtoxinine-β-lactam-insensitive leaf glutamine synthetases." *Plant Physiol.*, 88, 333–39.

190. Thompson, C. J., N. R. Movva, R. Tizard, R. Crameri, J. E. Davies, M. Lauwerys, and J. Botterman. 1987. "Characterization of the herbicide-resistance gene *bar* from *Streptomyces hygroscopicus.*" *EMBO J.*, 6, 2519–23.

191. Wohlleben, W., W. Arnold, I. Broer, D. Hillemann, E. Strauch, and A. Pühler. 1988. "Nucleotide sequence of the phosphinothricin *N*-acetyltransferase gene from *Streptomyces viridochromogenes* Tü494 and its expression in *Nicotiana tabacum.*" *Gene*, 70, 25–37.

192. Bartsch, K., and C. C. Tebbe. 1989. "Initial steps in the degradation of phosphinothricin (glufosinate) by soil bacteria." *Appl. Environ. Microbiol.*, 55, 711–16.

193. Spencer, T. M., W. J. Gordon-Kamm, R. J. Daines, W. G. Start, and P. G. Lemaux. 1990. "Bialaphos selection of stable transformants from maize cell culture." *Theor. Appl. Genet.*, 79, 625–31.

194. Gordon-Kamm, W. J., T. M. Spencer, M. L. Mangano, T. R. Adams, R. J. Daines, W. G. Start, J. V. O'Brien, S. A. Chambers, W. R. Adams, Jr., N. G. Willetts, T. B. Rice, C. J. Mackey, R. W. Krueger, A. P. Kausch, and P. G. Lemaux. 1990. "Transformation of maize cells and regeneration of fertile transgenic plants." *The Plant Cell*, 2, 603–18.

195. de Block, M., J. Botterman, M. Vandewiele, J. Dockx, C. Thoen, V. Gosselé, N. R. Movva, C. Thompson, M. van Montagu, and J. Leemans. 1987. "Engineering herbicide resistance in plants by expression of a detoxifying enzyme." *EMBO J.*, 6, 2513–18.

196. De Greef, W., R. Delon, M. De Block, J. Leemans, and J. Botterman. 1989. "Evaluation of herbicide resistance in transgenic crops under field conditions." *Bio/Technology*, 7, 61–64.

197. Hilton, J. L., P. C. Kearney, and B. N. Ames. 1965. "The mode of action of the

herbicide 3-amino-1,2,4-triazole: inhibition of an enzyme in histidine biosynthesis." *Arch. Biochem. Biophys.*, 112, 544–47.

198. Kidd, G. L., and S. R. Gross. 1984. "Specific regulatory interconnection between the leucine and histidine pathways of *Neurospora crassa*." *J. Bacteriol.*, 158, 121–27.
199. Klopotowski, T., and A. Wiater. 1965. "Synergism of amitrole and phosphate on the inhibition of yeast IGP dehydrase." *Arch. Biochem. Biophys.*, 112, 562–66.
200. Heim, D. R., and I. M. Larrinua. 1989. "Primary site of action of amitrole in *Arabidopsis thaliana* involves inhibition of root elongation but not of histidine or pigment biosynthesis." *Plant Physiol.*, 91, 1226–31.
201. Davies, M. E. 1971. "Regulation of histidine biosynthesis in cultured plant cells: Evidence from studies on amitrole toxicity." *Phytochemistry*, 10, 783–88.
202. Hilton, J. L. 1969. "Inhibitions of growth and metabolism of 3-amino-1,2,4-triazole (amitrole)." *J. Agric. Food Chem.*, 17, 182–98.
203. McWhorter, C. G., and J. L. Hilton. 1967. "Alterations in amino acid content caused by 3-amino-1,2,4-triazole." *Physiol. Plant.*, 20, 30–40.
204. Margoliash, E., A. Novogrodsky, and A. Schejter. 1960. "Irreversible reaction of 3-amino-1,2,4-triazole and related inhibitors with the protein catalase." *Biochem. J.*, 74, 339–50.
205. Baumann, G., and G. Günther. 1975. "The influence of 3-amino-1,2,4-triazole (amitrole) on the acid-soluble nucleotides and some other phosphorus compounds in oat seedlings." *Biochem. Physiol. Pflanzen*, 167, 339–410.
206. Dörfling, P., W. Dümmler, and D. Mücke. 1970. "The occurrence of coproporhyrin in cultures of *Poteriochromonas stipitata* after incubation with 3-amino-1,2,4-triazole." *Experientia*, 26, 728.
207. Nakamoto, H., M. S. B. Ku, and G. E. Edwards. 1982. "Inhibition of C_4 photosynthesis by (benzamidooxy)acetic acid." *Photosyn. Res.*, 3, 293–305.
208. John, R. A., A. Charteris, and L. J. Fowler. 1978. "The reaction of amino-oxyacetate with pyridoxyl phosphate-dependent enzymes." *Biochem. J.*, 171, 771–79.
209. Moore, R. E., W. P. Niemczura, O. C. H. Kwok, and S. S. Patil. 1984. "Inhibitors of ornithine carbamoyltransferase from *Pseudomonas syringae* pv. *phaseolicola*: Revised structure of the phaseolotoxin." *Tetra. Lett.*, 25, 3931–34.
210. Turner, J. G. 1986. "Effect of phaseolotoxin on the synthesis of arginine and protein." *Plant Physiol.*, 80, 760–67.
211. Turner, J. G., and R. E. Mitchell. 1985. "Association between symptom development and inhibition of ornithine carbamoyltransferase in bean leaves treated with phaseolotoxin." *Plant Physiol.*, 79, 468–73.
212. Giovanelli, J., L. Owens, and S. Mudd. 1971. "Mechanisms of inhibition of β-cystathionase by rhizobitoxine." *Biochim. Biophys. Acta*, 227, 671–84.
213. Nishino, T., and S. Murao. 1983. "Isolation and some properties of an aspartate aminotransferase inhibitor, gostatin." *Agric. Biol. Chem.* 47, 1961–66.
214. Piryns, I., S. Vernaillen, and M. Jacobs. 1988. "Inhibitory effects of aspartate-derived amino acids and aminoethylcysteine, a lysine analog, on the growth of sorghum seedlings; relation with three enzymes of the aspartate pathway." *Plant Sci.*, 57, 93–101.

Chapter 14

Herbicides with Auxin Activity

14.1 AUXIN AND HERBICIDE STRUCTURES

The natural auxin, indole-3-acetic acid (IAA), is synthesized in plant tissue mainly from tryptophan. The size of the tryptophan pool may be up to three (or more) orders of magnitude larger than the IAA pool. It is therefore difficult to differentiate between spontaneous conversion to IAA and the enzymatic and microbial rates of IAA synthesis [1]. The pool of free cellular IAA is, in turn, strongly affected by oxidation as well as rapidly reversible and slowly reversible (''slow release'') conjugations, resulting in regulated intracellular and intercellular IAA pool sizes. In corn seed, for example, IAA esters exceed the free IAA pool by a factor of 100, whereas in seedling tissue this factor is approximately 10. Free auxin is transported exclusively basipetally in plant tissues, which is another factor modulating the pool of free auxin available in a particular tissue.

A large number of herbicides are active auxins, as judged by their growth-promoting effects in cell cultures, specific tissue systems (e.g., the ''growing'' coleoptile), and in intact plants [2]. A selection of auxin herbicides is presented in Figure 14.1. They have been separated into two groups, one with an oxygen bridge between an aromatic substituent and a carboxylic acid, the other with a carboxyl group directly attached to the aromatic ring. Of course, all of these molecules have the free carboxyl group that is required for auxin transport and activity, but the variability of the distance to and the substituents at the aromatic ring system is quite large. Attempts to find a charge distribution common to all active auxins by

Common name	Structure
Phenoxy- and Pyridoxy-carbonic acids:	
2, 4-D	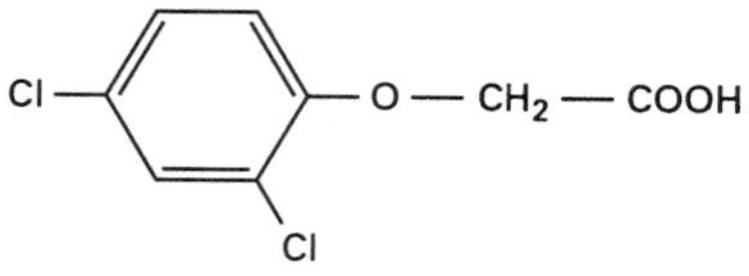 Cl, Cl; —O — CH_2 — COOH
2, 4-DB	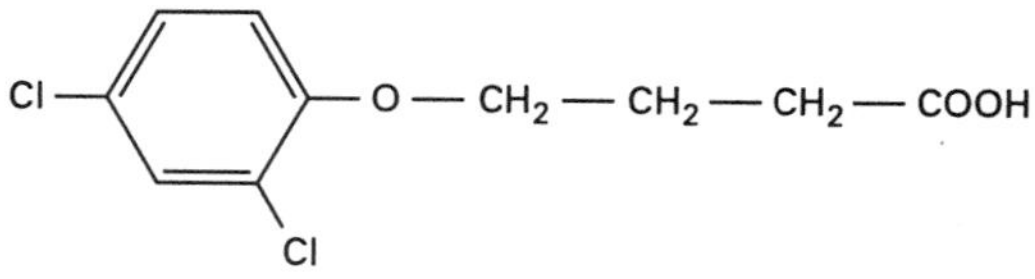 Cl, Cl; —O — CH_2 — CH_2 — CH_2 — COOH
2, 4-DP, dichlorprop	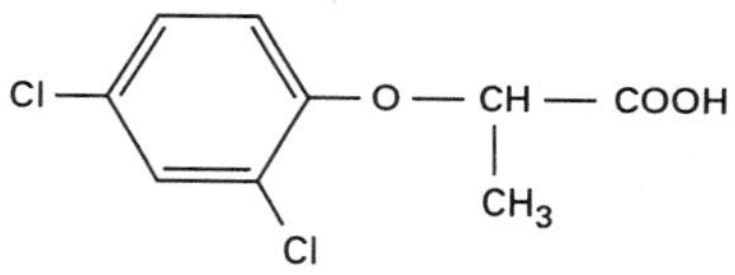 Cl, Cl; —O — CH(CH_3) — COOH
MCPA	Cl, CH_3; —O — CH_2 — COOH
2, 4, 5-T	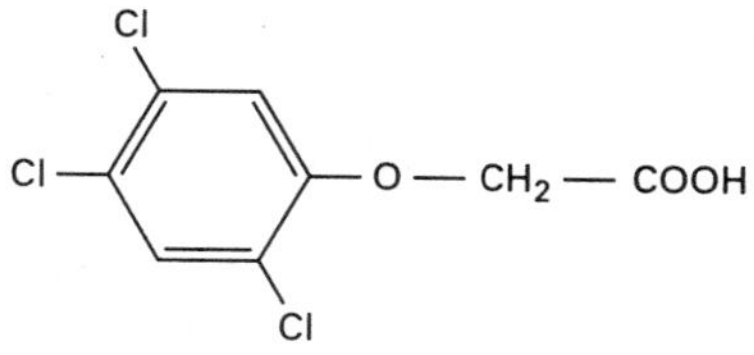 Cl, Cl, Cl; —O — CH_2 — COOH
Fluroxypyr	F, N, Cl, H_2N, Cl; —O — CH_2 — COOH

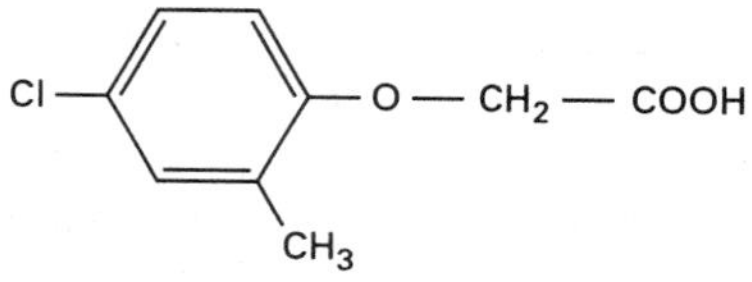

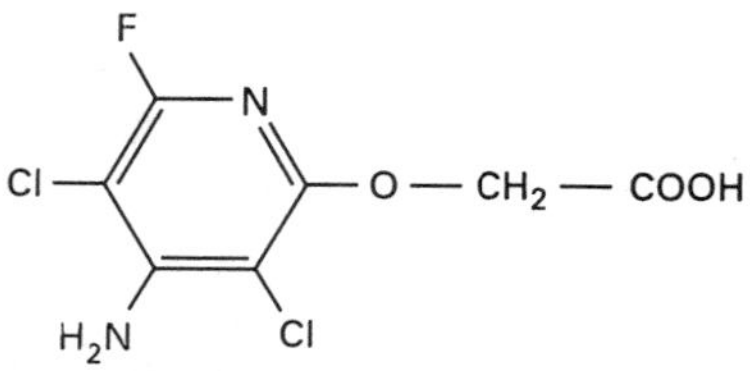

Figure 14.1 Structures of some herbicides with auxin activity.

Common name	Structure
Triclopyr	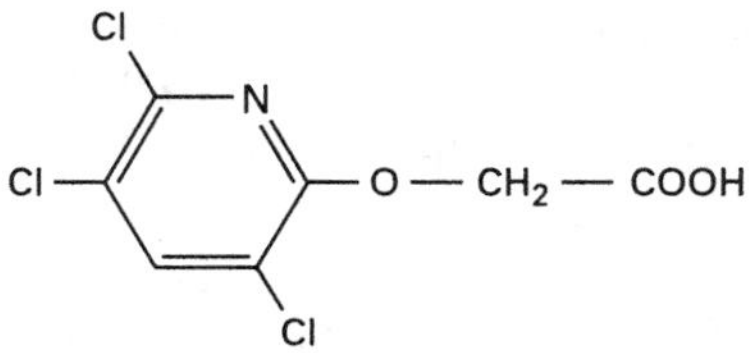
Benzoic acids and analogues:	
Chloramben	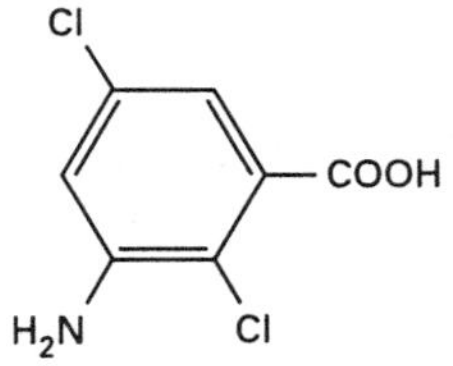
Clopyralid	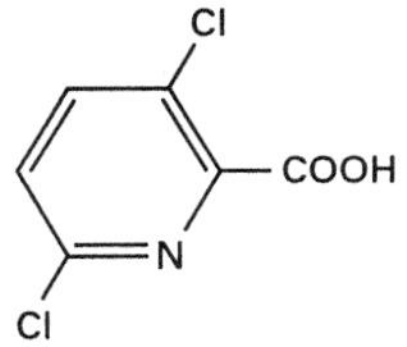
Dicamba	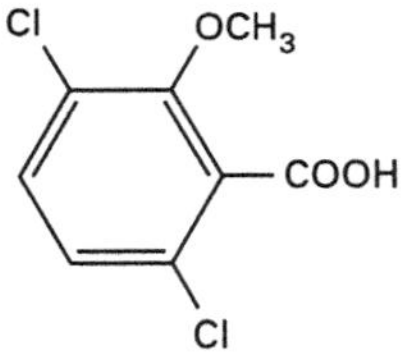
Picloram	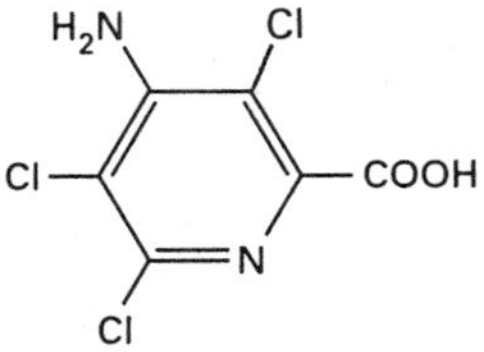
Quinmerac (= BAS 518H)	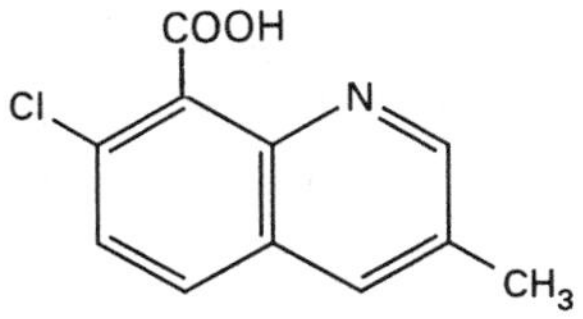

Figure 14.1 (Continued).

molecular orbital calculations have been only partially successful, because several auxin-active molecules do not follow the earlier established rules, and vice versa [3–5]. The herbicide 2,4-DB is a precursor molecule that is transformed into the active auxin herbicide 2,4-D by the well-known catabolic process of β-oxidation [6]. Quinmerac is one of two new auxin herbicides recently developed [2]. The closely similar quinchlorac (3,7-dichloro-8-quinoline carboxylic acid, BAS 514H), however, has an additional mechanism of herbicidal action, leading to wilting and chlorotic leaves in *Echinochloa crus-galli* by an as yet unidentified mechanism [7].

Auxin herbicides are used mainly in grasses (cereal grain crops), coniferous trees (tree nurseries), and certain legumes that are generally quite resistant [8]. They also have utility in pastures, waterways and, in combination with other herbicides, in industrial areas for total vegetation control. Dicotyledonous plants, including woody plants, are mostly very sensitive and respond to the application of auxin herbicides by aberrant growth, necroses, desiccation, and eventual die-back.

Two aspects of auxin behavior in plants, transport and hormone action, have already been mentioned. These aspects are also important for the action of the auxin herbicides. However, some molecules, like 2,3,5-triiodobenzoic acid (TIBA) and α-naphthylphthalamic acid (NPA), bind to an auxin transport site but are not transported [9, 10]. This means that these auxin analogues act as auxin transport inhibitors. They have no herbicidal activity and should not be confused with the active auxins. In effect, they can show anti-auxin activity in certain systems, such as dormancy release (from apical dominance) of axillary buds in NPA-treated bean plants [11]. Several reviews on auxin and auxin herbicide action, in particular on phenoxyacetic acid derivatives, have been written during recent years [3, 8, 12, 13].

14.2 HOW AUXINS WORK

Auxins are plant hormones; that is, they regulate plant cell growth and differentiation in cooperation with other plant hormones. The molecular mechanism of auxin action is still not known in detail, and the elucidation of this and refinement of current hypotheses are the focus of intense ongoing research. An important facet of auxin action is the sensitivity of the tissue [14, 15, 16]. Large differences have been found between different tissues, such as roots, buds, active meristems, or callus cells, and between tissues at different physiological stages of growth. In a growing plant, at any one specific time, one type of tissue may respond to auxin, whereas others may not. The type of response may be positive or negative, inhibitory or stimulatory, depending on the auxin concentration, and can be modulated further by the simultaneous presence of other plant hormones.

The most rapid and most obvious effect induced by auxin and auxin herbicides is growth by cell elongation [9, 17, 18] (as opposed to growth by cell division). Scheme 14.1 lists some early auxin effects and some present-day thoughts on auxin action, arranged in a possible cause-effect sequence. One intensely studied topic is

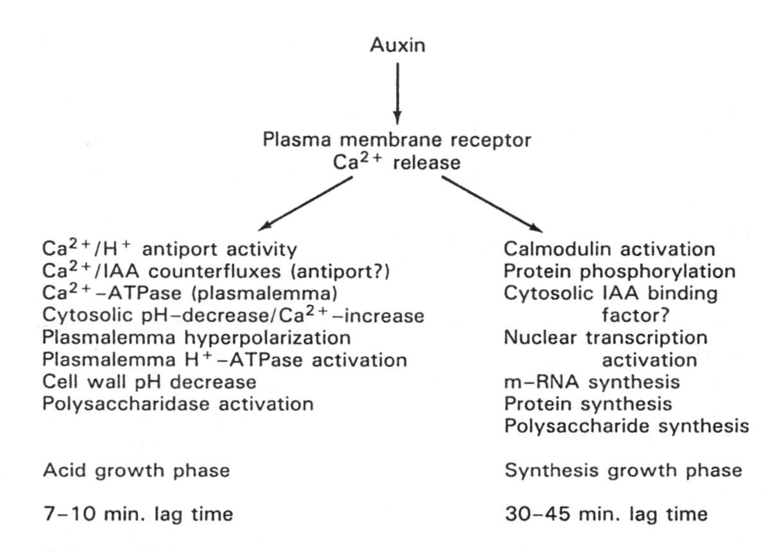

Scheme 14.1 A concept of the mechanism(s) of auxin-induced growth [9]

the nature and sequence of events occurring in tissue between auxin application and the rapid growth response 7–10 min later. The long-lasting growth response, starting after about 30–45 min, is thought to be supported by a different mechanism involving gene activation and the synthesis of specific enzymes required for the increased growth rate.

Auxins are thought to bind to specific receptors on the external surface of the plasmalemma and to induce a cascade of events involving Ca^{2+} as a second messenger [9, 19, 20, 21]. Two distinctly different types of membrane (auxin binding) carriers have been described; an "uptake carrier" which is evenly distributed along the surface of the plasmalemma, and an "efflux carrier" which is located primarily at the lower end of the cell. In cooperation, these two carriers are thought to mediate polar (downward) auxin transport [9]. It is the efflux carrier that is inhibited by the auxin transport inhibitors TIBA and NPA. There are, however, indications that this binding does not occur at the native auxin binding site, but at a separate site on this protein [17]. Auxin carrier systems have been studied in considerable detail. They are found in different membrane fractions and are also known as "auxin-binding membrane proteins." One of these membrane-bound proteins might also be responsible for the induction of the rapid growth response (and also of the delayed growth response) portrayed in Scheme 14.1. The uptake/efflux carrier concept is also known as the "chemiosmotic polar diffusion hypothesis," involving IAA accumulation from the acidic apoplasmic space by electroneutral symport of the auxin or auxin herbicide anion and H^+.

The details of the cause-and-effect sequence started after binding of an auxin herbicide to a plasmalemma auxin-binding protein are not known. The second messenger concept involving Ca^{2+} suggests a rapid and transient Ca^{2+} increase in the cytoplasm [22–24]. The cytosolic Ca^{2+}-concentration is usually around 0.1 to 1 μM, whereas the Ca^{2+}-concentration in the cell wall, endoplasmic reticulum (ER), and vacuole may be as high as 1 mM. The stimulus amplification could therefore easily be brought about by an influx through membrane Ca^{2+} gates. The sequestering of Ca^{2+} into subcellular compartments by an H^+/Ca^{2+} antiport system might cause a transient cytosolic pH decrease. Simultaneously, a rapid and transient unspecific plasmalemma depolarization (5 to 12 min) and a subsequent (15 to 20 min) auxin specific hyperpolarization (inside negative) can be measured. In summary, a number of complex, fast, and transient effects after auxin application to certain tissues leads to the eventual activation of a plasmalemma ATPase that exports protons and acidifies the cell wall space.

The situation in intact plants might differ in some details, such as the actual auxin concentration in the cellular, extracellular, and subcellular spaces, from that observed in excised tissues (e.g., coleoptiles), but the mechanism of the response leading to the growth stimulation is thought to be similar. There appear to be profound differences, however, between monocotyledonous and dicotyledonous species. The results and suggestions presented in Scheme 14.1 have been obtained primarily from studies with coleoptiles, and might differ considerably from those obtained using other tissues, in particular those from dicotyledonous plants.

The lower cell wall pH eventually initiates the acid growth phase via an activation of specific acid polysaccharidases. The actual driving force that leads to the increase in cellular volume by stretching the weakened cell wall matrix is, of course, cell turgor pressure. However, it has also become clear in recent years that the fast growth response cannot be explained entirely by the acid growth effect. Additional modifying (either supporting or interfering) systems must also be involved. Accordingly, the list of affected physiological and biochemical systems given in Scheme 14.1 can serve as no more than a guide to some of our present-day thoughts. The continued presence of IAA leads to growth stimulation over longer time spans, that is, hours and days. The rapid acid growth effect must therefore be supplemented by a tuning of the cellular metabolism to the needs of an extended growth response. The signal chain that is thought to trigger the responses leading to the expression of additional nuclear genes may also be started by the second messenger Ca^{2+} (Scheme 14.1). Interestingly, auxin-regulated cell elongation can be inhibited by calmodulin-binding drugs [22]. After calmodulin activation and specific membrane protein phosphorylation, a signal may be produced that diffuses from the plasmalemma to the nucleus. Soluble auxin-binding proteins have, for example, been found in tobacco cell cultures and in pea epicotyls. The actual amount depends on the growth stage [25–27]. Increased and specific m-RNA synthesis has also been documented, for example in tobacco cell suspension cultures [28]. The conceptual presentation given in Scheme 14.1 envisages a "master switch" which, upon binding IAA or an auxin herbicide, triggers two different cascades of cause-effect sequences with two different final target compartments and responses. On the other hand, it has been suggested that several different auxin-binding proteins might have evolved in higher plants by a "paralogous evolution" from a primitive algal Ca^{2+} gate [9]. At least three different functional auxin-binding proteins have so far been suggested: the uptake carrier, the efflux carrier, and a soluble auxin receptor. For the induction of other auxin-controlled or auxin-induced responses, such as specific morphogenesis and differentiation, the presence of additional receptors ("auxin binding proteins") might have to be invoked. Multifunctional control of morphogenesis in cooperation with other plant hormones, as is the case in root and shoot morphogenesis, may require even more complex response systems.

14.3 METABOLISM OF AUXINS AND AUXIN HERBICIDES

Artificial (herbicidal) auxins as well as the natural auxin IAA are rapidly and extensively conjugated with amino acids in plant tissue [1]. The amino acids that are commonly bound to the auxins are glutamic acid and aspartic acid. Other compounds to which auxin may be bound include myo-inositol, rhamnose, galactose, arabinose, and glucose, or several of these sugars (or sugar alcohols). The amount of conjugates formed can be quite large, and may exceed the concentration of free IAA by a factor of 10 to 100, as previously mentioned. The ubiquity of IAA conjugates in the plant kingdom is well established, and the general interpretation of these conjugates is that they serve as slow release buffers in the control of the free IAA concentration [29]. The situation is somewhat similar in the case of the herbicidal auxin molecules. In one study, for example, the free internal concentration of 2,4-D in soybean root callus cultures was maintained at about 4 nmoles 2,4-D per gram fresh weight by controlled conversion of the excess 2,4-D into amino acid conjugates [30]. In a study with different 2,4,5-T-amino acid conjugates, it has been shown that these are all biologically active; that is, they can be hydrolyzed in the tissue, yielding the active auxin herbicide [31]. Interestingly, the conjugates with aspartate and glutamate were only weakly active, supporting the function of these conjugates as slow release forms of endogenous auxin. Similar results have been obtained with 1-naphthaleneacetic acid and its amino acid conjugates [32].

When considering the many different known metabolic conversions of auxin herbicides, a more complex picture is obtained. The phenoxy herbicides 2,4-D and MCPA are depicted in Scheme 14.2 as examples of the metabolic behavior of auxin herbicides in general. Besides conjugation reactions, oxidation, and hydroxylation, binding to the lignin component of the cell wall must also be considered [1, 33–36]. Contrasting with conjugation, the latter metabolic conversions are true detoxifications because of their irreversibility. There is a tendency toward more metabolism via hydroxylation in wheat and barley (tolerant) and more conjugation to amino acids in soybean (susceptible) [37, 38]. Thus, metabolism in tolerant grasses is primarily to irreversible detoxification products, whereas that in suscep-

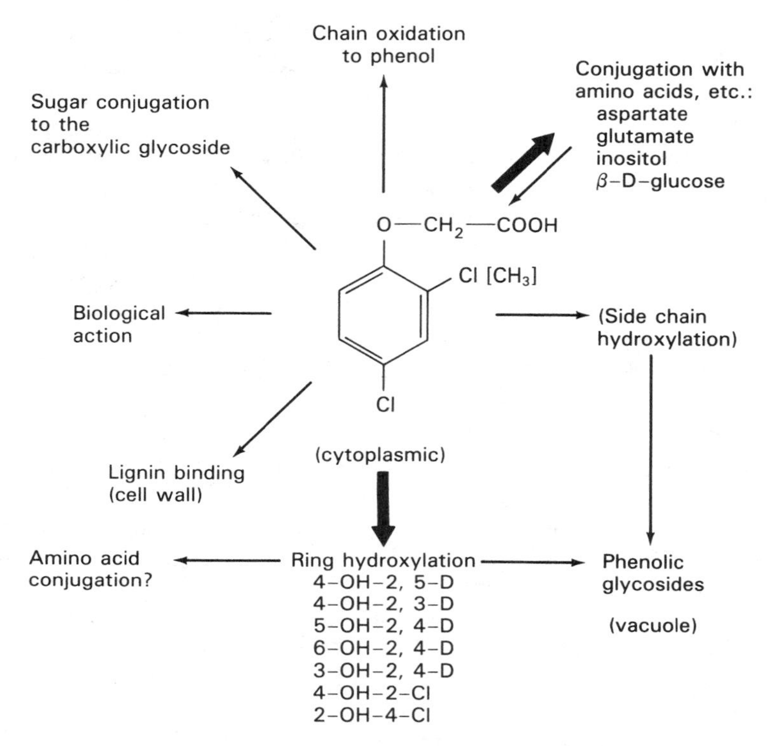

Scheme 14.2 Reversible and irreversible metabolic conversions of the auxin herbicides 2,4-D and MCPA.

tible dicotyledonous species is primarily to reversible conjugates. However, it is not known to what extent these different pathways of herbicide metabolism contribute to the selectivity of auxin herbicides.

The situation is somewhat similar for the auxin herbicide triclopyr; predominantly aspartate conjugates have been found in the sensitive weeds *Stellaria media* and *Chenopodium album*, compared to predominantly glycoside esters in the tolerant crops wheat and barley [39]. Interestingly, the different types of metabolites are deposited into different subcellular compartments: the phenolic glycosides of the hydroxylated herbicides are transported into the vacuole, and the amino acid conjugates are excreted into the cell wall space [40, 41]. Oxidative and decarboxylating metabolic reactions, for example by peroxidases, complement the detoxification reactions. There is, however, an indication that IAA conjugates are fairly immune to peroxidative attack [1].

The hydroxylating monooxygenases (see Chapter 6) catalyze a broad spectrum of hydroxylations, involving all the positions available in the aromatic ring and substituent. The hydroxylation in the 4-position involves transfer of the chlorine into either position 3 or (more commonly) position 5 by an NIH-shift reaction mechanism [42, 43]. The hydroxyl groups introduced into the ring positions or the methyl side chain [44] are extensively conjugated to glucose and other sugars and, possibly, to amino acids also. Fairly high concentrations of MCPA and 2,4-D (10-500 μM) have been shown to stimulate oxygenation in aged potato slices [45, 46].

14.4 SELECTIVE ACTION OF AUXIN HERBICIDES

Selective control of broadleaved weeds in cereal grain crops was one of the most successful early uses of herbicides, and has made the phenoxyacetic acid herbicides one of the most widespread and important herbicide groups. Profound differences in the metabolism of the auxin herbicides between monocotyledonous (tolerant) and dicotyledonous (sensitive) species have already been mentioned, and may be an important factor in the selective herbicide action.

Further support for a metabolic basis for tolerance comes from several studies of tolerant dicotyledonous plants [40, 47–50]. A *Lotus corniculatus* biotype that developed 2,4-D tolerance after repeated applications of this auxin herbicide rapidly inactivates 2,4-D by hydroxylation, glycosylation, and insoluble residue formation [47]. Increased side chain breakdown has also been reported. The ester forms of most auxin herbicides are more efficaceous than the acid forms when applied to the foliage because they are more readily absorbed through plant cuticles and cell membranes (Chapters 3 and 4). Thus, de-esterification may lead to reduced herbicide entry into plant cells (particularly if it occurs in the cell wall region) and, thus, to reduced herbicide action.

Tolerant cell cultures can be obtained relatively easily by growing the cultures on elevated concentrations of auxin herbicides (e.g., 40 mg l^{-1} 2,4-D) [51]. Tolerant cell cultures selected on phytotoxic concentrations of one auxin herbicide are frequently also tolerant of other auxin-type herbicides. For example, cross-tolerance has been shown for 2,4-D, 2,4,5-T, and 2,4-DB in *Trifolium repens* cell suspension cultures after selection with any one of these auxins [52], or for 2,4-D, IAA, and NAA in tobacco cell cultures after selection with 2,4-D [53]. Where the basis for the increased tolerance has been studied, both lower rates of uptake and higher rates of hydroxylation and conjugate formation have been found to contribute to the tolerance. However, the increased rates of detoxification cannot always explain the observed level of tolerance. An unspecified difference in metabolic activity between tolerant and sensitive lines has also been suggested as the basis of the observed tolerance [54].

The insufficient potential of metabolic differences to explain the selective behavior completely has led to many alternative suggestions over the years, but none of these hypotheses has been confirmed. Differences in morphology between monocotyledonous and dicotyledonous plants, differences in growth physiology and auxin responsiveness, and differences in nuclear volume have all been observed. An interesting recent example of 2,4-D resistance is an *Arabidopsis tha-*

liana mutant that also has an altered growth habit. The horizontal growth characteristics and altered plant morphology of this mutant have led the authors to suggest that a site-of-action mutant may have been selected [55]. However, the precise mechanism of resistance has not yet been determined.

14.5 THE PHYTOTOXIC ACTION OF AUXIN HERBICIDES

The phytotoxic action of auxin-type herbicides occurs largely as a result of their ability to mimic the activity of endogenous auxin. As repeatedly discussed in other parts of this chapter, tolerance of auxin herbicides is mainly correlated with their metabolism. Plants normally maintain controlled levels of IAA in their different tissues, each being the result of the relative rates of synthesis (if any), import, export, degradation, and reversible as well as irreversible conjugation in the particular tissue. The main role in maintaining auxin concentration in plant tissue seems to be played by the reversible conjugation reactions. The conjugating enzymes and the specific hydrolases should, therefore, be regulated independently, and are probably located in different compartments of the cell. The main reason that an auxin-active chemical acts as a herbicide is probably a lack of control over its intracellular concentration. The tissue concentration of auxin-active herbicides therefore becomes excessively high and, consequently, the interaction with other plant hormones in the regulation of plant metabolism and morphogenesis is disturbed.

The herbicidal actions of auxin herbicides are, therefore, principally those associated with excessively high auxin concentrations in the tissue. Scheme 14.3 lists some of the important responses that occur during the course of herbicidal action, and that serve to mediate it. The first response is preparatory for cell and tissue growth and includes reserve mobilization, increased rates of protein and RNA synthesis, and depolymerization and degradation of wall xyloglucans and arabinogalactans [56]. Growth effects follow, and can be detected morphologically as stem and petiole elongation and curling, stem and petiole thickening, and differentiation of new cells and organs (e.g., meristems and roots) [57]. Since the normal control mechanisms that limit the availability of IAA to the physiologically correct cells, tissues, and time intervals do not operate in the case of the auxin herbicide molecules, the ongoing and unlimited mobilization of metabolic reserves for excessive growth eventually leads to the loss of cellular functions, cellular integrity, and repair capacities. Loss of membrane functions such as semipermeability and compartmental separation causes autolysis, desiccation, and eventual disintegration of the tissue.

On an ultrastructural level, the most sensitive organelle is the chloroplast. Chloroplast swelling precedes a general loss of membrane integrity in the nucleus, plasmalemma, tonoplast, and so on [58, 59]. The interplay with other plant hormones can delay or modify the herbicidal activity, but cannot profoundly affect its occurrence. For example, the inhibition of ethylene biosynthesis with aminoethox-

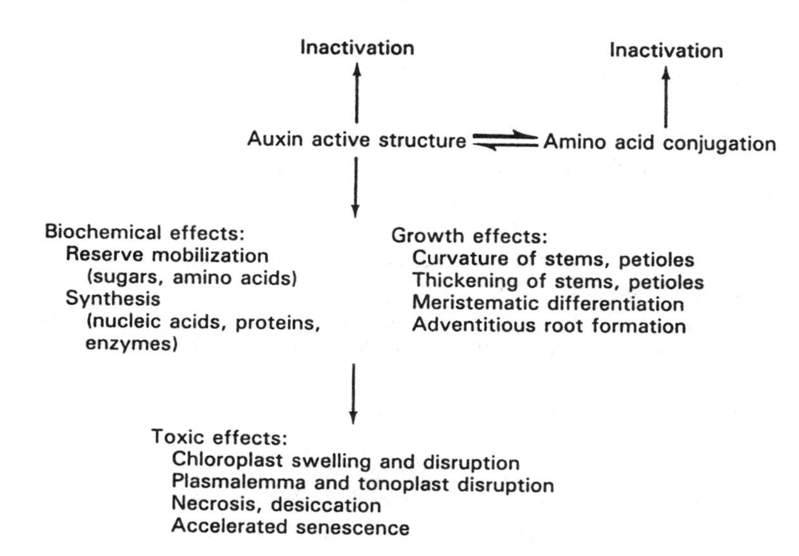

Scheme 14.3 Sequence of biochemical and physiological effects leading to the phytotoxic action of auxin herbicides.

yvinylglycine (AVG) has been shown to delay the development of morphological changes induced by picloram and clopyralid in sensitive species [60]. This result suggests that the massive release of ethylene that occurs soon after application of auxin-type herbicides to sensitive species is closely related to the morphological effects also induced. Cytokinins inhibit the conjugation of 2,4-D in cultured soybean cells, but do not affect growth of the cells [61].

REFERENCES

1. Reinecke, D. M., and R. S. Bandurski. 1987. "Auxin biosynthesis and metabolism." In *Plant Hormones and their Role in Plant Growth and Development,* P. J. Davies, ed. Dordrecht, Netherlands: Martinus Nijhoff Publishers, pp. 24–42.
2. Berghaus, R., and B. Wuerzer. 1987. "The mode of action of the herbicidal quinolinecarboxylic acid, quinmerac (BAS 518 H)." In Proc. Brit. Crop Prot. Conf.—Weeds, Vol. 3. Thornton Heath, U.K.: BCPC Publ., pp. 1091–96.
3. Katekar, G. F. 1979. "Auxins: on the nature of the receptor site and molecular requirements for auxin activity." *Phytochemistry,* 18, 223–33.
4. Farrimond, J. A., M. C. Elliott, and D. W. Clack. 1980. "Auxin structure/activity relationships: benzoic acids and phenols." *Phytochemistry,* 19, 367–71.
5. Farrimond, J. A., M. C. Elliott, and D. W. Clack. 1981. "Auxin structure/activity relationships: aryloxyacetic acids." *Phytochemistry,* 20, 1185–90.
6. McComb, A. J., and J. A. McComb. 1978. "Differences between plant species in their

ability to utilize substituted phenoxybutyric acids as a source of auxin for tissue culture growth." *Plant Sci. Lett.*, 11, 227–32.

7. Berghaus, R., and B. Würzer. 1987. "The mode of action of the new experimental herbicide quinchlorac (BAS 514H)." In Proc. 11th Asian-Pacific Weed Sci. Soc. Conf., Weed Sci. Soc. of the Rep. of China, Taipei, Taiwan, pp. 81–87.
8. Council for Agricultural Science and Technology (CAST). 1975. "The phenoxy herbicides." *Weed Sci.*, 23, 253–63.
9. Hertel, R. 1983. "The mechanism of auxin transport as a model for auxin action." *Z. Pflanzenphysiol.*, 112, 53–67.
10. Brummel, D. A., and J. L. Hall. 1987. "Rapid cellular responses to auxin and the regulation of growth." *Plant Cell Envir.*, 10, 523–43.
11. Tamas, I. A. 1987. "Hormonal regulation of apical dominance." In *Plant Hormones and their Role in Plant Growth and Development,* P. J. Davies, ed. Dordrecht, Netherlands: Martinus Nijhoff Publishers pp. 393–410.
12. Black, C. C., and G. A. Buchanan. 1980. "How herbicides work—the phenoxyacetic acids and related herbicides." *Weeds Today,* 11, 13–15.
13. Pillmoor, J. B., and J. K. Gaunt. 1981. "The behaviour and mode of action of the phenoxyacetic acids in plants." *Progress in Pestic. Biochem.*, 1, 147–218.
14. Trewavas, A. J. 1982. "Growth substance sensitivity: the limiting factor in plant development." *Physiol. Plant.*, 55, 60–72.
15. Davies, P. J. 1987. "The plant hormone concept: transport, concentration, and sensitivity." In *Plant Hormones and their Role in Plant Growth and Development,* P. J. Davies, ed. Dordrecht, Netherlands: Martinus Nijhoff Publishers, pp. 12–23.
16. Nissen, P. 1985. "Dose responses of auxins." *Physiol. Plant.*, 65, 357–74.
17. Rubery, P. H. 1987. "Auxin transport." In *Plant Hormones and their Role in Plant Growth and Development,* P. J. Davies, ed. Dordrecht, Netherlands: Martinus Nijhoff Publishers, pp. 341–62.
18. Taiz, L. 1984. "Plant cell expansion: regulation of cell wall mechanical properties." *Annu. Rev. Plant Physiol.*, 35, 585–657.
19. Cross, J. W. 1985. "Auxin action: the search for the receptor." *Plant Cell Envir.*, 8, 351–59.
20. Venis, M. A. 1987. "Auxin-binding proteins in maize: purification and receptor function." In *Molecular Biology of Plant Growth Control,* J. E. Fox, M. Jacobs, eds. New York: A. R. Liss, Inc. pp. 219–28.
21. Napier, R. M., and M. A. Venis. 1990. "Receptors for plant growth regulators: recent advances." *J. Plant Growth Regul.*, 9, 113–26.
22. Hepler, P. K., and R. O. Wayne. 1985. "Calcium and plant development." *Annu. Rev. Plant Physiol.*, 36, 397–439.
23. Gross, J., and M. Sauter. 1987. "Are auxin and calcium movements in corn coleoptiles linked processes?" *Plant Sci.*, 49, 189–98.
24. Evans, M. L., K.-H. Hasenstein, C. L. Stinemetz, and J. J. McFadden. 1987. "Calcium as a second messenger in the response of roots to auxin and gravity." In *Plant Hormones and their Role in Plant Growth and Development,* P. J. Davies, ed. Dordrecht, Netherlands: Martinus Nijhoff Publishers, pp. 361–70.

25. Elliott, M. C., A. M. O. O'Sullivan, J. F. Hall, G. M. Robinson, J. A. Lewis, D. A. Armitage, H. M. Bailey, R. D. J. Barker, K. R. Libbenga, and A. M. Mennes. 1987. "Plant cell division—the roles of IAA and IAA binding protein." In *Plant Hormones and their Role in Plant Growth and Development,* P. J. Davies, ed. Dordrecht, Netherlands: Martinus Nijhoff Publishers, pp. 245–55.

26. Jacobsen, H.-J., and K. Hajek. 1987. "Growth-stage dependent occurrence of soluble auxin-binding proteins in pea." In *Plant Hormones and their Role in Plant Growth and Development,* P. J. Davies, ed. Dordrecht, Netherlands: Martinus Nijhoff Publishers, pp. 257–66.

27. Libbenga, K. R., H. J. van Telgen, A. M. Mennes, P. C. G. van der Linde, and E. J. van der Zaal. 1987. "Characterization and function analysis of a high-affinity cytoplasmic auxin-binding protein." In *Plant Hormones and their Role in Plant Growth and Development,* P. J. Davies, ed. Dordrecht, Netherlands: Martinus Nijhoff Publishers, pp. 229–43.

28. van der Zaal, E. J., A. M. Mennes, and K. R. Libbenga. 1987. "Auxin-induced rapid changes in translatable m-RNAs in tobacco cell suspension." *Planta,* 172, 514–19.

29. Hangarter, R. P., and N. E. Good. 1981. "Evidence that IAA conjugates are slow release sources of free IAA in plant tissues." *Plant Physiol.,* 68, 1424–27.

30. Davidonis, G. H., R. H. Hamilton, and R. O. Mumma. 1978. "Metabolism of 2,4-dichlorophenoxyacetic acid in soybean root callus and differentiated soybean root cultures as a function of concentration and tissue age." *Plant Physiol.,* 62, 80–83.

31. Davidonis, G. H., M. Arjmand, R. H. Hamilton, and R. O. Mumma. 1979. "Biological properties of amino acid conjugates of 2,4,5-trichlorophenoxyacetic acid." *J. Agric. Food Chem.,* 27, 1086–88.

32. Smulders, M. J. M., E.T.W.M. van de Ven, A. F. Croes, and C. J. Wullems. 1990. "Metabolism of 1-naphthanleneacetic acid in explants of tobacco: evidence for release of free hormone from conjugates." *J. Plant Growth Regul.,* 9, 27–34.

33. Davidonis, G. H., R. H. Hamilton, and R. O. Mumma. 1980. "Evidence for the conversion of 2,4-D amino acid conjugates to free 2,4-D." *Plant Physiol.,* 66, 537–40.

34. Cohen, J. D., and R. S. Bandurski. 1982. "Chemistry and physiology of the bound auxins." *Annu. Rev. Plant Physiol.,* 33, 403–30.

35. Feung, C. S., S. L. Loerch, R. H. Hamilton, and R. O. Mumma. 1978. "Comparative metabolic fate of 2,4-dichlorophenoxyacetic acid in plants and plant tissue cultures." *J. Agric. Food Chem.,* 26, 1064–67.

36. Hutber, G. N., E. I. Lord, and B. C. Loughman. 1978. "The metabolic fate of phenoxyacetic acids in higher plants." *J. Exp. Bot.,* 29, 619–29.

37. Scheel, D., and H. Sandermann. 1981. "Metabolism of 2,4-dichlorophenoxyacetic acid in cell suspension cultures of soybean (*Glycine max* L.) and wheat (*Triticum aestivum* L.). I. General results." *Planta,* 152, 248–52.

38. Scheel, D., and H. Sandermann. 1981. "Metabolism of 2,4-dichlorophenoxyacetic acid in cell suspension cultures of soybean (*Glycine max* L.) and wheat (*Triticum aestivum* L.). II. Evidence for incorporation into lignin." *Planta,* 152, 253–58.

39. Lewer, P., and W. J. Owen. 1987. "Triclopyr: an investigation of the basis for its species-selectivity." Proc. Brit. Crop Prot. Conf.—Weeds, Vol. 1. Thornton Heath, U.K.: BCPC Publ., pp. 353–60.

40. Schmidt, R., and H. Sandermann. 1982. "Specific localization of β-glucoside conjugates of 2,4-dichlorophenoxyacetic acid in soybean vacuoles." *Z. Naturforsch.*, 37c, 772–77.
41. Davidonis, G. H., R. H. Hamilton, and R. O. Mumma. 1982. "Evidence for compartmentalization of conjugates of 2,4-dichlorophenoxy-acetic acid in soybean callus tissue." *Plant Physiol.*, 70, 939–42.
42. Bärenwald, G., B. Schneider, and H.-R. Schütte. 1987. "Metabolism of the herbicide 2-(2,4-dichlorophenoxy)-propionic acid (dichlorprop) in barley (*Hordeum vulgare*)." *Z. Naturforsch.*, 42c, 486–90.
43. Drinkwine, A. D., and J. R. Fleeker. 1981. "Metabolism of 2,5-dichloro-4-hydroxyphenoxyacetic acid in plants." *J. Agric. Food Chem.*, 29, 763–66.
44. Cole, D. J., and B. C. Loughman. 1985. "Factors affecting the hydroxylation and glycosylation of (4-chloro-2-methylphenoxy)acetic acid in *Solanum tubersum* tuber tissue." *Physiol. Vég.*, 23, 879–86.
45. Adele, P., D. Reichart, J. P. Salaün, I. Benveniste, and F. Durst. 1981. "Induction of cytochrome P-450 monooxygenase activity by 2,4-dichlorophenoxyacetic acid in higher plant tissue." *Plant Sci. Lett.*, 22, 39–46.
46. Cole, D. C., and B. C. Loughman. 1982. "Metabolism of phenoxyacetic acid in potato (*Solanum tuberosum* L.) tuber slices." *Plant Sci. Lett.*, 27, 289–98.
47. Davis, C., and D. L. Linscott. 1986. "Tolerance of birdsfoot trefoil (*Lotus corniculatus*) to 2,4-D." *Weed Sci.*, 34, 373–76.
48. Hallmen, U. 1975. "Translocation and complex formation of root-applied 2,4-D and picloram in susceptible and tolerant species." *Physiol. Plant.*, 34, 266–72.
49. Smith, A. E. 1985. "Differential response of *Trifolium* species to 4-(2,4-dichlorophenoxy)butyric acid treatments." *Physiol. Plant.*, 65, 124–28.
50. Sanders, G. E., and K. E. Pallett. 1987. "Comparison of the uptake, movement and metabolism of fluroxypyr in *Stellaria media* and *Viola arvensis*." *Weed Res.*, 27, 159–66.
51. Davidonis, G. H., R. H. Hamilton, and R. O. Mumma. 1982. "Metabolism of 2,4-D-resistant soybean callus tissue." *Plant Physiol.*, 70, 104–07.
52. Oswald, T. H., A. E. Smith, and D. V. Phillips. 1977. "Herbicide tolerance developed in cell suspension cultures of perennial white clover." *Can. J. Bot.*, 55, 1351–58.
53. Nakamura, C., M. Nakata, M. Shioji, and H. Ono. 1985. "2,4-D resistance in a tobacco cell culture variant: cross-resistance to auxins and uptake, efflux, and metabolism of 2,4-D." *Plant Cell Physiol.*, 26, 271–80.
54. Nakamura, C., N. Mori, M. Nakata, and H. Ono. 1986. "2,4-D resistance in a tobacco cell culture variant. II. Effects of 2,4-D on nucleic acid and protein synthesis and cell respiration." *Plant Cell Physiol.*, 27, 243–51.
55. Estelle, M. A., and C. Somerville. 1987. "Auxin-resistant mutants of *Arabidopsis thaliana* with an altered morphology." *Mol. Gen. Genet.*, 206, 200–06.
56. Nishitani, K., and Y. Masuda. 1981. "Auxin-induced changes in the cell wall structure: Changes in the sugar compositions, intrinsic viscosity and molecular weight distributions of matrix polysaccharides of the epicotyl cell of *Vigna angularis*." *Physiol. Plant.*, 52, 482–94.

57. Sanders, G. E., and K. E. Pallett. 1987. "Physiological and ultrastructural changes in *Stellaria media* following treatment with fluroxypyr." *Ann. Appl. Biol.*, 111, 385–98.
58. Ayling, R. D. 1976. "Ultrastructural changes in leaf and needle segments treated with herbicides containing picloram." *Weed Res.*, 16, 301–04.
59. Bretherton, G., and N. D. Hallam. 1979. "The movement of 2,4,5-trichlorophenoxyacetic acid into the leaves of *Rubus procerus* P. J. Muell. and its effect on chloroplast ultrastructure." *Weed Res.*, 19, 307–13.
60. Hall, J. C., P. K. Bassi, M. S. Spencer, and W. H. Vanden Born. 1985. "An evaluation of the role of ethylene in herbicidal injury induced by picloram or clopyralid in rapeseed and sunflower plants." *Plant Physiol.*, 79, 18–23.
61. Montague, M. J., R. K. Enns, N. R. Siegel, and E. G. Jaworski. 1981. "Inhibition of 2,4-D conjugation to amino acids by treatment of cultured soybean cells with cytokinins." *Plant Physiol.*, 67, 701–04.

Chapter 15

Other Sites of Herbicide Action

15.1 INTRODUCTION

Although normal growth and development of higher plants is dependent on thousands of enzymatic, energy transfer, and membrane function processes, the molecular targets of the majority of commercial herbicide action are represented by only a very small fraction of these. In fact, this book has thus far dealt with only about ten certain molecular sites of herbicides and these are the sites of action of the preponderance of commercial herbicides (Table 15.1). This chapter will deal with several sites of lesser importance, known to be the primary sites of action of commercial herbicides, herbicides currently under development, and herbicides that have been or are being considered for development.

In addition to the primary sites for herbicides mentioned in previous chapters and listed in Table 15.1, there are other target sites of these herbicides that are of lesser importance. These sites might be termed secondary sites. Some of these target sites will be discussed briefly in this chapter.

15.2 INHIBITION OF CARBON ASSIMILATION AND CARBOHYDRATE SYNTHESIS

Although electron transfer processes of photosynthesis have been the most successfully exploited sites of action for commercial herbicides, carbon fixation has not been found to be a viable site for herbicide action. In the case of ribulose

TABLE 15.1 MOLECULAR SITES OF ACTION OF HERBICIDES MENTIONED IN CHAPTERS 7–14.

Process	Molecular Site	Herbicide Class
Photosynthesis		
	D-1	*s*-triazines
		substituted ureas
		carboxanilides
		as-triazinones
		uracils
		hydroxybenzonitriles
		biscarbamates
	PSI	bipyridiliums
		heteropentalenes
Tetrapyrrole synthesis		
	Protoporphyrinogen oxidase	*p*-nitro-diphenylethers
		oxadiazoles
		cyclic imides
		phenyl pyrazoles
Carotenoid synthesis		
	Phytoene desaturase	pyridazinones
		fluridone
		m-phenoxybenzamides
		4-dihydroxypyridines
Amino acid synthesis		
	EPSP synthase	glyphosate
	Acetolactate synthase	imidazolinones
		sulfonylureas
		triazolopyrimidines
	Glutamine synthetase	glufosinate and analogs
Cell division		
	Tubulin	dinitroanilines
		phosphoric amides
Lipid synthesis		
	Acetyl CoA carboxylase	cyclohexanediones
		aryl-propanoic acids

1,5-bisphosphate carboxylase/oxygenase (Rubisco), this may be because the amount of herbicide needed to saturate the relatively large number of target site molecules would be too large. Rubisco accounts for 20 percent to 35 percent of the soluble protein of green higher plants. In fact, the concentration of RuBP carboxylase catalytic sites in the chloroplast far exceeds the normal concentration of carbon dioxide. Rubisco is not a very efficient enzyme in vivo because of the low CO_2 concentration in the chloroplast and the competition of molecular oxygen for the CO_2 binding site [1].

Many inhibitors of RuBP-carboxylase are known [2]. For example, 2- and 4-carboxylarabinitol-1,5-bisphosphate are potent, almost irreversible inhibitors of

RuBP-carboxylase [3, 4]. Their structures are very similar to the six-carbon reaction intermediate of RuBP-carboxylase. A potent structurally related inhibitor, 2-carboxyarabinitol-1-phosphate, is an endogenous inhibitor that accumulates at night and under low light conditions [5]. Evidence is accumulating that this inhibitor is important in regulation of in vivo RuBP carboxylase activity [6].

Other inhibitors of enzymes of C_3 carbon assimilation are known, but have not proven to be commercially viable herbicides. For instance, iodoacetolphosphate inhibits glyceraldehyde-3-phosphate dehydrogenase at nanomolar concentrations [7], and inhibits CO_2 fixation in isolated chloroplasts at micromolar levels.

Inhibition of carbon assimilation enzymes found primarily in C_4 plants has been one basis for the search for selective herbicides. Pyruvylphosphate dikinase (PPDK) and phosphoenolpyruvate carboxylase (PEP carboxylase) are both strongly inhibited by methyl-2,4-diketo-*n*-pentanate [8, 9]. Other compounds with less activity have also been found. It has been concluded that discovery of compounds with specificity for these enzymes is likely to be difficult because of the biochemical ubiquity of the substrates. The growth regulator glyphosine (Figure 15.1), an analog of glyphosate (Figure 13.2), is a weak inhibitor of PEP carboxylase [10]. PEP carboxylase is much more strongly inhibited by 3-mercaptopicolinic acid (I_{50} = 10 μM) [11]. However, this compound does not block carbon assimilation effectively in C_4 plants. The PEP carboxylase inhibitor 3,3-dichloro-2-(dihydroxyphosphinoylmethyl)-propenoate selectively inhibits photosynthesis in C_4 versus C_3 species [12]. At 1 mM, this compound inhibits photosynthesis by 79 percent to 98 percent in a range of seven C_4 species, but only 12 percent to 46 percent in four C_3 species. PEP carboxylase of C_4 grasses is almost completely inhibited by 4 mM 2,4-D and two close analogs [13]. The PEP carboxylases of C_3 plants are affected much less. Because 2,4-D is not a grass herbicide, it is likely that this effect at the molecular level is unimportant in its herbicidal action.

A number of formulated herbicides have been shown to markedly inhibit PEP carboxylase; however, the effect is lost when only the active herbicidal compound is tested [14]. The PEP carboxylase-inhibiting activity was found to reside in a formulation constituent, dodecylbenzenesulfonic acid, an anionic detergent, which has an I_{50} on PEP carboxylase of about 15 μM.

There is some evidence that the mechanism of action of methylarsonic acid (Figure 15.1) is through inhibition of malic enzyme in C_4 plants [15]. Apparently, methylarsonate is photosynthetically reduced to form sulfhydryl group reagents, including arsenomethane. Arsenomethane reaction with $NADP^+$-malic enzyme, an enzyme especially sensitive to sulfhydryl reagents, results in rapid accumulation of malic acid in C_4 plants such as johnsongrass (*Sorghum halepense*). Inhibition of this enzyme and the accumulation of malic acid results in cessation of carbon fixation which, in bright light, causes rapid photooxidative damage (see Chapter 9).

Aminooxyacetate (Figure 15.1) is a potent inhibitor of transaminases and other pyridoxal phosphate-requiring enzymes [16], and is an inhibitor of aspartate-

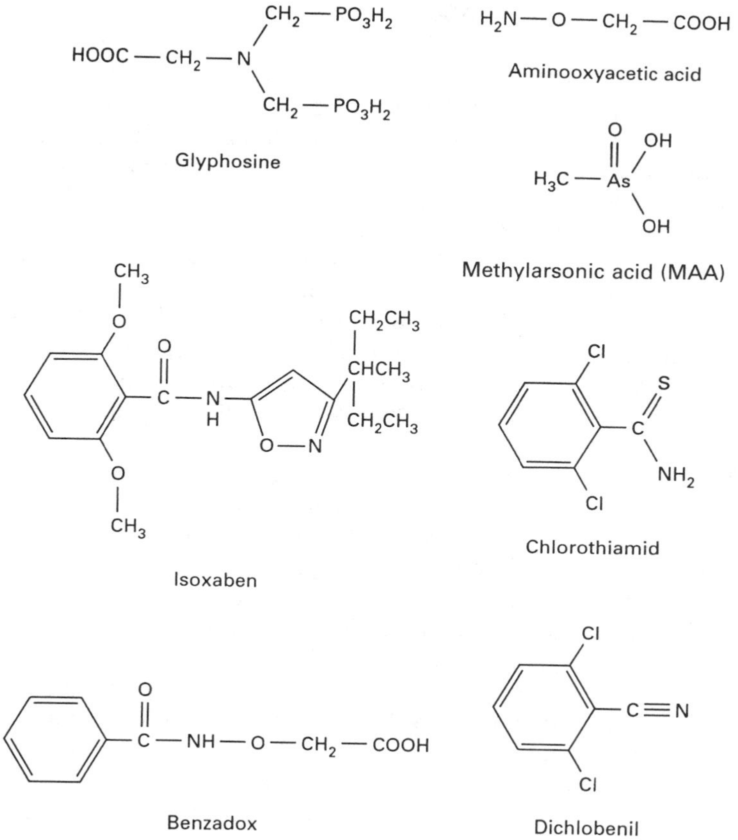

Glyphosine

Aminooxyacetic acid

Methylarsonic acid (MAA)

Isoxaben

Chlorothiamid

Benzadox

Dichlobenil

Figure 15.1 Structures of some carbon metabolism-inhibiting herbicides.

dependent photosynthetic oxygen evolution in cells of C_4 plants [17]. It is extremely phytotoxic and has been patented as a herbicide [18]. However, its general mechanism of action and subsequent mammalian toxicity would preclude it from commercialization. Benzadox (Figure 15.1), an analog of aminooxyacetate, effectively controls C_4 weeds, apparently due to its metabolism to aminooxyacetate [19].

Hadacidin (Figure 18.1), a *Penicillium* product, inhibits starch synthesis [20] and adenylosuccinate synthetase [21]. It is a weak inhibitor of plant growth [21, 22]. The more complex streptothricin-like compound, SF-701 (Figure 18.2), is a much more potent inhibitor of starch synthesis [23]. At about 2 μM it completely inhibits starch synthesis in *Panicum crus-galli*. As a foliar spray, it is strongly herbicidal to this weed at about 20 μM, while having no detectable effect on rice.

15.3 INHIBITION OF CELLULOSE SYNTHESIS

Cellulose biosynthesis appears to be specifically inhibited by dichlobenil (Figure 15.1) [24–27], although the precise molecular site remains to be determined. The herbicide has rapid in vivo effects on cellulose synthesis at micromolar concentrations, while having little or no effect on other physiological parameters [24], including noncellulosic polysaccharide synthesis [25–29]. The hydroxylation products of dichlobenil are quite potent uncouplers [30] (see Sections 15.8 and 15.11); however, dichlobenil itself is active on celluose synthesis at lower concentrations than are the metabolites on photosynthetic or respiratory phosphorylation. Labeling of cotton fiber proteins with a photoaffinity ^{14}C-dichlobenil analog results in specific labeling of uncharacterized 18 and 12 kD proteins in cotton and tomato, respectively [27]. In cotton, the amount of labeled 18 kD protein increases substantially with the onset of secondary cell wall synthesis. Since dichlobenil has no effect on in vitro synthesis of β-glucans by plant membrane preparations [28], the 18 kD protein may be a regulatory component of the cellulose synthesis complex.

The herbicide chlorthiamid (Figure 15.1) is converted to dichlobenil as the result of transformations in soil [31] or photochemical action [32]. Thus, it presumably has the same mechanism of action as dichlobenil.

Isoxaben (Figure 15.1) inhibits synthesis of acid-insoluble cell wall components (presumably cellulose) in intact *Arabidopsis thaliana* plants (I_{50} = 10 nM) about 40 times more strongly than dichlobenil [33]. Its effect is much stronger on cell wall biosynthesis than on biosynthesis of lipids, nucleic acids, or protein. Furthermore, dichlobenil and isoxaben were the only two herbicides of six (the others were amitrole, fluridone, ethalfluralin, and chlorsulfuron) to have rapid effects on incorporation of glucose into acid-insoluble cell wall materials. These data indicate that isoxaben is an even more powerful inhibitor of cell wall biosynthesis than dichlobenil.

15.4 INHIBITION OF FOLIC ACID SYNTHESIS

Folic acid or its coenzyme form, tetrahydrofolic acid, serves as an intermediate carrier of hydroxymethyl, formyl, or methyl groups in many enzymatic reactions such as the synthesis of amino acids, purines, and pyrimidines. Thus, a constant supply of this metabolite is essential for the health of the plant. The phytotoxicities of sodium sulfanilic acid and sulfanilamide (a sulfa drug) (Figure 15.2) are thought to be through inhibition of folate synthesis, because 4-aminobenzoate (Figure 15.2) reverses the phytotoxicity [34–38], just as it does in microbes affected by sulfanilamide [39]. Sulfanilate and sulfanilamide are obvious substrate analogs of *p*-aminobenzoate for 7,8-dihydropteroate synthase (Figure 15.2). Selectivity and chlorosis caused by sulfanilate in sensitive species have been correlated with effects on folate content [40]. In asparagus (insensitive), sulfanilate is not metabolized, whereas in groundnut (sensitive), a large fraction of the compound is conju-

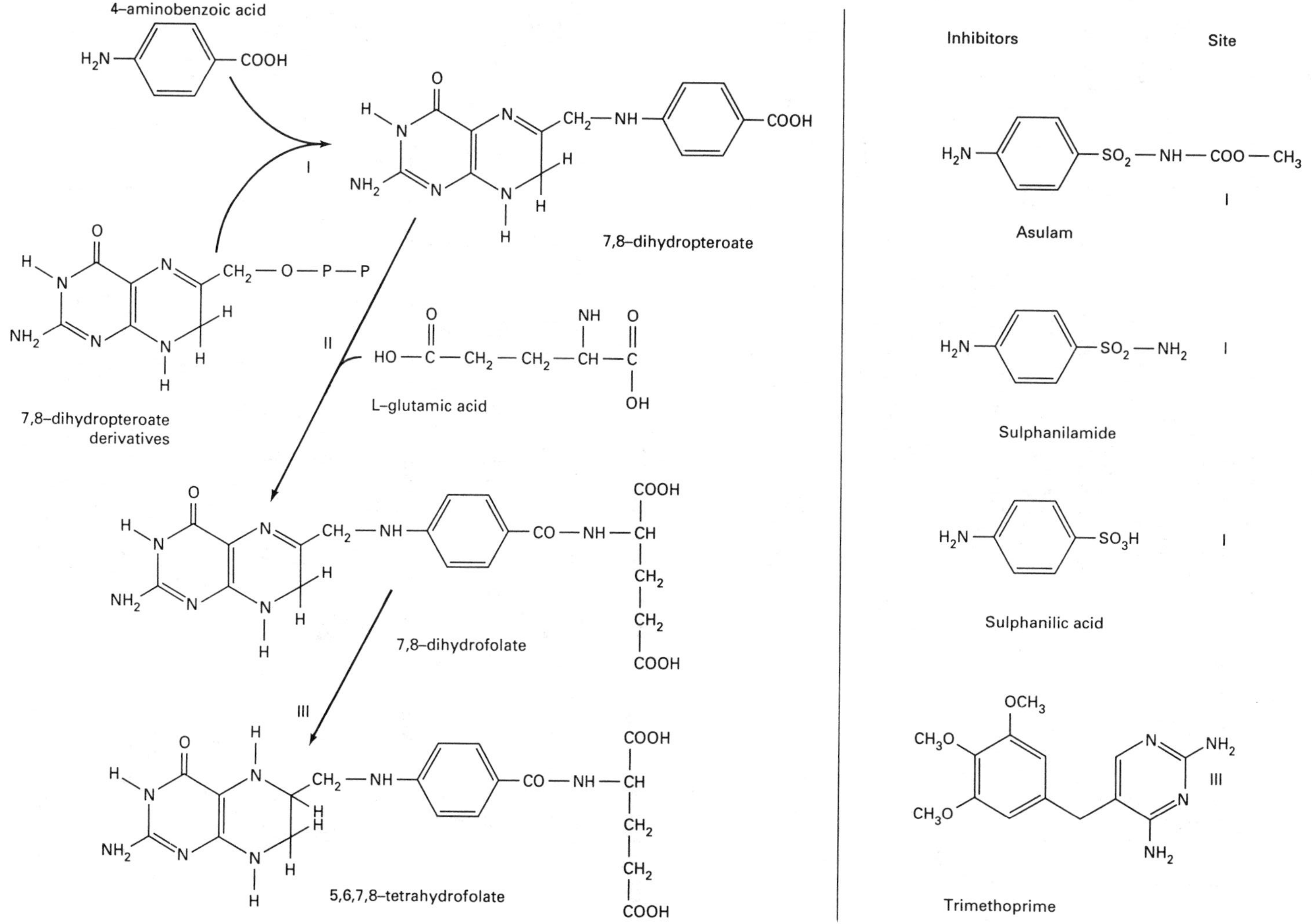

4-aminobenzoic acid

7,8-dihydropteroate derivatives

7,8-dihydropteroate

L-glutamic acid

7,8-dihydrofolate

5,6,7,8-tetrahydrofolate

Inhibitors	Site
Asulam	I
Sulphanilamide	I
Sulphanilic acid	I
Trimethoprime	III

Figure 15.2 The folate synthesis pathway and structures of some folate synthesis inhibitors.

gated to form a compound that behaves as a folate analog in some respects, thereby presumably inhibiting folate synthesis [41].

The herbicide asulam (Figure 15.2) apparently has the same mechanism of action as sulfanilate. Asulam-inhibited growth of carrot cell suspension cultures or wheat seedling roots can be reversed with addition of *p*-aminobenzoate or folate to the media [42, 43]. Asulam-inhibited elongation of flax and snapdragon (*Antirrhinum majus* L.) roots in culture is completely reversed by addition of only 4.5 μM folate to the media [44]. Reductions in fresh weight of wheat, wild oat, flax, and *Stellaria media* L. caused by foliar application of asulam are reduced by half or more by simultaneous spraying with *p*-aminobenzoate [44]. Total folates are reduced in asulam-treated plants, but the reduction is less in plants sprayed with asulam plus *p*-aminobenzoate [45]. Sulfanilamide and asulam inhibit 7,8-dihydropteroate synthase activity equally in wheat seedling extracts [44], with approximately 70 percent inhibition at 100 μM. This site of action is currently more convincing than the proposed site of action in microtubule function (section 10.3.5).

Inhibitors of dihydrofolate reductase (Figure 15.2), such as methotrexate (4-amino-10-methylfolic acid), aminopterin (4-aminofolic acid), and trimethoprime (Figure 15.2), are also quite effective plant growth inhibitors [46–48]. None of these compounds are used as herbicides.

15.5 INHIBITION OF GIBBERELLIN SYNTHESIS

Several herbicides apparently have direct effects on enzymes of the gibberellic acid synthesis pathway (Figure 15.3). However, this is not thought to be the primary mechanism of action of any of these compounds. In most cases, the effects of these herbicides on lipid synthesis (Chapter 11) are considered to be more important to their phytotoxicity. Wilkinson has found that EPTC and metolachlor block conversion of geranylgeranyl pyrophosphate (GGPP) to *ent*-kaurene, and alachlor, mefluidide, and CDAA block *ent*-kaurene conversion to *ent*-kaurenoic acid [49–54]. Diallate inhibits several reactions between GGPP and *ent*-kaurenoic acid [55]. As discussed earlier (Chapter 8), clomazone apparently causes inhibition of gibberellic acid synthesis by inhibition of GGPP formation [56]. Application of exogenous gibberellic acid does not reverse the herbicidal effects of alachlor, mefluidide, CDAA, or alachlor [57–60], but it can reverse some of the growth-inhibiting effects of clomazone in dark- or light-grown plants [56, 61].

Many highly effective inhibitors of gibberellic acid are used or have been considered for use as plant growth regulators [60, 62]. However, these compounds have not been used as herbicides. Apparently, the slow effect and the inability to kill the plant precludes specific gibberellin synthesis inhibitors from consideration as herbicides. Wilkinson [63] concluded that gibberellic acid synthesis-inhibiting herbicides would be most effective during seedling development, when growth is dependent on gibberellic acid-regulated mobilization of seed storage reserves.

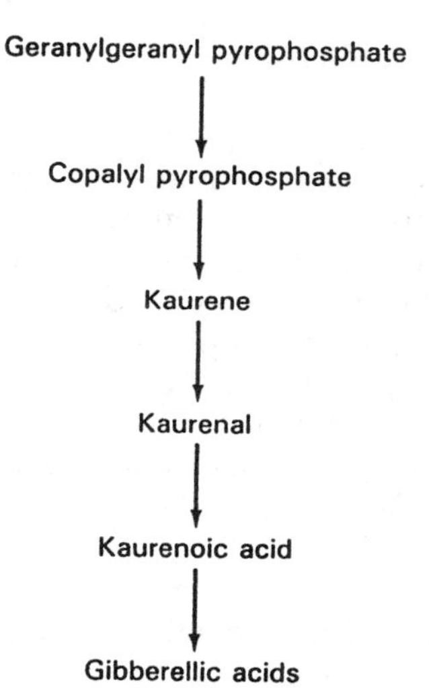

Figure 15.3 The gibberellic acid synthesis pathway.

Still, he did not ascribe the mechanism of action of metolachlor to inhibition of gibberellic acid biosynthesis alone.

15.6 INHIBITION OF LIGNIFICATION AND PHENOLIC COMPOUND SYNTHESIS

A very large percentage of the carbon assimilated by plants if funneled through the shikimate pathway (Figure 13.1), eventually becoming secondary phenolic compounds (e.g., flavonoids, tannins), particularly lignin. The first enzyme involved in production of secondary compounds derived from phenylalanine and tyrosine is phenylalanine ammonia-lyase (PAL). Theoretically, specific inhibition of this enzyme should have the most profound effects on production of all aromatic secondary compounds. Although early searches for herbicides that might act through such a mechanism were unsuccessful [64], several specific inhibitors of PAL, such as α-aminooxy-β-phenylpropionic acid [65] and R-(1-amino-2-phenylethyl)phosphonic acid [66], and of cinnamate-4-hydroxylase, such as 1-aminobenzo-triazole [67], have subsequently been found. However, these compounds are relatively poor growth inhibitors. Specific inhibitors of lignification, *N*(*o*-hydroxyphenyl) and *N*(*o*-aminophenyl)sulfinamoyl tertiobutryl acetate, which act by inhibition of cinnamyl alcohol dehydrogenase, stimulate growth by preventing lignification and, thus, keep cell walls elastic for longer periods [68]. A triazine [4-amino-6-methyl-3-phenylamino-1,2,4-triazin-5(4*H*)-one] inhibits coniferyl alcohol oxidation and the generation of hydrogen peroxide, two of the reactions required for lignification [69]. Eventually, these inhibitors could cause plant death; however, the effects are too slow for consideration as sites of action for herbicides. Secondary effects of herbicides on lignification are discussed in Chapter 16.

15.7 EFFECTS ON NITROGEN UTILIZATION

No herbicides are known to act by inhibiting nitrate or nitrite reductase. However, the nonselective herbicide sodium chlorate is bioactivated to the highly phytotoxic chlorite ion by the reductive action of nitrate reductase [70, 71]. The chlorite ion can then inactivate nitrate reductase; however, the exact mode of action of the chlorite ion is unknown. Plants with nonfunctional nitrate reductase are resistant to chlorate and, thus, chlorate has been utilized to select for nitrate reductase mutants (e.g., [72]). Since chlorite production is dependent on nitrate reductase, chlorate toxicity is maximal under conditions that favor high nitrate reductase levels (e.g., grown on a nitrogen source other than ammonia) or high in vivo nitrate reductase activity (e.g., under high light levels). As might be expected, nitrate antagonizes chlorate as a herbicide.

Nitrate reductase would probably be a poor choice for a molecular target of a herbicide because nitrate-deficient plants die very slowly and fertilization with other nitrogen sources would preclude use of the herbicide. Nitrite reductase might be a better choice because inhibition of this enzyme under conditions in which nitrate is being reduced would result in accumulation of phytotoxic levels of nitrite. Several researchers have hypothesized that nitrite accumulation plays a major role in the mode of action of photosynthesis-inhibiting herbicides. This is discussed in detail in Chapter 16.

Urease, the enzyme that converts urea to ammonia and carbon dioxide, is inhibited by substituted urea herbicides [73]. The K_i of these compounds ranges from 2 to 28 μM for neburon and fenuron, respectively. However, the primary site of all of these herbicides is inhibition of photosystem II electron transport (Chapter 7), and there are no physiological data to suggest that this inhibitory site plays a role in their herbicidal action.

15.8 INHIBITION OF PHOTOPHOSPHORYLATION

Direct effects on photophosphorylation can be through uncoupling (i.e., a protonophore) or through effects on the coupling factor complex (CF_0 plus CF_1) of the thylakoid [74]. In the first case, the proton gradient generated by splitting of water and maintained by thylakoid membranes is dissipated through protonophore activity. In the second case, the ATPase activity of CF_1 is usually inhibited. Compounds that induce these effects in the chloroplast are likely to affect analogous processes in the mitochondrion (Section 15.11). No commercial herbicides are known to affect these processes directly as part of their primary mechanism of action. However, some herbicide classes, such as certain diphenylethers (e.g., [75]) can inhibit photophosphorylation through inhibition of CF_1 ATPase. At much higher concentrations than required for inhibition of electron transport, several photosynthetic electron transport inhibitors also uncouple cyclic photophosphorylation (Figure 15.4A).

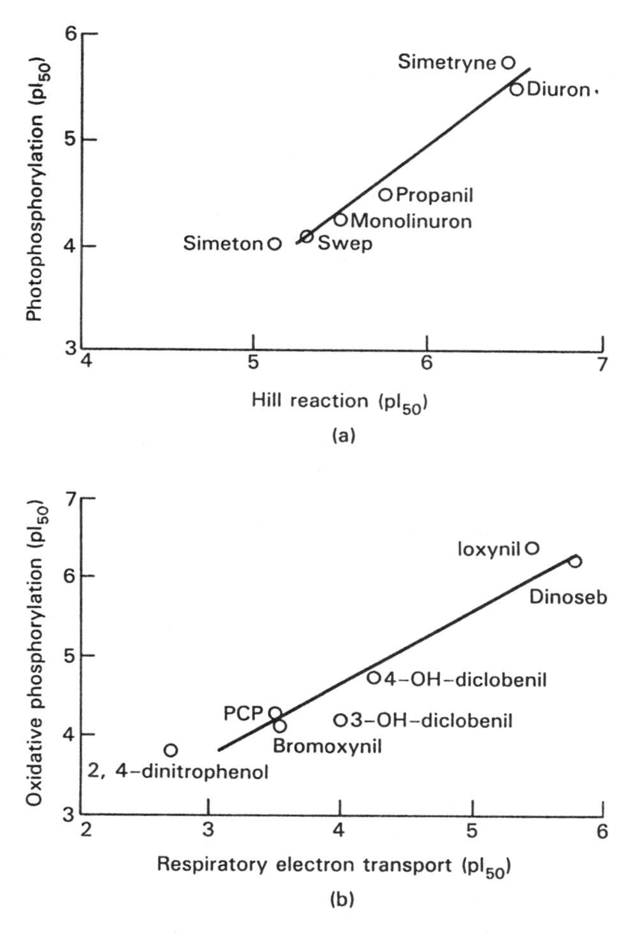

Figure 15.4 Comparison of pI_{50} values of photosynthetic (A) and respiratory (B) electron transport inhibition with those of uncoupling of photophosphorylation (A) and oxidative phosphorylation (B) for different herbicides. Taken from [76, 77].

At least one herbicide discovery effort has focused on uncoupling of photophosphorylation as a site of action [78]. A series of aryloxyalkylamines were examined for their uncoupling activity and this activity was compared with herbicidal activity. Generally, increases in lipophilicity correlated with uncoupling activity; however, uncoupling activity did not correlate well with herbicidal activity, and the most potent herbicide discovered did not act by uncoupling photophosphorylation. In another study, the photophosphorylation-uncoupling activity of several aliphatic amines, 2-aryl-amino-1,4,5,6-tetrahydropyrimidines, and alkylated *N*-aryl-1,2-ethanediamines, all correlated well with the lipophilicity of the compounds [79]. Nineteen members of the latter chemical group were all more potent as uncouplers of photophosphorylation than they were as photosynthetic electron transport inhibitors.

Targeting chloroplast coupling factor function rather than the coupling of photophosphorylation might hold more promise for linking an in vitro effect with a whole plant response, because uncoupling is generally due to a very general effect on membrane permeability, whereas an in vitro effect on coupling factor function would be more specific. An example of a herbicidal compound with very specific effects on coupling factor function is tentoxin (see Chapter 18).

15.9 EFFECTS ON PLASMALEMMA AND TONOPLAST FUNCTION

The plasmalemma and tonoplast control the water, ionic, nutrient, and, to some extent, xenobiotic status of the plant cell. Many membrane proteins and lipids interact in complex and, in most cases, still unknown mechanisms to control these functions. No herbicides have been shown to have as their primary mechanism of action a direct inhibition of plasmalemma or tonoplast function. It could be argued that plasmalemma or tonoplast function is a poor choice for targeting a herbicide because of the potentially large number of actual sites per cell that would have to be affected and the relatively greater likelihood for mammalian toxicity. Many herbicides can directly affect plasmalemma functions at sufficiently high concentrations, especially lipophilic compounds. The topic of herbicide effects on the plasmalemma and tonoplast has been reviewed previously [80, 81]. Since many herbicides are quite lipophilic and, thus, preferentially partition into membranes, it is not surprising that at high concentrations many of these compounds cause some plasmalemma dysfunction. As one would expect, effects on plasmalemma functions often correlate well with lipophilicity (e.g., [82]), just as lipophilicity usually correlates with other target sites that are membrane-bound. The synthetic herbicides with auxin-like activities apparently bind plasmalemma auxin receptors. Since this is reviewed in Chapter 14, it will not be addressed here.

Most reported studies on herbicide effects on membrane function simply describe the effects of the herbicide on cell or liposome leakage of electrolytes or an easily monitored plant pigment such as betacyanin (e.g., [83]). The effects are generally not demonstrated to be the result of a direct effect on the plasmalemma. In relatively few cases have direct effects of herbicides on membrane activities such as plasmalemma ATPase been determined. For example, the effects of alachlor and barban on plasmalemma permeability have been linked to direct effects on plasmalemma ATPase activities [84].

Balke [81] categorized herbicides that directly affect the plasmalemma as: (1) permeability alterers; (2) transport inhibitors; (3) alterers of membrane enzymatic activities; and (4) those that affect hormonal and environmental regulation of membrane function. Table 15.2 provides a partial listing of herbicides that have been reported to act on the plasmalemma through each of these mechanisms. In many cases, effects on one of these parameters will affect the others, and lack of an

TABLE 15.2 EXAMPLES OF HERBICIDES THAT AFFECT PLASMALEMMA FUNCTION BY VARIOUS MECHANISMS.

Mechanism	Herbicide	Reference
Permeability alteration		
Permeability to:		
electrolytes	2,4-D	85
	simazine	85
amino acids	diclofop-methyl	86
^{32}P-solute	metolachlor	87
Transport inhibition		
Mineral ion:		
potassium	nitrofen	88
potassium	chlorsulfuron	89
sulfate	atrazine	90
phosphate	propanil	91
ammonium	2,4-D	92
Hydrogen ion	dinitramine	93
	chlorpropham	92
Membrane enzyme activity inhibition		
Enzyme:		
ATPase	dinitramine	94
glucan synthase	dinitramine	94
Effect on hormonal or environmental regulation:		
antiauxin activity	chlorfenprop-methyl	95
phototropic response	acifluorfen	96

effect on a particular parameter is not a good predictor of the effect on a process controlled by that parameter. For example, several herbicides that have little or no effect on plasmalemma ATPase activity have profound effects on the proton gradient that is maintained by that ATPase activity [93]. Diclofop-methyl and diclofop increase the proton permeability of tonoplasts by a mechanism unrelated to ATP formation [97]. The two enantiomers of diclofop-methyl are equally active. Because phytotoxicity of this herbicide is stereospecific, the relationship between this effect and phytotoxicity is not clear.

Aryl-propanoic acid herbicides (e.g., diclofop) and phenoxy alkanoic acid herbicides (e.g., 2,4-D) antagonize some membrane effects of each other [98, 99] (discussed further in Chapter 11). However, this mutual antagonism is not good proof of a plasmalemma-related function mechanism of action for aryl-propanoic acid herbicides. The antagonism might be explained by the fact that they are structural analogs [98]. Whether phenoxy alkanoic acids can protect acetyl-CoA carboxylase (see Chapter 11) from aryl-propanoates has not been reported.

Several herbicides rapidly depolarize the electrical differential across the plasmalemma (see [99–102]). This effect can be caused by several mechanisms. Loss of membrane polarization will result in inability to take up ions actively and will ultimately lead to leakage of electrolytes. Conversely, leakage of electrolytes can lead to loss of membrane polarization. In the case of sethoxydim, an inhibitor of acetyl-CoA carboxylase (Chapter 11), only concentrations (> 0.2 mM) of the herbicide, well above those required for herbicidal activity, are effective in causing membrane depolarization [102]. Furthermore, there is little difference between susceptible and resistant species of *Poa* in this response to the herbicide. Similarly, there is no difference in sethoxydim-inhibited alanine uptake between the two species and, again, this effect is observed only at high herbicide concentrations. In vivo inhibition of plasmalemma-bound H^+-ATPase by sethoxydim has also been found, but no in vitro effects have been reported. The effect of sethoxydim on the activity of this ATPase is the same on sethoxydim-susceptible and -tolerant *Poa* species, as well as *Festuca* species. However, a membrane-bound redox system of the more susceptible species of each genus is more inhibited by sethoxydim than that of its more tolerant counterpart [103]. A difference in the effect between susceptible and tolerant species has only been detected at very high (0.4 mM) sethoxydim concentrations. It is unlikely that an effect at such high concentrations could play an important role in the mode of action of this herbicide.

Herbicide formulations often have surfactants and other additives that have profound effects on plant membranes [104]. Therefore, the use of commercial formulations can give especially misleading results in studies to determine the effects of herbicides on membrane functions. In other herbicide mechanism of action studies on whole cells or with membrane preparations, formulation constituents are likely to confound the results. Structure-activity studies of formulation constituent effects on membrane function have been conducted (e.g., [105]); however, there is no evidence that this type of information has been used in designing herbicide/additive formulations.

15.10 INHIBITION OF POLYAMINE SYNTHESIS

Polyamines are intimately involved in cellular processes that are critical to plants and animals, including DNA replication and cell division (see [106, 107]). Six or seven enzymes are required for their biosynthesis, and no herbicides are known to act directly on any of these enzymes. Inhibitors of several of the enzymes of the pathway are known [108]. Neither ornithine decarboxylase (ODC) nor arginine decarboxylase (ADC) inhibitors effectively inhibit polyamine synthesis in vivo, apparently because of a dual synthetic pathway [109]. Even treating plants with inhibitors of both ODC and ADC does not decrease polyamine levels, possibly due to a third pathway. Reducing spermidine to undetectable levels was not lethal to the plants in this study, and the authors concluded that reduction of all polyamine levels below 1 nmole per gram of fresh weight would probably not be lethal. Cinmethylin (Figure 18.3) at 5 μM strongly inhibits growth and weakly inhibits accumulation of putrescine in pea roots [110]. However, treating cinmethylin-treated plants with exogenous putrescine does not alleviate the herbicidal effect of cinmethylin [110]. In studies with *Phaseolus vulgaris*, putrescine partially reverses the growth-inhibiting effect of canavanine, an arginine decarboxylase inhibitor [111]. Thus, the cinmethylin effect on growth may be unrelated to inhibition of putrescine synthesis.

Chlorsulfuron inhibits spermidine accumulation in mitotic tissues of root tips [112]. Chlorsulfuron is a more effective spermidine accumulation inhibitor than are the specific spermidine synthesis inhibitors methylglyoxal-bis(guanylhydrazone) and cyclohexylamine. These specific inhibitors cause putrescine accumulation and, in these treatments, inhibited spermidine accumulation could be reversed with exogeneous spermidine. Neither effect was observed in chlorsulfuron treatments, suggesting that the effect could be due to enhanced spermidine degradation. How this effect relates to ALS inhibition (see Chapter 13) is not clear.

Other herbicides can cause dramatic increases in polyamine levels in plants [113]; however, there is no indication that this effect is any more than a stress reaction. Apparently, the exceedingly low requirement for polyamines and multiple pathways in plants make polyamine synthesis a poor choice of sites for herbicide discovery.

15.11 EFFECTS ON RESPIRATION AND CATABOLIC CARBON METABOLISM

Many herbicides directly affect mitochondrial respiration [114]. However, no currently marketed herbicides have this as their primary site of action. Respiratory inhibitors may inhibit this process either by inhibiting mitochondrial electron transport, uncoupling oxidative phosphorylation, or by directly interfering with the mitochondrial coupling factor. In some cases, a compound can inhibit more than one of these processes directly. Any protonophore or other compound that will dissipate the proton gradient required for mitochondrial ATP formation via the mitochondrial coupling factor ATPase will uncouple oxidative phosphorylation. Some phenolic herbicides (e.g., hydroxybenzonitriles such as bromoxynil, or nitrophenols such as dinoseb) and certain NH-acidic herbicides (e.g., benzimidazoles such as fluromidine) are effective uncouplers. Most of the photophosphorylation uncouplers mentioned in Section 15.8 that affect membrane permeability also uncouple mitochondrial respiration. At higher concentrations, most of them also inhibit electron transport, which is the opposite pattern from that observed with the relative effects of photosynthetic electron transport inhibitors on uncoupling of cyclic photophosphorylation (Figure 15.4). All commercial herbicides that are oxidative phosphorylation uncouplers are much more effective as inhibitors of photosynthetic electron transport or some other primary site of action. In many cases, the effect of the herbicide on mitochondrial function, whether electron transport or coupled phosphorylation, has been attributed to effects on membrane properties such as membrane fluidity (e.g., [115]).

Several herbicides have been discovered that interfere with glycolysis or the

Krebs (tricarboxylic acid) cycle. Pyruvate dehydrogenase is a critical enzyme in linking glycolysis with the Krebs cycle in that it produces acetyl-CoA from pyruvate formed from the Krebs cycle:

$$\text{Pyruvate} + \text{NAD}^+ + \text{CoA·SH} \rightarrow \text{acetyl-CoA} + \text{NADH} + \text{H}^+ + \text{CO}_2 \text{ TPP}$$

The first step of the reaction involves combining pyruvate with the cofactor thiamine pyrophosphate (TPP). Pyruvate dehydrogenase is a component of a multienzyme complex consisting of pyruvate dehydrogenase, lipoate acetyltransferase, and dihydrolipoyl dehydrogenase. Biorational design of herbicides as analogs of the pyruvate-TPP complex has been attempted [116]. A series of acyl phosphinate analogs has been found to inhibit pyruvate dehydrogenase and to be herbicidal. Herbicidal activity generally correlates with enzyme sensitivity, and pyruvate levels are greatly elevated in treated plants. To date, none of the compounds has been developed commercially. Acetolactate synthase (ALS) has both pyruvate and TPP binding sites and has been one the most successfully exploited molecular targets of herbicides (Chapter 13). Although at least some ALS inhibitors have a binding site that overlaps the second pyruvate (or ketobutyrate) binding site, their binding site is not equivalent to either a substrate or cofactor binding site [117]. The herbicide EPTC and the herbicide safener dichlormid have been reported to have strong effects on pyruvate dehydrogenase and 2-oxoglutarate dehydrogenase [118]. The I_{50} values for EPTC on the two enzymes from wheat mitochondria are in the 1 pM range.

Haloxyfop (Table 11.1) inhibits both pyruvate and α-ketoglutarate dehydrogenases with K_i values of 1 to 10 and 1 mM, respectively [119]. With such low activity at these sites, it is unlikely that either site plays a significant role in the mechanism of action. Acetyl-CoA carboxylase is apparently the only significant molecular site of action for this herbicide (Chapter 11).

The experimental herbicide UKJ72J (2-ethylamino-4-amino-5-thiomethyl-6-chloropyrimidine) is a potent inhibitor of mitochondrial succinate oxidation [120, 121]. The I_{50} for inhibition of succinate oxidation varies widely between plant species (9 to 2300 μM). However, no data are available to correlate these values with herbicidal efficacy. This compound is more than 20-fold more effective on succinate oxidation of potato mitochondria than of rat liver mitochondria. The exact mechanism of inhibition of succinate oxidation has not been determined.

15.12 EFFECTS ON STEROL SYNTHESIS

Sterols and their derivatives are essential membrane components, and some may also act as hormones. Thus, a normal sterol component is required for optimal plant growth and development. Several sterol synthesis inhibitors have been very successful agricultural fungicides; however, none has been developed as a herbicide [122]. The potential of plant sterol synthesis inhibitors as growth regulators has been discussed [123]. Recently, enantiomers of triadimentol and paclobutrazol, both fungicides, have been suggested as herbicides [124]. Diclobutrazol, a related fungicide, retards germination and growth of wheat seedlings [125]. These compounds and structurally related compounds cause the accumulation of α-methyl-sterols. The mode of action of these compounds as fungicides is inhibition of a P-450 cytochrome-dependent monooxygenase that removes the C-14αmethyl group during sterol biosynthesis [122]. In plants, this class of compounds also inhibits *ent*-kaurene oxidase, another P-450 cytochrome oxidase-dependent enzyme that is involved in gibberellic acid synthesis [126].

REFERENCES

1. Pierce, J. 1988. "Prospects for manipulating the substrate specificity of ribulose bisphosphate carboxylase/oxygenase." *Physiol. Plant.*, 72, 690–98.
2. Black, C. C. 1986. "Effects of herbicides on photosynthesis." In *Weed Physiology. Vol. II. Herbicide Physiology*, S. O. Duke, ed. Boca Raton, Fla.: CRC Press, Inc., pp. 1–36.
3. Pierce, J., T. J. Andrews, and G. H. Lorimer. 1986. "Reaction intermediate partitioning by ribulose bisphosphate carboxylases with differing substrate specificities." *J. Biol. Chem.*, 261, 10248–256.
4. Schloss, J. V. 1989. "Modern aspects of enzyme inhibition with particular emphasis on reaction-intermediate analogs and other potent, reversible inhibitors." In *Target Sites of Herbicide Action*, P. Böger and G. Sandmann, eds. Boca Raton, Fla.: CRC Press, Inc., pp. 165–245.
5. Beck, E., R. Scheibe, and J. Reiner. 1989. "An assessment of the rubisco inhibitor. 2-caboxyarabinitol-1-phosphate and D-hamamelonic acid 2^1-phosphate are identical compounds." *Plant Physiol.*, 90, 13–16.
6. Servaites, J. C. 1990. "Inhibition of ribulose 1,5-bisphosphate carboxylase/oxygenase by 2-carboxyarabinitol-1-phosphate." *Plant Physiol.*, 92, 867–70.
7. Usada, H., and G. E. Edwards. 1981. "Inhibition of photosynthetic carbon metabolism in isolated chloroplasts by iodoacetal phosphate." *Plant Physiol.*, 67, 854–58.
8. Lawyer, A. L., S. R. Kelley, and J. I. Allen. 1987. "Use of pyruvate-phosphate dikinase as a target for herbicide design: analysis of inhibitor specificity." *Z. Naturforsch.*, 42c, 834–36.
9. Lawyer, A. L., S. R. Kelley, and J. I. Allen. 1987. "Pyruvate phosphate dikinase from maize: Purification, characterization, analysis of inhibitors, and its use for biorational design of herbicides." In *Progress in Photosynthesis, Vol. 3*, J. Biggens, Dordrecht, The Netherlands: Martinus Nijhoff, pp. 831–34.
10. Podestá, F. E., D. H. González, and C. S. Andreo. 1987. "Glyphosine inhibits maize leaf phosphoenolpyruvate carboxylase." *Plant Cell Physiol.*, 28, 375–78.
11. Ray, T. B., and C. C. Black. 1976. "Inhibition of oxalacetate decarboxylation during C_4 photosynthesis by 3-mercaptopicolinic acid." *J. Biol. Chem.*, 251, 5824–26.
12. Jenkins, C. L. D. 1989. "Effects of the phospho*enol*pyruvate carboxylase inhibitor

3,3-dichloro-2-(dihydroxyphosphinoylmethyl)propenoate on photosynthesis. C_4 selectivity and studies on C_4 photosynthesis." *Plant Physiol.,* 89, 1231–37.

13. Crétin, C., J. Vidal, P. Gadal, S. Tabache, and B. Loubinoux. 1983. "Differential inhibition of phosphoenol-pyruvate carboxylases by 2,4-dichlorophenoxyacetic acid and two newly synthesized herbicides." *Phytochemistry,* 22, 2661–64.
14. Weidner, M., and N. Burchartz. 1978. "Inhibition of phosphoenolpyruvate carboxylase by formulated herbicides and anionic detergents." *Biochem. Physiol. Pflanzen,* 173, 381–89.
15. Knowles, F. C., and A. A. Benson. 1983. "The mode of action of a herbicide. Johnsongrass and methanearsonic acid." *Plant Physiol.,* 71, 235–40.
16. John, R. A., A. Charteris, and L. J. Fowler. 1978. "The reaction of amino-oxyacetate with pyridoxal phosphate-dependent enzymes." *Biochem. J.,* 171, 771–79.
17. Shieh, Y.-J., M. S. B. Ku, and C. C. Black. 1982. "Photosynthetic metabolism of aspartate in mesophyll and bundle sheath cells isolated from *Digitaria sanguinalis* (L.) Scop., a $NADP^+$-malic enzyme C_4 plant." *Plant Physiol.,* 69, 776–80.
18. Hoagland, R. E., and S. O. Duke. 1982. "Effects of glyphosate on metabolism of phenolic compounds VIII. Comparison of the effects of aminooxyacetate and glyphosate." *Plant Cell Physiol.,* 23, 1081–88.
19. Nakamoto, H., M. S. B. Ku, and G. E. Edwards. 1982. "Inhibition of C_4 photosynthesis by (benzmidooxy)acetic acid." *Photosyn. Res.,* 3, 293–305.
20. Kida, T., and H. Shibai. 1985. "Inhibition by hadacidin, duazomycin A, and other amino acid derivatives of *de novo* starch synthesis." *Agric. Biol. Chem.,* 49, 3231–37.
21. Hatch, M. D. 1967. "Inhibition of plant adenylosuccinate synthetase by hadacidin and the mode of action of hadacidin and structurally related compounds on plant growth." *Phytochemistry,* 6, 115–19.
22. Gray, R. A., G. W. Gauger, E. L. Delayney, E. A. Kaczka, and H. B. Woodruff. 1964. "Hadacidin, a new plant-growth inhibitor produced by fermentation." *Plant Physiol.,* 39, 204–08.
23. Kida, T., T. Ishikawa, and H. Shibai. 1985. "Isolation of two streptothricin-like antibiotics. Nos. 6241-A and B, as inhibitors of *de novo* starch synthesis and their herbicidal activity." *Agric. Biol. Chem.,* 49, 1839–44.
24. Delmer, D. P. 1987. "Celluose biosynthesis." *Annu. Rev. Plant Physiol.,* 38, 259–90.
25. Hogetsu, T., H. Shiboaka, and M. Shimokoriyama. 1974. "Involvement of cellulose synthesis in actions of gibberellin and kinetin on cell expansion. 2,6-dichlorobenzonitrile as a new cellulose-synthesis inhibitor." *Plant Cell Physiol.,* 15, 389–93.
26. Montezinos, D., and D. P. Delmer. 1980. "Characterization of inhibitors of celluose synthesis in cotton fibers." *Planta,* 148, 305–11.
27. Delmer, D. P., S. M. Read, and G. Cooper. 1987. "Identification of a receptor protein in cotton fibers for the herbicide 2,6-dichlorobenzonitrile." *Plant Physiol.,* 84, 415–20.
28. Blaschek, W., U. Semler, and F. Franz. 1985. "The influence of potential inhibitors on the *in vivo* and *in vitro* cell-wall β-glucan biosynthesis in tobacco cells." *J. Plant Physiol.,* 120, 457–70.
29. Pillonel, C. H., and H. Meier. 1985. "Influence of external factors on callose and cellulose synthesis during incubation *in vitro* of intact cotton fibres with [^{14}C] sucrose." *Planta,* 165, 76–84.

30. Moreland, D. E., G. G. Hussey, and F. S. Hussey. 1974. "Comparative effects of dichlobenil and its phenolic alteration products on photo- and oxidative phosphorylation." *Pestic. Biochem. Physiol.,* 4, 356–64.

31. Milborrow, B. W. 1965. "Effects of rate of formation of 2,6-dichlorobenzonitrile on its toxicity to plants." *Weed Res.,* 5, 332–42.

32. Bahadir, M., S. Nitz, H. Parlar, and F. Korte. 1979. "On chemical problems of the mechanism of action of chlorothiobenzamides." *J. Agric. Food Chem.,* 27, 815–18.

33. Heim, D. R., J. R. Skomp, E. E. Tschabold, and I. M. Larrinua. 1990. "Isoxaben inhibits the synthesis of acid insoluble cell wall materials in *Arabidopsis thaliana.*" *Plant Physiol.,* 93, 695–700.

34. Lin, K.-H., and Y. B. Chen. 1985. "Mode of action of sodium sulfanilate for control of groundnut rust (*Puccinia arachidis* Speg) and its toxicide to host plant (*Arachis hypogaea* Linn.)." *Pestic. Biochem. Physiol.,* 23, 205–11.

35. Bonner, J. 1942. "A reversible growth inhibition of isolated tomato roots." *Proc. Natl. Acad. Sci. U.S.A.,* 28, 321–24.

36. Boll, W. G. 1955. "Inhibition of excised tomato roots by sulfanilimide and its reversal by a *p*-aminobenzoic acid and n-pterorylglutamic acid." *Plant Physiol.,* 30, 161–68.

37. Brian, P. W. 1944. "Effect of *p*-aminobenzoic acid on the toxicity of *p*-aminobenzene-sulphonamide to higher plants and fungi." *Nature,* 153, 83–84.

38. Scholl, R. L., and K. M. Dennison. 1978. "Sensitivity of cultured tissue of *Arabidopsis thaliana* races to sulfanilamide." *Physiol. Plant.,* 43, 321–25.

39. Woods, D. D. 1940. "The relation of *p*-aminobenzoic acid to the mechanism of the action of sulfanilimide." *Brit. J. Exp. Pathol.,* 21, 74–90.

40. Zhang, L.-H., and K.-H. Lin. 1988. "Mechanisms of selective action of sodium sulfanilate on plants." *Pestic. Biochem. Physiol.,* 32, 11–16.

41. Lin, K.-H., and L.-H. Zhang. 1988. "Mechanisms of differential inhibitory effects of sodium sulfanilate of folic acid biosynthesis in plants." *Pestic. Biochem. Physiol.,* 32, 17–24.

42. Killmer, J. L., J. M. Widholm, and F. W. Slife. 1980. "Antagonistic effect of *p*-aminobenzoate or folate on asulam [methyl (4-aminobenzensulphonyl carbamate)] inhibition of carrot suspension cultures." *Plant Sci. Lett.,* 19, 203–08.

43. Veerasekaran, P., R. C. Kirkwood, and E. W. Parnell. 1981. "Studies of the mechanism of action of asulam in plants. I. Antagonistic interaction of asulam and 4-aminobenzoic acid." *Pestic. Sci.,* 12, 325–29.

44. Mugnier, J. 1988. "Behaviour of herbicides in dicotyledonous roots transformed by *Agrobacterium rhizogenes.* I. Selectivity." *J. Exp. Bot.,* 39, 1045–56.

45. Veerasekaran, P., R. C. Kirkwood, and E. W. Parnell. 1981. "Studies of the mechanism of action of asulam in plants. II. Effect of asulam on the biosynthesis of folic acid." *Pestic. Sci.,* 12, 330–38.

46. Crosti, P. 1981. "Effect of folate analogues on the activity of dihydrofolate reductase and on the growth of plant organisms." *J. Exp. Bot.,* 32, 717–23.

47. Suzuki, N., and K. Iwai. 1970. "The occurrence and properties of dihydrofolate reductase in pea seedlings." *Plant Cell Physiol.,* 11, 199–208.

48. Mohammed, A. M. S., K. Al-Chalabi, and S. A. Abood. 1989. "Effect of folate analogues on the activity of dihydrofolate reductase and callus growth of sunflower." *J. Exp. Bot.*, 40, 701–06.
49. Wilkinson, R. E. 1981. "Metolachlor [2-chloro-*N*-(2-ethyl-6-methylphenyl)-*N*-(2-methoxy-1-methylethyl)acetamide] inhibition of gibberellin prescursor biosynthesis." *Pestic. Biochem. Physiol.*, 16, 199–205.
50. Wilkinson, R. E. 1982. "Alachlor influence on sorghum growth and gibberellin precursor biosynthesis." *Pestic. Biochem. Physiol.*, 17, 177–84.
51. Wilkinson, R. E. 1982. "Mefluidide inhibition of sorghum growth and gibberellin precursor biosynthesis." *J. Plant Growth Regul.*, 1, 85–94.
52. Wilkinson, R. E. 1983. "Gibberellin precursor biosynthesis inhibition by EPTC and reversal by R-25788." *Pestic. Biochem. Physiol.*, 19, 321–29.
53. Wilkinson, R. E. 1985. "CDAA inhibition of kaurene oxidation in etiolated sorghum coleoptiles." *Pestic. Biochem. Physiol.*, 23, 19–23.
54. Wilkinson, R. E. 1982. "Metolachlor [2-chloro-*N*-(2-ethyl-6-methylphenyl)-*N*-(2-methoxy-1-methylethyl)acetamide] inhibition of gibberellin precursor biosynthesis." *Pestic. Biochem. Physiol.*, 16, 199–205.
55. Wilkinson, R. E. 1986. "Diallate inhibition of gibberellin biosynthesis in sorghum coleoptiles." *Pestic. Biochem. Physiol.*, 25, 93–97.
56. Sandmann, G., and P. Böger. 1986. "Interconversion of prenyl pyrophosphates and subsequent reactions in the presence of FMC 57020." *Z. Naturforsch.*, 42c, 803–07.
57. Chang, T. C., H. V. Marsh, and P. H. Jennings. 1975. "Effect of alachlor on *Avena* seedlings: Inhibition of growth and interaction with gibberellic acid and indoleacetic acid." *Pestic. Biochem. Physiol.*, 5, 323–29.
58. Truelove, B., D. E. Davis, and C. G. P. Pillai. 1977. "Mefluidide effects on growth of corn (*Zea mays*) and the synthesis of protein by cucumber (*Cucumis sativus*) cotyledon tissue." *Weed Sci.*, 25, 360–63.
59. Leavitt, J. R. C., and D. Penner. 1978. "Potential antidotes against acetanilide herbicide injury to corn (*Zea mays*)." *Weed Res.*, 18, 281–86.
60. Rademacher, W. 1989. "Gibberellins: metabolic pathways and inhibitors of biosynthesis." In *Target Sites of Herbicide Action,* P. Böger and G. Sandmann, eds. Boca Raton, Fla.: CRC Press, Inc., pp. 127–45.
61. Vencill, W. K., K. K. Hatzios, and H. P. Wilson. 1989. "Growth and physiological responses of normal, dwarf, and albino corn (*Zea mays*) to clomazone treatments." *Pestic. Biochem. Physiol.*, 35, 81–88.
62. Miki, T., Y. Kamiya, M. Fukazawa, T. Ichikawa, and A. Sakurai. 1990. "Sites of inhibition by a plant-growth regulator, 4′-chloro-2-(α-hydroxybenzyl)-isonicotinanilide (inabenfide), and its related compounds in the biosynthesis of gibberellins." *Plant Cell Physiol.*, 31, 201–06.
63. Wilkinson, R. E. 1988. "Consequences of metolachlor induced inhibition of gibberellin biosynthesis in sorghum seedlings." *Pestic. Biochem. Physiol.*, 32, 25–37.
64. Jangaard, N. O. 1974. "The effect of herbicides, plant growth regulators and other compounds on phenylalanine ammonia-lyase activity." *Phytochemistry*, 13, 1769–75.
65. Amrhein, N., and K. H. Gödeke. 1977. "α-Aminooxy-β-phenylpropionic acid, a potent inhibitor of L-phenylalanine ammonia-lyase *in vitro* and *in vivo*." *Plant Sci. Lett.*, 8, 313–17.

66. Reichart, D., A. Simon, F. Durst, J. M. Mathews, and P. R. Ortiz de Montellano. 1982. "Autocatalytic inactivation of plant cytochrome P-450 enzymes: selective inactivation of cinnamic acid 4-hydroxylase from *Helianthus tuberosus* by 1-aminobenzotriazole." *Arch. Biochem. Biophys.*, 216, 522–29.
67. Waldmüller, T., and H. Grisebach. 1987. "Effects of R-(1-amino-2-phenylethyl)phosphonic acid on glyceollin accumulation and expression of resistance to *Phytopthora megasperma* f.sp. *glycinea* in soybean." *Planta*, 172, 424–30.
68. De Jaehger, G., and N. Boyer. 1987. "Specific inhibition of lignification in *Bryonia dioica*. Effects on thigmomorphogenesis." *Plant Physiol.*, 84, 10–11.
69. Muñoz, R., A. Martinez-Martinez, A. Ros-Barceló, and M. A. Pedreño. 1990. "Effect of 4-amino-6-methyl-3-phenylamino-1,2,4-triazin-5(4*H*)-one on the lignification process catalyzed by peroxidase from lupin (*Lupinus albus*)." *Pestic. Sci.*, 28, 283–89.
70. Solomonson, L. P., and B. Vennesland. 1972. "Nitrate reductase and chlorate toxicity in *Chlorella vulgaris* Beijerinck." *Plant Physiol.*, 50, 421–24.
71. Hofstra, J. J. 1977. "Chlorate toxicity and nitrate reductase activity in tomato plants." *Physiol. Plant.*, 41, 65–69.
72. Doddema, H., and H. Otten. 1979. "Uptake of nitrate by mutants of *Arabidopsis thaliana*, disturbed in uptake or reduction of nitrogen. III. Regulation." *Physiol. Plant.*, 45, 339–46.
73. Cervelli, S., P. Nannipieri, G. Giovannini, and A. Penna. 1975. "Jack bean urease inhibition by substituted ureas." *Pestic. Biochem. Physiol.*, 5, 221–25.
74. McCarty, R. E. 1980. "Delineation of the mechanism of ATP synthesis in chloroplasts: use of uncouplers, energy transfer inhibitors, and modifiers of coupling factor 1." *Methods Enzymol.*, 69, 719–28.
75. Huchzermeyer, B., and A. Loehr. 1983. "Effects of nitrofen on chloroplast coupling factor-dependent reactions." *Biochim. Biophys. Acta*, 724, 224–29.
76. Nishimura, K., T. Kawata, K. Asada, and M. Nakajima. 1975. "Effect of Hill reaction inhibitors on photoreduction of ferricyanide and cyclic photophosphorylation by spinach-chloroplasts." *Agric. Biol. Chem.*, 39, 867–72.
77. Fedtke, C. 1982. *Biochemistry and Physiology of Herbicide Action*. Berlin: Springer-Verlag, 202 pp.
78. Wright, B. J., A. C. Baillie, K. Wright, J. R. Dowsett, and T. M. Sharpe. 1980. "Synthesis of potential herbicides designed to uncouple photophosphorylation." *Phytochemistry*, 19, 61–65.
79. van den Berg, G., and M. Brandse. 1984. "Effects of N-aryl-N′,N′-dialkyl-1,2-ethanediamines on ATP formation in chloroplast. QSAR of amine uncouplers." *Z. Naturforsch.*, 39c, 107–14.
80. Scalla, R., and C. Gauvrit. 1984. "Action of herbicides on plant cell membranes." In *Membranes and Compartmentalization in the Regulation of Plant Functions*, A. M. Boudet et al., eds. Oxford: Clarendon Press, pp. 301–21.
81. Balke, N. E. 1985. "Herbicide effects on membrane functions." In *Weed Physiology Vol. II Herbicide Physiology*, S. O. Duke, ed. Boca Raton, Fla.: CRC Press, pp. 113–39.
82. Bujtás, C., T. Cserháti, E. Cseh, Z. Illés, and Z. Szigeti. 1987. "Effect of some benzonitrile esters on the K^+ uptake of wheat seedlings: quantitative structure-activity relationships." *Biochem. Physiol. Pflanzen*, 182, 465–71.

83. O'Leary, N. F., J. T. O'Donovan, and G. N. Prendeville. 1980. "Effect of diclofop methyl and 2,4-D alone and in combination on leaf cell membrane permeability of wild oats and barley." *Can. J. Plant Sci.*, 60, 773–75.

84. Watson, M. C., P. G. Bartels, and K. C. Hamilton. 1980. "Action of selected herbicides and Tween 20 on oat (*Avena sativa*) membranes." *Weed Sci.*, 28, 122–27.

85. O'Brien, M. C., and G. N. Prendeville. 1979. "Effect of herbicides on cell membrane permeability in *Lemna minor*." *Weed Res.*, 19, 331–34.

86. Towne, C. A., P. G. Bartels, and J. L. Hilton. 1978. "Interaction of surfactant and herbicide treatments on single cells of leaves." *Weed Sci.*, 26, 182–88.

87. Mellis, J. M., P. Pillai, D. E. Davis, and B. Truelove. 1982. "Metolachlor and alachlor effects on membrane permeability and lipid synthesis." *Weed Sci.*, 30, 399–404.

88. Blein, J.-P. 1982. "Action of some herbicides on growth, respiration, plasmalemma integrity, and proton extrusion of *Acer pseudoplatanus* cells II. Amides, diphenyl ethers, nitriles, phenols, triazines, and uracils." *Pestic. Biochem. Physiol.*, 17, 156–61.

89. de Agazio, M., and M. C. Giardina. 1987. "Inhibition of fusicoccin-stimulated K^+/H^+ transport in root tips from maize seedlings pretreated with chlorsulfuron." *Plant Cell Environ.*, 10, 229–32.

90. Ferrari, G., S. Nardi, G. Cacco, and G. Dell'Agnola. 1981. "Sulfate transport of excised roots as an index of genotype response to herbicides." *Physiol. Plant.*, 52, 29–32.

91. Ingle, M., and B. J. Rogers. 1961. "Some physiological effects of 2,2-dichloropropionic acid." *Weeds*, 9, 264–72.

92. Zsoldos, F., and E. Haunold. 1982. "Influence of 2,4-D and low pH on potassium, ammonium and nitrate uptake by rice roots." *Physiol. Plant.*, 54, 63–68.

93. Ratterman, D. M., and N. E. Balke. 1988. "Herbicidal disruption of proton gradient development and maintenance by plasmalemma and tonoplast vesicles from oat root." *Pestic. Biochem. Physiol.*, 31, 221–36.

94. Travis, R. L., and W. G. Woods. 1977. "Studies on the mechanism of action of dinitramine. Effect on soybean root plasma membrane." *Plant Physiol.*, 60, 54–57.

95. Andreev, G. K., and N. Amrhein. 1976. "Mechanism of action of the herbicide 2-chloro-3-(4-chlorophenyl)propionate and its methyl ester: interaction with cell responses mediated by auxin." *Physiol. Plant.*, 37, 175–82.

96. Leong, T.-Y., and W. R. Briggs. 1982. "Evidence from studies with acifluorfen for participation of a flavin-cytochrome complex in blue light photoreception for phototropism of oat coleoptiles." *Plant Physiol.*, 70, 875–81.

97. Ratterman, D. M., and N. E. Balke. 1989. "Diclofop-methyl increases the proton permeability of isolated oat-root tonoplast." *Plant. Physiol.*, 91, 756–65.

98. Duke, S. O., and W. H. Kenyon. 1988. "Polycyclic alkanoic acids." In *Herbicides—Chemistry, Degradation, and Mode of Action, Vol. 3,* P. C. Kearney and D. D. Kaufman, eds. New York: Marcel Dekker, pp. 71–116.

99. Shimabukuro, R. H., W. C. Walsh, and J. P. Wright. 1989. "Effect of diclofop-methyl and 2,4-D on transmembrane proton gradient: a mechanism for their antagonistic interaction." *Physiol. Plant.*, 77, 107–14.

100. Linsel, G., I. Dahse, and E. Müller. 1983. "Electrophysiological evidence for the herbicidal mode of action of phosphinic esters." *Physiol. Plant.*, 73, 77–84.

101. Wright, J. P., and R. H. Shimabukuro. 1987. "Effects of diclofop and diclofop-methyl on membrane potentials of wheat and oat coleoptiles." *Plant Physiol.*, 85, 188–93.

102. Weber, A., E. Fischer, H. S. von Branitz, and U. Lüttge. 1988. "The effects of the herbicide sethoxydim on transport processes in sensitive and tolerant grass species I. Effects on the electrical membrane potential and alanine uptake." *Z. Naturforsch.*, 43c, 249–56.

103. Weber, A., and U. Lüttge. 1988. "The effects of the herbicide sethoxydim on transport processes in sensitive and tolerant grass species II. Effects on membrane-bound redox systems in plant cells." *Z. Naturforsch.*, 43c, 257–63.

104. McWhorter, C. G. 1985. "The physiological effects of adjuvants on plants." In *Weed Physiology. Vol. II. Herbicide Physiology*, S. O. Duke, ed. Boca Raton, Fla.: CRC Press, Inc., pp. 141–58.

105. Bujtás, C., T. Cserháti, and Z. Szigeti. 1988. "Effect of some nonionic tenzides on potassium influx in roots of wheat seedlings." *Biochem. Physiol. Pflanzen*, 183, 277–81.

106. Smith, T. A. 1985. "Polyamines." *Annu. Rev. Plant Physiol.*, 36, 117–43.

107. Galston, A. W., and R. K. Sawhney. 1990. "Polyamines in plant physiology." *Plant Physiol.*, 94, 406–10.

108. Christ, M., H. Felix, and J. Harr. 1989. "Inhibitors influencing plant enzymes of the polyamine biosynthetic pathway." *Z. Naturforsch.*, 44c, 49–54.

109. Felix, H., and J. Harr. 1989. "Influence of inhibitors of polyamine biosynthesis on polyamine levels and growth of plants." *Z. Naturforsch.*, 44c, 55–58.

110. DiTomaso, J. M., and S. O. Duke. 1991. "Is polyamine biosynthesis a possible site of action of cinmethylin or artemisinin?" *Pestic. Biochem. Physiol.*, 39, 158–67.

111. Palavan-Ünsal, N. 1987. "Polyamine metabolism in the roots of *Phaseolus vulgaris*. Interaction of the inhibitors of polyamine biosynthesis with putrescine in growth and polyamine biosynthesis." *Plant Cell Physiol.*, 28, 565–72.

112. Giardina, M. C., and S. Carosi. 1990. "Effects of chlorsulfuron on polyamine content in maize seedlings." *Pestic. Biochem. Physiol.*, 36, 229–36.

113. DiTomaso, J. M., T. L. Rost, and F. M. Ashton. 1988. "Herbicide-induced diamine accumulation in pea roots. The effect of napropamide on polyamine levels." *Plant Cell Physiol.*, 29, 1367–72.

114. Moreland, D. E. 1985. "Effect of herbicides on respiration." In *Weed Physiology Vol. II Herbicide Physiology*, S. O. Duke, ed. Boca Raton, Fla.: CRC Press, pp. 37–61.

115. Gauvrit, C. 1984. "An alteration in membrane fluidity can explain the various effects of the herbicides benzoylpropethyl and flampropisopropyl on plant mitochondria." *Physiol. Vég.*, 22, 57–66.

116. Baillie, A. C., K. Wright, B. J. Wright, and C. G. Earnshaw. 1988. "Inhibitors of pyruvate dehydrogenase as herbicides." *Pestic. Biochem. Physiol.*, 30, 103–12.

117. Schloss, J. V. 1990. "Acetolactate synthase, mechanism of action and its herbicide binding site." *Pestic. Sci.*, 29, 283–92.

118. Wilkinson, R. E., and T. H. Oswald. 1987. "*S*-Ethyl dipropylthiocarbamate (EPTC) and 2,2-dichloro-N,N-di-2-propenylacetamide (dichlormid) inhibitions of synthesis of acetyl-coenzyme A derivatives." *Pestic. Biochem. Physiol.*, 28, 38–43.

119. Cho, H.-Y., J. M. Widholm, and F. W. Slife. 1988. "Haloxyfop inhibition of the

pyruvate and the α-ketoglutarate dehydrogenase complexes of corn (*Zea mays* L.) and soybean (*Glycine max* [L.] Merr.)." *Plant Physiol.*, 87, 334–40.

120. Gauvrit, C. 1983. "The experimental herbicide UKJ72J is an inhibitor of succinate oxidation in plant mitochondria." *FEBS Lett.*, 158, 222–24.
121. Gauvrit, C. 1985. "Inhibition of succinate oxidation by the herbicide UKJ72J." *Phytochemistry*, 24, 2181–83.
122. Burden, R. S., D. T. Cooke, and G. A. Carter, 1989. "Inhibitors of sterol biosynthesis and growth in plants and fungi." *Phytochemistry*, 28, 1791–1804.
123. Burden, R. S., G. A. Carter, T. Clark, D. T. Cooke, S. J. Croker, A. H. B. Deas, P. Hedden, C. S. James, and J. R. Lenton. 1987. "Comparative activity of the enantiomers of triadimenol and paclobutrazol as inhibitors of fungal growth and plant sterol and gibberellin biosynthesis." *Pestic. Sci.*, 21, 253–67.
124. Burden, R. S., C. S. James, D. T. Cooke, and N. H. Anderson. 1987. "C-14 demethylation in phytosterol biosynthesis—A new site for herbicidal activity." Brit. Crop Protect. Conf.—Weeds, 13, 171–78.
125. Khalil, I. A., and E. I. Mercer. 1990. "Effect of diclobutrazol on the growth and sterol and photosynthetic pigment content of winter wheat." *Pestic. Sci.*, 28, 271–81.
126. Shive, J. B., Jr. and H. D. Sisler. 1976. "Effects of ancymidol (a growth retardant) and triarimol (a fungicide) on the growth, sterols, and gibberellins of *Phaseolus vulgaris* (L.)." *Plant Physiol.*, 57, 640–44.

Chapter 16

Secondary Physiological Effects of Herbicides

16.1 INTRODUCTION

After the primary site of a herbicide is affected, a series of related biochemical and physiological events occur sequentially. These events are often interconnecting and relate to each other in complex ways so that connecting secondary events to a primary effect is sometimes quite difficult. By the time a plant is killed by a herbicide, virtually every biochemical and physiological process of the plant will be affected. Association of primary effects with secondary effects can be especially difficult when the time between treatment and the development of symptoms is relatively long, allowing significant involvement of secondary effects in the demise of the plant. Literature on the mode of action of herbicides is full of papers that confuse secondary and primary effects.

A careful study of the sequence of physiological effects of a herbicide can eliminate some secondary effects from consideration as primary effects. However, to our knowledge, this method has never produced the critical piece of information that has led to the discovery of a primary molecular target site of a herbicide.

Primary effects of secondary importance are sometimes mistakenly identified as secondary effects. In a plant that is killed by a herbicide, action at a primary site of action of secondary importance will probably play an insignificant role. However, following a sublethal dose in plants that are resistant at the targeted primary site (e.g., crops genetically engineered to be herbicide-resistant), action at other sites could cause profound physiological changes of agricultural importance.

Similarly, in plants that are tolerant of a particular herbicide due to metabolic alteration of the herbicide, metabolites of the herbicide can have significant physiological effects. In most cases, differentiation between secondary effects resulting from primary effects and primary effects of secondary importance is quite difficult. Both possibilities should be considered when studying secondary effects of herbicides.

Other chapters in this volume have dealt to some extent with secondary effects of the herbicides covered in those chapters. Therefore, this chapter will deal only with important secondary effects not covered previously. These include secondary effects on nitrogen metabolism, secondary metabolism, hormone levels and sensitivities to hormones, and chloroplast organization and function. Secondary effects can be important in nontarget plants, including crops. Sublethal effects of herbicides influence metabolic processes that affect plant growth and development, interactions with pathogens and other pests, responses to the physical environment, and mineral nutrition of the plant. These aspects of secondary effects of herbicides will also be discussed in this chapter.

16.2 EFFECTS ON NITROGEN METABOLISM

The chloroplast enzyme, nitrite reductase, is dependent on photosynthetic electron flow to ferredoxin, which is used by the enzyme as a reductant. Inhibitors of photosynthetic electron transport therefore stop or reduce the reduction of nitrite to ammonium ion. If nitrate continues to be reduced to nitrite in the cytoplasm, toxic levels of nitrite and nitrous acid and, in soybeans, free radicle NO_X will accumulate rapidly [1]. At least part of the generation of NO_X is enzymatic [2]. In nitrate-fertilized plants, photosynthetic inhibitor herbicides cause rapid accumulation of high nitrite concentrations and high levels of NO_X evolution via this mechanism [1, 3–6]. The effect is so pronounced that an assay for nitrite accumulation has been suggested as a rapid screen for photosynthetic inhibitors [5].

Nitrite accumulation can be caused in darkness by compounds such as respiratory inhibitors [1, 6, 7] that cause an increase in NADH availability for nitrate reductase. For example, 2,4-dinitrophenol causes nitrite levels in soybean leaves to increase from undetectable levels to 5 μmoles nitrite per gram of fresh weight within 90 min of treatment in darkness [6]. Such compounds can synergize the effects of photosynthesis inhibitors on NO_X evolution and nitrite accumulation [1, 7] (Figure 16.1). Compounds that do not directly inhibit photosynthesis will eventually inhibit photosynthesis through secondary effects and cause nitrite accumulation as a tertiary effect. For instance, although the primary mechanism of action of thiobencarb is probably on lipid synthesis (Chapter 11), its ability to cause nitrite accumulation in barnyardgrass (*Echinochloa crus-galli*) leaves is associated with its inhibition of photosynthesis and more pronounced inhibition of in vitro nitrite reductase activity than of nitrate reductase [8].

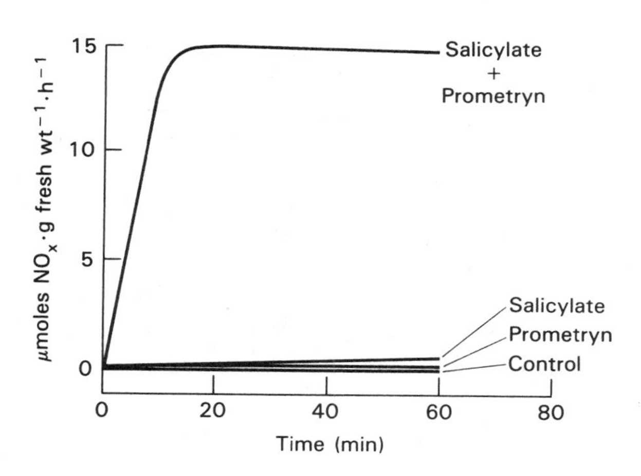

Figure 16.1 NO_x emission rates by soybean leaves treated with salicylate, prometryn, or a combination of salicylate and prometryn [1].

In soybean cotyledons, photobleaching of the chloroplast caused by inhibition of carotenoid synthesis by norflurazon results in very high levels of extractable nitrate reductase activity [9, 10] (Figure 16.2). The isoform of the enzyme found at elevated levels is more like the isoform normally found in the root than in green tissues; it is not the constitutive nitrate reductase isoform of soybeans with which NO_X generation from nitrate has been found to be associated [2]. In alfalfa, norflurazon-induced bleaching is associated with large increases in certain

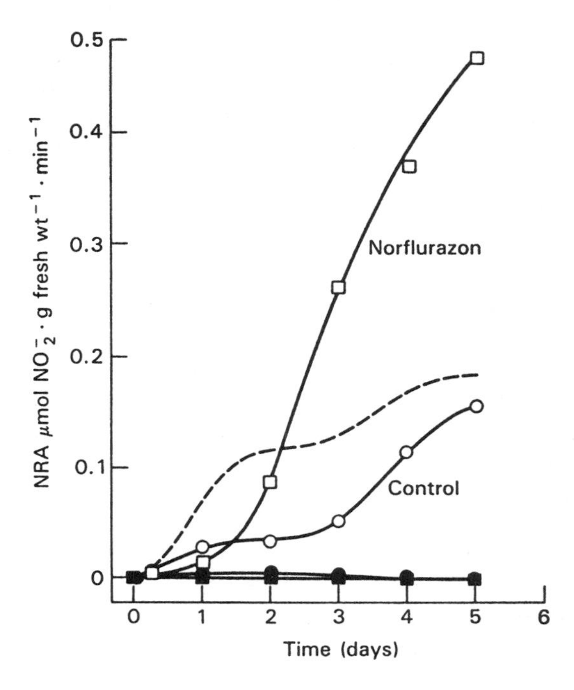

Figure 16.2 Superinduction of NADH nitrate reductase activity (NRA) (pH 7.5) in photobleached cotyledons of norflurazon-treated soybean seedlings. Three-day-old, dark-grown plants were exposed to white light at zero time. Closed symbols are the dark controls. The dotted line represents nitrate reductase of the untreated seedlings assayed at its pH maximum. The pH maximum of the photobleached tissues remained at 7.5, whereas that of the control was closer to 6.5 [9].

amylases [11]. The mechanism of increases in extraplastidic enzymes in tissues of plants treated with bleaching herbicides is not known. In some species (e.g., *Sinapis alba* and barley), carotenoid synthesis inhibitors generally cause a complete absence or loss of extractable nitrate reductase (e.g., [12, 13]). The effect on extractable nitrate reductase is associated with the photobleaching effect of the herbicide, because the effect is much less under nonphotobleaching, far-red light [13]. The mechanisms of bleaching herbicide effects on nitrate reductase are not known; however, some forms of nitrate reductase apparently require a plastidic factor for induction of the nuclear gene that codes for them [14]. Differences among species in the effects of norflurazon on expression of genes coding for nitrate reductase may be due to differences in both the isoforms that are controlled by the plastid and those that are not.

Not all of the inhibition of development of nitrate reductase activity may be associated with photobleaching of the plastid. The nonbleaching pyridazinone herbicides Sandoz 9785 or BASF 13-338 cause very pronounced reductions in both nitrite reductase and nitrate reductase of green barley leaves [15]. This effect may be due to the activity of these compounds as photosynthesis inhibitors. Many effects of photosynthetic electron transport inhibitors on nitrate reductase and nitrogen metabolism of plants have been reported [16]. These effects are often conflicting and difficult to interpret.

PS II inhibitors can inhibit nitrate reductase induction in greening tissues of some species (e.g., [17]); in other species, however, no effect has been found [18, 19]. In green tissues, PS II-inhibiting herbicides can inhibit nitrate-induced increases in extractable nitrate reductase [20]. In greening [17] or green tissues [20], supplying exogenous sugars can alleviate the inhibition of nitrate reductase activity by the PS II-inhibiting herbicide. However, in roots of both wheat and cucumber, large simazine-induced reductions in extractable nitrate reductase activity could not be alleviated by supplying the plants with exogenous sugars [20]. This diversity in responses is not understood.

In green tissues, PS II-inhibiting herbicides have generally been reported to have no effect or to increase extractable nitrate reductase (e.g., [21–24]). Increased protein yields that are associated with the increases in nitrate reductase in crops treated with sublethal levels of triazine herbicides have led some to propose sublethal levels of triazines as growth regulators [25].

The mechanism of increased nitrate reductase by PS II inhibitors is not clear. One might expect that plants would have a mechanism to lower nitrate reductase activity when photosynthesis is reduced. Ries [25] concluded that simazine increased nitrate reductase levels in plants by increasing nitrate concentrations in the plant. That this is due to a membrane effect seems unlikely because PS II inhibitors of widely varying structural groups all increase nitrate reductase activity. Therefore, if they all increase nitrate uptake, the effect is likely to be a secondary effect of inhibition of PS II, the site of action that they all share.

An alternative hypothesis is that herbicides that reduce cellular ATP levels cause nitrate leakage from vacuoles, resulting in enhanced nitrate reductase levels and activities [24]. No herbicides have been reported to have a significant direct effect on nitrate reductase. One might expect that a comparison of the effects of PS II inhibitors on nitrate reductase in resistant and susceptible biotypes with little or no other differences in their genome should resolve the question of whether the effect of PS II inhibitors on nitrate reductase is a secondary effect of inhibition of photosynthesis or is an effect unrelated to photosynthesis. Using atrazine-resistant and susceptible biotypes of *Chenopodium album,* Lawrence et al. [26] found that, three days after treatment with atrazine, nitrate content in the susceptible biotype increased to levels almost threefold those of untreated plants; no increase was observed in the resistant biotype. However, both in vitro and in vivo nitrate reductase activity levels were dramatically reduced in both biotypes. The authors concluded that the effect on nitrate reductase activity was due to an effect unrelated to inhibition of photosynthesis. Other potential physiological differences between the biotypes were not considered. No studies with triazine-resistant and susceptible biotypes have been conducted in which the herbicide has been found to increase nitrate reductase. The mechanism(s) of induction of nitrate reductase by PS II inhibitors remain(s) unresolved.

Glufosinate (see Chapter 13) inhibits nitrate reductase by inhibiting uptake of nitrate [27]. Uptake of nitrate as well as other anions is inhibited due to direct depolarization of the plasmalemma and to the accumulation of ammonium ion and the resulting alkalinization of the cytoplasm and uncoupling of electron transport. Nitrate reductase activity can be directly inhibited by high ammonium ion levels; however, reductions in nitrate reductase activity in glufosinate-treated alfalfa occur before ammonia levels are greatly increased [28].

Other reports of herbicide effects on nitrate and/or nitrite reductase activities exist (e.g., [29, 30]). However, in these reports the mechanism of the effect is even more obscure than with herbicides that inhibit chloroplast development or photosynthesis. In a whole plant, decreased transpiration caused by a herbicide can limit nitrate uptake and consequently reduce nitrate-induced nitrate reductase [31].

Herbicides may also affect the nitrogen metabolism of plants through effects on nitrogen fixation, either by effects on the plant or on the bacterium [32, 33]. In most cases, the evidence indicates that reductions in nodulation and/or fixation by existing nodules are due to effects on the plant rather than on the bacterium [32–35]. Only dinoseb has been found to have a direct effect on nitrogenase activity [35]. In some cases, herbicides have been reported to increase nitrogen fixation. For example, cyanazine, simazine, and trifluralin have been shown to increase acetylene reduction activity (=nitrogenase) in *Lupinus albus* [33]. In this study, the two triazines also increased grain yield and seed weight.

16.3 EFFECTS ON SECONDARY METABOLISM

Secondary products of plants are compounds with no direct role in primary metabolism and for which there is no known short-term requirement by the plant. Tens of thousands of these compounds have been isolated and identified, and hundreds of thousands of them are thought to exist. Most of the compounds that have been identified are products of the shikimate (Figure 13.1) or the terpenoid pathways (Figure 8.1); however, secondary products of many chemical classes exist in plants. These compounds are thought to function primarily in interactions of plants with pathogens, herbivores, and other plants and as structural components of cell walls such as lignins. Stress of any type can have profound effects on the biosynthesis of these compounds [36]. Although the effects of herbicides on secondary metabolism have not been studied in detail, the information we have indicates that the effects that occur may be important in agroecosystems [37]. Both herbicide-caused reductions and increases in content of secondary compounds in plants can be cause for concern. For instance, 2,4,5-T markedly increases poisonous alkaloid content of some plants [38]. Such an effect could be quite important in pasture or rangeland [39]. Maleic hydrazide can cause significant decreases in nicotine content of tobacco [40], an effect that could be interpreted as either desirable or undesirable. When a slow-acting herbicide is used to kill a cover crop or as a harvest aid, changes induced in secondary metabolism of the target plants can influence the potential toxicity of the plant residues to the next crop. Weston and Putnam [41] found glyphosate-killed *Agropyron repens* tissue to be significantly less phytotoxic to several legume crops than were untreated tissues.

Several types of stress, such as those caused by wounding [42] or by plant pathogens [43], will generally cause dramatic increases in the levels of certain phenolic compounds in plants. Repair of wounds requires sealing off the affected area with phenolic polymers such as lignin as well as prevention of infection with antimicrobial phenolics and terpenoids (phytoalexins). Plants have a repertoire of inducible phytoalexins with which to combat pathogens. The effects of herbicidal damage are not so clear-cut because the type and mechanism of damage varies considerably among herbicides and affected species, and with herbicide concentration and/or method of application with specific herbicides used on particular species. In a survey of the effects of 16 herbicides on synthesis of anthocyanin in soybean hypocotyls, half of the herbicides (amitrole, diclofop-methyl, DSMA, fluridone, maleic hydrazide, nitralin, perfluidone, and 2,4-D) had no significant effect, while the other half (atrazine, fenuron, metribuzin, norflurazon, paraquat, propanil, propham, and TCA) significantly reduced the anthocyanin content [44]. There was a rough correlation between the effect on anthocyanin content and the effect on extractable phenylalanine ammonia-lyase (PAL) activity. PAL is the first enzyme in the production of secondary phenolic compounds, in that it deaminates either phenylalanine or tyrosine to form cinnamic or *p*-coumaric acids, respectively.

The effects of most herbicides on synthesis of secondary products of the shikimic acid pathway are probably very indirect; however, there are two groups of herbicides for which the mechanism is more completely understood. The effects of glyphosate on the synthesis of phenolic secondary compounds have been relatively well documented. By inhibiting EPSPS of the shikimic acid pathway (Figure 13.1), glyphosate blocks synthesis of aromatic amino acids and all phenolic compounds derived from them, including flavonoids [45], lignins [46], cinnamic acids [47], and flavans [48]. Reported effects on synthesis of specific phenolic compounds are numerous and are summarized in detail in two reviews [37, 49]. Unlike other herbicides that have a more indirect effect on phenolic biosynthesis [44], glyphosate generally causes an increase in extractable PAL activity concomitantly with the decrease in synthesis of secondary phenolic compounds—probably because of decreased feedback inhibition of the enzyme [49]. In addition to the block at the EPSPS site, there is evidence that glyphosate also blocks synthesis of flavonoids at a later step in the pathway [50]. The large increase in shikimate caused by inhibition of EPSPS and deregulation of the shikimate pathway (see Chapter 13) results in large increases of certain benzoic acids that are apparently derived directly from shikimate [47, 51, 52] (Figure 16.3). Content of tannins derived from benzoic acids can also be increased by glyphosate [53]. Benzoic acids can also be derived by side-chain degradation of cinnamic acids in some or all plant species, and this is apparently the principal route of synthesis in some species [54]. The effect on benzoic acid content is relatively much greater than that on the content of aromatic amino acid-derived phenolic compounds. For example, the content of protocatechuic acid can be increased by more than 100-fold in certain tissues of pigweed (*Amaranthus retroflexus*) and soybean [51] (Figure 16.3), whereas the content of aromatic amino acid-derived compounds is seldom reduced by more than 50 percent by glyphosate [37]. One reason that glyphosate-induced reductions in aromatic amino acid-derived phenolic compounds is less than might be expected is that this pathway has access to aromatic amino acids formed by protein degradation and in storage pools in vacuoles, as well as freshly formed amino acids [55]. The level of reliance of secondary compound synthesis on the two pools can vary between tissues, resulting in differences in the effects of glyphosate on biosynthesis of these compounds from the two pools.

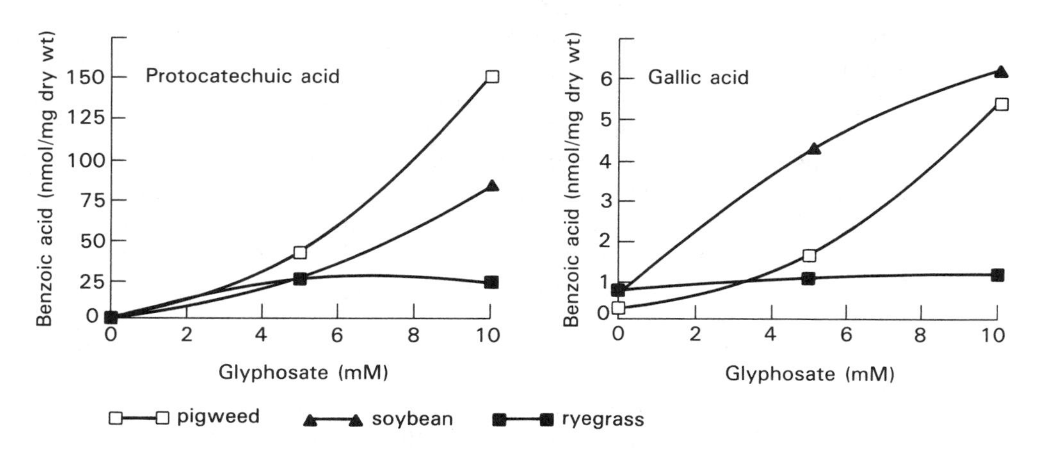

Figure 16.3 Effects of different concentrations of glyphosate on protcatechuic acid and gallic acid from three plant species [51].

In a few cases glyphosate has been reported to increase levels of phenolic compounds in plants. In some of these cases (e.g. [56]) the apparent increases may be an artifact of the assay procedure. For example, a general method of estimating total phenolics, such as a procedure based on reaction with Folin-Ciocalteau reagent, may not be accurate because of qualitative changes caused by the herbicide. However, there are reports of glyphosate induction of specific phenolic compounds such as glyceollins in soybean [57]. In this study, levels of glyceollin precursors were reduced by glyphosate, resulting in a net reduction in total carbon in glyceollins plus their precursors.

At sublethal glyphosate levels, the effects on phenolic compound levels in metabolite-exporting tissues are transitory. However, the effects can remain pronounced in metabolic sinks such as fruits or flowers [52]. This may be due to the fact that glyphosate itself is very phloem mobile and accumulates in these and other carbohydrate sinks (see Chapter 5). Thus, the effects on secondary products apparently can be maintained only in the presence of the herbicide.

The significance of the effects of glyphosate on secondary metabolism may be great. There are reports of herbicide effects on crop or plant tissue susceptibility to pathogens [58–61] that may be due entirely or in part to herbicide effects on secondary metabolism. For example, the increase in susceptibility of raspberry tissues to the fungus *Didymella applanata* is negatively correlated with the amount of phenolic compounds in the tissues [62]. Depending on whether the herbicide increases or decreases synthesis of phytoalexins, susceptibility can be decreased or increased, respectively. Since glyphosate blocks synthesis of phenolic phytoalexins, one would expect this herbicide to enhance susceptibility to pathogens. There is evidence to indicate that this is often the case.

In one study, bean (*Phaseolus vulgaris*) plants treated with glyphosate died when grown in nonsterile soil, but survived in sterile soil [63]. The effect was demonstrated to be due to differences in susceptibility to root-infesting pathogens. Similar results have been obtained with *Sinapis alba*, apple, wheat, and corn [64]. The major pathogens infecting glyphosate-treated wheat and beans are *Pythium* and *Fusarium* spp. Glyphosate-treated *Agropyrons repens* [65] and flax [66] have been shown to be rapidly infected by fungi. Glyphosate significantly increases infections of several weed species with *Fusarium* spp. [67]. In addition, glyphosate has been shown to reimpose susceptibility of a genetically resistant tomato variety to *Fusarium* crown and root rot [68]. This effect was associated with reduced incorporation of phenolic compounds into structures normally formed in response to the pathogen. However, exogenous phenylalanine did not alleviate the effect of glyphosate.

In other studies, glyphosate-reduced phytoalexin synthesis has been correlated with increased susceptibility of several species to fungal pathogens. For instance, the susceptibility of *Phaseolus vulgaris* to *Colletotrichum lindemuthianum* is increased by glyphosate and this increase can be partly explained by the reduction in phenolic phytotalexins by glyphosate [69]. Glyphosate-reduced medicarpin in lucerne [70] and glyceollin in soybeans [71, 72] reduces the resistance of these crops to fungal infection.

The significance and importance of the dramatic increase in shikimate and benzoic acids caused by glyphosate is not known. Whether these compounds play a significant role in the mechanism of action of glyphosate is unclear; however, these compounds are phytotoxic and they accumulate to abnormally high concentrations in the most affected tissues. Whether shikimate is carcinogenic is not clear from the literature. A compound from bracken fern (*Pteridium aquilinum*) that is carcinogenic and mutagenic to mice was identified as shikimate [73]; however, the effects could not be reproduced in rats [74]. Later, an International Agency for Cancer Research Working Group study concluded that the mouse studies were inadequate to conclude that shikimate is a carcinogen or mutagen [75]. Jacobsen et al. [76] found no evidence of shikimate mutagen activity in the Ames test. If the high levels of shikimate in glyphosate-treated plants were a mammalian health hazard, glyphosate use in pastures, as a harvest aid, as a growth regulator, or on glyphosate-tolerant crops could be a source of concern.

At sublethal levels, the photobleaching diphenylether herbicides cause dramatic changes in levels of secondary compounds of plants, including both phenolic and terpenoid compounds (Figure 16.4). Sublethal concentrations of acifluorfen cause a more than 25-fold increase in levels of a phenolic amine derivative of ferulic acid in spinach leaves [77, 78]. Content of flavonoid phytoalexins increases greatly in legumes sprayed with 5 ppm acifluorfen (Figure 16.4) [79–81]. Since these crops are largely tolerant of acifluorfen due to metabolic detoxification [82], the use of photobleaching diphenylethers to kill weeds in these crops may provide an additional benefit of inducing resistance to pathogens. Similarly, some agricultural fungicides may act by activating the crop's defense mechanisms [37, 71]. Concentrations of other secondary compounds such as medicarpin in broad bean, xantho-

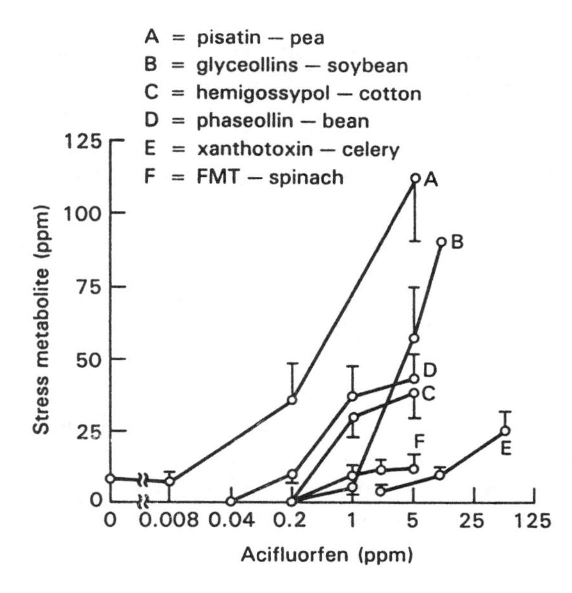

Figure 16.4 Leaf content of six secondary plant metabolites in six different crop species at various treatment levels of acifluorfen after a 48-h exposure period. FMT = *N*-feruloyl-3-methoxytyramine. Redrawn from [79].

toxin in celery, and the terpenoid hemigossypol in cotton are also greatly increased by acifluorfen [79]. The mechanism of induction of phenolic compound production by photobleaching herbicides is not known; however, PAL activity is induced by these herbicides [78, 80, 83, 84], and abiotic phytoalexin induction has been linked to the generation of toxic oxygen species [85].

The chloroacetamides are the only other herbicide group for which there is documentation of a profound effect on phenolic metabolism. Early reports described drastically reduced lignification in alachlor- and metolachlor-treated plant tissues [86, 87]. Later, Molin et al. [88] reported that micromolar concentrations of alachlor strongly inhibit anthocyanin (I_{50} = 20 μM) and lignin (I_{50} = 28 μM) synthesis in mesocotyl sections of sorghum. Other chloroacetamides have similar effects, but alachlor is the most active. The effect is fully reversible by removal of the herbicide from the tissue. The kinetics of alachlor-induced accumulation of *p*-coumaric and ferulic acids are similar to those of accumulation of anthocyanin in untreated tissues [89]. Apparently, the mechanism involves inhibition of incorporation of cinnamic acids into flavonoids and other more complex phenolic compounds. Low concentrations of alachlor reduce extractable *p*-coumarate:CoA ligase, the first enzyme of flavonoid synthesis, by 75 percent, although there is no effect on the enzyme in vitro [89]. Because biosynthesis of both anthocyanin and lignin require CoA, and alachlor can interact directly with CoASH, it has been suggested that the effect of chloroacetamides on anthocyanin and lignin synthesis could be due to reduced CoA availability [88].

More recently, the herbicidally active *cis* forms of 2-halo-*N*-methyl-*N*-phenylacetamides were found to have I_{50} values for anthocyanin biosynthesis of 4 and 13 μM for the iodo and bromo forms, respectively [90]. The *trans* forms are not herbicidal, nor do they inhibit anthocyanin synthesis. These data suggest a close connection between effects on anthocyanin synthesis and the primary herbicidal site of action.

There are many other reports of effects of herbicides on the secondary metabolism of plants in which there is no obvious connection between the primary site of action and the effect. For example, the sulfonylurea herbicide chlorsulfuron greatly increases anthocyanin content in soybean [91] and total hydroxycinnamic acid levels in sunflower [92] seedlings. In sunflower, *p*-coumaric and caffeic acid levels are elevated to more than 30-fold those of untreated seedlings after 3 days of treatment. In both cases, chlorsulfuron increases extractable PAL activity. Although ethylene evolution is also stimulated by chlorsulfuron, ethylene alone does not induce synthesis of phenolic compounds. The mechanism of these effects has not been determined. Whether other ALS inhibitors (see Chapter 13) have a similar effect has not been reported.

A number of purely phenomenological reports of herbicide effects on levels of secondary compounds, including terpenoids, alkaloids, and phenolics, can be found [37]. However, there is no clear understanding of these effects from a mechanism of action or ecological standpoint. Nor is it known whether or not these effects play a significant role in influencing secondary metabolism of nontarget plants.

16.4 EFFECTS ON HORMONE LEVELS

Due to their hormonal activities, auxin herbicides such as 2,4-D (see Chapter 14) cause symptoms (such as rapid and abnormal growth) that are characteristic of abnormal hormone balances. The interference of aryl-propanoates with auxin action is discussed in Sections 11.3 and 17.2. Other herbicides that have mechanisms of action unrelated to hormones can also induce effects similar to those caused by application of hormones or hormone inhibitors. For example, glyphosate, which exerts its action through disruption of the shikimate pathway (Chapter 13), disrupts apical dominance at sublethal doses [93, 94]. Under some circumstances, photobleaching diphenylethers also break apical dominance [93]. Such effects could be due to direct disruption of the apical meristem(s), to indirect effects on endogenous hormone levels, to hormone-like effects of the herbicide or one or more of its metabolites, or to herbicide-induced changes in sensitivity to endogenous hormones. Herbicides that are preferentially translocated to apical meristems (e.g., glyphosate), where their effects are initially much more pronounced than in vegetative tissues, can be expected to interfere with apical dominance. However, in the case of glyphosate, this interpretation may be overly simplistic.

Synthesis of tryptophan, an aromatic amino acid and precursor of auxin, is directly inhibited by glyphosate (Chapter 13). Whether the tryptophan pool for indole-3-acetic acid (IAA) synthesis is limited by glyphosate is not known. Although no studies have directly measured the effect of glyphosate on endogenous IAA biosynthesis, two studies that have directly measured IAA content in glyphosate-treated plant tissues have shown increased IAA levels (Table 16.1). Cañal et al. [96] associated the increase in IAA in glyphosate-treated yellow nutsedge (*Cyperus esculentus*) leaf tissues with the increases in benzoic acids [47, 96]. The content of gentisic acid, an inhibitor of IAA oxidase, increased about 5-fold and the IAA content increased 2.5-fold [96]. The normal level of gentisic acid in the tissue reduced the activity of commercial IAA oxidase (horse radish peroxidase with IAA as a substrate) by 18 percent, whereas the levels in glyphosate-treated tissues reduced the activity by 95 percent. In other species and tissues, there is direct and indirect evidence that glyphosate reduces the concentrations of IAA in affected tissues. In soybean and pea internodes, glyphosate reduces free levels of endogenous IAA [94]. Exogenous IAA will partially reverse the growth-retarding effects of glyphosate on tobacco callus cultures [104]. In this system and in internodes of soybean and pea seedlings, the glyphosate-treated tissues metabolize exogenously supplied IAA more rapidly than do control tissues [94, 105–107]. Increased metabolism of IAA in glyphosate-treated tissues has been attributed to changes in IAA oxidase due to altered phenolic compound content [94, 105]. IAA oxidase activity can be stimulated or inhibited by different phenolic compounds, and the in vivo activity will be influenced by many different phenolic compounds [108, 109]. Thus, the conflicting results with different species and systems could be due to glyphosate causing different qualitative changes in phenolic compounds in these species. Glufosinate is more active than glyphosate in promoting IAA metab-

TABLE 16.1 EFFECTS OF VARIOUS HERBICIDES ON ENDOGENOUS HORMONE LEVELS IN PLANTS
Note that many of these effects were observed long after the herbicide affected the primary site of action.

Herbicide (treatment concentration or μg/seedling)	Species	Hormone	Percent increase (+) or reduction (−)	Reference
Glyphosate (0.5 mM)	Pea	IAA	−51	94
Glyphosate (0.5 mM)	Napier grass	IAA	+67	95
		ABA	+97	
Fluridone (0.3 mM)	Napier grass	IAA	+23	95
		ABA	−98	
Glyphosate (10 mM)	Yellow nutsedge	IAA	+150	96
Glyphosate (1 mM)	Tobacco	Ethylene	−55	96
Glyphosate (0.2 mM)	Amer. germander	IAA	−80	97
Chlorsulfuron (10 μg)	Sunflower	Ethylene	+2200	98
Paraquat (1 mM)	Bean	Ethylene	+300	99
Picloram (250 μg)	*Brassica napus*	Ethylene	+600	100
Norflurazon (50 μM)	Cotton	ABA	−70	101
Norflurazon (13 μM)	Pearl millet	ABA	−91	102
Mefluidide (20 ppm)	Corn	ABA	+400	103

olism, whereas glyphosine and aminomethylphosphonic acid, a metabolic product of glyphosate, are only mildly effective [107].

A positive correlation among plant species has been found between rates of IAA metabolism in untreated tissues and tolerance of glyphosate [110] (Figure 16.5). The authors hypothesized that plants that normally have high rates of IAA metabolism might be less dependent on IAA and, thus, less affected by reductions in IAA caused by glyphosate. There is also evidence that IAA is more subject to metabolism by conjugation in glyphosate-treated than in control tissues [106, 107]. Phenolic compounds can also influence conjugation of IAA. For example, 2,6-

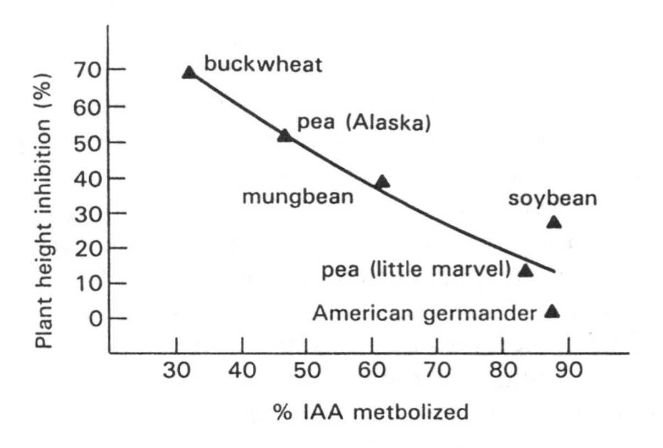

Figure 16.5 Relationship between capacity to metabolize IAA and sensitivity to glyphosate among six plant species. IAA metabolism is given as percent of radiolabeled IAA taken up that was metabolized with 4 h. Growth inhibition is percent inhibition of increase in plant height 8 days after treatment with 1 mM glyphosate. Data taken from [110].

dihydroxyacetophenone, a phenolic inhibitor of conjugation of IAA with L-aspartate and L-glutamate [111], almost completely prevents glyphosate-induced growth inhibition and IAA metabolism in American germander (*Teucrium canadense* L.) [97]. IAA levels are only partially returned to normal by 2,6-dihydroxyacetophenone in this system. The comparative effectiveness of this phenol with other phenols in preventing metabolism of IAA was reported to vary among species. Therefore, differences in the effect of glyphosate on IAA levels among species can be due to differences in the absolute and relative rates of IAA oxidase and IAA conjugation activities, as well as differences in the amounts and types of phenolic compound changes caused by glyphosate and differences in the effects of these phenolic compounds on each of the routes of metabolism. Furthermore, transport of IAA is differentially affected by different phenolic compounds [112], and is severely inhibited by glyphosate in corn coleoptiles and cotton seedling hypocotyls [113]. The mechanism of this effect has not been determined. The effect of glyphosate on IAA levels is just one example of the complex and sometimes conflicting secondary effects of herbicides on plant hormones.

Hormones interact in complex ways and herbicides usually affect more than one hormone. In addition to its effects on IAA, glyphosate also has strong effects on generation of ethylene. Ethylene synthesis has been reported to be increased [114], little affected [29, 115], or decreased [96, 107, 115] by glyphosate in different systems. In some cases, IAA levels influence the effect of glyphosate on ethylene generation [115]. Enhanced ethylene synthesis is a general stress symptom [116] that accompanies treatment by many herbicides with different primary sites of action [e.g., [98–100, 117]) (Table 16.1). In some cases, the ethylene levels induced by the herbicide may be phytotoxic to the plant. For example, treatment with aminoethoxyvinylglycine, an inhibitor of ethylene synthesis, before treatment with picloram or clopyralid prevents herbicide-induced ethylene synthesis and delays the development of herbicide-induced morphological effects [100] (Chapter 14).

Other herbicides affect the levels of certain plant hormones. Inhibitors of carotenoid biosynthesis, such as fluridone and norflurazon, can also inhibit abscisic acid (ABA) biosynthesis [95, 101, 102] (Table 16.1), perhaps because some ABA is formed as an oxidation product of carotenoids [118]. Stress can induce ABA synthesis and some stress symptoms can be explained by the increases in ABA [119]. Fluridone can inhibit the development of these symptoms [119]. Herbicides can also increase ABA levels, presumably due to a general stress effect. For example, mefluidide increases ABA levels in corn (Table 16.1). The effects of herbicides on biosynthesis of gibberellic acid are discussed in section 15.4.

16.5 EFFECTS ON CHLOROPLAST ORGANIZATION AND FUNCTION

The secondary effects of herbicides with primary sites of action on photosynthesis or chloroplast pigment synthesis on the organization and function of the chloroplast have been discussed in the chapters covering those herbicides. The mechanisms of these effects are generally straightforward to deduce, although at sublethal concentrations these effects can become subtle and complex. For instance, at sublethal concentrations PS II inhibitors often cause chloroplast development that is similar to that of biotypes resistant to the same herbicide (e.g., [120]). In both cases, the plants have chloroplast organization similar to that of shade-adapted plants. The main differences are in lipid composition rather than in proteins [121]. This topic is discussed in more detail in Chapter 7. Herbicides that have no direct effect on pigment synthesis or photosynthesis eventually disrupt the chloroplast. Although these effects are usually many steps separated from the initial site of action, they often result in the first visual symptoms of phytotoxicity. We will give several examples.

The primary site of action of glyphosate is the shikimate pathway (Chapter 13); however, chlorosis, especially of new growth, is one of the first visual symptoms of this slow-acting herbicide. The relationship of this chlorosis to inhibition of aromatic amino acid synthesis is unclear. Glyphosate strongly inhibits synthesis of chlorophyll [122] and its precursor δ-aminolevulinic acid (ALA) [123] (Figure 16.6). Incorporation of glutamate, 2-ketoglutarate, and glycine into ALA is inhibited; however, incorporation of ALA into chlorophyll is unaffected [123]. Because glyphosate phytotoxicity can be partially overcome by supplying 2-ketoglutarate [124], an ALA precursor, inhibited ALA synthesis may play a significant role in the mode of action of glyphosate. Reduced 2-ketoglutarate levels in glyphosate-treated plants may be the result of uncontrolled flow of carbon into the shikimate pathway due to its deregulation by EPSPS inhibition. This theory is discussed in detail in Chapter 13. The plastids of developing tissues treated with glyphosate begin to swell relatively rapidly and this effect parallels accumulation of shikimate [125], even though shikimate is not compartmentalized in the plastid (Chapter 13). The thylakoids of chloroplasts of green soybean leaves treated with glyphosate become spiral-shaped, and starch and stroma lamellae are much less abundant than in untreated plants [126]. The interrelationships between inhibition

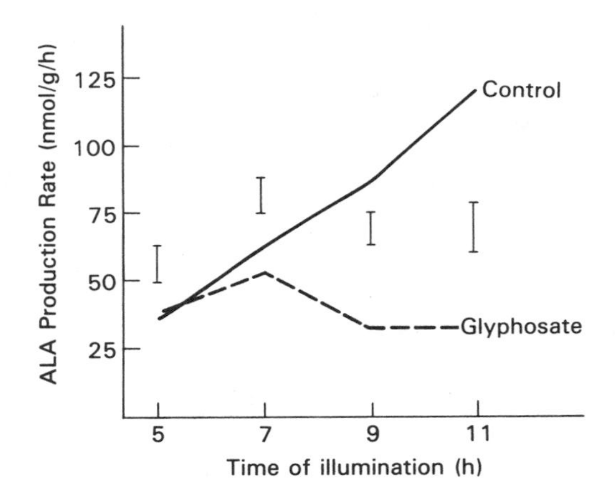

Figure 16.6 Accumulation of ALA in control or glyphosate-treated (1 mM), 7-day-old etiolated barley shoots during subsequent illumination. Plants were exposed to glyphosate for 8 h in darkness before exposure to light. Taken from [123].

of chlorophyll synthesis, shikimate accumulation, and ultrastructural effects on the chloroplast of glyphosate-treated plants are still unclear.

Although the primary mechanism of action of cyclohexanedione herbicides is now known to be inhibition of fatty acid synthesis by inhibition of acetyl-CoA carboxylase (Chapter 11), earlier research suggested that these herbicides inhibited chloroplast biogenesis [127, 128]. In barley seedlings, sethoxydim reduces chloroplast division, and lowers thylakoid stacking and relative amounts of thylakoid membrane [127]. Plastids at all stages of development are affected, although photosynthesis is not directly affected [128]. Chloroplast biogenesis and differentiation were concluded to be major targets for sethoxydim in this early work.

Chloroplast structure and development are also strongly affected by 2,4-D [129]. In one study, chloroplasts developed abnormally during 24 h of light after treatment of etiolated radish cotyledons with 2,4-D; the chloroplasts were smaller and thylakoids were much more tightly appressed than in untreated plants. The photosynthetic unit size was reduced by 25 percent in treated plants, and the maximum turnover rate of PS II in chloroplasts from treated plants was less than half that in untreated plants [130]. However, photophosphorylation efficiency was unaffected by 2,4-D and it was concluded that there was no deterioration of the membranes. All of these effects were considered to be secondary. The ultrastructural effects on chloroplasts of other herbicides with auxin activity are summarized by Bartels [131]. In general, the mechanisms of these effects are unknown.

16.6 SUMMARY

This chapter has provided some examples of secondary effects of herbicides and has pointed out the potential importance of secondary effects. Since most of the literature on the mode of action of herbicides deals with secondary effects, the sampling in this chapter represents only a small fraction of the available literature on secondary effects of herbicides. Any herbicide can eventually affect any physiological process of a plant, and many scientists have taken advantage of this fact as the basis for experimentation and publication. Connecting secondary effects to primary effects and determination of whether a secondary effect is meaningful or important is much more difficult than demonstrating the effect. Still, secondary effects are extremely important in the chain of events that eventually lead to death of target organisms, and can also be important in alteration of the physiology of plants that are not killed by the herbicide.

REFERENCES

1. Klepper, L. A. 1988. "Synergistic levels of NO_x emissions from soybean leaves caused by a combination of salicylic acid and photosynthetic inhibitor herbicides." *Pestic. Biochem. Physiol.*, 32, 173–79.

2. Dean, J. V., and J. E. Harper. 1988. "The conversion of nitrite to nitrogen oxide(s) by the constitutive NAD(P)H-nitrate reductase enzyme from soybean." *Plant Physiol.*, 88, 389–96.
3. Klepper, L. A. 1974. "A mode of action of herbicides: inhibition of the normal process of nitrite reductase." Nebraska Agric. Exp. Stn. Res. Bull. 259, 1–42.
4. Klepper, L. A. 1975. "Inhibition of nitrate reduction by photosynthetic inhibitors." *Weed Sci.*, 23, 188–90.
5. Fedtke, C. 1977. "Formation of nitrite in plants treated with herbicides that inhibit photosynthesis." *Pestic. Sci.*, 8, 152–56.
6. Klepper, L. A. 1976. "Nitrite accumulation within herbicide-treated leaves." *Weed Sci.*, 24, 533–35.
7. Klepper, L. A. 1979. "Effects of certain herbicides and their combinations on nitrate and nitrite reduction." *Plant Physiol.*, 64, 273–75.
8. Prakash, T. R., R. S. Murthy, and P. M. Swamy. 1989. "Influence of thiobencarb on nitrate reductase, nitrite reductase and DCPIP photoreduction in rice and *Echinochloa crus-galli* (L.) (barnyardgrass)." *Weed Res.*, 19, 427–32.
9. Duke, S. O., K. C. Vaughn, and S. H. Duke. 1982. "Effects of norflurazon (San 9789) on light-increased extractable nitrate reductase activity in soybean [*Glycine max* (L.) Merr.] seedlings." *Plant Cell Environ.*, 5, 155–62.
10. Kakefuda, G., S. H. Duke, and S. O. Duke. 1983. "Differential light induction of nitrate reductases in greening and photobleached soybean seedlings." *Plant Physiol.*, 73, 56–60.
11. Saeed, M., and S. H. Duke. 1990. "Chloroplastic regulation of apoplastic α-amylase activity in pea seedlings." *Plant Physiol.*, 93, 131–40.
12. Deane-Drummond, C. E., and C. B. Johnson. 1980. "Absence of nitrate reductase activity in San 9789 bleached leaves of barley seedlings (*Hordeum vulgare* cv. Midas)." *Plant Cell Environ.*, 3, 303–08.
13. Schuster, C., R. Oelmüller, R. Bergfeld, and H. Mohr. 1988. "Recovery of plastids from photooxidative damage: significance of a plastidic factor." *Planta*, 174, 289–97.
14. Schuster, C., S. Schmidt, and H. Mohr. 1989. "Effect of nitrate, ammonium, light and a plastidic factor on the appearance of multiple forms of nitrate reductase in mustard (*Sinapsis alba* L.) cotyledons." *Planta*, 177, 74–83.
15. Rao, R. K., R. M. Mannon, A. Gnanam, and S. Bose. 1988. "Inhibition of nitrate and nitrite reductase induction in wheat by Sandoz 9785." *Phytochemistry*, 27, 685–88.
16. Duke, S. H., and S. O. Duke. 1984. "Light control of extractable nitrate reductase activity in higher plants." *Physiol. Plant.*, 62, 485–93.
17. Sihag, R. K., S. Guha-Mukherjee, and S. K. Sopory. 1979. "Effect of ammonium, sucrose, and light on the regulation of nitrate reductase level in *Pisum sativum*." *Physiol. Plant.*, 45, 281–87.
18. Nasrulhaq-Boyce, A., and O. T. Jones. 1977. "The light-induced development of nitrate reductase in etiolated barley shoots: an inhibitory effect of laevulinic acid." *Planta*, 137, 77–84.
19. Vijayaraghavan, S. J., S. K. Sopory, and S. Guha-Mukherjee. 1979. "Role of light in the regulation of nitrate reductase level in wheat (*Triticum aestivum*)." *Plant Cell Physiol.*, 20, 1251–62.

20. Buczek, J., J. Borkowski, and I. Jarzynska. 1986. "The dependence of nitrate reductase activity on the level of soluble sugars in wheat and cucumber roots growing in the presence of simazine, in light or in darkness." *Acta Soc. Bot. Pol.,* 55, 589–600.

21. Aslam, M., and R. C. Huffaker. 1973. "Effect of DCMU, simazine and atrazine on nitrate reductase in *Hordeum vulgare* in vitro and in vivo." *Physiol. Plant.,* 28, 400–04.

22. Tweedy, J. A., and S. K. Ries. 1967. "Effect of simazine on nitrate reductase in corn." *Plant Physiol.,* 42, 280–82.

23. Mohandas, S., W. Wallace, and D. J. D. Nicholas. 1978. "Effects of atrazine on the assimilation of inorganic nitrogen in cereals." *Phytochemistry,* 17, 1021–28.

24. Soares, M. I. M., S. H. Lips, and C. F. Cresswell. 1985. "Regulation of nitrate and nitrite reduction in barley leaves." *Physiol. Plant.,* 64, 492–500.

25. Ries, S. K. 1976. "Subtoxic effects on plants." In *Herbicides—Physiology, Biochemistry, and Ecology, Vol. 2,* L. J. Audus, ed. New York: Academic, pp. 313–44.

26. Lawrence, J. M., R. J. Foster, and H. E. Herrick. 1980. "Reduction of nitrate and nitrite in lambsquarters (*Chenopodium album*) biotypes resistant and susceptible to atrazine toxicity." *Plant Physiol.,* 65, 984–89.

27. Trogisch, G. D., H. Köcher, and W. R. Ullrich. 1989. "Effects of glufosinate on anion uptake in *Lemna gibba* G. 1." *Z. Naturforsch.,* 44c, 33–38.

28. Lacuesta, M., B. González-Moro, C. González-Murua, and A. Muñoz-Rueda. 1990. "Temporal study of the effect of phosphinothricin on the activity of glutamine synthestase, glutamine dehydrogenase and nitrate reductase in *Medicago sativa* L." *J. Plant Physiol.,* 136, 410-414.

29. Cole, D. J., J. C. Caseley, and A. D. Dodge. 1983. "Influence of glyphosate on selected plant processes." *Weed Res.,* 173–83.

30. Muñoz-Rueda, A., C. González-Murua, F. L. Hernando, and M. Sánchez-Diaz. 1986. "Effects of glyphosate (N-(phosphonomethyl)-glycine) on water potential, and activities of nitrate and nitrite reductase and aspartate aminotransferase in lucerne and clover." *J. Plant Physiol.,* 123, 107–15.

31. Muñoz-Rueda, A., C. González-Murua, J. L. Becerril, C. Arreseigor, and M. Sánchez-Díaz. 1986. "The effect of asulam on water potential and nitrate reduction." *Plant Sci.,* 46, 21–27.

32. Goring, C. A. I., and D. A. Laskowski. 1982. "The effects of pesticides on nitrogen transformations in soils." *Agronomy,* 22, 689–720.

33. Pozuelo, J. M., M. Fernandez-Pascual, M. M. Lucas, and M. R. De Felipe. 1989. "Effect of eight herbicides from five different chemical groups on nitrogen fixation and grain yield in *Lupinus albus* L. grown in semi-arid zones." *Weed Res.,* 29, 419–25.

34. Moorman, T. B. 1989. "A review of pesticide effects on microorganisms and microbial processes related to soil fertility." *J. Prod. Agric.,* 2, 14–23.

35. Lal, R., and S. Shivaji. 1988. "Effects of pesticides on asymbiotic nitrogen fixation." In *Pesticides and Nitrogen Cycle, Vol. III,* R. Lal and S. Lal, eds. Boca Raton, Fla.: CRC Press, Inc., pp. 1–46.

36. DiCosmo, F., and G. H. N. Towers. 1984. "Stress and secondary metabolism in cultured plant cells." *Rec. Adv. Phytochem.,* 18, 97–175.

37. Lydon, J., and S. O. Duke. 1989. "Pesticide effects on secondary metabolism of higher plants." *Pestic. Sci.,* 25, 361–73.
38. Williams, M. C., and E. H. Cronin. 1963. "Effect of silvex and 2,4,5-T on alkaloid content of tall larkspur." *Weeds,* 11, 317–19.
39. Williams, M. C., and L. F. James. 1983. "Effects of herbicides on the concentration of poisonous compounds in plants: A review." *Amer. J. Vet. Res.,* 44, 2420–22.
40. Birch, E. C., and L. S. Vickery. 1961. "The effect of maleic hydrazide on certain chemical constituents of flue-cured tobacco." *Can. J. Plant Sci.,* 41, 170–75.
41. Weston, L. A., and A. R. Putnam. 1985. "Inhibition of growth, nodulation, and nitrogen fixation of legumes by quackgrass." *Crop Sci.,* 561–65.
42. Dyer, W. E., J. M. Henstrand, A. K. Handa, and K. M. Herrmann. 1989. "Wounding induces the first enzyme of the shikimate pathway in Solanaceae." *Proc. Natl. Acad. Sci. USA.* 86, 7370–73.
43. Harborne, J. B. 1986. "The role of phytoalexins in natural plant resistance." Amer. Chem. Soc. Symp. Ser., 296, 22–35.
44. Hoagland, R. E., and S. O. Duke. 1983. "Relationships between phenylalanine ammonia-lyase activity and physiological responses of soybean (*Glycine max*) seedlings to herbicides." *Weed Sci.,* 31, 845–52.
45. Holländer, H., and N. Amrhein. 1980. "The site of the inhibition of the shikimate pathway by glyphosate. I. Inhibition by glyphosate of phenylpropanoid synthesis in buckwheat (*Fagopyrum esculentum* Moench)." *Plant Physiol.,* 66, 823–29.
46. Saltveit, M. E., Jr. 1988. "Postharvest glyphosate application reduces toughening, fiber content, and lignification of stored asparagus spears." *J. Amer. Soc. Hort. Sci.,* 113, 569–72.
47. Cañal, M. J., R. S. Tamès, and B. Fernández. 1987. "Effects of glyphosate on phenolic metabolism in yellow nutsedge leaves." *Physiol. Plant.,* 69, 627–32.
48. Ishikura, N., S. Teramoto, Y. Takeshima, and S. Mitsui. 1986. "Effects of glyphosate on the shikimate pathway and regulation of phenylalanine ammonia-lyase in *Cryptomeria* and *Perilla* cell suspension cultures." *Plant Cell Physiol.,* 27, 677–84.
49. Duke, S. O. 1985. "Effects of glyphosate on metabolism of phenolic compounds." In *The Herbicide Glyphosate,* E. Grossbard and D. Atkinson, eds. London: Butterworths, pp. 75–91.
50. Laanest, L. 1987. "Incorporation of exogenous tyrosine and phenylalanine into c-glycosylflavones in glyphosate-treated barley seedlings." *Eesti NSV Tead. Akad. Toim., Biol.,* 36, 204–09.
51. Lydon, J., and S. O. Duke. 1988. "Glyphosate induction of elevated levels of hydroxybenzoic acids in higher plants." *J. Agric. Food Chem.,* 36, 813–18.
52. Becerril, J. M., S. O. Duke, and J. Lydon. 1989. "Glyphosate effects on shikimate pathway products in leaves and flowers of velvetleaf (*Abutilon theophrastic* Medic.)." *Phytochemistry,* 28, 695–99.
53. Kakhniashvili, Kh. A., N. N. Sikharulidze, G. G. Gogoladze, and M. Sh. Giauri. 1989. "Effect of sym-triazines and Roundup on biochemical characteristics and quality of tea." *Subtrop. Kul't.,* 2, 111–14.
54. Heide, L., H. G. Floss, and M. Tabata. 1989. "Incorporation of shikimic acid into *p*-hydroxybenzoic acid in *Lithospernum erythrorhizon* cell cultures." *Phytochemistry,* 28, 2643–45.

55. Margna, J., T. Vainjärv, and L. Laanest. 1989. "Different L-phenylalanine pools available for the biosynthesis of phenolics in buckwheat seedling tissues." *Phytochemistry,* 28, 469–75.
56. Morandi, D. 1989. "Effect of xenobiotics on endomycorrhizal infection and isoflavonoid accumulation in soybean roots." *Plant Physiol. Biochem.,* 27, 697–701.
57. Falco, J. M., L. Vilanova, and J. Segura. 1989. "Effects of glyphosate on phenolic compounds and free amino acids in oat seedlings." *Agrochimica,* 33, 166–73.
58. Altman, J., and C. L. Campbell. 1977. "Effects of herbicides on plant diseases." *Annu. Rev. Phytopath.,* 15, 361–86.
59. Altman, J. 1983. "Impact of herbicides on plant diseases." In *Ecology and Management of Soilborne Plant Pathogens,* C. A. Parker, A. D. Rovira, K. J. Moore, P. T. W. Wong, and J. F. Kollmorgan, eds. St. Paul, Minn.: Amer. Phytopath. Soc., pp. 227–31.
60. Altman, J., and A. D. Rovira. 1989. "Herbicide-pathogen interactions in soil-borne root diseases." *Can. J. Plant Pathol.,* 11, 166–72.
61. Katan, J., and Y. Eshel. 1973. "Interactions between herbicides and plant pathogens." *Res. Rev.,* 45, 145–77.
62. Kozlowska, M., Z. Krzywanski, and U. Taszek. 1988. "Effect of 2,4-D and benzyladenine on colonization of red raspberry callus tissue with *Didymella applanata* (Niessl) Sacc. and on accumulation of phenolics." *Acta Physiol. Plant.,* 10, 25–30.
63. Johal, G. S., and J. E. Rahe. 1984. "Effect of soil-borne plant pathogenic fungi on the herbicidal action of glyphosate on bean seedlings." *Phytopathology,* 74, 950–55.
64. Rahe, J. E., C. A. Lévesque, and G. S. Johal. 1990. "Synergistic role of soil fungi in the herbicidal efficacy of glyphosate." Amer. Chem. Soc. Symp. Ser., 439, 260–75.
65. Lynch, J. M., and D. J. Penn. 1980. "Damage to cereals caused by decaying weed residues." *J. Sci. Food Agri.,* 31, 321–24.
66. Brown, A. E., and H. S. S. Sharma. 1984. "Production of polysaccharide-degrading enzymes saprophytic fungi from glyphosate-treated flax and their involvement in retting." *Ann. Appl. Biol.,* 105, 65–74.
67. Lévesque, C. A., J. E. Rahe, and D. M. Eaves. 1987. "Effect of glyphosate on *Fusarium* spp.: its influence on root colonization of weeds, propagule density in the soil, and crop emergence." *Can. J. Microbiol.,* 33, 354–60.
68. Brammall, R. A., and V. J. Higgins. 1988. "The effect of glyphosate on resistance of tomato to *Fusarium* crown and root rot disease and on formation of host structural defensive barriers." *Can. J. Bot.,* 66, 1547–55.
69. Johal, G. S., and J. E. Rahe. 1988. "Glyphosate, hypersensitivity and phytoalexin accumulation in the incompatible bean anthracnose host-parasite interaction." *Physiol. Molec. Plant Pathol.,* 32, 267–81.
70. Latunde-Dada, A. O., and J. A. Lucas. 1985. "Involvement of the phytoalexin medicarpin in the differential response of callus lines of lucerne (*Medicago sativa*) to infection by *Verticillium albo atrum*." *Physiol. Plant Pathol.,* 26, 31–42.
71. Ward, E. W. B. 1984. "Suppression of metalaxyl activity by glyphosate. Evidence that host defense mechanisms contribute to metalaxyl inhibition of *Phytophthora megasperma* F Sp. *glycinea* in soybeans." *Physiol. Plant Pathol.,* 25, 381–86.
72. Holliday, M. J., and N. T. Keen. 1982. "The role of phytoalexins in the resistance of

soybean leaves to bacteria: effect of glyphosate on glyceollin accumulation." *Phytopathology,* 72, 1470–74.

73. Evans, I. A., and M. A. Osman. 1974. "Carcinogenicity of bracken and shikimic acid." *Nature* (London), 250, 348–49.
74. Hirono, I., K. Fushimi, and N. Matsubara. 1977. "Carcinogenicity test of shikimic acid in rats." *Toxicol. Lett.,* 1, 9–10.
75. IARC. 1986. *IARC Monographs on the Evaluation of the Carcinogenic Risk of Chemicals to Humans,* Vol. 40. Lyon, France: International Agency for Cancer Research, pp. 47–66.
76. Jacobsen, L. B., C. L. Richardson, and H. G. Floss. 1978. "Shikimic acid and quinic acid are not mutagenic in the Ames assay." *Lloydia,* 41, 450–52.
77. Suzuki, T., I. Holden, and J. E. Casida. 1981. "Diphenyl ether herbicides remarkably elevate the content in *Spinacea oleracea* of (*E*)-3-(4-hydroxy-3-(4-hydroxy-3-methoxyphenyl)-*N*-[2-(4-hydroxy-3-methoxyphenyl)ethyl]-2-propenamide." *J. Agric. Food Chem.,* 29, 992–95.
78. Kömives, T., and J. E. Casida. 1982. "Diphenyl ether herbicides: effects of acifluorfen on phenylpropanoid biosynthesis and phenylalanine ammonia-lyase activity in spinach." *Pestic. Biochem. Physiol.,* 18, 191–96.
79. Kömives, T., and J. E. Casida. 1983. "Acifluorfen increases the leaf content of phytoalexins and stress metabolites in several crops." *J. Agric. Food Chem.,* 31, 751–55.
80. Cosio, E. G., G. Weissenböck, and J. W. McClure. 1985. "Acifluorfen-induced isoflavonoids and enzymes of their biosynthesis in mature soybean leaves." *Plant Physiol.,* 78, 14–19.
81. Rubin, B., D. Penner, and A. W. Saettler. 1983. "Induction of isoflavonoid production in *Phaseolus vulgaris* L. leaves by ozone, sulfur dioxide and herbicide stress." *Environ. Toxicol. Chem.,* 2, 295–306.
82. Frear, D. S., H. R. Swanson, and E. R. Mansager. 1983. "Acifluorfen metabolism in soybean: diphenylether bond cleavage and the formation of homoglutathione, cysteine, and glucose conjugates." *Pestic. Biochem. Physiol.,* 20, 299–310.
83. Sato, R., E. Nagano, H. Oshio, and K. Kamoshita. 1987. "Diphenylether-like physiological and biochemical actions of S-23141." *Pestic. Biochem. Physiol.,* 28, 194–200.
84. Hikawa, M. 1984. "The mode of action of chlomethoxyfen." *Pestic. Sci.,* 15, 562–70.
85. Epperlein, M. M., A. A. Noronha-Dutra, and R. N. Strange. 1986. "Involvement of the hydroxyl radical in the abiotic elicitation of phytoalexins in legumes." *Physiol. Molec. Plant Pathol.,* 28, 67–77.
86. Hickey, J. S., and W. A. Krueger. 1974. "Alachlor and 1,8-naphthalic anhydride effects on corn coleoptiles." *Weed Sci.,* 22, 250–52.
87. Ebert, E. 1980. "Herbicidal effects of metolachlor (2-chloro-*N*-[2-ethyl-6-methylphenyl]-*N*-[2-methoxy-1-methylethyl]acetamide) at the cellular level in sorghum." *Pestic. Biochem. Physiol.,* 13, 227–36.
88. Molin, W. T., E. J. Anderson, and C. A. Porter. 1986. "Effects of alachlor on anthocyanin and lignin synthesis in etiolated sorghum (*Sorghum bicolor* (L.) Moench) mesocotyls." *Pestic. Biochem. Physiol.,* 25, 105–11.

89. Lee, T. C., W. T. Molin, and C. A. Porter. 1986. "The effects of alachlor on phenolic acid metabolism in buckwheat [*Fagopyrum tataricum* (L.) Gaertn.] hypocotyls." *WSSA Abstr.*, 26, 81.

90. Molin, W. T., C. A. Porter, J. P. Chupp, and K. Naylor. 1990. "Differential inhibition of anthocyanin synthesis in etiolated sorghum (*Sorghum bicolor* (L.) Moench) mesocotyls by rotomeric 2-halo-*N*-methyl-*N*-phenylacetamides." *Pestic. Biochem. Physiol.*, 36, 277–80.

91. Suttle, J. C., and D. R. Schreiner. 1982. "Effects of DPX-4189 (2-chloro-N-(4-methyl-6-methyl-1,3,5-triazin-2-yl)aminocarbonyl)benzenefulfonamide) on anthocyanin synthesis, phenylalanine ammonia lyase activity and ethylene production in soybean hypocotyls." *Can. J. Bot.*, 60, 741–45.

92. Suttle, J. C., H. C. Swanson, and D. R. Schreiner. 1983. "Effect of chlorsulfuron on phenylpropanoid metabolism in sunflower seedlings." *J. Plant Growth Regul.*, 2, 1443–45.

93. Parker, C. 1975. "Effects on the dormancy of plant organs." In *Herbicides—Physiology, Biochemistry, Ecology, Vol. 1*. L. J. Audus, ed. New York: Academic, pp. 165–90.

94. Lee, T. T. 1984. "Release of lateral buds from apical dominance by glyphosate in soybean and pea seedlings." *J. Plant Growth Regul.*, 3, 227–35.

95. Rajasekaran, K., M. B. Hein, and I. K. Vasil. 1987. "Endogenous abscisic acid and indole-3-acetic acid and somatic embryogenesis in cultured leaf explants of *Pennisetum purpureum* Schum. Effects *in vivo* and *in vitro* of glyphosate, fluridone, and paclobutrazol." *Plant Physiol.*, 84, 47–51.

96. Cañal, M. J., R. S. Tamès, and B. Fernàndez. 1987. "Glyphosate-increased levels of indole-3-acetic acid in yellow nutsedge leaves correlated with gentisic acid levels." *Physiol. Plant.*, 71, 384–88.

97. Lee, T. T., and A. N. Staratt. 1989. "Phenol-glyphosate interaction: effects on IAA metabolism and growth of plants." In *Adjuvants and Agrochemicals Vol. I, Mode of Action and Physiological Activity*, P. N. P. Chow, C. A. Grant, A. M. Hinshalwood, and E. Simundson, eds. Boca Raton, Fla.: CRC Press, pp. 35–40.

98. Suttle, J. C. 1983. "Effect of chlorsulfuron on ethylene evolution from sunflower seedlings." *J. Plant Growth Regul.*, 2, 31–35.

99. Weckx, J., J. Vangronsveld, and M. Van Poucke. 1989. "Effect of paraquat on ethylene biosynthesis by intact green *Phaseolus vulgaris* seedlings." *Physiol. Plant.*, 75, 340–45.

100. Hall, J. C., P. K. Bassi, M. S. Spencer, and W. H. Vanden Born. 1985. "An evaluation of the role of ethylene in herbicidal injury induced by picloram or clopyralid in rapeseed and sunflower plants." *Plant Physiol.*, 79, 18–23.

101. Stegink, S. J., and K. C. Vaughn. 1988. "Norflurazon (SAN-9789) reduces abscisic acid levels in cotton seedlings: a glandless isoline is more sensitive than its glanded counterpart." *Pestic. Biochem. Physiol.*, 31, 269–75.

102. Henson, I. E. 1984. "Inhibition of abscisic acid accumulation in seedling shoots of pearl millet (*Pennisetum americanum* [L.] Leeke) following induction of chlorosis by norflurazon." *Z. Pflanzenphysiol.*, 114, 35–43.

103. Zhang, C.-L., R. H. Li, and M. L. Brenner. 1986. "Relationship between mefluidide treatment and abscisic acid metabolism in chilled corn leaves." *Plant Physiol.*, 81, 699–701.

104. Lee, T. T. 1980. "Characteristics of glyphosate inhibition of growth in soybean and tobacco callus cultures." *Weed Res.*, 20, 365–69.

105. Lee, T. T. 1982. "Promotion of indole-3-acetic acid oxidation by glyphosate in tobacco callus tissue." *J. Plant Growth Regul.*, 1, 37–48.

106. Lee, T. T. 1982. "Mode of action of glyphosate in relation to metabolism of indole-3-acetic acid." *Physiol. Plant.*, 54, 289–94.

107. Lee, T. T., T. Dumas, and J. J. Jevnikar. 1983. "Comparison of the effects of glyphosate and related compounds on indole-3-acetic acid metabolism and ethylene production in tobacco callus." *Pestic. Biochem. Physiol.*, 20, 354–59.

108. Lee, T. T., A. N. Starratt, and J. J. Jevnikar. 1982. "Regulation of enzymic oxidation of indole-3-acetic acid by phenols: structure-activity relationships." *Phytochemistry*, 21, 517–23.

109. Grambow, H. J., and B. Langenbeck-Schwich. 1983. "The relationship between oxidase activity, peroxidase activity, hydrogen peroxide, and phenolic compounds in the degradation of indole-3-acetic acid in vitro." *Planta*, 157, 131–37.

110. Lee, T. T., and T. Dumas. 1985. "Effect of glyphosate on indole-3-acetic acid metabolism in tolerant and susceptible plants." *J. Plant Growth Regul.*, 4, 29–39.

111. Lee, T. T., and A. N. Staratt. 1986. "Inhibition of conjugation of IAA with amino acids by 2,6-dihydroxyacetophenone in *Teucrium canadense*." *Phytochemistry*, 25, 2457–61.

112. Stenlid, G. 1976. "Effects of flavonoids on the polar transport of auxins." *Physiol. Plant.*, 38, 262–66.

113. Baur, J. R. 1979. "Effect of glyphosate on auxin transport in corn and cotton tissues." *Plant Physiol.*, 63, 882–86.

114. Abu-Irmaileh, B. E., L. S. Jordan, and J. Kumamoto. 1979. "Enhancement of CO_2 and ethylene production and cellulase activity by glyphosate in *Phaseolus vulgaris*." *Weed Sci.*, 27, 103–06.

115. Lee, T. T., and T. Dumas. 1983. "Effect of glyphosate on ethylene production in tobacco callus." *Plant Physiol.*, 72, 855–57.

116. Yang, S. F., and N. E. Hoffman. 1984. "Ethylene biosynthesis and its regulation in higher plants." *Annu. Rev. Plant Physiol.*, 35, 155–89.

117. Kenyon, W. H., S. O. Duke, and K. C. Vaughn. 1985. "Sequence of herbicidal effects of acifluorfen on ultrastructure and physiology of cucumber cotyledon discs." *Pestic. Biochem. Physiol.*, 24, 240–50.

118. Creelman, R. A. 1989. "Abscisic acid physiology and biosynthesis in higher plants." *Physiol. Plant.*, 75, 131–36.

119. Cammue, B. P. A., W. F. Broekaert, J. T. C. Kellens, N. V. Raikhel, and W. J. Peumans. 1989. "Stress-induced accumulation of wheat germ agglutinin and abscisic acid in roots of wheat seedlings." *Plant Physiol.*, 91, 1432–35.

120. Vaughn, K. C., and S. O. Duke. 1984. "Ultrastructural alterations to chloroplasts in triazine-resistant weed biotypes." *Physiol. Plant.*, 62, 510–20.

121. Grenier, G., L. Proteau, J.-P. Marier, and G. Beaumont. 1987. "Effects of a sublethal concentration of atrazine on the chlorophyll and lipid composition of chlorophyll-protein complexes of *Lemna minor*." *Plant Physiol. Biochem.*, 25, 409–13.

122. Kitchen, L. M., W. W. Witt, and C. E. Rieck. 1981. "Inhibition of chlorophyll accumulation by glyphosate." *Weed Sci.*, 29, 513–16.

123. Kitchen, L. M., W. W. Witt, and C. E. Rieck. 1981. "Inhibition of δ-aminolevulinic acid synthesis by glyphosate." *Weed Sci.*, 29, 571–77.

124. Killmer, J. L., J. M. Widholm, and F. W. Slife. 1981. "Reversal of glyphosate inhibition of carrot cell culture growth by glycolytic intermediates and organic and amino acids." *Plant Physiol.*, 68, 1299–1302.

125. Mollenhauer, C., C. C. Smart, and N. Amrhein. 1987. "Glyphosate toxicity in the shoot apical region of the tomato plant I. Plastid swelling is the initial ultrastructural feature following *in vivo* inhibition of 5-enolpyruvylshikimic acid 3-phosphate synthase." *Pestic. Biochem. Physiol.*, 29, 55–65.

126. Vaughn, K. C., and S. O. Duke. 1986. "Ultrastructural effects of glyphosate on *Glycine max* seedlings." *Pestic. Biochem. Physiol.*, 26, 56–65.

127. Lichtenthaler, H. K. 1984. "Chloroplast biogenesis, its inhibition and modification by new herbicide compounds." *Z. Naturforsch.*, 39c, 492–99.

128. Lichtenthaler, H. K., and D. Meier. 1984. "Inhibition by sethoxydim of chloroplast biogenesis, development and replication in barley seedlings." *Z. Naturforsch.*, 39c, 115–22.

129. Nadakavukaren, M., and D. McCracken. 1977. "Effect of 2,4-dichlorophenoxyacetic acid on the structure and function of developing chloroplasts." *Planta*, 137, 65–69.

130. McCracken, D. A., D. R. Ort, and M. Nadakavukaren. 1981. "The effects of 2,4-dichlorophenoxyacetic acid altered chloroplast development on photosynthesis." *Physiol. Plant.*, 52, 285–91.

131. Bartels, P. G. 1985. "Effects of herbicides on chloroplast and cellular development." In *Weed Physiology, Vol. II. Herbicide Physiology*, S. O. Duke, ed. Boca Raton, Fla.: CRC Press, Inc., pp. 63–90.

Chapter 17

Herbicide Interactions with Herbicides, Synergists, and Safeners

17.1 DEFINITIONS AND CALCULATIONS

When mixing two or more herbicides, the joint action may be equal to, more, or less than expected. The problem in examining interactions is that there is no universally accepted reference model for the expected response of herbicide mixtures. To begin with, a reference model should inform the experimenter of the level of response to expect when the compounds in the mixture behave and act independently. This situation is called "independent joint action," and the corresponding reference models have been termed "additive dose models" [1].

Let us first consider the special case where one compound in a mixture of two has no effect when applied alone. This situation is very simple: we can use the term "synergistic interaction" or "antagonistic interaction" when the joint effect is significantly more or less than the effect of the active compound applied alone. Preferentially, the synergistic or antagonistic interaction should be shown not only for one or two mixture concentrations, but the dose response curve of the active compound should be obtained in the presence of different added amounts of the "inactive" compound [2]. In accordance with terminology used in the insecticide and pharmaceutical fields, "inactive" compounds that synergize or antagonize an "active" material should be called synergists or antagonists, respectively. In the herbicide field the term "safener" (instead of antidote or antagonist) is commonly used, and will also be adopted here. As an example of the former definition, the herbicidally inactive insecticide carbaryl acts as a synergist when applied in mixture with propanil [3]. Similarly, the herbicide tridiphane synergizes the herbicidal activity of atrazine in *Setaria faberi* [4, 5]. In these and other cases, inhibition of degradation of the herbicidally active component by the herbicidally inactive component is the basis for the synergism. The synergist may behave as described in one system, but the same compound may exert other and/or additional physiological or toxicological effects in other systems. Many compounds act differently in different systems, and the type of action must therefore always be defined with reference to the system under study; a knowledge of the mode of action is helpful in defining a specific situation or system. Synergists and safeners that fit the definitions given above are discussed in more detail in Sections 17.3 and 17.4.

Several approaches or rationales have been suggested to "prove" synergism or antagonism of mixtures of active materials. The following discussion will consider only mixtures of two active materials, which constitute the vast majority of cases. Since there is no universal agreement with respect to the applicable reference systems, and since different proposed reference systems can differ considerably [1, 6], we will use the terms "more active" or "less active" than expected from the specific reference model (one exception from this rule will be discussed in connection with Figure 17.3). Many of the herbicide combinations that have been found to be more active than expected have been adopted in patent specifications. Such herbicide combinations must fulfill the requirements of "unobviousness" and "novelty" in order to convince patent officers [7]. Novelty is usually not a problem, but unobviousness is often difficult to "prove."

The most simple reference model states:

Definition 1: the combined effect equals the sum of the effects of each of the agents when used alone, that is:

$$E_e = Y_1 + Y_2$$

where E_e is the expected effect and Y_1 and Y_2 are the individual effects/responses. This model means that the dose-response function is linear and starts out at zero effect. If E_o (the observed effect) is significantly higher than E_e, than the increased joint action would be higher than expected and, consequently, "unobvious." However, a linear dose-response relationship, as anticipated with this model, is seldom obtained. Only when data at low effect levels are considered (e.g., 20% effect at concentration x, 40% effect at concentration 2x) might such a relationship be obtained. The linear dose-response relationship that is adopted in Definition 1 is certainly not true for wider concentration ranges. A second reference model avoids this problem:

Definition 2: the combined effect of two compounds equals the effect of one compound when tested at the additive dose:

$$\begin{aligned} E_e(x_1S_1 + x_2S_2) &= E_o(x_1 + x_2)S_1 \\ \text{or } &= E_o(x_1 + x_2)S_2 \end{aligned}$$

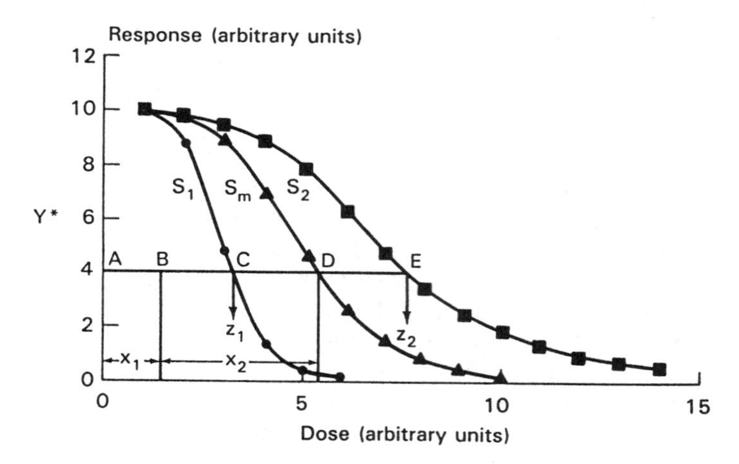

Figure 17.1 Calculation of the joint action of a mixture S_m of two compounds S_1 and S_2 from the known dose-response curves for S_1 and S_2 according to Morse [1]. z_1 and z_2 represent the dose equivalents of S_1 and S_2 required for the response Y*; x_1 and x_2 are the doses of S_1 and S_2 in the mixture S_m.

where x_1 and x_2 are the dose equivalents of the compounds S_1 and S_2, respectively, and E_e and E_o are the corresponding expected and observed effects. This approach gives reasonable values when S_1 and S_2 both have similar activity. In this case, the reference model gives an expected percentage effectiveness or expected response along a sigmoidal dose-response curve (Figure 17.1). However, when one compound is active at lower doses than the other, testing at the additive dose $(x_1 + x_2)$ will obviously lead to erroneous results.

One way of avoiding these problems can be found in the next reference model:

Definition 3: the combined action at a given response level is obtained by mixing the individual compounds according to their relative effectiveness at this response level.

This definition describes the method for obtaining a joint action curve S_m as shown in Figure 17.1 [1]. The position of the joint action curve (D) at a certain response level Y^* is given by

$$R = n_2 z_1 / n_1 z_2$$

where $R = CD/DE$, $0 \leq n_i \leq 1$ (i = 1, 2; $n_1 + n_2 = 1$) are the fractions of the mixing partners, and z_i (i = 1, 2) are the doses of the compounds S_1 and S_2 required individually for the same response level. If one dose in a mixture (x_1) is given at a certain response level, the dose of the second compound (x_2) is calculated according to the relative effectiveness of S_1 and S_2 (z_2/z_1) at that response level, that is:

$$x_2 = \frac{z_2}{z_1} (z_1 - x_1)$$

This approach gives a joint action curve that lies entirely between S_1 and S_2 and reflects the relative effectiveness of the mixing partners at varying response levels.

Definitions 1–3 describe additive dose models using different dose-response relationships for the summation of individual actions. From a linear additive effect (Definition 1) via a nonlinear additive effect of two compounds with equal effectiveness (Definition 2), we come to a more generally applicable calculation of expected joint action curves by assuming additive effects according to the relative effectiveness of the individual active ingredients (Definition 3).

It is generally advisable to fit mathematical functions of the form $E_0 = f(x)$ to the individual effect curves as well as to the joint action curve. Different functions have been suggested and calculated for some standard examples [8–10]. A procedure has also been proposed for the derivation of a mathematical function for the joint action curve from the parameters of the individual curve functions [8]. This procedure has been subject to additional suggestions [1].

Besides the additive dose models described so far, one "multiplicative survival model" [1] must be mentioned here because of its wide use and general acceptance (e.g., in patent specifications). The reasoning is as follows: if the effect of a compound S_1 is $Y_1\%$ and that of S_2 is $Y_2\%$, then in a mixture, S_2 can only act on the remaining response which is not inhibited by S_1, that is:

$$E_e = Y_1 + (100 - Y_1) \frac{Y_2}{100}$$

Rearrangement gives Limpel's formula [11], also known as the Colby formula [12]:

Definition 4: the combined action of two active materials is given by

$$E_e = Y_1 + Y_2 - \frac{(Y_1 Y_2)}{100} \qquad [Y_1, Y_2 = \% \text{ effect}]$$

or, if Y_1 and Y_2 are expressed as % of control, then

$$E_e = \frac{Y_1 Y_2}{100} \qquad [Y_1, Y_2 = \% \text{ of control}]$$

In Figure 17.2a, curve 1 is calculated according to Limpel's formula for a hypothetical compound that is mixed with itself. The calculation starts with 5% effect at dose 1. Then at dose 2 the effect will be (5+5–0.25)%; further development gives curve 1 with an I_{50}-concentration of 13.5. Curve 2 is a sigmoidal function with the same I_{50}-concentration and similar slope. Curve 1 cuts curve 2 several times and becomes steeper at higher concentration (i.e., at 50 to 100 dosage units). This is indeed the case, because Limpel's formula is basically a quadratic function which, by definition, cannot accurately describe a sigmoidal dose-response relationship. The quadratic character of curve 1 becomes obvious in Figure 17.2b where the

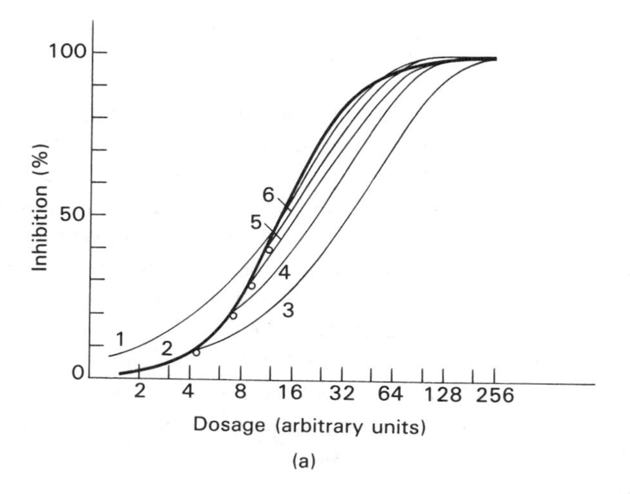

(a)

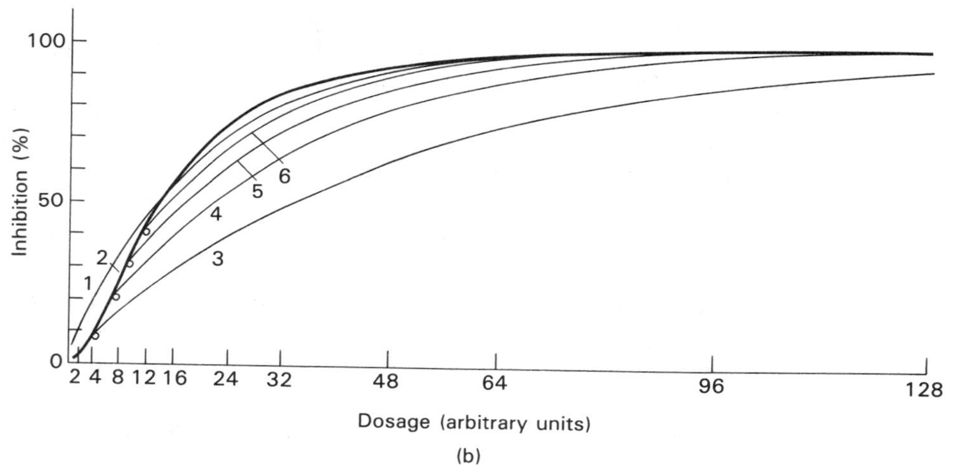

(b)

Figure 17.2 Plot of percent inhibition vs. dosage in semilogarithmic (a) and linear (b) scaling. Curve 1 is a dose-response curve for a hypothetical compound mixed with itself and calculated according to Limpel's formula [11]. The starting point is 5% inhibition at dose 1. Curve 2 is a sigmoidal dose-response curve calculated according to [8] with a I_{50}-dosage and average steepness matching curve 1. Curves 3 to 6 again represent hypothetical mixtures of a compound with itself calculated according to Limpel's formula; the individual starting points are chosen at inhibition levels of 8, 20, 30, and 40% effect. The expected inhibition in the mixture approaches the sigmoidal dose-response curve only when calculated from individual (single) inhibition values above 40–50%. A related steepness of the single dose-response curve 2 is additionally required if the Limpel formula is to have predictive value.

dosage is plotted on a linear scale. The sigmoidal curve in Figure 17.2a has been calculated by a function given by Rummens [8]:

$$Y = 100 - \frac{100}{\frac{(X)^B}{C} + 1}$$

with the parameters set at $B = 2$ and $C = 13.5$ (same I_{50}-concentration as in curve 1). As pointed out by Rummens, in the case of two compounds S_1 and S_2 with different effectiveness, Limpel's formula yields a joint action curve that is located outside the area between the curves S_1 and S_2 (compare with Figure 17.1). Of course, this cannot occur in an additive interaction. When a mixture of two compounds S_1 and S_2 with equal effectiveness is considered (which can be calculated as if a compound were mixed with itself), curves 3 to 6 in Figure 17.2 are obtained. The approximation of a sigmoidal function becomes better when comparing the compounds at increasing effectiveness, that is, in the upper portion of the sigmoidal curve. In this portion both curves have a positive curvature, which is not the case in the lower part (Figure 17.2b). The conclusion is, therefore, that Limpel's formula can roughly describe joint herbicidal actions only for compounds of about equal effectiveness, for situations of more than about 40 percent effect of the individually applied mixing partners, and for individual dose-response curves roughly following a quadratic relationship at these levels of effectiveness (i.e., Y increasing from 50% to 75% effect when doubling the concentration).

It may be difficult to detect an effect more or less than expected by direct comparison with a calculated joint action curve. One method that shows deviations from the expected joint action more directly is the bologram method (Figure 17.3). The isobole is the line that connects points of equal expected response, that is, all combinations of doses x_1 and x_2 of the mixing partners S_1 and S_2 for which the

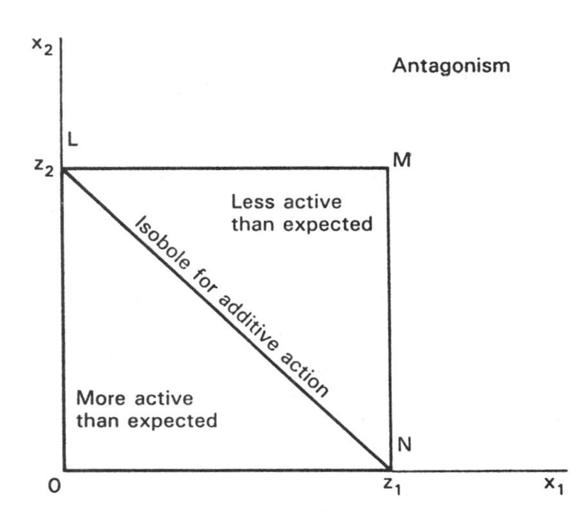

Figure 17.3 Bologram for mixtures of S_1 and S_2 in concentrations of x_1 and x_2, respectively. The isobole line represents the levels (x_1 and x_2) of the two components required to produce the measured response Y*, provided S_1 and S_2 are additive. Designations of different types of joint action are given.

expected response Y^* is equal (line CE in Figure 17.1 and line LN in Figure 17.3). The isobole cuts the axes at z_1 and z_2, reflecting the individual doses of S_1 and S_2 required at the same response level Y^* when applied alone. The line LN will be linear when independent joint action of two compounds with the same mode of action is considered, that is, when free competition at the same site of action can be expected. In the case of multiplicative survival models, however, the isoboles are nonlinear [1].

The greatest accuracy can be obtained with 50 percent isoboles, representing half-maximal action. Half-maximal means the middle of the close-to-linear portion of the sigmoidal dose-response curve. These values can be obtained with the greatest accuracy, partly because there is less variation in normal populations close to mean [13]. Similarly, isoboles for effects between 30 percent and 70 percent can be obtained from the close-to-linear portion of the dose response curve; higher or lower effect values may be less accurate and reflect uncertainties about the correct fit of the sigmoidal dose-response curve.

If a combination of doses x_1 and x_2 required for an observed response is not located on the isobole expected for additive combined action at that response level, three different situations can be considered (Figure 17.3): location within the triangle LON means that the combination is more active than expected, whereas location within LMN means that the combination is less active than expected [14]. When, in the combination (x_1, x_2) at the considered dose-response level, either x_1 exceeds z_1 or x_2 exceeds z_2, or both, true antagonism can be concluded. In that case, in which either x_1 or x_2, or both, exceed the individual dose required for the same level of effect, the measured effect value Y^* for the combination is lower than the lowest effect value measured for the individual compounds S_1 and S_2 when tested at the concentration x_1 and x_2, respectively (e.g., [15]).

17.2 HERBICIDE-HERBICIDE INTERACTIONS

A selection of reported interactions between herbicidally active compounds is provided in Table 17.1. The individual examples are presented in five groups of more-or-less related possible mechanisms. Group VI contains examples of unknown and/or unique suggested possible mechanisms of interaction.

The antagonistic herbicides (2) in Group I are inhibitors of cell division and cell growth—in particular the dinitroanilines, which affect microtubule function (see Chapter 10). Sublethal herbicidal activity of these compounds reduces growth of the root system; this results in a decrease in transpiration rate, which in turn is the basis for the uptake and mobility of the xylem-mobile, photosynthesis-inhibiting herbicides (1) in Group I (Chapter 7). Root uptake and mobility of the "acting" herbicides (1) is therefore decreased because the root system of the treated plants develops poorly in the presence of the "interacting" herbicides (2). The result is decreased herbicidal action of these combinations.

The reverse is true for the herbicide combinations in Group II. The interacting herbicides (2) of this group are known to affect lipid synthesis and, in particular, wax deposition on the surface of treated leaves, leading to an increased transpiration rate (Chapter 11). Xylem-mobile herbicides may therefore be taken up and transported to the leaf more effectively in these plants. Leaf-applied herbicides may also be taken up more effectively through the altered cuticle on the leaf surface.

The herbicide combinations listed in Groups III, IV, and V show decreased or increased herbicidal actions that can be correlated with increased or decreased rates of detoxification. The interacting herbicides (2) in Group III exhibit auxin activity, except bromoxynil, which shows only a weak effect [23]. The acting herbicides (1) of Group III must first be hydrolyzed to the free acids in order to be herbicidally active in the plant (see Chapter 6). These acids can either be irreversibly detoxified by aryl hydroxylation and subsequent conjugation, as in wheat [39], or they can be reversibly detoxified by conjugation of the carboxyl group of the free acid herbicide, as in oat and wild oat [40]. These conjugation reactions in wild oat are strongly increased when an auxin-type herbicide (2) is applied in combination with an aryl-propanoic acid herbicide (1). The conjugates formed are complex and appear to change over time [27]. Although the interacting herbicides (2) in Group III are active auxins, not all herbicides with auxin activity induce negative interactions when combined with herbicides (1). More specifically, the "pyridine auxin herbicides" picloram and fluroxypyr cause very little or no interaction [26, 27]. The molecular mechanism(s) leading to increased conjugation(s) of herbicides (1) are not clear. Since only certain subgroups of auxin-active herbicides induce this interaction, it could be speculated that a specific auxin-triggered induction system, that is, an auxin-induced auxin conjugation system, becomes functional; this auxin conjugation system must have structure-activity requirements differing from those of the auxin receptor protein discussed in Chapter 14.

The interactions listed in Groups IV and V in Table 17.1 are dealt with in detail in Sections 17.4 and 17.3. The reason that they are included here is that some herbicidally active compounds can apparently also act by mechanisms similar to those of safeners (Section 17.4) and synergists (Section 17.3). Of course, true safeners and synergists have, by definition, no herbicidal activity of their own.

In general, the miscellaneous reports of increased or decreased herbicidal activity listed in group VI have not been studied in much detail. Also, the magnitude of the effect is sometimes not very pronounced. One example of this type is given in Figure 17.4a. The individual concentrations required for 50 percent effect of the atrazine-alachlor combinations deviate from the additive dose straight line, indicating a more-than-additive interaction in these combinations (compare with Figure 17.3). The more-than-additive interaction in Figure 17.4a and the less-than-additive interaction in Figure 17.4c contradict the group behavior shown in Table 17.1 (Groups I and II). Moderately "synergistic" or "antagonistic" interactions obviously depend very much on the specific experimental conditions, and may vary considerably.

The special case of one herbicidally inactive mixing partner is reproduced in Figure 17.4b. Sun 11E oil strongly increases the activity of atrazine, without showing any herbicidal activity of its own. The isobole line therefore never reaches

TABLE 17.1 HERBICIDE-HERBICIDE INTERACTIONS. The herbicidal activity of herbicide 1 is either positively (+) or negatively (−) influenced by the simultaneous application of herbicide 2 (compound 2).

Herbicide 1 (acting)	Herbicide 2 (interacting)	Plant(s)	+/− Possible mechanism(s)	Reference
Group I				
Metribuzin, atrazine, simazine, prometryne, linuron, buthidazole	Trifluralin, pendimethalin, oryzalin, napropamide, nitralin, alachlor	Soybean, tomato, pea, corn, weeds, etc.	(−) Reduction of root system by 2; reduced uptake and transport of 1	16 17 18 19
Group II				
Desmedipham, atrazine, 2, 4-D, diquat	Ethofumesate, TCA, diallate	Pea, sugar beet	(+) Reduction of leaf wax deposition by 2; increased root and foliar uptake and transport of 1	20 21
Group III				
Diclofop-methyl, benzoylprop-ethyl, flamprop esters	2,4-D MCPA, dicamba, bromoxynil, IAA, 2,3,6-TBA	Wild oat, oat, corn	(−) Increased rate of conjugation of free acid of 1	22 23 24 25 26 27–29
Group IV				
Bensulfuron (=DPX-84). CDAA	Thiobencarb, dimepiperate (=MY-93), CDAA	Rice	(−) Increased rate of detoxification of 1 in the presence of 2	15 30

Group V				
Atrazine	Tridiphane	*Setaria*, etc.	(+) Inhibition of detoxification of 1 by 2	4
Group VI (miscellaneous)				
Atrazine	Alachlor	*Echinochloa*	(+) Unknown	13
Trifluralin	Alachlor	*Ipomoea*	(+) Unknown	31
Glyphosate	Simazine, atrazine	Corn, beans	(−) Physical binding of 1 in spray solution of 2	32
Glyphosate	2,4-D, dicamba	*Convolvulus arvensis*	(+) Increased uptake and transport of 1	33
Ioxynil	Mecoprop	*Stellaria media*	(+) Unknown	34
Mefluidide	Bentazon	Soybean	(−) Unknown	35
Mefluidide	Bentazon	Rice	(+) Unknown	
Ethofumesate	Metamitron	Oat	(−) Unknown	14
Lenacil	Metamitron	Oat	(+) Unknown	
Lenacil	Pyrazon	Oat	(+) Unknown	
Barban	Flamprop-methyl	Wild oat	(+) Increased absorption	36
Barban	2,4-D 2,4,5-T	Wild oat	(−) Counteraction of DNA synthesis	37
Bentazon, glyphosate	Gibberellic acid	Bean, *Cirsium arvense*	(+) Increased sensitivity of GA-stimulated plants	38

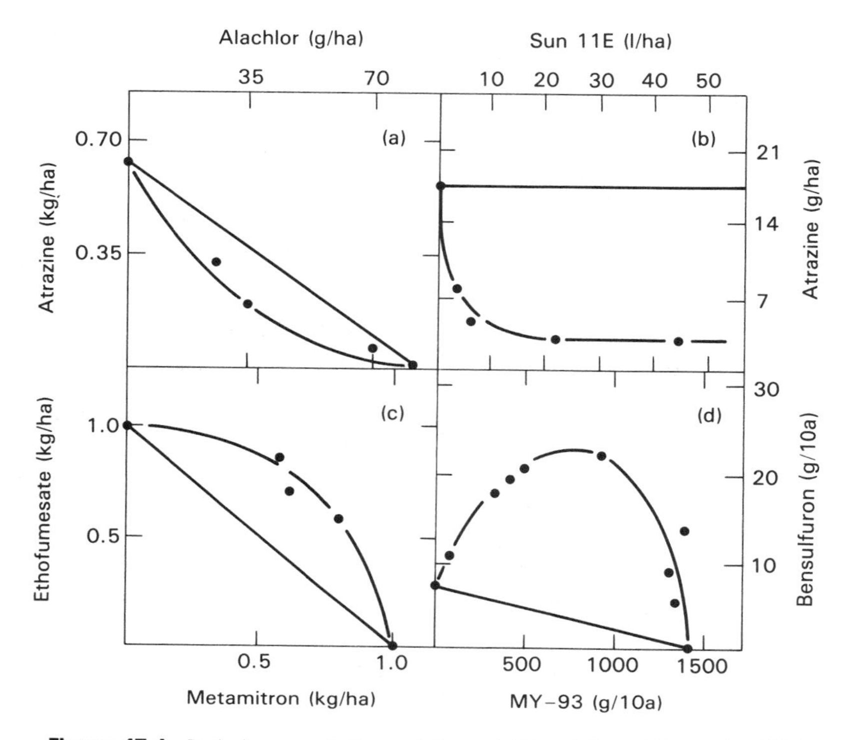

Figure 17.4 Isobole presentations of the activities of four different herbicide mixtures in an additive dose model: (a) and (b) 50% growth inhibition in *Echinochloa crus-galli* [13]; (c) 50% growth inhibition in oats [14]; (d) 10% growth suppression in rice shoots [15]. MY-93 = dimepiperate = *S*-(1-methyl-1-phenylethyl)-1-piperidine carbothioate; Sun 11E = 85% insecticidal spray oil and 15% emulsifier.

the Sun 11E oil axis. Antagonistic herbicidal interactions are shown in Figures 17.4c and 17.4d. Whereas only a moderately negative interaction is shown in example c, strong antagonistic interaction is demonstrated in example d (see Sections 17.1 and 17.4).

17.3 SYNERGISTS FOR HERBICIDAL ACTION

17.3.1 Tridiphane: Inhibition of Glutathione-S-Transferases

Tridiphane (Figure 17.5) is a postemergence grass herbicide used in corn and sorghum. The herbicidal activity of this compound is not very pronounced, but in combination with atrazine, a synergistic interaction in the control of panicoid grasses (*Panicum miliaceum, Setaria faberi, Setaria italica,* and *Digitaria sangui-*

Compound	Structure
1-Aminobenzotriazole (ABT)	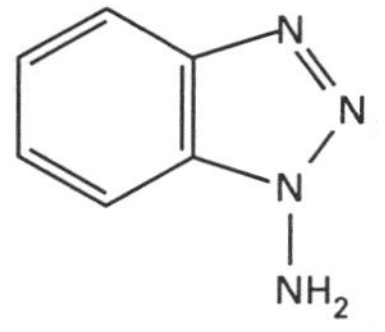 N, N, N, NH_2
Piperonyl butoxide (PBO)	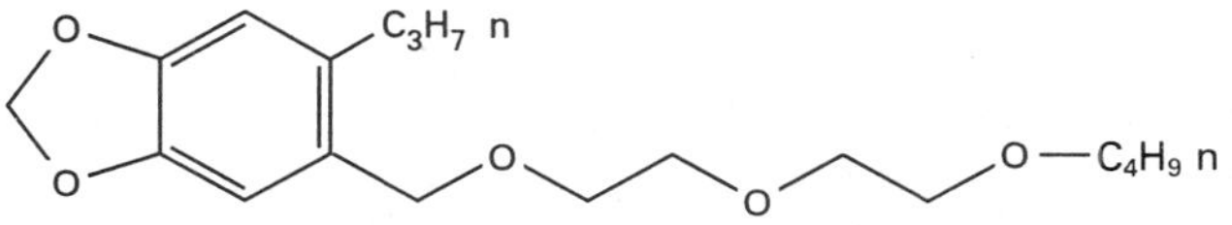 O, O, C_3H_7 n, O, O, O—C_4H_9 n
Carbaryl	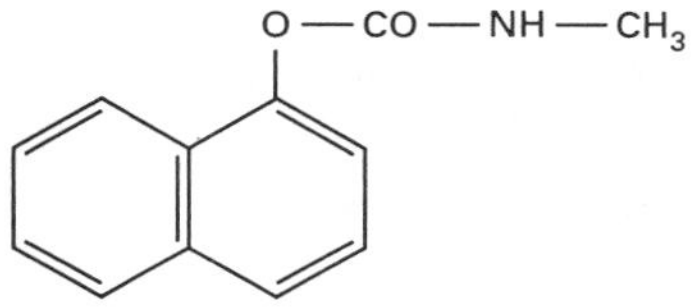 O—CO—NH—CH_3
Carbofuran	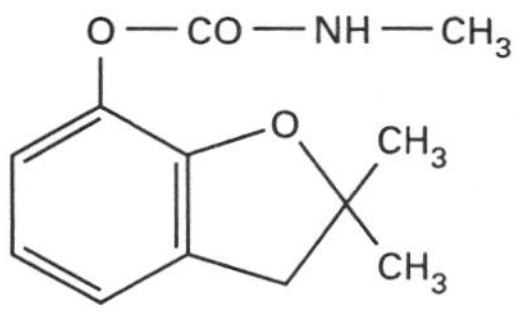 O—CO—NH—CH_3, O, CH_3, CH_3
Tridiphane	Cl_3C—H_2C, O, C—CH_2, Cl, Cl
Edifenphos	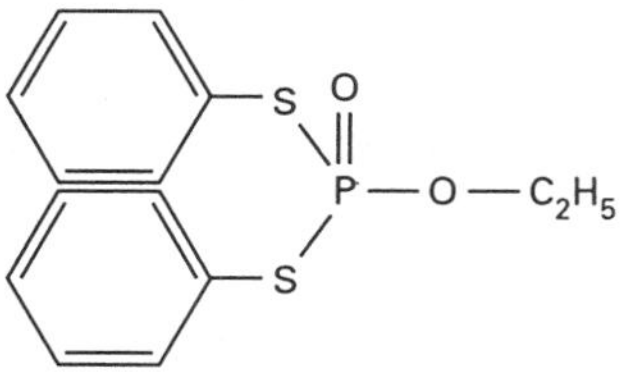 S, O, P—O—C_2H_5, S
Dietholate (R-33865)	H_5C_2—O, S, P—O, H_5C_2—O
Picolinic acid *t*-butyl amide (PABA)	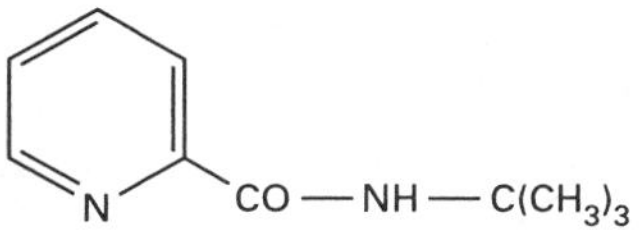 N, CO—NH—$C(CH_3)_3$

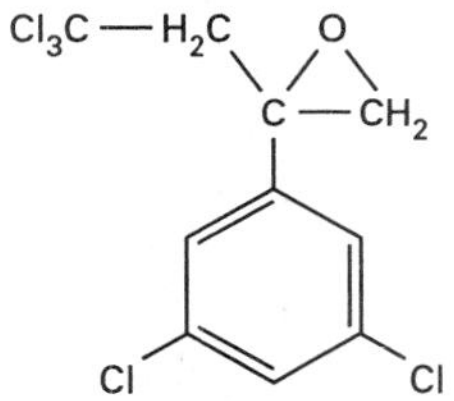

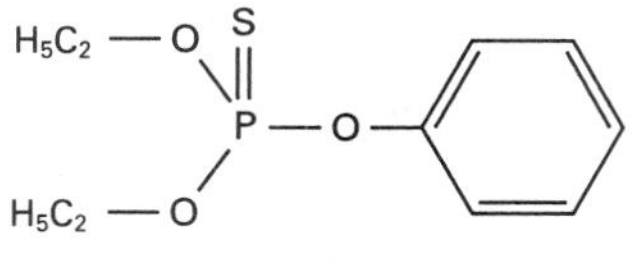

Figure 17.5 Herbicide synergists and related inhibitors of enzymes that detoxify herbicides.

nalis) can be observed [4, 41]. This interaction is of particular importance because the herbicidal action of atrazine on the panicoid grasses is very poor. The basis of the synergistic interaction is inhibition of atrazine detoxification. Although a significant (i.e., field-applicable) synergistic interaction is usually obtained only with atrazine, synergism has also been reported for the combinations of tridiphane with EPTC or alachlor in corn and *Panicum miliaceum* [41], and with cyanazine in *Setaria faberi* [4, 42].

In the presence of 560 g ha^{-1} tridiphane, the atrazine dose for 50 percent growth reduction of *Setaria faberi* is reduced to 0.45 kg ha^{-1} as compared to 4.7 kg ha^{-1} when atrazine is applied alone [43]. The synergistic interaction can be further enhanced by split applications; when atrazine is applied at 1.68 kg ha^{-1} between 1 and 10 days after treatment with tridiphane at 0.56 kg ha^{-1} (both applications given postemergence in the greenhouse), control of *Panicum miliaceum* increases to greater than 95 percent from only 40 percent for the simultaneous treatment [44]. More detailed studies have shown that the inhibition of atrazine conjugation after tridiphane treatment is only complete after a 15–20 h lag phase [43]. The reason for this lag phase lies in the conversion of tridiphane to the tridiphane-glutathione conjugate, which appears to be the actual in vivo inhibitor of glutathione-*S*-transferase [5] (Figure 17.6). The increased activity after split applications can be interpreted as follows: differences in uptake and distribution among the two compounds (tridiphane and atrazine) might occur, but the main reason for the lower activity of simultaneous applications appears to be the delayed inhibition of atrazine conjugation.

When discussing the activity and inhibition of glutathione-*S*-transferases (GSTs), some enzymological peculiarities should be kept in mind. These include the following [5]:

1. Rates of conjugation of different substrates can differ considerably; for example, the model substrate CDNB (1-chloro-2,4-dinitrobenzene) is conjugated three orders of magnitude faster than atrazine (3000 x) or tridiphane (1500 x) by the isolated enzyme.
2. Sensitivity toward (i.e., inhibition by) different glutathione conjugates (GS-conjugates) or other inhibitors differs among different GST isozymes.
3. Substrate specificities of the different GSTs (K_m for the herbicide or other substrate to be conjugated) differ.
4. Herbicide concentration in the tissue is important.
5. Glutathione concentration is of similar importance.
6. In vivo stability of the glutathione conjugates determines the duration of the inhibition.
7. The number and relative amounts of GST isozymes present in the tissue may differ in the characteristics described in 1–3.

An additional factor that influences the behavior of a particular tissue is the age of the plant. Whereas the levels of GST and GSH both decrease in maturing

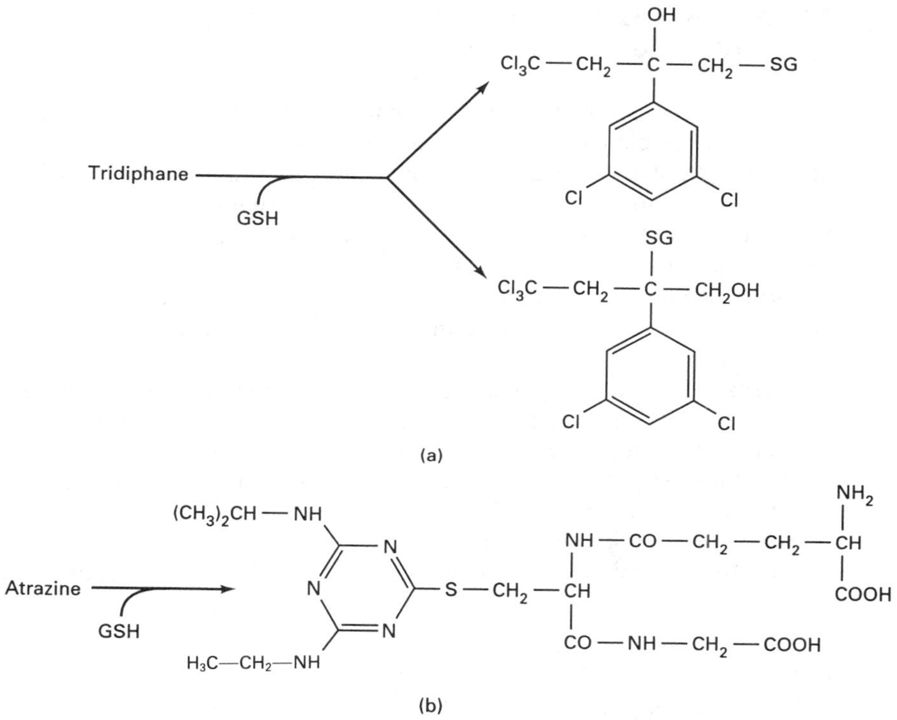

Figure 17.6 Conversion of tridiphane into two diastereomeric glutathione conjugates (a) and conversion of atrazine into its glutathione conjugate (b) by glutathione-S-transferase.

corn, they increase in maturing *Setaria faberi*. In addition, the tridiphane-GS conjugate is stable in *Setaria faberi* but unstable in corn. Tridiphane-GS is also a more potent inhibitor of the *Setaria faberi* enzyme (K_i = 2 μM) than the corn enzyme (K_i = 8 μM). For inhibition of GST by tridiphane, a K_i of 5 μM has been reported for both plant species [45]. However, it should be made clear that in in vitro studies the inhibition is caused by tridiphane itself, because the conjugate concentration does not reach inhibitory levels. The GS-conjugate of tridiphane is synthesized in two diastereomeric forms (Figure 17.6). Since most of the GST preparations used thus far have not been extensively purified, and may therefore contain several isozymes, specific K_i and K_m values for inhibitors and substrates are only preliminary estimates [46–48]. A constitutive GST I and a safener-inducible GST II have been described in corn [48]. The tridiphane-GS conjugate inhibits a variety of GST enzymes with several different substrates, and has been reported to be a competitive inhibitor with respect to GSH [5].

Tridiphane-glucosulfonate conjugates have also been reported to be very effective inhibitors of GST enzymes [49]. Other conversions and inhibitions might also be of relevance, therefore, in the tridiphane/atrazine synergism.

17.3.2 Aminobenzotriazole and Other Inhibitors of Monooxygenases

Monooxygenases are important detoxifying enzymes in all eucaryotic organisms [50]. In higher plants, monooxygenases are largely restricted to the germination and seedling stages. In growing plants they are found only at specific developmental stages and in certain tissues (e.g., in expanding leaves) [51]. It has been speculated that this is a specific adaptation against herbivores, since most endogenous metabolic waste molecules are not detoxified but are simply deposited in the vacuole. Some monooxygenases, in particular cinnamic acid-4-hydroxylase, which is required by growing plants, are generally present in developing tissue.

Much like glutathione-*S*-transferases (Sections 6.3.1, 17.3.1, 17.4), monooxygenases occur in living tissues as multiple isozymes with different substrate specificities [51, 52]. Furthermore, the isozymes also differ with respect to the type of monooxygenation that they catalyze (Section 6.2.2). Besides the cytochrome P-450 monooxygenases, other related enzymes occur in plant tissues [53]. This discussion is restricted to the so-called P-450 enzymes. Of the different P-450 monooxygenases known (see Chapter 6), the following three have been demonstrated to be inhibited by 1-aminobenzotriazole during herbicide detoxification [54]:

1. two-step *N*-demethylation
2. hydroxylation at 4′-alkylphenyl
3. aromatic ring hydroxylation

In the reaction mechanism of monooxygenation, one oxygen atom of the O_2 molecule is introduced into the substrate molecule, while the second oxygen is reduced to water by NADPH.

Of the monooxygenase inhibitors that have been tested for inhibition of herbicide detoxification, 1-aminobenzotriazole (ABT, Figure 17.5) has been the most successful. ABT has been shown to inhibit 4′-methylphenyl oxygenation and the second *N*-demethylation of chlortoluron and isoproturon in wheat [54, 55]. The *N*-demethylation appears to be less sensitive in *Veronica persica* [55]. The first *N*-demethylation is not a detoxification, since the mono-methyl analogues of chlortoluron and isoproturon are still phytotoxic. ABT also inhibits the metabolic breakdown of chlortoluron by *N*-demethylation and benzylalcohol formation in cell suspension cultures of maize and cotton [56]. The 4′-ring hydroxylation of phenoxyacetic acid in wheat is also inhibited by ABT [57]. Resistance to chlortoluron and isoproturon in *Alopecurus myosuroides* populations can be overcome by addition of ABT to these herbicides [58]. Cross-resistance to the herbicides terbutryne, diclofop-methyl, chlorsulfuron, and pendimethalin is also reversed by ABT. The generally accepted conclusion is that synergism or resistance reversal by an ABT-herbicide combination indicates the involvement of P-450 monooxygenase enzymes in the detoxification pathway/resistance mechanism of the herbicide.

Other monooxygenase inhibitors that have shown some synergism or enhanced herbicidal activity in combination with different herbicides include: the insecticide synergists piperonyl butoxide (PBO, Figure 17.5) [59] and 3-(2,4-dichlorophenoxy)-1-propyne [56]; the gibberellin biosynthesis inhibitors and growth regulators tetcyclacis and paclobutrazol [60]; a large number of azol fungicides that inhibit sterol biosynthesis (e.g., triadimenol, imazalil, clotrimazol) [61]; and the antioxidant *N*-(2-(2-oxo-1-imidazolidinyl)ethyl)-*N*-phenylurea (EDU) [59]. However, no monooxygenase inhibitor is yet known that increases the herbicidal activity on a difficult-to-control weed without simultaneously increasing the phytotoxicity toward the crop plant. Nevertheless, it should be possible to find a commercially useful synergist that fulfills these requirements, either by not inhibiting the relevant monooxygenase isozymes in the crop plant, or because of metabolic instability of the monooxygenase inhibitor in the crop plant. In addition, the compound should not by itself be phytotoxic to the crop plant.

ABT and PBO are suicide substrates that irreversibly bind to cytochrome P-450 and therefore inhibit its enzymatic activity [62]. The mechanism of bonding has been described for PBO as π-σ-bonding, after the enzymatically catalyzed formation of the carbene structure [50]:

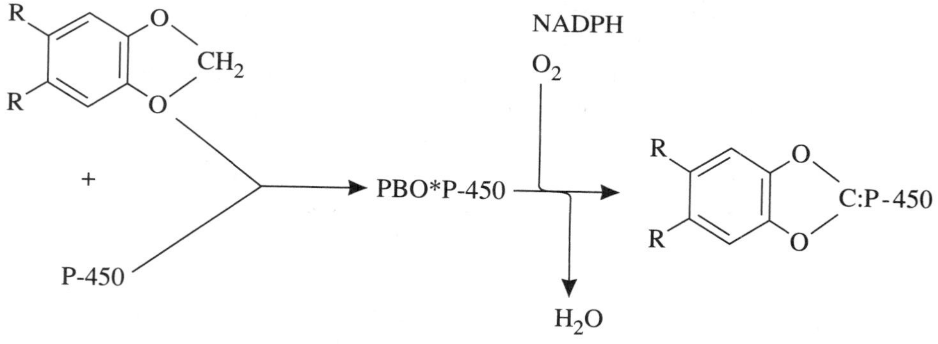

In addition, the stable complex that is formed can no longer bind carbon monoxide.

17.3.3 Picolinic acid *tert*-Butylamide: Inhibition of Metribuzin Degradation

The morningglories (*Ipomoea* species) are widespread and important weeds in the soybean growing areas of the United States. About 12 *Ipomoea species,* including the closely related *Pharbitis purpurea* and several *Ipomoea* subspecies, have been listed as economically important [63]. Of these, *Ipomoea hederacea* and *Pharbitis purpurea* are particularly common and widespread, but cannot be adequately controlled by several of the most important soybean herbicides. The weak action of metribuzin on these two morningglories can successfully be overcome by mixing with the synergist picolinic acid *tert*-butylamide (PABA; Figure 17.5).

The synergism between PABA and metribuzin is particularly strong in *Ipomoea hederacea* and *Pharbitis purpurea*, in which the control of field populations is increased from 40–50 percent to 80–90 percent by a 1 : 1 combination [2]. Since the recommended metribuzin application rate varies locally, the PABA application rate must be adjusted in a similar manner. In the greenhouse application of 500 g ha^{-1} PABA plus 50 g ha^{-1} metribuzin increases the herbicidal action against *Ipomoea hederacea* from 0 percent to 100 percent control. Tolerance of soybean plants to metribuzin is not affected by the combination with PABA. PABA also shows no herbicidal or phytotoxic activity at concentrations up to 2 kg ha^{-1} when applied alone. Other weed species that respond positively to a combination of PABA with metribuzin include *Cassia obtusifolia*, *Datura stramonium*, and *Portulaca oleracea*, but the synergism is less strong in these cases. Furthermore, increased herbicidal activity is generally observed when PABA is combined with herbicides that inhibit photosystem II.

Interestingly, PABA and other compounds that synergize metribuzin also synergize the activity of some insecticides [2]. Conversely, the organophosphate soil insecticides, phorate, disulfoton, and terbufos have been shown to promote metribuzin phytotoxicity in soybeans when applied in combination with metribuzin [64]. This observation reflects the considerable structural variation in metribuzin synergists, some of which (e.g., the insecticides mentioned above) do not discriminate between soybeans and *Ipomoea*.

The basis for the synergistic action of PABA with metribuzin appears to be inhibition of metribuzin degradation (Figure 17.7) [65]. In *Ipomoea hederacea*

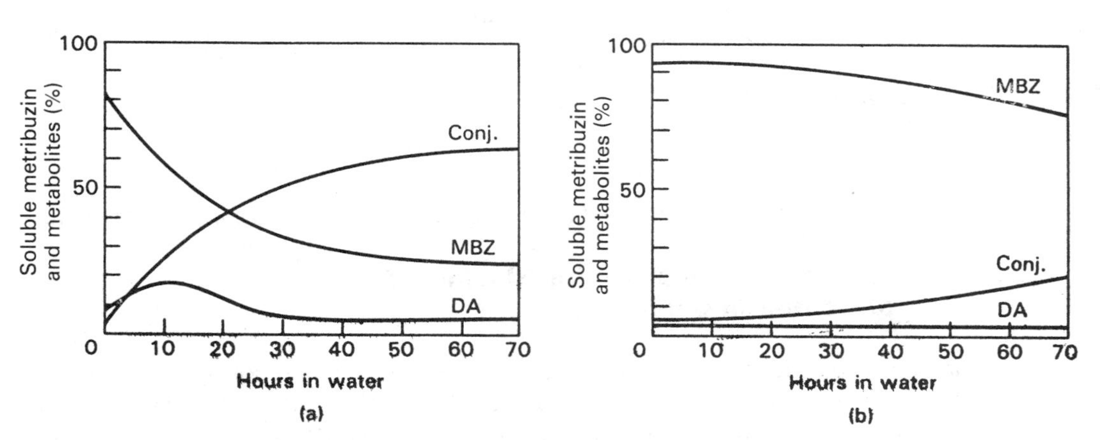

Figure 17.7 Kinetic changes of metribuzin (MBZ), deaminated metribuzin (DA), and conjugates (Conj.) in *Ipomoea hederacea* cotyledons after a 3-h metribuzin pulse given via the roots. The plants were untreated (a) or had been pretreated with the synergist PABA (3.4 μM). The plants were then kept in water for up to 70 h after the metribuzin pulse [65].

plants that have not been treated with PABA, a rapid early rise and a subsequent decrease of the fraction of deaminated metribuzin in the soluble pool indicates a precursor-product relationship in the following order (Figure 17.7a):

$$\text{metribuzin} \longrightarrow \text{deaminated metribuzin} \longrightarrow \text{conjugation products}$$

It has been suggested that a similar reaction occurs in soybean plants [66, 67] (Chapter 6). In the presence of PABA, the degradation of metribuzin is strongly inhibited, and both deaminated metribuzin and conjugates are only slowly formed. Deaminated metribuzin does not accumulate to any sizeable extent (Figure 17.7b). When comparing these results with our knowledge of the degradation/conjugation of metribuzin in other weeds and crop plants, the possibility of direct metribuzin conjugation must be considered (Section 6.3.2). Since the deamination is obviously inhibited in the presence of PABA, the conjugate formation shown in Figure 17.7b might reflect the amount of direct metribuzin conjugation that is not inhibited by the synergist. The fraction of conjugate formation that is sensitive to PABA can therefore be attributed to the conjugation of deaminated metribuzin, which can no longer occur after application of the synergist.

In vitro deamination of metribuzin, which can be measured in isolated peroxisomes, is not sensitive to PABA [2]. The mechanism of peroxisomal metribuzin deamination has been shown to require reducing conditions and to liberate the exocyclic amino group as ammonia (Chapter 6). The requirement for reducing conditions can be met by ascorbate or glutathione. With this background knowledge it is particularly interesting that, after application of PABA, the in vivo level of ascorbate decreases in *Ipomoea* leaves for several days, depending on the PABA concentration. It seems, therefore, that the rate of metribuzin deamination in vivo in the presence of PABA may be limited by a decreased endogenous supply of a reductant, possibly ascorbate.

The decreased concentrations of ascorbate in the leaves of PABA-treated plants may also explain the low-level synergistic interaction observed with herbicidal inhibitors of photosystem II in general [2, 68]. It is well known that the phytotoxic mechanism of herbicides affecting photosynthesis includes different toxic oxygen species (Section 7.6, Chapter 9, Section 17.3.6). Ascorbate serves as one of several protectants against the phytotoxic action of activated oxygen species. Therefore, it may well be that the synergistic interaction of PABA and similarly acting compounds with other herbicides inhibiting photosynthesis is caused by a lack of protection, which in turn is caused by decreased ascorbate concentrations.

17.3.4 Carbamates and Phosphoric Esters: Inhibitors of Amidases and Related Enzymes

The interaction of herbicides with other agrochemicals has been extensively reviewed [69]. Among other types of interaction, the synergism of propanil in combination with many insecticides is of particular interest. A number of carbamate (carbaryl, carbofuran; Figure 17.5), phosphate (chlorfenvinfos, phosphamidon), thiophosphate (azinfos-methyl, diazinon, disyston, fensulfothion, malathion), and phosphonate (fonofos, trichlorfon) insecticides have been found to increase the phytotoxicity of propanil in rice and/or tomato plants [70–73]. Dry matter production and yield both may be decreased. Chlorinated hydrocarbon insecticides do not interact with the herbicidal activity of propanil.

Propanil is hydrolyzed in tolerant plants by an aryl acylamidase that has been partially purified from rice and some other species (Section 6.2.1). For the present discussion it is of particular interest that the hydrolysis of propanil is inhibited in rice plants that are pretreated with one of the synergistic insecticides [3, 74]. Moreover, a more detailed in vitro study with an enzyme extract has revealed a positive correlation between the inhibitory activities of a number of insecticides on this amidase and the phytotoxic activities of the respective insecticide-propanil combinations. It can be concluded, therefore, that the synergistic insecticides act by inhibiting the aryl acylamidase which hydrolyzes propanil in tolerant plants.

Other herbicides can also be synergized by the insecticides listed above; the degradation of linuron, chlorpropham, dicamba, pronamide, and pyrazon has been found to be inhibited in different leaf tissues by several of the insecticides listed above [70]. Disyston has been found to enhance the phytotoxic action of diuron in oat [6]. Several insecticides similarly enhance the phytotoxicity of bentazon in soybean and bean [75]. Although it is generally suggested that inhibition of herbicide degradation is the mode of synergistic interaction in these examples, it is usually not known which type of enzyme is involved. In some instances, at least, it may be an amidase similar to the propanil-hydrolyzing aryl acylamidase.

Inhibition of an amidase is also the mechanism of the synergistic interaction between the herbicide mefenacet [2-(2-benzothiazolyl-oxy)-*N*-methyl-*N*-phenyl-acetamide] and the thiophosphate fungicide edifenphos (Figure 17.5) or several of the insecticides listed above [76, 77]. In this case, the inhibition of mefenacet degradation appears to be more effective in the weeds *Echinochloa crus-galli* and *Echinochloa oryzicola* than in rice plants. The edifenphos-mefenacet combination can therefore be used for improved weed control at lower herbicide doses, and with increased herbicide selectivity. When measured in vitro, however, the amidases from rice plants and these weeds are equally sensitive to edifenphos (I_{50} = 32 nM).

17.3.5 Soil-Life Extenders

The inhibition of herbicide degradation, which forms the basis for the synergistic interactions described in Sections 17.3.1–17.3.4, can also be observed in the soil. Important examples are the so-called "thiocarbamate history soils" that have received repeated applications of thiocarbamate herbicides over several years, and have developed the ability to degrade the herbicide faster than untreated ("nonhistory") soils. This increased rate of thiocarbamate herbicide degradation can be inhibited by the soil-life extender R-33865 (Figure 17.5) [78, 79]. In two experiments, the halflife of EPTC in "EPTC history soils" was increased from 9 to 18 days and from 6 to 15 days when treated with R-33865 [79]. The chemical structure of R-33865 is remarkably similar to the insecticidal and fungicidal inhibitors of amide splitting in plants that were discussed in Section 17.3.4. Since thiocarbamates are also amides, and can consequently be hydrolyzed by amidase enzymes, inhibition of amidases by phosphoric esters and carbamates appears to be a very general and widespread mechanism of interaction with herbicide degradation. The effect of inhibition of herbicide degradation in the soil is, of course, prolonged life of the herbicide in the soil; hence the term "soil-life extender."

Another example is the increased soil life of CIPC and the resulting increased time span of control of the weed *Cuscuta* in the presence of PCMC (*p*-chlorophenyl-*N*-methylcarbamate) in the soil [80]. PCMC also inhibits the degradation of propanil in tomato leaf discs [70], again suggesting amidase inhibition by PCMC.

17.3.6 Interference with Mechanisms of Oxygen Toxicity

Herbicides that interfere with the photosynthetic electron transport chain either by inhibiting photosystem II (Chapter 7), or by accepting electrons from photosystem I (Section 8.5), induce their phytotoxic action through the activation of molecular oxygen. (The mechanisms of oxygen activation and of inactivation of the toxic oxygen species are described in Section 8.3 and Chapter 9.) Herbicides that interfere with different chloroplast pigment biosyntheses (carotenoids: Sections 8.1, 8.2; tetrapyrroles: Section 8.4) likewise exert their phytotoxic action via activated and highly reactive oxygen species.

For the present discussion, the mechanisms of inactivation of the toxic oxygen species (superoxide anion, singlet oxygen, hydroxyl radical) are of particular importance. In a paraquat-resistant biotype of *Conyza bonariensis,* found in Egypt in a location of heavy and frequent paraquat applications, the levels of superoxide dismutase (1.6x), ascorbate peroxidase (2.5x), and glutathione reductase (2.9x) were found to be increased on a soluble protein basis [81]. This finding led to the conclusion that the detoxification of toxic oxygen species (superoxide anion in the case of paraquat) provides plant tolerance toward the herbicide, and that an increase in the relevant detoxifying capacity can confer increased tolerance. However, this mechanism of tolerance has been questioned in *Conyza*. Other mechanisms may therefore contribute to the increased in vivo tolerance of the *Conyza* biotype (Section 8.5).

If increased superoxide dismutase activity confers increased herbicide tolerance on the tissue, then inhibition of the enzyme should lead to increased phytotoxicity. Table 17.2 shows that the copper chelator DDC (Na-diethyldithiocarbamate) interacts synergistically with the herbicides paraquat and atrazine. A similar synergistic interaction has been reported with acifluorfen, which also exerts its phytotoxic action via oxygen activation (Section 8.4).

Since the chloroplast superoxide dismutase contains bound copper and zinc, the mechanism of interaction of DDC may be the induction of copper deficiency

TABLE 17.2 SYNERGISTIC INTERACTION BETWEEN THE COPPER CHELATOR DDC (DIETHYLDITHIOCARBAMATE) AND THE OXIDANT GENERATING HERBICIDES PARAQUAT AND ATRAZINE IN ISOLATED *ASPARAGUS SPRENGERI* CELLS [82]

DDC	Chlorophyll content (% of control)			
	Control	Paraquat	Atrazine	
%		10 μM	30 μM	100 μM
0	100	107	95	92
0.02	118	11	40	8

after its efficient removal by complex formation. DDC can also reverse paraquat tolerance in the resistant *Conyza bonariensis* biotype described above, in agreement with a mechanism of resistance based on increased activity of the detoxifying enzyme; similarly, the cross-tolerance with acifluorfen can be overcome by simultaneous application of DDC. However, fairly high concentrations (0.5 percent spray or 0.02 percent in solution) of the chelator are required for efficient complex formation and synergism. Other complexing agents are 4-octyl-2-acetyl phenol for copper, and di-2-ethylhexyl-phosphoric acid for zinc. The strategies for the development of different herbicide synergists and the results obtained with the different strategies have been summarized in a recent review, with particular emphasis on the chelation mechanism [82].

A low-level, unspecific synergism with photosynthesis inhibitors and other photosynthesis-inhibiting herbicides has also been reported for the metribuzin synergist PABA, and for similarly acting compounds (Section 17.3.3) [68]. Since this synergist decreases the endogenous level of ascorbic acid in *Ipomoea hederacea,* a similar mechanism of synergism by enhanced oxygen toxicity may be responsible.

17.4 SAFENERS FOR HERBICIDES

Reviews on herbicide safeners have appeared steadily over the years [83–89], and very recent reviews that include the latest developments and ideas are also available [90, 91]. The safener compounds that are presently available will be described briefly in this section, followed by a more detailed discussion of some current ideas and developments in the safener mode of action field.

Herbicide safeners, as suggested by their name, are intended to increase the tolerance of crop plants to herbicides. The safening effect increases the margin of tolerance of the herbicide and allows for improved control of problem weeds with the herbicide-safener combination. The herbicide, when applied alone, would either not be sufficiently active on the weeds at lower dose levels, or would cause some crop injury at the higher dose levels required for satisfactory weed control.

Herbicide safeners are also referred to as antidotes, protectants, or antagonists; "safener" avoids much of the possible confusion with other compounds or types of activity, and is now the generally accepted term.

The structures of 11 safeners with agricultural use or use potential, and one experimental compound (OTC = R-2-oxo-4-thiazolidine carboxylic acid), are shown in Figure 17.8. The development of the safeners in Group I (dichloroacetamides and related structures) and Group II (cyometrinil and related structures) began early (NA in 1968 by Gulf Oil, dichlormid in 1971 by Stauffer, cyometrinil in 1976 by Ciba-Geigy), and is still being actively pursued. Development of safeners in Group III (safeners for wheat and rice) began more recently (with fenclorim in 1983) and has provided access to additional crop species for commercial safening.

The three groups in Figure 17.8 and Scheme 17.1 are not as strictly different as it might seem. For example, safeners in Group I and II also show some safening activity in wheat, and with other herbicides such as the sulfonylureas. These activities are, however, of minor importance, whereas the safeners in Group III have been developed and optimized for these new uses. Group III safeners therefore represent a considerable and important extension over the earlier safeners of Groups I and II. The same is true for the "main mechanisms of safening" given in Scheme 17.1. The respective entries refer to the main uses, but other safening

Name	Structure
Group I	
Dichloroacetamides and related structures	
Dichlormid (R-25, 788; DDCA)	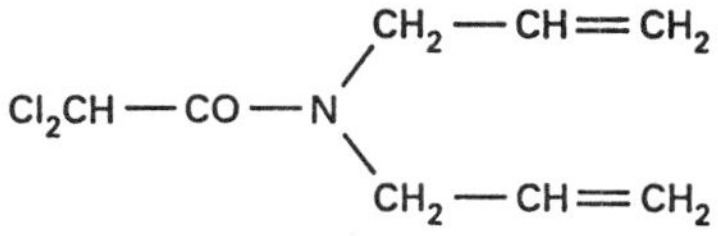 $Cl_2CH-CO-N(CH_2-CH{=}CH_2)_2$
Benoxacor (CGA-154, 281)	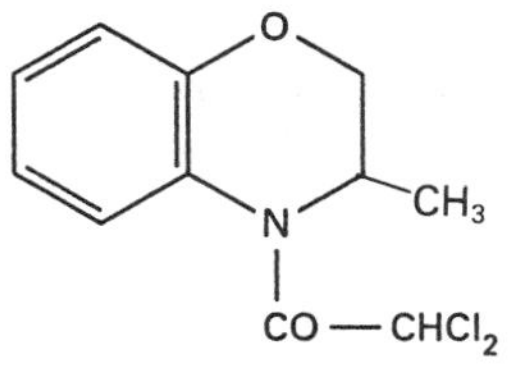 O; N; CH_3; $CO-CHCl_2$
MG-191	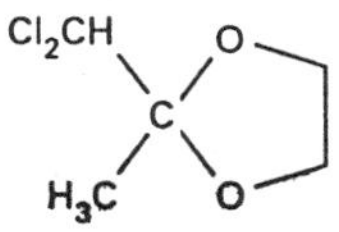 Cl_2CH; O; C; H_3C; O

Figure 17.8 Structures of herbicide safeners.

Name	Structure
Group II	
Cyometrinil and related structures	
NA (naphthalic anhydride)	O, O, O
Cyometrinil (CGA-43,089)	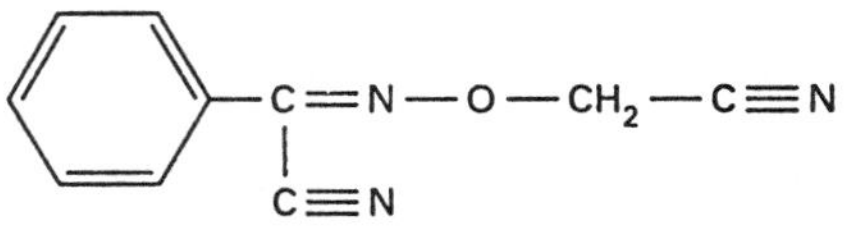 $C{=}N{-}O{-}CH_2{-}C{\equiv}N$; $C{\equiv}N$
Oxabetrinil (CGA-92, 194)	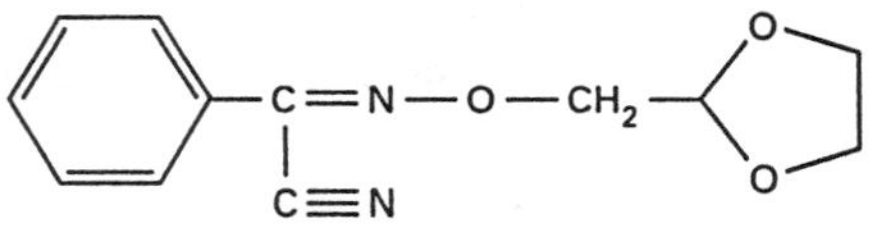 $C{=}N{-}O{-}CH_2$; $C{\equiv}N$; O, O
Fluxofenim (CGA-133, 205)	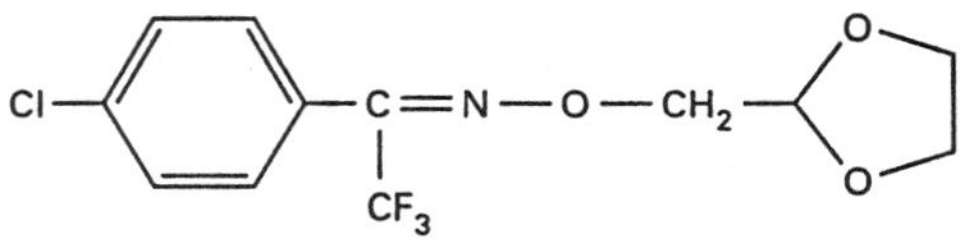 Cl; $C{=}N{-}O{-}CH_2$; CF_3; O, O
Flurazole (MON-4606)	$CH_2{-}O{-}CO$; CF_3; S, N; Cl
Group III	
Safeners for rice and wheat	
Fenclorim (CGA-123, 407)	N, N; Cl, Cl

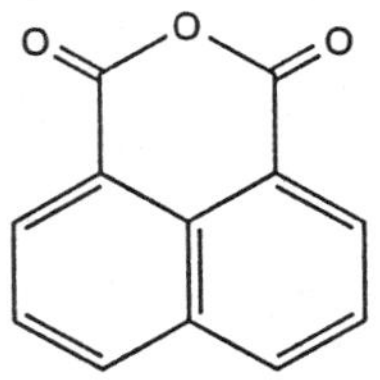

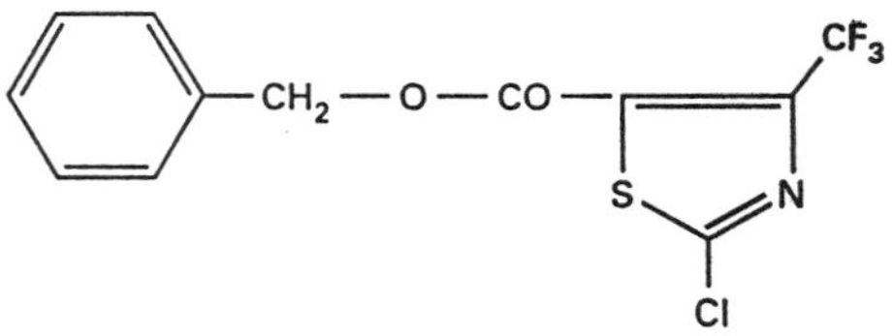

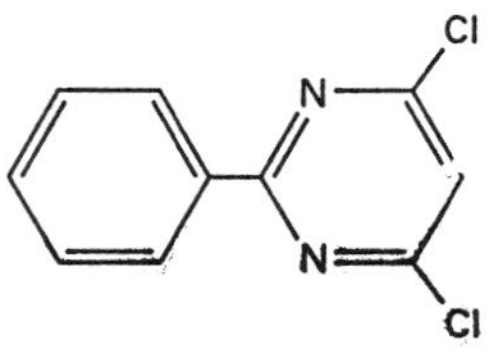

Figure 17.8 (Continued).

Name	Structure
Fenchlorazol-ethyl (HOE–70, 542)	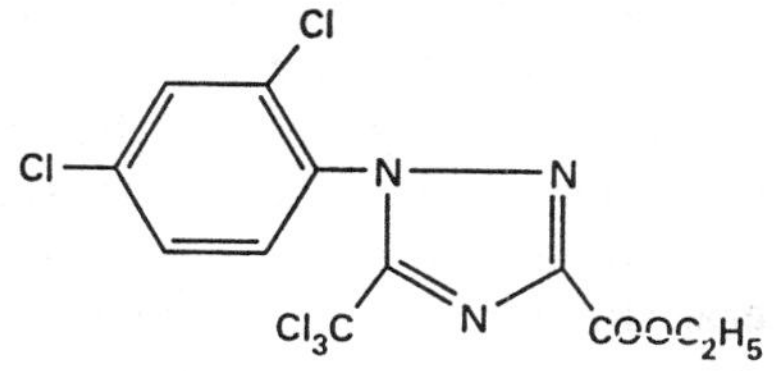
CGA–185, 072	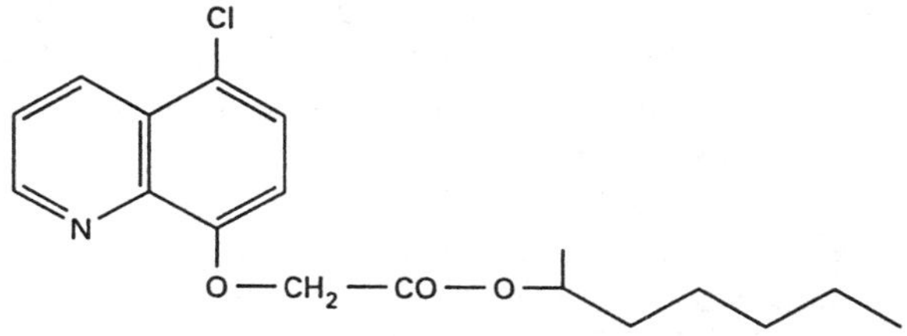
Other safening compounds	
OTC	

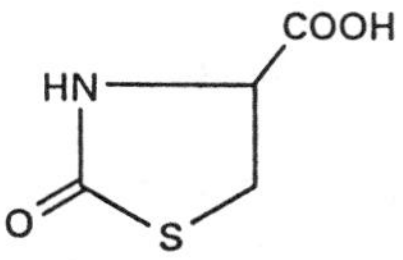

Figure 17.8 (Continued).

Scheme 17.1 MAIN APPLICATIONS AND USE PATTERNS OF THE THREE DIFFERENT SAFENER GROUPS PRESENTED IN FIGURE 17.8. MINOR USES AND SAFENING ACTIVITIES OF THESE COMPOUNDS ARE KNOWN WHICH CONNECT THE THREE DIFFERENT GROUPS.

	Group I	Group II	Group III
Main crop plants	Corn, sorghum	Corn, sorghum	Rice, wheat, corn
Main herbicides	Chloroacetamides, thiocarbamates	Chloroacetamides, thiocarbamates	Chloroacetamides, sulfonylureas, aryl-propanoic acids (fops)
Application form	Herbicide mix	Seed treatment	Herbicide mix
Dosage range (safener : herbicide or % seed treatment)	1 : 10–1 : 30	0.04–0.15%	1 : 2–1 : 4
Main mechanism of safening	Glutathione conjugation	Glutathione conjugation	Hydroxylation and glucosylation

mechanisms for the main uses and for possible additional uses may also be important.

The broad variety of possible safener-herbicide interactions are summarized in Figure 17.9, but the level of safening obtained is in most cases much lower than required for the commercial safening of these herbicides. The selection given in Figure 17.9 covers most of the important known cases of safening interactions. The level of safening observed with sulfonylureas in combination with the safeners NA (naphthalic anhydride), dichlormid, and oxabetrinil is also fairly high. Some general statements can be made with respect to safener chemistry: whereas NA is fairly unspecific and exerts safening interactions in a wide variety of crop plants, the other safeners are much more specific. In particular, dichlormid and other structurally related dichloroacetamides and α-chloropropionamides show strong safening interactions only with thiocarbamates and α-chloroacetamide herbicides. The most important message contained in Figure 17.9 can be formulated as follows: herbicide safeners show broad structural variability, but in general interact strongly only in the crop plants corn and sorghum, less so in rice and wheat, and only with herbicides from the thiocarbamate, α-chloroacetamide, arylpropanoic acid, and sulfonylurea groups. Other crop plants and herbicides from other structural groups do show safening effects, but the level of safening obtained is generally much lower.

The mode of action of safeners is now known to include stimulation of the pathway of detoxification via glutathione conjugation. This stimulation includes increased levels of reduced glutathione (GSH) and glutathione-*S*-transferase (GST), but many other parameters in plant growth and metabolism are also increased. The following is a list of the increases observed:

1. uptake of sulfate [92]
2. sulfate activation/ATP sulfurylase [92]
3. reduced glutathione [93, 94]
4. glutathione reductase [95]
5. microsomal and cytosolic glutathione-*S*-transferase [95, 96]
6. formation of herbicide-glutathione conjugates in vivo [97]
7. herbicide uptake [30, 97]
8. soluble protein content on a fresh weight basis [98]
9. activity of acetolactate synthase (ALS) [98–100]
10. herbicide monooxygenation [101, 102]
11. chlortoluron monooxygenation [60]
12. lipid synthesis from acetate [103]
13. glucosylation of OH-chlorimuron ethyl [104] and CGA 185,072 [105]
14. root and shoot growth [106, 107]

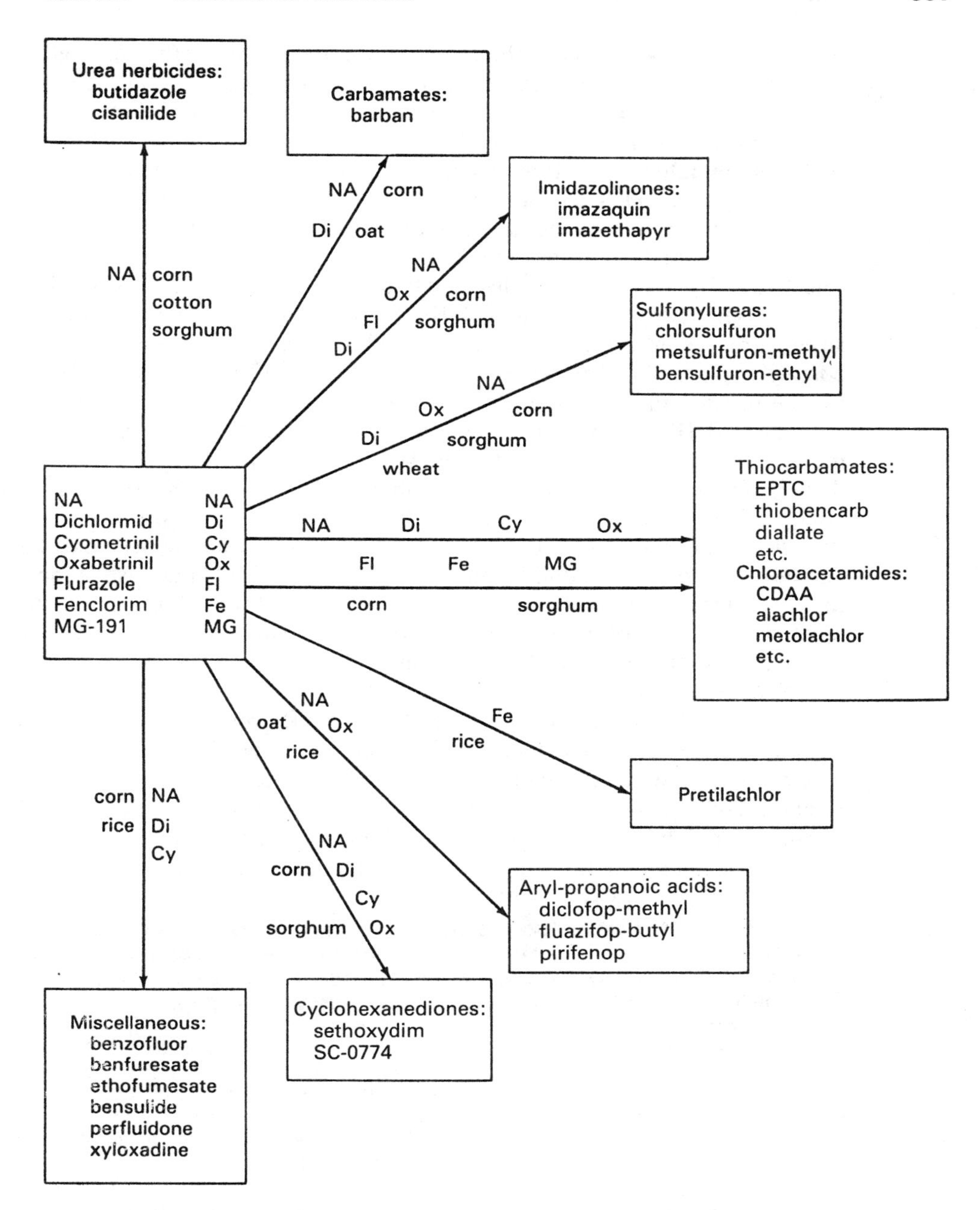

Figure 17.9 A selection of reported safener actions on herbicide toxicities in different crop plants.

When arranged in a metabolic sequence, the activity of the entire glutathione pathway increases, along with the amount of soluble protein and the activities of some soluble enzymes. The safener also seems to change the formation of membrane lipids and might thereby lead to increased herbicide uptake. The increase in GSH content after application of dichlormid or structurally similar safeners in corn is very strong (+200–400%), but may not be detected at all with other safeners and in other crop plants. The increase in GST activity seems to be of more widespread importance, especially when conjugation with the relevant herbicide is considered. The data in Table 17.3 clearly illustrate that it can be misleading to test GST activity with a model substrate such as CDNB (1-chloro-2,4-dinitrobenzene). In contrast, large increases in conjugation are observed with the herbicide metolachlor as the conjugation partner. As was pointed out in Section 17.3.1, several GST isozymes can be present in the tissue. After the application of safeners to corn, a newly induced GST II can be observed, but the constitutive GST I also increases [108]. A more detailed analysis has differentiated at least three GSTs that are different dimers of the monomeric polypeptide subunits GST A, GST B, GST C, and GST D [109]. Whereas GST I conjugates both atrazine and alachlor, GST II is more specific, and preferentially conjugates the α-chloroacetanilide herbicides alachlor and metolachlor. After application of dichlormid, a 3-4 fold increase of the m-RNA for the major GST A has been observed. The effect of the safener can therefore be located at the level of transcription. Moreover, a distinction can be made between cytosolic and microsomal GSTs; both increase after the application of dichlormid [95] or MG-191 [110].

In accordance with the increased levels of different GSTs, an increased rate of detoxification of metolachlor in safener-treated sorghum has been reported [97]. The proportional increases of the GSTs and the rates of detoxification are positively correlated with the safening activities of the different safeners. Further circumstantial evidence for safener action via increased GSH-conjugation can be seen in the fact that the highly safened thiocarbamate and α-chloroacetamide

TABLE 17.3 EFFECTS OF DIFFERENT SAFENERS ON REDUCED GLUTATHIONE CONTENT AND GLUTATHIONE-*S*-TRANSFERASE (GST) ACTIVITY WITH THE SUBSTRATES CDNB (1-CHLORO-2,4-DINITROBENZENE) AND METOLACHLOR IN EXTRACTS OBTAINED FROM UNTREATED AND SAFENER-TREATED SORGHUM SHOOTS [96].

Parameter	Safener applied			
	Oxabetrinil	Dichlormid	NA	Flurazole
Glutathione	Increase or decrease content (%)			
	+ 12	+ 31	− 6	+ 58
Glutathione-*S*-transferase activity with:	GST-activity (treated/control)			
CDNB	1.73	1.82	1.90	1.96
metolachlor	19.7	5.1	16.7	29.7

herbicides are detoxified by glutathione conjugation, and that GSTs in corn seedlings make up approximately 1–2 percent of the soluble protein. However, other researchers have been unable to find an increased rate of herbicide breakdown that can be correlated with the level of safening [91, 107, 111]. One difference between these experiments was the type of incubation; whereas the data with a positive response were obtained after incubation with tissue pieces for only a few hours, those incubations which often failed to show any increase of the herbicide detoxification rate in safener-pretreated plants were more practically oriented and were conducted over a longer time span. In summary, although safened corn plants clearly contain an increased capacity for glutathione conjugation of herbicides, it has not been possible thus far to obtain convincing and independent experimental evidence that this is the mechanism of safening for the different herbicide-safener combinations in Groups I and II under varying field conditions. The mechanism of safening via increased hydroxylation and increased glucosylation (Group III safeners, but also some "minor" activities of Group I and II safeners) [101, 104, 105] is, however, clearly related to increased herbicide metabolism.

The search for the mode of action of Group I and Group II safeners is severely hindered by our lack of knowledge on the mode of action of thiocarbamate and α-chloroacetamide herbicides [112]. There is indirect evidence that cytochrome P-450 monooxygenases are involved in the activity of these herbicides [113]. Several of the inhibitory activities observed (those in the biosynthesis of lipids, gibberellic acid, and lignins) might involve P-450 monooxygenases. There are some remarkable similarities between thiocarbamate and α-chloroacetamide herbicides, herbicide safeners, and some synergists that affect monoxygenases (Figure 17.10):

1. Several safeners and the herbicide metolachlor have been shown to act as insecticide synergists, similar to the monooxygenase inhibitor PBO (piperonyl butoxide) [114].
2. The thiocarbamate CDAA can act as its own safener [30].
3. Safeners increase tolerance toward sulfonylurea and arylpropanoic acid herbicides by inducing increased rates of herbicide monooxygenation [101, 105].
4. Thiocarbamate herbicides can act as safeners for sulfonylurea herbicides by inducing increased sulfonylurea monooxygenation [15, 115].
5. The suicide inhibitors of cytochrome P-450 monooxygenation, PBO and ABT (1-aminobenzotriazole) (see Section 17.3.2), decrease herbicidal activity in some systems [60, 113] but lead to an increase in other systems [54–57, 61].
6. PBO and ABT both induce increased concentrations of GSH in oat roots [116].
7. Dichlormid and cyometrinil, and also the monooxygenase inhibitor tetcyclacis, inhibit the monooxygenation of chlortoluron in wheat cell suspension cultures immediately after addition, but stimulate the oxygenation 2 days after addition. With ABT and paclobutrazol, partial reversal of the original inhibition is observed after 2 days [60].

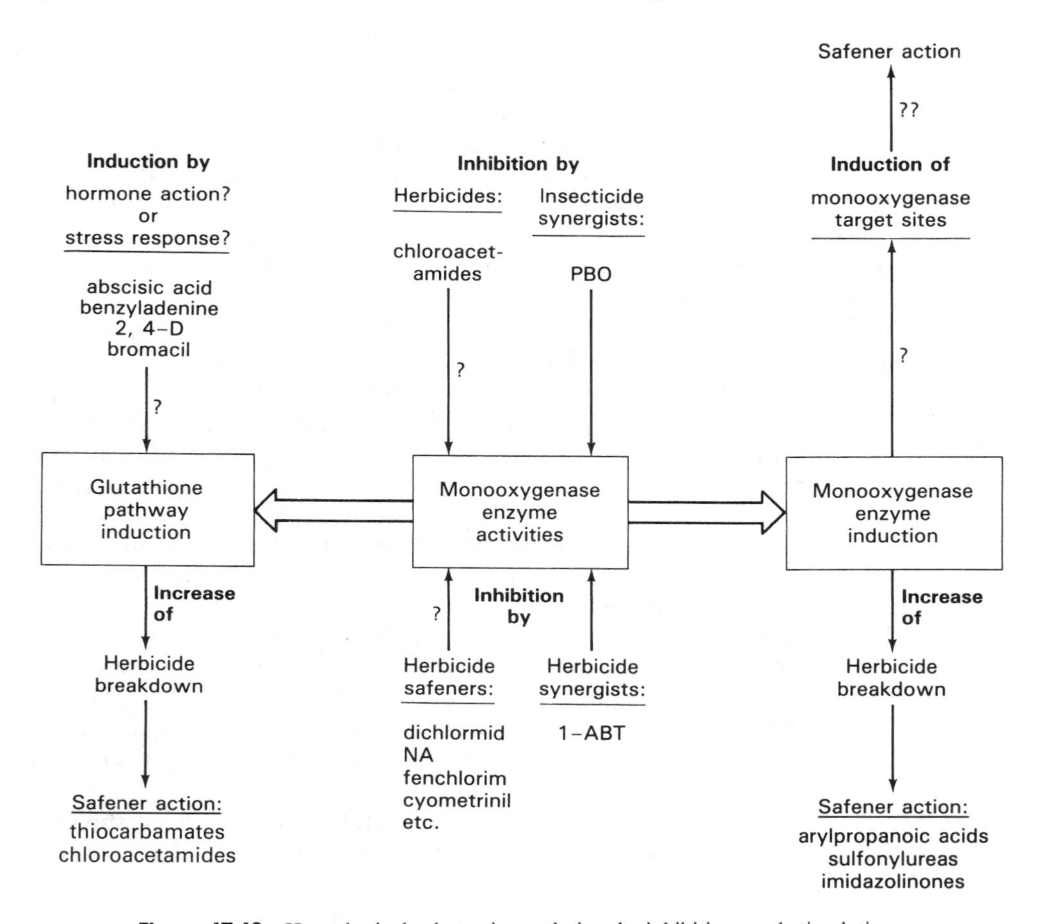

Figure 17.10 Hypothetical scheme interrelating the inhibitions and stimulations caused by safeners and synergists in the glutathione-S-transferase and monooxygenase pathways.

It is well known from extensive studies with tissues from insects and mammals that inhibitors of monooxygenases also act (after some delay) as inducers of monooxygenases [50]. When the information listed above under points 1 to 7 is investigated for an underlying pattern, trying to bring it into a logical context and arrange it in a physiological sequence, the scheme presented in Figure 17.10 can be deduced [116]. An important first step in explaining metabolic stimulations and inductions is to find a primary metabolic trigger or interference site. No information is presently available on primary effects of herbicide safeners on the metabolism of plants. We must therefore try to draw indirect conclusions from the available information. In looking for a mechanism that could possibly trigger a stimulating or inducing activity, there are two possibilities: either a "direct" stimulation, which would be a hormone-type action, or an "indirect" stimulation, which would be an endogenous (over)compensation reaction following original inhibition. Hormone-type inductions are well known and require a specific receptor on a regulatory protein. Such a protein is unlikely to recognize (bind) the many different safener structures. Regulatory adaptations after an original inhibition (disturbance) are, on the contrary, an integral part of plant life, and also good candidates for the action of safeners (Figure 17.11). One well-known example is the "shade-type adaptation" triggered in plants by sublethal photosynthesis inhibition (Chapter 16). Plants must constantly adapt to changing environmental conditions and should, to this end, have internal sensing systems that "measure" the important environmental parameters and subsequently trigger/control the required biochemical composition and metabolic activity.

It has been suggested that monooxygenase enzymes may be likely candidates for such a control system that is affected by safeners [113–116]. Monooxygenases are most important during early seedling life, which is also the time span that is relevant for safener action [51]. They are largely suppressed during cell vacuolization and tissue maturation. Their natural functions include secondary plant

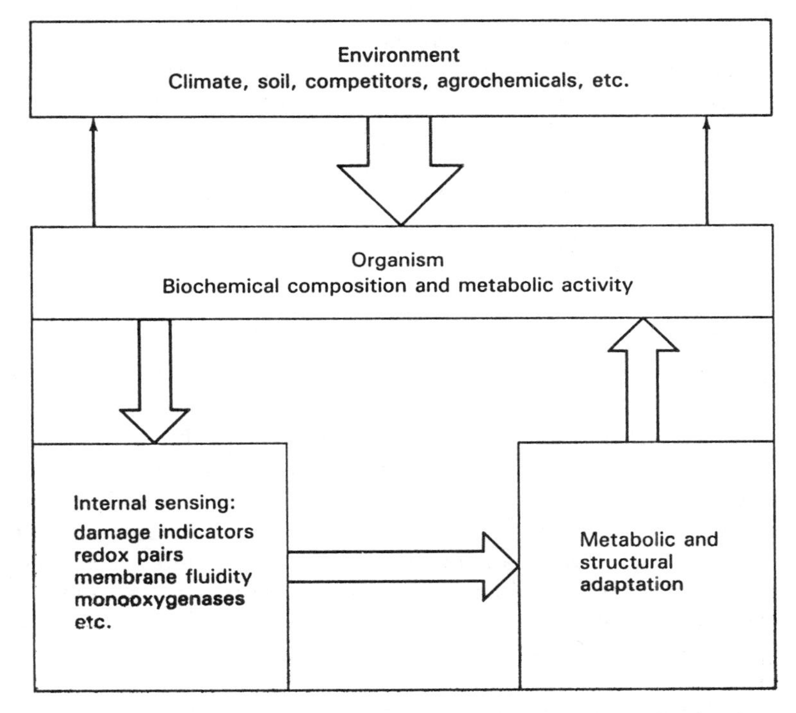

Figure 17.11 Scheme of the strategy of environmental adaptation in plants. Herbicide safeners appear to interfere with a step of the internal sensing system, most likely a monooxygenase enzyme.

metabolism (molecular specialization of flavonoids, lipids, etc.), xenobiotic detoxification (Section 6.2.2), plant hormone syntheses and degradations (gibberellic acid, abscisic acid), and lignification. Monooxygenases are a family of enzymes that are species- and tissue-specific, and might fulfill a control function through induction and adaptation upon original inhibition. However, it is not clear which type(s) of monooxygenase reaction(s) might be inhibited by safeners and cause the safening effect. The presently published cases of inhibition [60, 114] can of course only serve as models lending support to the hypothesis outlined here.

According to the relationships and responses described, it can now be understood how the same compound can be a synergist (inhibitory phase) or a safener (stimulatory phase; Figure 17.10). In any one tissue, however, not all the possible interactions may be realized because of the specific responses in different tissues and species. Monooxygenases frequently detoxify herbicides and other xenobiotics, but toxification has also been discussed, in the case of the thiocarbamates [88]. Inhibition of monooxygenase detoxification will cause synergism, as has been shown for ABT and a number of other monooxygenase inhibitors, including growth regulators and fungicides (Section 17.3.2). The delayed stimulatory response can be the basis for an increased rate of detoxification, as shown for sulfonylureas [101] and chlortoluron [60]. The fact that thiocarbamates and α-chloroacetamides can inhibit monooxygenases qualifies them as potential safeners, as has indeed been observed in some instances [30, 91]. The range of monooxygenases inhibited and the specific monooxygenases induced after the original inhibition is not known. However, it must be assumed that only a few monooxygenase enzymes are involved, and that different tissues and plant species react differently; the restriction of specific monooxygenase isozymes with specific inhibitor sensitivities to a particular tissue or plant species has been well documented [51, 52].

In the scheme and hypothesis that are presented here it is suggested that induction of the glutathione pathway is also caused by monooxygenase inhibition. This conclusion is supported by the many individual increases observed after safener applications, which suggest the existence of a ''master switch'' reaction, and also by the increased GSH content in plants treated with ABT and PBO [116]. The possibility of ''direct'' stimulation of the glutathione pathway by a hormone-type interaction cannot be excluded for some plant hormone-type inducers [117]. These compounds are not safeners, however. The increased capacity for glutathione conjugation in safener-treated sorghum and corn plants seems to explain the safener action in several reported cases, or at least to contribute significantly to the safener action. As already mentioned, however, there are several reports in which increased rates of herbicide conjugation or breakdown were not observed in safened plants [88, 89, 111]. This situation raises the question of whether the known herbicide safeners all have a similar mode of action. One way of avoiding the resulting contradictions is to consider a different and/or additional mechanism of safening. A tolerance mechanism by target site gene multiplication and/or increased target site gene expression has often been described in plants or tissue cultures selected for increased herbicide tolerance. For example, a 15- to 20-fold overproduction of EPSP synthase has been reported to convey glyphosate tolerance to selected cell lines of petunia and *Catharanthus* (see Section 13.2). A similar mechanism is found in glufosinate tolerance in alfalfa cell line selections [118]. If the induction and/or increase of monooxygenase enzymes were to include the herbicide target sites which, according to this hypothesis, are also monooxygenase enzymes, then increased tolerance would be the direct consequence.

Hendry [51] stated that "xenobiotics induce a cytochrome P-450 with narrow or broad specificity to structurally related xenobiotics." This response can explain the sometimes very close structural similarity between herbicides and safeners [30, 87]. It also explains why herbicides can sometimes be safeners, and it can be used as a guide in the synthesis of new safeners [111].

The experimental safener OTC (Figure 17.8) can be compared with the structurally similar thioproline (L-thiazolidine-4-carboxylic acid) and flurazole [119]. OTC and thioproline increase the thiol content of excised corn roots and act as safeners for the herbicides tridiphane and alachlor in sorghum [120]. Of the three safener compounds listed above, thioproline and flurazole are phytotoxic to sorghum roots above 10 μM.

REFERENCES

1. Morse, P. M. 1978. "Some comments on the assessment of joint action of herbicide mixtures." *Weed Sci.*, 26, 58–71.
2. Fedtke, C., G. Marzolph, W. Lunkenheimer, and W. Zeck. 1988. "Picolinic acid t-butylamide, a synergist for metribuzin." In *Proc. EWRS Symp., Factors affecting herbicidal activity and selectivity,* Wageningen, Netherlands: Ponsen & Looijen, pp. 133–38.
3. Matsunaka, S. 1968. "Propanil hydrolysis: inhibition in rice plants by insecticides." *Science,* 160, 1360–62.
4. Boydston, R. A. and F. W. Slife. 1987. "Postemergence control of giant foxtail (*Setaria faberi*) in corn (*Zea mays*) with tridiphane and triazine combinations." *Weed Sci.,* 35, 103–08.
5. Lamoureux, G. L., and D. G. Rusness. 1986. "Tridiphane (2-(3,5-dichlorophenyl)-2-(2,2,2-trichloroethyl)oxirane) an atrazine synergist: enzymic conversion to a potent glutathione-*S*-transferase inhibitor." *Pestic. Biochem. Physiol.*, 26, 323–42.
6. Nash, R. G. 1981. "Phytotoxic interaction studies—techniques for evaluation and presentation of results." *Weed Sci.,* 29, 147–55.
7. Richer, D. L. 1987. "Synergism—a patent view." *Pestic. Sci.,* 19, 309–15.
8. Rummens, F. H. A. 1975. "An improved definition of synergistic and antagonistic effects." *Weed Sci.,* 23, 4–6.
9. Streibig, J. C. 1981. "A method for determining the biological effect of herbicide mixtures." *Weed Sci.*, 469–73.
10. Pike, D. J., and A. M. Hasted. 1987. "Experimental design and response surface analysis of pesticide trials." *Pestic. Sci.,* 19, 297–307.

11. Limpel, L. E., P. H. Schuldt, and D. Lamont. 1962. "Weed control by dimethyl tetrachloroterephthalate alone and in certain combinations." *Proc. Northeastern Weed Control Conf.*, 16, 48–53.

12. Colby, S. R. 1967. "Calculating synergistic and antagonistic responses of herbicide mixtures." *Weeds*, 15, 20–22.

13. Akobundu, I. O., R. D. Sweet, and W. D. Duke. 1975. "A method of evaluating herbicide combinations and determining herbicide synergism." *Weed Sci.*, 23, 20–25.

14. Streibig, J. C. 1986. "Joint action of some root-absorbed herbicides in oats (*Avena sativa* L.)." *Weed Res.*, 26, 207–14.

15. Taketomi, I., F. Ishijuma, H. Yoshizawa, and H. Yamamura. 1986. "Study of a paddy herbicide, DPX-84M. Safener-action of MY-93 to DPX-84." *Weed Res.* (Japan), 31 Suppl., 105–6.

16. Ladlie, J. L., W. F. Meggitt, and D. Penner. 1977. "Effect of trifluralin and metribuzin combinations on soybean tolerance to metribuzin." *Weed Sci.*, 25, 88–93.

17. Ryan, G. F., R. N. Rosenthal, and R. L. Berger. 1981. "Napropamide and oryzalin effect on simazine tolerance of ornamental species." *Weed Sci.*, 29, 329–32.

18. O'Donovan, J. T., and G. N. Prendeville. 1976. "Interactions between soil-applied herbicides in the roots of some legume species." *Weed Res.*, 16, 331–36.

19. York, A. C., and F. W. Slife. 1981. "Interaction of buthidazole and acetanilide herbicides." *Weed Sci.*, 29, 461–68.

20. Duncan, D. N., W. F. Meggitt, and D. P. Penner. 1982. "Basis for increased activity from herbicide combinations with ethofumesate applied on sugar beet." *Weed Sci.*, 30, 195–200.

21. Davis, D. G., and K. E. Dusbabek. 1973. "Effect of diallate on foliar uptake and translocation of herbicides in pea." *Weed Sci.*, 21, 16–18.

22. Qureshi, F. A., and W. H. Vanden Born. 1979. "Interaction of diclofop-methyl and MCPA on wild oats." *Weed Sci.*, 27, 202–5.

23. O'Sullivan, P. A., and W. H. Vanden Born. 1980. "Interaction between benzoylprop-ethyl, flamprop-methyl or flamprop-isopropyl and herbicides used for broadleaved weed control." *Weed Res.*, 20, 53–57.

24. Todd, B. G., and E. H. Stobbe. 1980. "The basis for the antagonistic effect of 2,4-D on diclofop-methyl toxicity to wild oat (*Avena fatua*)." *Weed Sci.*, 28, 371–77.

25. Olson, W. O., and J. D. Nalewaja. 1981. "Antagonistic effects of MCPA on wild oat (*Avena fatua*) control with diclofop." *Weed Sci.*, 29, 566–71.

26. Taylor, H. F., M. P. C. Loader, and S. J. Norris. 1983. "Compatible and antagonistic mixtures of diclofop-methyl and flamprop-methyl with herbicides used to control broad-leaved weeds." *Weed Res.*, 24, 185–90.

27. Taylor, H. F., and M. P. C. Loader. 1984. "Metabolism of diclofop-methyl with reference to its interaction with other compounds." *Pestic. Sci.*, 15, 527–28.

28. Jacobsen, A., R. H. Shimabukuro, and C. McMichael. 1985. "Response of wheat and oat seedlings to root-applied diclofop-methyl and 2,4-dichlorophenoxyacetic acid." *Pestic. Biochem. Physiol.*, 24, 61–67.

29. Shimabukuro, R. H., W. C. Walsh, and R. A. Hoerauf. 1986. "Reciprocal antagonism between the herbicides, diclofop-methyl and 2,4-D, in corn and soybean tissue culture." *Plant Physiol.*, 80, 612–17.

30. Ezra, G., D. G. Rusness, G. L. Lamoureux, and G. R. Stephenson. 1985. "The effect of CDAA (*N,N*-diallyl-2-chloroacetamide) pretreatments on subsequent CDAA injury to corn (*Zea mays* L.)." *Pestic. Biochem. Physiol.*, 23, 108–15.

31. Prosch, S. D., and G. Kapusta. 1983. "Ivyleaf morningglory (*Ipomoea hederacea*) control with herbicides in soybeans (*Glycine max*)." *Weed Sci.*, 31, 23–27.

32. Appleby, A. P., and M. Somabhi. 1978. "Antagonistic effect of atrazine and simazine on glyphosate toxicity." *Weed Sci.*, 26, 135–39.

33. Flint, J. L., M. Barrett, and G. L. Olson. 1987. "Contribution of herbicide uptake and translocation to synergism in the response to glyphosate combinations with 2,4-D or dicamba in field bindweed. (*Convolvulus arvensis* L.)." *Abstr. Weed Sci. Soc. Amer.*, 27(183).

34. Merritt, C. R., B. O. Bartlett, and D. F. Wyatt. 1986. "Interaction of ioxynil and mecoprop when applied to *Stellaria media* (L.) Vill." *Weed Res.*, 26, 105–13.

35. Rao, S. R., and T. R. Harger. 1981. "Mefluidide-bentazon interactions on soybeans (*Glycine max*) and red rice (*Oryza sativa*)." *Weed Sci.*, 29, 208–12.

36. Sharma, M. P., F. A. Qureshi, and W. H. Vanden Born. 1982. "The basis for synergism between barban and flamprop on wild oat (*Avena fatua*)." *Weed Sci.*, 30, 147–52.

37. Chow, P. N. P., and H. F. Taylor. 1983. "Wild oat herbicide studies. 3. Physiological and biochemical bases for interaction of barban and growth regulator herbicides in wild oat." *J. Agric. Food Chem.*, 31, 575–78.

38. Sterrett, J. P., and R. H. Hodgson. 1983. "Enhanced response of bean (*Phaseolus vulgaris*) and Canada thistle (*Cirsium arvense*) to bentazon or glyphosate by gibberellin." *Weed Sci.*, 31, 396–401.

39. Shimabukuro, R. H., W. C. Walsh, and R. A. Hoerauf. 1979. "Metabolism and selectivity of diclofop-methyl in wild oat and wheat." *J. Agric. Food Chem.*, 27, 615–23.

40. Donald, W. W., and R. H. Shimabukuro. 1980. "Selectivity of diclofop-methyl between wheat and wild oat: growth and herbicide metabolism." *Physiol. Plant.*, 49, 459–64.

41. Ezra, G., J. H. Dekker, and G. R. Stephenson. 1985. "Tridiphane as a synergist for herbicides in corn (*Zea mays*) and proso millet (*Panicum miliaceum*)." *Weed Sci.*, 33, 287–90.

42. Burroughs, F. G. 1985. "Field and laboratory evaluations of the herbicidal efficacy, uptake and translocation of tridiphane and its effects on atrazine uptake, translocation, and metabolism." *Weed Abstr.*, 34(3097), 338.

43. Sheppard, B. R., P. J. McCall, L. E. Stafford, and P. S. Zorner. 1987. "Factors regulating the physiological interaction of tridiphane and atrazine in controlling panicoid grasses in corn." *Pestic. Sci.*, 19, 331–32.

44. Zorner, P. S. 1987. "In vivo kinetics of glutathione-*S*-transferase inhibition by tridiphane." *Weed Abstr.*, 36(3138), 366.

45. Boydston, R. A., and F. W. Slife. 1986. "Alteration of atrazine uptake and metabolism by tridiphane in giant foxtail (*Setaria faberi*) and corn (*Zea mays*)." *Weed Sci.*, 34, 850–58.

46. Frear, D. S., and H. R. Swanson. 1970. "Biosynthesis of *S*-(4-ethylamino-6-isopropylamino-2-*s*-triazino)glutathione: partial purification and properties of a glutathione-*S*-transferase from corn." *Phytochemistry*, 9, 2123–32.

47. Guddewar, M. B., and W. C. Dautermann. 1979. "Purification and properties of a glutathione-*S*-transferase from corn which conjugates *s*-triazine herbicides." *Phytochemistry,* 18, 735–40.
48. Mozer, T. J., D. C. Tiemeier, and E. G. Jaworski. 1983. "Purification and characterization of corn glutathione-*S*-transferase." *Biochemistry,* 22, 1068–72.
49. Zorner, P. S., L. D. Markley, L. E. Stafford, P. G. Ray, and J. M. Renga. 1986. "Physiological mechanism responsible for triazine synergism with pro-epoxides of tridiphane herbicide." 6th IUPAC Pestic. Chem. Congr., Abstr. No. 7D-09.
50. Hodgson, E. 1985. "Microsomal mono-oxygenases." In *Comprehensive Insect Physiology, Biochemistry, and Pharmacology, Vol. 11.* Oxford: Pergamon Press, pp. 226–321.
51. Hendry, G. 1986. "Why do plants have cytochrome P-450? Detoxification versus defense." *New Phytol.,* 102, 239–47.
52. O'Keefe, D. P., J. A. Romesser, and K. J. Leto. 1987. "Plant and bacterial cytochromes P-450: involvement in herbicide metabolism." In *Phytochemical Effects of Environmental Compounds,* J. A. Saunders, L. Kosak-Chaming, E. E. Conn, eds. Plenum Publ. Corp. pp. 151–73.
53. Butt, V. S., and C. J. Lamb. 1981. "Oxygenases and the metabolism of plant products." In *The Biochemistry of Plants, Vol. 7, Secondary Plant Products,* E. E. Conn, ed. New York: Academic, pp. 627–65.
54. Cabanne, F., D. Huby, R. Scalla, and F. Durst. 1987. "Effect of the cytochrome P-450 inactivator 1-aminobenzotriazole on the metabolism of chlortoluron and isoproturon in wheat." *Pestic. Biochem. Physiol.,* 28, 371–80.
55. Gonneau, M., B. Pasquette, F. Cabanne, and R. Scalla. 1988. "Metabolism of chlortoluron in tolerant species: possible role of cytochrome P-450 mono-oxygenases." *Weed Res.,* 28, 19–25.
56. Cole, D. J., and W. J. Owen. 1987. "Influence of monooxygenase inhibitors on the metabolism of the herbicides chlortoluron and metolachlor in cell suspension cultures." *Plant Sci.,* 50, 13–20.
57. Gonneau, M., B. Pasquette, F. Cabanne, R. Scalla, and B. C. Loughman. 1987. "Transformation of phenoxyacetic acid and chlortoluron in wheat, barren brome, cleavers, and speedwell. Effects of an inactivator of monooxygenases." Proc. Brit. Crop Prot. Conf.—Weeds, Vol. 1. Thornton Heath, U.K.: BCPC Publ., pp. 329–36.
58. Kemp, M. S., and J. C. Caseley. 1987. "Synergistic effects of 1-aminobenzotriazole on the phytotoxicity of chlortoluron and isoproturon in a resistant population of blackgrass (*Alopecurus myosuroides*)." Proc. Brit. Crop Prot. Conf.—Weeds, Vol. 3. Thornton Heath, U.K.: BCPC Publ., pp. 895–99.
59. Rubin, B., J. R. C. Leavitt, D. Penner, and A. W. Saettler. 1980. "Interactions of antioxidants with ozone and herbicide stress." *Bull. Environ. Contam. Toxicol.,* 25, 623–29.
60. Canivenc, M. C., B. Cagnac, F. Cabanne, and R. Scalla. 1988. "Manipulation of chlortoluron fate in wheat cells." In Proc. EWRS Symp. Factors Affecting Herbicidal Activity and Selectivity. Wageningen, Netherlands: Ponsen & Looijen, pp. 115–20.
61. Kemp, M. S., L. W. Newton, and J. C. Caseley. 1988. "Synergistic effects of some P-450 oxidase inhibitors on the phytotoxicity of chlortoluron in a resistant population

of blackgrass (*Alopecurus myosuroides*)." In Proc. EWRS Symp. Factors Affecting Herbicidal Activity and Selectivity. Wageningen, Netherlands: Ponsen & Looijen, pp. 121–26.

62. Reichhardt, D., A. Simon, F. Durst, J. M. Mathews, and P. R. Ortiz de Montellano. 1982. "Autocatalytic inactivation of plant cytochrome P-450 enzymes: selective inactivation of cinnamic-4-hydroxylase from *Helianthus tuberosus* by 1-aminobenzotriazole." *Arch. Biochem. Biophys.*, 216, 522–29.
63. Oliver, D. 1985. "Identification and control of morning glory species in soybeans." *Weeds Today*, 16, 17–20.
64. Hammond, R. B. 1983. "Phytotoxicity of soybean caused by the interaction of insecticide-nematicides and metribuzin." *J. Econ. Entomol.*, 76, 17–19.
65. Klamroth, E. E., C. Fedtke, and W. C. Kühbauch. 1989. "Mechanism of synergism between metribuzin and MZH 2091 on ivyleaf morningglory (*Ipomoea hederacea*)." *Weed Sci.*, 37, 517–20.
66. Fedtke, C. 1986. "Selective metabolism of triazinone herbicides." *Pestic. Sci.*, 17, 65–66.
67. Fedtke, C. 1991. "Deamination of metribuzin in tolerant and susceptible soybean (*Glycine max*) cultivars." *Pestic. Sci.*, 31, 175–83.
68. Fedtke, C., and R. H. Strang. 1990. "Synergistic activity of the herbicide safener dichlormid with herbicides affecting photosynthesis." *Z. Naturforsch.*, 45c, 565–67.
69. Hatzios, K. K., and D. Penner. 1985. "Interactions of herbicides with other agrochemicals in higher plants." *Rev. Weed Sci.*, 1, 1–63.
70. Chang, F.-Y., L. W. Smith, and G. R. Stephenson. 1971. "Insecticide inhibition of herbicide metabolism in leaf tissues." *J. Agric. Food Chem.*, 19, 1183–86.
71. Bowling, C. C., and H. R. Hudgins. 1966. "The effect of insecticides on the selectivity of propanil in rice." *Weeds*, 14, 94–95.
72. Bowling, C. C., and W. T. Flinchum. 1968. "Interaction of propanil with insecticides applied as seed treatments on rice." *J. Econ. Entomol.*, 61, 67–69.
73. El-Refai, A. R., and M. Mowafy. 1973. "Interaction of propanil with insecticides absorbed from soil and translocated into rice plants." *Weeds*, 21, 246–48.
74. Yukimoto, M., and M. Oda. 1973. "Phytotoxicity on rice plant of herbicide propanil in combination with carbamate insecticides." *Weed Res.* (Japan), 16, 28–32.
75. Campbell, J. R., and D. Penner. 1982. "Enhanced phytotoxicity of bentazon with organophosphate and carbamate insecticides." *Weed Sci.*, 30, 324–26.
76. Goto, T., A. Kamachi, and H. Hayakawa. 1987. "Herbicidal composition." *Japanese Patent Application Sho* 62-85880, 8.12.
77. Krauskopf, B., C. Fedtke, and D. Feucht. 1989. "Uptake, translocation, and metabolism of mefenacet in rice and *Echinochloa*, and the mechanism of the synergistic interaction with edifenphos." Proc. 4th Mediterranean EWRS Symp., Problems of Weed Control in Fruit, Horticultural Crops, and Rice. pp. 386–93.
78. Miaullis, B., G. J. Nohynek, and F. Pereiro. 1982. "R-33865: A novel concept for extended weed control by thiocarbamate herbicides." Proc. Brit. Crop Prot. Conf.—Weeds, Vol. 1. Croydon, U.K.: BCPC Publ., pp. 205–10.
79. Obrigawitch, T., F. W. Roeth, A. R. Martin, and R. G. Wilson. 1982. "Addition of R-33865 to EPTC for extended herbicide activity." *Weed Sci.*, 30, 417–22.

80. Dawson, J. H. 1972. "Inhibitor of microbial enzyme prolongs dodder control with chlorpropham." *Weed Sci.*, 20, 465–67.

81. Shaaltiel, Y., and J. Gressel. 1987. "Biochemical analysis of paraquat resistance in *Conyza* leads to pinpointing synergists for oxidant generating herbicides." In *Pesticide Science and Biotechnology,* R. Greenhalgh, T. R. Roberts, eds. Oxford: Blackwell Scientific Publ., pp. 183–86.

82. Gressel, J., and Y. Shaaltiel. 1988. "Biorational herbicide synergists." In *Biotechnology for Crop Protection,* ACS Symp. Series. 379, 4–24.

83. Pallos, F. M., J. E. Casida, eds. 1978. *Chemistry and Action of Herbicide Antidotes.* New York: Academic, 171 pp.

84. Blair, A. M., C. Parker, and L. Kasasian. 1976. "Herbicide protectants and antidotes—a review." *PANS,* 22, 65–74.

85. Hatzios, K. K. 1984. "Biochemical and physiological mechanisms of action of herbicide antidotes." In *Biochemical and Physiological Mechanisms of Herbicide Action,* S. O. Duke, ed. Southern Section, American Society of Plant Physiology. pp. 7–30.

86. Fedtke, C. 1985. "Einsatzmöglichkeiten und Wirkungsweise von Herbizid-Antidots." *Z. PflKrankh. PflSchutz.,* 92, 654–64.

87. Parker, C. 1983. "Herbicide antidotes—a review." *Pestic. Sci.,* 14, 40–48.

88. Hatzios, K. K. 1983. "Herbicide antidotes: development, chemistry, and mode of action." *Adv. Agron.,* 36, 265–316.

89. Stephenson, G. R., and G. Ezra. 1983. "Herbicide antidotes—a new era in selective chemical weed control." In *Plant Growth Regulating Chemicals, Vol. II,* L. G. Nickell, ed. Boca Raton, Fla.: CRC Press, pp. 193–231.

90. Hatzios, K. K., and R. E. Hoagland, eds. 1989. *Crop Safeners for Herbicides—Development, Uses, and Mechanisms of Action.* San Diego: Academic Press. 400 pp.

91. Stephenson, G. R., and G. Ezra. 1987. "Chemical approaches for improving herbicide selectivity and crop tolerance." *Weed Sci.,* 35 (Suppl. 1), 24–27.

92. Adams, C. A., E. Blee, and J. E. Casida. 1983. "Dichloroacetamide herbicide antidotes enhance sulfate metabolism in corn roots." *Pestic. Biochem. Physiol.,* 19, 350–60.

93. Lay, M. M., and J. E. Casida. 1976. "Dichloroacetamide antidotes enhance thiocarbamate sulfoxide detoxification by elevating corn glutathione-*S*-transferase activity." *Pestic. Biochem. Physiol.,* 6, 442–56.

94. Breaux, E. J., J. E. Patonella, and E. F. Sanders. 1987. "Chloroacetanilide herbicide selectivity: analyses of glutathione and homoglutathione in tolerant, susceptible, and safened seedlings." *J. Agric. Food Chem.,* 35, 474–78.

95. Kömives, T., V. A. Kömives, M. Balazs, and F. Dutka. 1985. "Role of glutathione-related enzymes in the mode of action of herbicide antidotes." Proc. Brit. Crop Prot. Conf.—Weeds, Vol.3. Croydon, U.K.: BCPC Publ., pp. 1155–62.

96. Gronwald, J. W., E. P. Fuerst, C. V. Eberlein, and M. A. Egli. 1987. "Effect of herbicide antidotes on glutathione content and glutathione-*S*-transferase activity of sorghum shoots." *Pestic. Biochem. Physiol.,* 29, 66–76.

97. Fuerst, E. P., and J. W. Gronwald. 1986. "Induction of rapid metabolism of metolachlor in sorghum (*Sorghum bicolor*) shoots by CGA-92194 and other antidotes." *Weed Sci.,* 34, 354–61.

98. Barret, M. 1989. "Protection of corn (*Zea mays*) and Sorghum (*Sorghum bicolor*) from imazethapyr toxicity with antidotes." *Weed Sci.*, 37, 293–301.

99. Polge, N. D., A. D. Dodge, and J. C. Caseley. 1987. "Biochemical aspects of safener action: effects on glutathione, glutathione-*S*-transferase and acetohydroxyacid synthetase in maize." Proc. Brit. Crop Prot. Conf.—Weeds, Vol. 3. Thornton Heath, U.K.: BCPC Publ., pp. 1113–20.

100. Rubin, B., and J. E. Casida. 1985. "R-25788 effects on chlorsulfuron injury and acetohydroxyacid synthase activity." *Weed Sci.*, 33, 462–68.

101. Sweetser, P. B. 1985. "Safening of sulfonylurea herbicides to cereal crops: mode of herbicide antidote action." Proc. Brit. Crop Prot. Conf.—Weeds, Vol. 3. Croydon, U.K.: BCPC Publ., pp. 1147–54.

102. Fonné-Pfister, R., and K. Kreuz. 1990. "Ring-methyl hydroxylation of chlortoluron by an inducible cytochrome-P450-dependent enzyme from maize." *Phytochemistry*, 29, 2793–96.

103. Ezra, G., J. Gressel, and H. M. Flowers. 1983. "Effects of the herbicide EPTC and the protectant DDCA on incorporation and distribution of 2-14C-acetate into major lipid fractions of maize cell suspension cultures." *Pestic. Biochem. Physiol.*, 19, 225–34.

104. Lamoureux, G. L., and D. G. Rusness. 1991. "The effect of BAS 145,138 safener on chlorimuron ethyl metabolism and toxicity in corn (*Zea mays* L.)." *Z. Naturforsch.*, 46c, 882–86.

105. Kreuz, K., J. Gaudin, J. Stingelin, and E. Ebert. 1991. "Metabolism of the aryloxyphenoxypropanoate herbicide, CGA 184927, in wheat, barley, and maize: differential effects of the safener, CGA 185072." *Z. Naturforsch.*, 46c, 901–05.

106. Jackson, L. A., G. Kapusta, and J. H. Yopp. 1985. "Early growth effects of flurazole as a safener against acetochlor in grain sorghum." *Weed Sci.*, 33, 740–45.

107. Ekler, Z., and G. R. Stephenson. 1987. "Comparative action of three herbicide antidotes in corn and sorghum." Abstr. Weed Sci. Soc. Amer. 27(181).

108. Shah, D. M., C. M. Hironka, R. C. Wiegand, E. I. Harding, G. G. Krivi, and D. C. Tiemeier. 1986. "Structural analysis of a maize gene coding for glutathione-*S*-transferase involved in herbicide detoxification." *Plant Molec. Biol.*, 6, 203–11.

109. Wiegand, R. C., D. M. Shah, T. J. Mozer, E. I. Harding, J. Diaz-Collier, C. Saunders, E. G. Jaworski, and D. C. Tiemeier. 1986. "Messenger RNA encoding a glutathione-*S*-transferase responsible for herbicide tolerance in maize is induced in response to safener treatment." *Plant Molec. Biol.*, 7, 235–43.

110. Kömives, T., and F. Dutka. 1988. "Biochemical mode of action of the safener MG-191." In Proc. EWRS Symp. Factors Affecting Herbicidal Activity and Selectivity. Wageningen, Netherlands: Ponsen & Looijen, pp. 381–85.

111. Jackson, L. A., J. H. Yopp, and G. Kapusta. 1986. "Absorption and distribution of flurazole and acetochlor in grain sorghum." *Pestic. Biochem. Physiol.*, 25, 373–80.

112. Fuerst, E. P. 1987. "Understanding the mode of action of the chloroacetamide and thiocarbamate herbicides." *Weed Technol.*, 1, 270–77.

113. Fedtke, C. 1987. "Physiological activity spectra of existing graminicides and the new herbicide 2-(2-benzothiazolyl-oxy)-*N*-methyl-*N*-phenylacetamide (mefenacet)." *Weed Res.*, 27, 221–28.

114. Ketchersid, M. L., F. W. Plapp, and M. G. Merkle. 1985. "Sorghum (*Sorghum bicolor*) seed safeners as insecticide synergists." *Weed Sci.*, 33, 774–78.

115. Yuyama, T., P. B. Sweetser, R. C. Ackerson, and S. Takeda. 1986. "Study of a new paddy herbicide, DPX-F5384. Safening action of thiocarbamate herbicides." *Weed Res.* (Japan) 31(Suppl.)101–02.

116. Fedtke, C., and A. Trebst. 1987. "Advances in understanding herbicide modes of action." In *Pesticide Science and Biotechnology*, R. Greenhalgh, T. R. Roberts, eds. Oxford: Blackwell Scientific Publ., pp. 161–68.

117. Fedtke, C. 1981. "Wirkung von Herbiziden und Herbizid-Antidots auf den Glutathion-Gehalt in Mais- und Sojabohnenwurzeln." *Z. PflKrankh. PflSchutz., Sonderheft*, IX, 141–46.

118. Fraley, R., G. Kishore, C. Gasser, S. Padgette, R. Horsch, S. Rogers, G. Della-Cioppa, and D. Shah. 1987. "Genetically engineered herbicide tolerance—technical and commercial considerations." Proc. Brit. Crop Prot. Conf.—Weeds, Vol. 2. Thornton Heath, U.K.: BCPC Publ., pp. 463–71.

119. Hilton, J. L., and P. Pillai. 1988. "Thioproline protection of crops against herbicide toxicity." *Weed Technol.*, 2, 72–76.

120. Hilton, J. L., and P. Pillai. 1986. "L-2-Oxothiazolidine-4-carboxylic acid protection against tridiphane toxicity." *Weed Sci.*, 34, 669–75.

Chapter 18

Naturally Occurring Chemicals as Herbicides

18.1 INTRODUCTION

The search for more cost-effective, efficacious, selective, and environmentally safe herbicides has led to new emphasis on several underutilized herbicide discovery strategies. These include utilization of natural products from plants and microorganisms. Natural products offer a huge number of chemical structures and, since many secondary compounds are presumed to have a role in interspecies interactions, many have potential as leads for new herbicides. These chemical compounds are often exotic and difficult to synthesize; however, they may offer leads to more easily synthesized compounds with similar activity. Many of them are highly water soluble compounds (e.g., small peptides) quite unlike the types of compounds that pesticide chemists traditionally synthesize for evaluation. The general topic of natural products as herbicides has been the focus of previous reviews [1, 2], and plant [3–7] and microbial products [8–18] as herbicides have been reviewed separately.

Plants and microorganisms produce many secondary products with biological activity, and each has been the source of patented herbicides. The strategies and rationales with each source are somewhat different. Thus, they will be considered separately in this chapter.

18.2 COMPOUNDS DERIVED FROM PLANTS

Tens of thousands of secondary products of plants have been identified and there are estimates that hundreds of thousands of these compounds exist. There is growing evidence that most of these compounds are involved in the interaction of plants with other species—primarily the defense of the plant against plant pests. Thus, these secondary compounds represent a large reservoir of chemical structures with biological activity. This resource is largely untapped for use as herbicides. Inhibition of plant growth and induction of phytotoxic symptoms by certain plants and their residues is a well-established phenomenon [19–21]. In searching for potential herbicides from plants, screening of compounds known to function in plant-plant interactions is a logical strategy. All plants produce secondary compounds that are phytotoxic to some degree. However, in only relatively few cases has it been clearly established that particular compounds (allelochemicals) provide the producing species with a competitive advantage over other species that are less tolerant to the compound. Many claims of allelopathy are unproven.

18.2.1 Phenolic Compounds from Plants

Due to their ubiquity and moderate phytotoxicity, common cinnamic and benzoic acids, such as ferulic and caffeic acid, have been claimed to be important in allelopathy (e.g., [22]) and in apparent phytotoxicity of crop residues (e.g., [23]). However, significant effects have generally only been demonstrated at relatively high concentrations and/or in bioassays in petri dishes or in hydroponic or sand culture [3, 24]. In soil, these compounds are largely inactive, even at quite high concentrations [24, 25]. Phenolic compounds can be inactivated by sorption to soil [26, 27] and by rapid metabolism to nonphytotoxic compounds by soil microbes [28, 29]. Even juglone (Figure 18.1), a phenolic quinone with a very high level of phytotoxicity, is rapidly degraded by a common soil microbe [29]. Natural phenolic compounds are generally more phytotoxic when foliage-applied than soil-applied [30]. At a rate of 11.2 kg/ha, the benzoic acid, salicylic acid, and the coumarin, umbelliferone, are more phytotoxic as a foliar spray than as preplant incorporated or preemergence herbicides. However, at this rate, there is no difference in phytotoxicity of *p*-benzoic acid applied as a foliar spray or to soil. At this concentration, these compounds are only mildly herbicidal.

There are at least two other reasons that make it doubtful that these compounds cause phytotoxicity in agroecosystems: in many of the studies reporting effects, harsh alkaline extraction of plant residues or soils has generated phenolic acids by hydrolysis of phenolic glucosides and other bound phenolic forms [31]; and, although phenolic acids have been reported to act synergistically (e.g., [22]), more careful studies indicate that they act only additively or antagonistically as plant phytotoxins [32, 33].

Many mechanisms of action have been attributed to simple phenolic acids. As discussed earlier (Chapter 16), these compounds can influence metabolism of IAA.

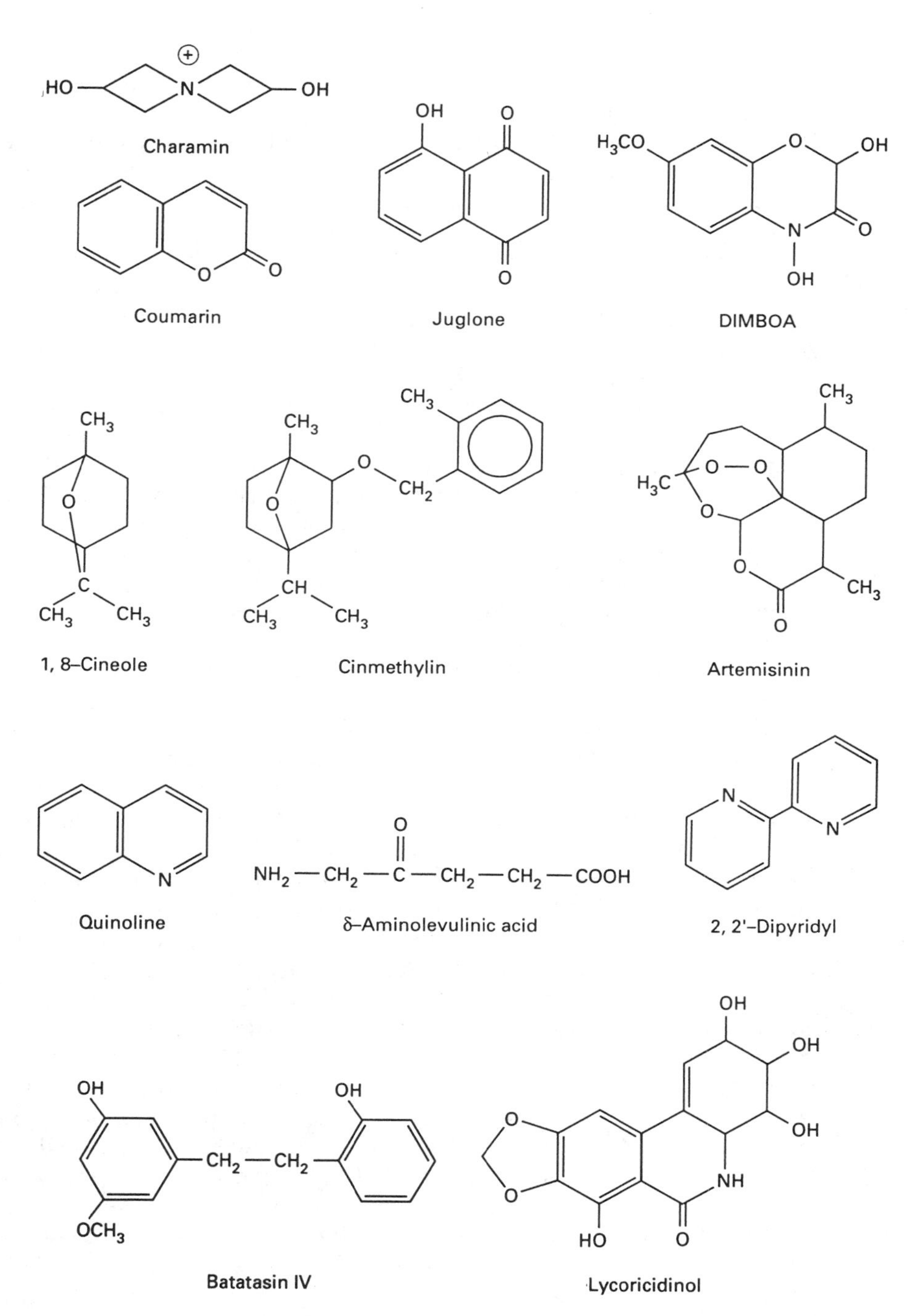

Figure 18.1 Structures of phytotoxic compounds from plants, a herbicide derived from a plant-derived compound (cinmethylin), and a herbicide synergist (2,2′-dipyridyl) of a plant-derived compound.

They can also inhibit photophosphorylation [34], mitochondrial respiration [35, 36], potassium uptake [37], calcium uptake [36], and many other physiological processes [1]. Each compound seems to have a multitude of effects at sufficiently high concentrations. Whether these effects are secondary or primary is often not known.

Simple phenolic acids have no potential as herbicides. Several halogen-substituted benzoic acid derivatives such as dicamba are widely used as herbicides; the commercial examples of these compounds all have auxin-like activity (Chapter 14). It is likely that halogen-substituted cinnamic and benzoic acids have been fully explored for herbicidal potential.

The coumarins are generally significantly more phytotoxic than common phenolic acids or other phenolic compounds [1, 3]. Coumarin itself (Figure 18.1) is the most phytotoxic of the coumarins in most bioassays (e.g., [38]), and some substituted coumarins have been patented as herbicides. Coumarin inhibits microfibril formation [38] and both oxidative and photosynthetic phosphorylation [39].

Some phenolic quinones such as juglone (Figure 18.1) are also highly phytotoxic when compared to other phenolic compounds. Several plant-derived naphthoquinones are as phytotoxic to velvetleaf as juglone [40]. Juglone is a potent inhibitor of plant mitochondrial respiration [41].

The hydroxamic acids of the benzoxazinone-type (4-hydroxy-1,4-benzoxanzin-3-ones), found in many species of the Gramineae, have a high level of biological activity [42], including herbicidal activity [43]. Phytotoxic benzoxazinones derived from plant tissue include 2,4-dihydroxy-1,4(2H)-benzoxazin-3-one (DIBOA) and its breakdown product 2(3H)-benzoxazolinone (BOA), as well as methoxylated DIBOA (DIMBOA, Figure 18.1) [44]. Symptoms of herbicidal injury of these compounds resemble those of photosynthetic inhibitors [43] and these compounds contain the $-N-C=O$ group of many PSII inhibitors (see Chapter 7). However, DIMBOA is only weakly effective (I_{50} = 1 mM) as an inhibitor of coupled photosynthetic electron transport [45]. It has no effect on uncoupled electron transport, but inhibits the ATPase activity of CF_1. In soil, BOA can be transformed by microbes to the diazoperoxide, 2,2-oxo-1,1′-azobenzene, a compound as much as 10-fold more phytotoxic than DIBOA [46].

The phytotoxicity of the flavonoids is low, similar to that of the phenolic acids and, like the phenolic acids, they have been reported to affect many different physiological processes in plants, including oxidative phosphorylation, protoplasmic streaming, photophosphorylation, Ca^{2+}-dependent ATPase, and NADH dehydrogenase [1, 3, 47]. However, unlike the phenolic acids, no halogen-substituted analogs are used (or patented?) as herbicides. A significant number of known or putative phytoalexins are flavonoids. There is evidence that these compounds are often phytotoxic, leading to death of both the pathogen that induces their production and the toxin-producing cells. These compounds are generally only produced in response to infection, and only at the site of infection. Batatasin IV (Figure 18.1), a dihydrostilbene (phenolic analogs of flavonoids), completely stops growth of sorghum roots at a concentration of 10 μM [48]. The related dihydrostilbene from liverworts, lunularic acid, is also a growth inhibitor with effects similar to those of abscisic acid [49]. Of the flavonoids and related compounds, phytoalexins might be the most promising leads for herbicide discovery.

The phloroglucinol derivative, homograndinol, is a growth inhibitor and strong photosynthesis inhibitor. With an I_{50} of 2 μM, it is almost as active as atrazine in inhibiting the Hill reaction in isolated chloroplasts [50].

Plants probably avoid autotoxicity caused by hydroxyphenolic compounds by attachment of acyl, glycosyl, or other groups to the hydroxyl groups of these compounds. This modification is generally a requirement for movement of the compound into the vacuole where most of these compounds are compartmentalized.

18.2.2 Terpenoids from Plants

Secondary compounds from the terpenoid pathway are generally more phytotoxic than phenolic compounds [2]. In some of the most obvious cases of plant/plant allelopathy, terpenoids are involved. For example, the monoterpenoid 1,8-cineole (Figure 18.1) from sagebrush is largely responsible for the lack of vegetation around this species [51]; it is one of the more phytotoxic plant-derived terpenoids [52]. Little is known of the mechanism of action of 1,8-cineole other than that it inhibits respiration [53]. A derivative of 1,8-cineole, with the common name of cinmethylin (Figure 18.1), was advanced to the final phases of commercial development as a herbicide [54]; however, it has not been marketed as of this writing. Cinmethylin is a growth inhibitor with good activity on most grassy weeds and some dicotyledonous weeds. It prevents entry of meristematic cells into mitosis, but the mechanism of action is unknown [55].

Toxaphene, a mixture of nearly 200 chlorinated camphene derivatives, was sold as a herbicide and an insecticide, but was removed from the market in the United States in 1982 by the U.S. Environmental Protection Agency. It had limited use for control of sicklepod in soybeans before it was banned [56]. Although some of the insecticidal compounds of the mixtures were identified, the phytotoxic constituents were never characterized.

Many other terpenoids, particularly sesquiterpenoid lactones, have been demonstrated to have some degree of phytotoxin activity [1, 2, 55, 57]. For instance, the antimalarial sesquiterpenoid lactone, artemisinin (Figure 18.1), was found to inhibit plant growth as effectively as the herbicide cinmethylin [58]. In many respects, its effects are quite similar to those of cinmethylin. Many herbicides increase putrescine synthesis as a general stress effect (see Chapter 15); however, both artemisinin and cinmethylin can reduce free pools of putresine in plant tissues [59]. Exogenously applied putresine will not relieve the growth-retarding effects of either herbicide [58, 59], suggesting that putrescine biosynthesis is not the primary site of action of artemisinin. Artemisinin also lowers putrescine levels in red blood cells and reduces infection by malarial *Plasmodium* [60].

Sesquiterpenoid lactones with an α-methylene-λ-lactone moiety can react with nucleophilic groups of other molecules, such as the thiol group of proteins. Often, small molecules with a thiol group, such as cysteine or glutathione, can reduce or eliminate the phytotoxicity of sesquiterpenoids containing this moiety [52, 61].

Podolactone E, a diterpenelactone, is an effective greening inhibitor, reducing chlorophyll accumulation in barley seedlings by more than 90 percent at 10 μM [62]. At this concentration it also greatly inhibits δ-aminolevulinic acid (ALA) synthesis.

Several important plant hormones and growth regulators (gibberellins, abscisic acid, brassinolides) are products of the terpenoid pathway. Analogs of these compounds have been studied, but none are used as herbicides.

18.2.3 Alkaloids from Plants

Many plant-derived alkaloids are potent inhibitors of mitosis (see Chapter 10 for details on the mechanism of action) [63, 64]. However, most of these have complex structures and, in many cases, are more toxic to mammals than plants. Caffeine, a less complex alkaloid, has been field-tested as a herbicide and was found to provide selective weed control in some crops, but only at extremely high application rates [30, 65]. It reduces extractable amylase activity of the more susceptible *Amaranthus spinosus,* while having no effect on this enzyme from tolerant *Phaseolus mungo* [66]. The reduction was not an in vitro effect on the enzyme. The alkaloid lycoricidinol (Figure 18.1) is a potent inhibitor of ALA synthesis [62] and lycorine is an inhibitor of ascorbate synthesis [67]. Certain *N*-acyl derivatives of loline, a pyrrolizidine alkaloid found in biotically stressed tall fescue (*Festuca arundinacea* Schreb.), are plant growth and seed germination inhibitors with only limited selectivity and about the same herbicidal activity as juglone [68].

One group of herbicides has been developed from a plant alkaloid—the quinolines, including quinchlorac (Chapter 14). These compounds are based on the nicotine precursor quinoline (Figure 18.1), a derivative of tryptophan.

18.2.4 Phytodynamic Compounds from Plants

Plants produce many photodynamic compounds, including quinones, porphyrins, furanocoumarins, thiophenes, and polyacetylenes [69, 70]. These compounds are strongly phytotoxic in the presence of light, provided they can get into the plant cell. These compounds are unlikely to be developed as herbicides because, in the presence of light, they are toxic to all living organisms. However, any plant can be induced to generate phytotoxic levels of photodynamic porphyrin compounds by treating the plant with both ALA (Figure 18.1), a natural porphyrin precursor, and 2,2′-dipyridyl (DP) (Figure 18.1), a synthetic compound. This combination of compounds has been proposed both as a herbicide [71–73] and as an insecticide [74]. Relatively little has been published on its efficacy and mechanism of action.

ALA alone can act as a herbicide; however, compounds that prevent complete conversion to chlorophyll synergize its activity. For example, the effective level of ALA can be reduced considerably by spraying it in combination with DP, which prevents the conversion of Mg-protoporphyrin IX monomethyl ester to protochlorophyllide [71, 75]. A further effect of DP is inhibition of heme synthesis by its iron-chelating properties. Heme is a feedback inhibitor of ALA synthesis [76], so that DP stimulates ALA synthesis via the same mechanism that inhibition of heme synthesis by protoporphyrinogen oxidase inhibitors affects ALA synthesis [77] (Chapter 8). Thus, DP further increases availability of ALA for porphyrin synthesis, as well as qualitatively affecting the porphyrins that accumulate (Table 18.1). Based on the qualitative changes, the more-than-additive effect of ALA and DP could be due to: (a) one or more protochlorophyllide precursors being more herbicidal than protochlorophyllide, (b) photodynamic synergism between tetrapyrroles, or (c) synergism of ALA and DP not based on their effects on tetrapyrroles.

The first possibility seems probable. Although the relative phytotoxicity of each of the chlorophyll intermediates is unknown, one might expect protochlorophyllide to be less phytotoxic because it can be converted to chlorophyll and it is the only intermediate that normally accumulates to high levels. At extremely high levels, however, there may not be sufficient enzymatic capacity to phototransform it to chlorophyllide or to phytyllate the resulting chlorophyllide to chlorophyll. In normal etiolated plants, phototransformation is generally complete after less than a minute of sunlight and phytyllation is complete within 30 min after this. Excess ALA due to ALA plus DP treatment, however, may overload the system with protochlorophyllide.

In the initial work on this herbicide combination, a long (17 h) postspray dark period was required for porphyrins to accumulate sufficiently for herbicide damage to occur upon exposure to light [71]. The herbicidal effect was age- and species-

TABLE 18.1 EFFECTS OF ALA AND 2,2′ DIPYRIDYL (DP), ALONE OR IN COMBINATION, ON PORPHYRIN ACCUMULATION AND HERBICIDAL DAMAGE. Cucumber seedlings (6-day-old) were assayed for porphyrins after being sprayed with the herbicides and incubated in darkness for 17 h. Herbicidal damage was assessed 10 days after porphyrins were assayed and during which they were exposed to greenhouse light conditions. PChlide = protochlorophyllide, MgPPIXME = Mg-protoporphyrin IX monomethyl ester, PPIX = protoporphyrin IX. Adapted from [71].

Treatment	Porphyrins (nmoles/100 mg protein)				Herbicidal damage (%)[a]
	PChlide	MgPPIXME	PPIX	Total	
Control	17.3	0.6	0.0	17.9	0
5 mM ALA	100.7	1.6	0.0	101.3	30
15 mM 2,2′-Dipyridyl	24.0	12.3	2.6	38.9	10
5 mM ALA + 15 mM DP	121.1	26.3	8.1	155.5	80

[a] % dead cotyledons

dependent and there was not always a good correlation between porphyrin accumulation and herbicidal damage. In some species, accumulation of porphyrins did not result in significant herbicidal damage; generally, dicots were more sensitive than monocots. The differential sensitivity between species was hypothesized to be due to differences in both the extent of porphyrin synthesis and to the nature of the accumulated tetrapyrroles [72, 73]. However, differential tolerance to reactive toxic oxyen species among plant species could also be involved, as has been hypothesized to explain species differences in susceptibility to commerial herbicides that cause protoporphyrin IX to accumulate [78]. A difference in uptake of ALA and DP can explain the differential effects of this combination on *Stellaria media* L. (susceptible) and *Rosa chinensis* (tolerant) [79]. Monocots could also be more tolerant of ALA than dicots because the accumulated porphyrins degrade more rapidly in light and more of the ALA accumulates as porphobilinogen, a nonphotodynamic compound [80].

Neither ALA nor the combination of ALA with other compounds affecting porphyrin synthesis has been marketed as a herbicide. However, several classes of commercial herbicides have recently been shown to act by causing target plant species to accumulate phytotoxic levels of protoporhyrin IX, a photodynamic chlorophyll and heme precursor (see Chapter 8). Thus, a natural product, not the synthetic herbicide, is the acutely toxic compound in these cases. Application of protoporphyrin IX alone to plant tissues, however, is not effective even when the cuticle barrier is eliminated [81], perhaps because it does not reach the proper cellular location(s) in sufficient quantity.

18.2.5 Miscellaneous Compounds

Some compounds produced by plants are difficult to categorize based on the metabolic pathway from which they are derived. In some cases, they contain constituents from more than one pathway. Also, contrary to popular belief, plants, especially marine algae, produce many halogenated compounds with biological activity [82, 83]. For example, the marine green alga *Cladophora fascicularis* produces a polybrominated diphenylether [84] with a structure similar to that of herbicidal diphenylethers. It has bactericidal activity against several species. It is unlikely to have photobleaching herbicide activity like the synthetic photobleaching diphenylethers, because one of the phenyl rings is not substituted properly (although the other is); however, it has not been tested for herbicidal activity. Very similar brominated diphenyl ether compounds are found in sponges of the genus *Dysidea* that contain large populations of cyanophyte algae [85]. Two of these compounds also have bactericidal activity. The marine alga *Chara gobularis* produces charamin (Figure 18.1), a powerful photosynthesis inhibitor [86].

The IAA analog 5-hydroxyindole-3-acetic acid apparently accounts for a significant portion of the allelopathic properties of quackgrass (*Agropyron repens* L.) [87]. At low doses, it stimulates growth of quackgrass and corn, while inhibiting growth of the roots of red kidney bean plants. Whether this represents a general monocot/dicot selectivity is not known. These are but a few examples of the many plant-derived compounds reported to be phytotoxic.

18.2.6 Problems with Plant-Derived Compounds

Without modification, plant-produced phytotoxins are generally only weakly active compared to commercial herbicides. Most known allelochemicals would have to be applied at rates of more than 10 kg/ha to achieve significant weed control (e.g., [30]), whereas most of the newer commercial herbicides achieve acceptable weed control at levels two to three orders of magnitude lower. Weak herbicidal activity is not unexpected, because production of highly phytotoxic compounds would lead to strong autotoxicity unless the producing plant develops metabolic or physical mechanisms to cope with its own phytotoxins. Some allelochemicals are toxic to the producing species, and this autotoxicity has been implicated in plant species shifts (succession) in abandoned agricultural fields. Microbial conversion of relatively nonphytotoxic compounds in the soil to highly phytotoxic derivatives may play a role in this process.

Plants have been much more successfully exploited as sources of insecticides and fungicides [3, 5, 7, 88]. This is probably due to several factors. The selection pressure exerted by pathogens and herbivores has probably been more acute and intense than that exerted by plant competitors. Plants can effectively compete with each other in ways other than poisoning each other with compounds potentially harmful to themselves and then having to cope with autotoxicity. Pathogens and herbivores have many potential physiological and biochemical sites of action for pesticides that the plant does not share, reducing the potential for autotoxicity. Plant/plant competition is a process that can span relatively long time periods, whereas the defense against pathogens or herbivores must usually be much more rapid to be effective. A chemical defense can meet the need for a rapid response.

Germination is usually much less sensitive to phytotoxins than is growth or other functions of growing plants [1, 3, 57]. Unfortunately, much of the data on allelopathic or suspected allelopathic compounds are confined to effects on germination. Many of the reported effects on growth are insufficient to judge their potential as herbicides.

18.2.7 Discovery and Development of Plant-Derived Compounds

The discovery process for natural product herbicides is more complicated than that for synthetic herbicides. Traditionally, new herbicides have been discovered by synthesis, bioassay, and evaluation. If the compound is sufficiently promising, quantitative structure-activity relationship-based synthesis of analogues is used to optimize desirable herbicidal properties (see Section 1.6).

The discovery process with natural compounds is complicated by several factors. First, the amount of purification initially conducted is a variable for which there is no general rule. Isolation and chemical characterization of herbicidally active compounds from plants can be a major effort compared to synthesizing a new compound. Furthermore, secondary products of plants are generally isolated in relatively small amounts compared to the amounts of synthesized chemicals available for screening for herbicide activity. Therefore, bioassays requiring very small amounts of material are often helpful, if not essential, in screening natural products from plants. Numerous published methods for assaying small amounts of compounds for pesticidal and biological activities are available in the allelochemical and natural product literature. However, as mentioned above, many of these bioassays are not good indicators of herbicidal potential. At some point in the discovery process, structural identification is required. This step can be quite difficult for some natural products. Finally, synthesis of the compound and analogues must be considered. This is generally much more difficult than identification. Despite these difficulties, modern instrumental analysis and other improved methods are reducing the difficulty, cost, and time involved in each of the above steps.

Many factors must be considered in the decision to develop and market a pesticide. An early consideration is the patentability of the compound. As with any synthetic compound, a patent search must be done for natural compounds also. Prior publication of the pesticidal properties of a compound could cause patent problems. Compared to synthetic compounds, there is a plethora of published information on the biological activity of natural products. For this reason, patenting synthetic analogues with no mention of the natural source of the chemical family might be wiser than patenting the natural product in some situations.

The toxicological and environmental properties of the compound must be considered. The fact that a compound is a natural product does not ensure that it is safe. The most toxic mammalian poisons known are natural products, and many of these are plant products. However, evidence is strong that natural products generally have a much shorter half life in the environment than synthetic herbicides. In fact, the relatively short environmental persistence of natural products may be a problem in some cases, because many herbicides must have some residual activity in order to be effective. As with pyrethroids, chemical modification can increase persistence. Agreement on a proper balance between adequate residual activity and a sufficiently short environmental half life to avoid the problems associated with persistence can be more difficult than achieving the desired level of persistence.

After promising biological activity is discovered, extraction of larger amounts of the compound for more extensive bioassays can be considered. Also, analogues of the compound should be made by chemical alteration of the compound and/or by chemical synthesis. Structural manipulation could lead to improved activity, selectivity, or toxicological properties, altered environmental effects, or discovery of an active compound that can be economically synthesized. This has been the case with several plant-derived compounds that have been used as a template for commercial pesticides (e.g., pyrethroids).

Before a decision is made to produce a natural pesticide for commercial use, the most cost-effective means of production must be found. Although this is a crucial question in considering the development of any pesticide, it is even more complex and critical with natural products. At present, the question of whether a plant product will be produced by biosynthesis and purification or by traditional chemical synthesis is heavily weighted in favor of industrial synthesis.

There are at least three advantages to the herbicide industry of synthetic production of herbicides. First, they have invested heavily in facilities for this approach and in personnel trained in disciplines oriented toward these methods. Retooling facilities and personnel is costly and disruptive. Second, in addition to the patent for use, patents for chemical synthesis often further protect the investment that a company makes in development of a pesticide. Many natural products are so complex that the cost of chemical synthesis would be prohibitive. Even so, more economically synthesized analogues with adequate or even superior biological activity may be possible. This is the third advantage of synthetic production—versatility in production of analogs. Thus, biosynthesis is currently not a likely choice for commercial production. However, there are a growing number of biosynthetic technologies that may eventually make biosynthesis a viable option.

To extract a compound from field-grown plants is not a viable option, even if production were optimized by genetic, cultural, and chemical methods. An alternative is to produce the compound in cell culture. With these methods, cell lines that produce higher levels of the compound can be selected rapidly. However, genetic stability of such traits has been a problem in cell culture production of secondary products. Cells that produce and accumulate massive amounts of possibly autotoxic secondary compounds are obviously at a metabolic disadvantage, and are thus selected against under many cell or tissue culture conditions. Alternative techniques, such as an immobilized cell column (bioreactor), that continuously removes secondary products can increase production by decreasing feedback inhibition of synthesis, reducing autotoxicity, and possibly increasing genetic stability. Still, economical production by plant cell culture of compounds with values of hundreds or even thousands of dollars per kilogram is very difficult [89].

A problem with cell culture is that these cells are undifferentiated, and undifferentiated cells normally produce very small amounts of secondary compounds compared to differentiated cells. Production of herbicides with this technique is not likely in the near future. Other culture methods that optimize production can also be utilized. For example, supplying inexpensively synthesized metabolic precursors can greatly enhance biosynthesis of many secondary products. Also, plant growth regulators, elicitors, and metabolic blockers can be used to increase production. Root culture of roots transformed by *Agrobacterium rhizogenes* to form "hairy roots" in bioreactors can take advantage of the higher level of secondary compound synthesis in differentiated cells while also taking advantage of the bioreactor methods worked out for immobilized cell culture [90]. The greatly increased surface area caused by the "hairy root" morphology of these roots lends itself to substrate/product exchange.

Genetic engineering and biotechnology may allow for the production of plant-derived secondary products by gene transfer to microorganisms and production by fermentation. This concept is attractive because of the existing fermentation technology for production of secondary products. However, it may be prohibitively difficult for complex secondary products in which several genes control the conversion of several complex intermediates to the desired product.

18.3 COMPOUNDS FROM MICROBES

Microbes have thus far been a much more lucrative source of new herbicides than plants. Microbes can be used directly for weed control as biocontrol agents or as sources of phytotoxins that can be used directly or as a template for new herbicides. Living, microbial herbicides can be bacterial or fungal (mycoherbicides). Considerable effort is being expended in the development of mycoherbicides.

18.3.1 Mycoherbicides

Using a plant pathogen itself as a biological herbicide is an option that is very attractive from an environmental standpoint. Two mycoherbicides of this type have been commercially developed and marketed in the United States, and others are under development [91–93] (Table 18.2). However, chemical herbicides, including those based on microbially produced toxins, offer several advantages over mycoherbicides. First, their storage requirements are less restrictive. Since mycoherbicides are living materials, they have relatively narrow environmental requirements during storage compared to a toxin. Also, the shelf life of phytotoxins is generally much longer than that of mycoherbicides, so that unused material from one year can be used the next year or even later. Use of mycoherbicides is generally limited by the restricted environmental conditions required for infection. Since toxins require less restricted conditions for efficacy, their environmental

TABLE 18.2 MYCOHERBICIDES THAT ARE CURRENTLY AVAILABLE OR UNDER DEVELOPMENT. This list does not include all products under development at the time of this writing.

Commercial name	Pathogen	Target weed species
BioMal	*Colletotrichum gloeosporoides*	Round-leaved mallow *(Malva pusilla)*
Casst	*Alternaria cassia*	Sicklepod *(Cassia obtusifolia)*
Collego	*Colletotricum gloeosporoides*	Northern jointvech *(Aeschynomene virginica)* in rice and soybean
DeVine	*Phytophthora palmivora*	Stranglervine *(Morrenia odorata)* in citrus and orchards
Velgo	*Colletotrichum coccoides*	Velvetleaf *(Abutilon theophrasti)*

window for use is quite wide compared to mycoherbicides. Furthermore, many mycoherbicides are incompatible with other herbicides, adjuvants, and tank mix additives with which toxins would not interact. The application and formulation technology for mycoherbicides is in its infancy, whereas existing technology can be used with microbial toxins. One of the biggest disadvantages of mycoherbicides is their extreme selectivity. Although some microbe-produced phytotoxins are extremely selective, others have highly desirable ranges of selectivity. An important potentially negative aspect of mycoherbicides for industry is that they have the potential of being too successful; that is, they could persist in the field and effectively eradicate the target organism for several years, eliminating the need for continued product purchases. Finally, there is the possibility of escape of mycoherbicides to nontarget organisms.

Comparatively little effort has been made to solve these problems, and eventually many of them will be overcome. For instance, the requirement for a wet microclimate over a period of several hours for infectivity can be met with slow-drying, invert emulsion formulations [94]. Manipulation of the genetics of the pathogen or the formulation in which it is sprayed has the potential of favorably altering shelf life, infectivity, virulence, and host range.

The mechanisms of action of mycoherbicides have not been extensively studied and will not be dealt with in this chapter. From plant pathology research on the physiology and biochemistry of pathogen infections of crops, we know that fungal phytotoxins often play a strong role in infectivity. There is considerable interest in using these compounds as templates for new herbicides.

Phytotoxins can be found in nonphytopathogenic microorganisms as well as pathogenic microbes, and those from plant pathogens can be divided into host-specific and non-host-specific toxins. There are rationales for choosing to survey each of these categories of phytotoxins for potential herbicidal utility.

18.3.2 Compounds from Nonphytopathogens

As with medicinal antibiotics, the most successful examples of microbial toxins for herbicidal use are from nonphytopathogenic soil microbes such as the Actinomycetes, particularly the genus *Streptomyces*. These microorganisms are generally much more readily isolated and cultured than plant pathogenic or saprophytic microbes. Furthermore, their secondary products are often less structurally complicated (e.g., fewer chiral centers) and sometimes have a more favorable range of affected weed species than phytotoxins from plant pathogens.

Antibiotics such as chloramphenicol, cycloheximide, and streptomycin, from soil microbes, have nonselective phytotoxic effects. A problem in screening soil microorganisms for phytotoxins is rediscovery of these common antibiotics [17, 18, 95]. Rapid analytical assays for these compounds must be available to eliminate them early in the discovery process. Despite these problems, one research group screening soil microbes for new phytotoxins reported that of 50 compounds promising enough for structural determination, 28 percent were novel compounds and another 16 percent were known antibiotics not previously reported to be herbicidal [17].

The most successful class of herbicides from this class are the oligopeptide glutamine synthetase (GS) inhibitors bialaphos, glufosinate (also called phosphinothricin) (Figure 13.9), and phosalacine from *Streptomyces hygroscopicus, S. viridochromogenes,* and *Kitasatosporia phosalacinea,* respectively (see Chapter 13). An unrelated GS inhibitor, oxetin, is derived from an unidentified *Streptomyces* species [96]. The GS-inhibiting properties of glufosinate were known before it was developed for herbicide use. Glufosinate is now sold as an ammonium ion salt in Europe, and is under development in other parts of the world. Structure-activity research has resulted in only a few active analogs, and all of these have been demonstrated to be metabolized to glufosinate in the plant [97]. For this reason, glufosinate was chosen as the compound with optimum herbicidal properties. Bialaphos is produced by fermentation and sold as a commercial herbicide in Japan, whereas glufosinate is being synthesized industrially. These herbicides are nonselective and are rapidly degraded in soil [98, 99]. The mechanism of action of these compounds is dealt with in detail in Chapter 13.

Gostatin (Figure 18.2), a *S. sumanensis* product, inhibits aspartate aminotransferase from wheat; however, it is more active against this enzyme from pig [100]. Its herbicidal activity has not been determined. Anisomycin (Figure 18.2), another *Streptomyces* sp. product, has selective herbicidal properties against barnyardgrass (*Echinochloa crus-galli*) and crabgrass (*Digitaria sanguinalis*), while having no activity on several crop species, including rice [101]. Methoxyphenone (Figure 18.2), a synthetic rice herbicide, was derived from anisomycin. Methoxyphenone acts as a photosynthetic pigment synthesis inhibitor [102].

The herbicidins, a group of adenine nucleoside analogs from *Streptomyces saganonenis,* show promise as herbicides for control of dicot weeds in rice. The herbimycins and other ansamitocins from *S. hygroscopicus* have been patented as rice herbicides. Another product of *S. hygroscopicus,* geldanamycin, is structurally related to the herbimycins and is an effective plant growth inhibitor [103, 104]. Nigericin, a polyether antibiotic from *S. hygroscopicus,* is also a potent plant growth inhibitor [103, 104]. Doses as low as 0.3 kg/ha cause visual damage to velvetleaf within 1 to 2 days after foliar application [104]. It is a photophosphorylation inhibitor [105] as well as a potassium ionophore [106].

Homoalanosine (Figure 18.2) is an amino acid from *Streptomyces galilaeus* that effectively controls certain monocot weeds in rice [107]. After foliar application, appearance of symptoms is slow and the axillary buds and roots, rather than the leaves, show symptoms of herbicide effects. Its growth-retarding activity on bacteria is inhibited by L-aspartic acid and L-glutamic acid, indicating that it may act as an inhibitor of the synthesis or utilization of these amino acids.

Toyocamycin (Figure 18.2), a pyrrolopyridine nucleoside from a species of *Streptomycin,* is a nonselective phytotoxin [101]. SF 2494 (5′-*O*-sulfamoyltubercidin) (Figure 18.2), a product of *Streptomcyces mirabilis,* is a nonselective herbi-

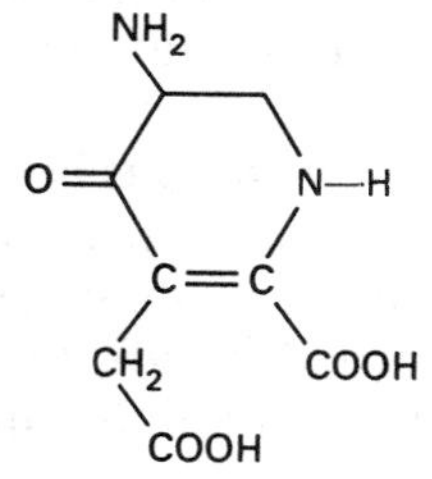

Gostatin

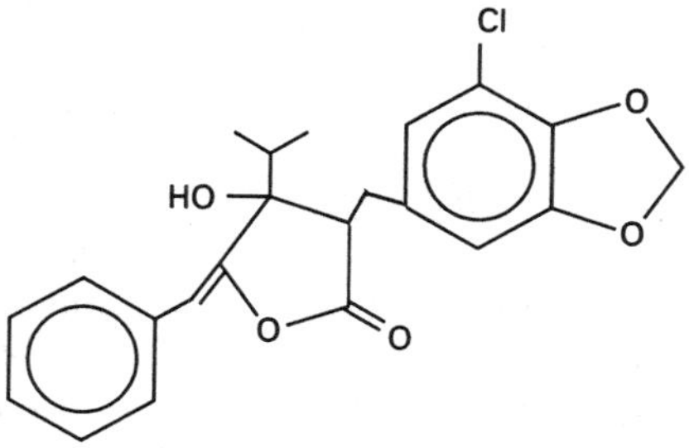

Cyanobacterin

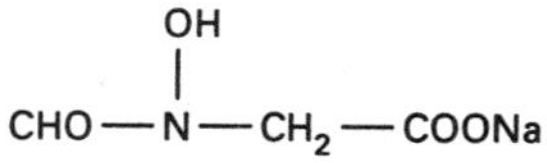

Hadacidin

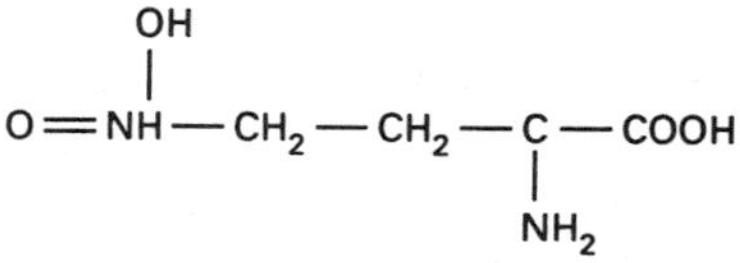

Homoalanosine

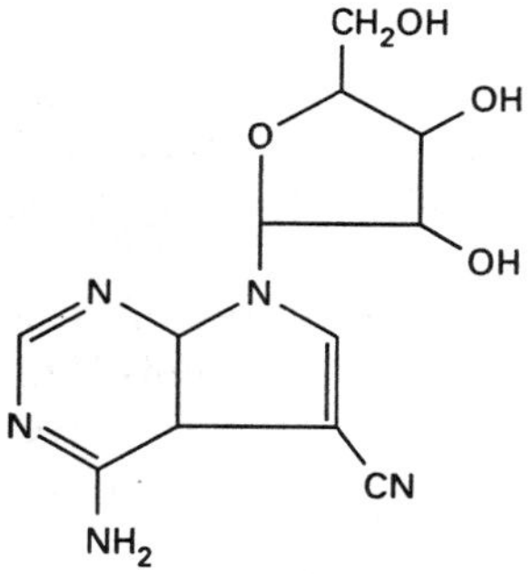

Toyocamycin

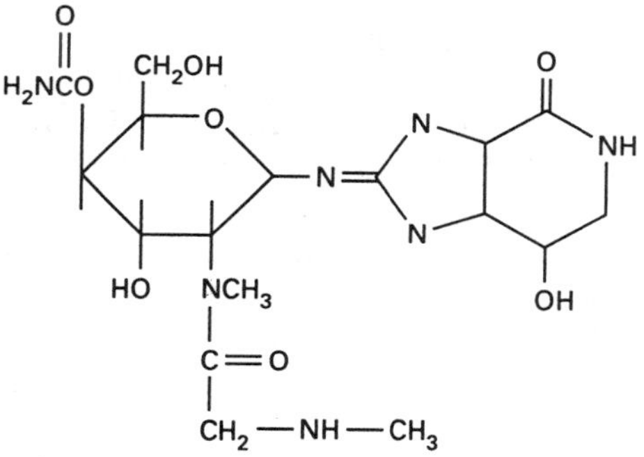

SF 701

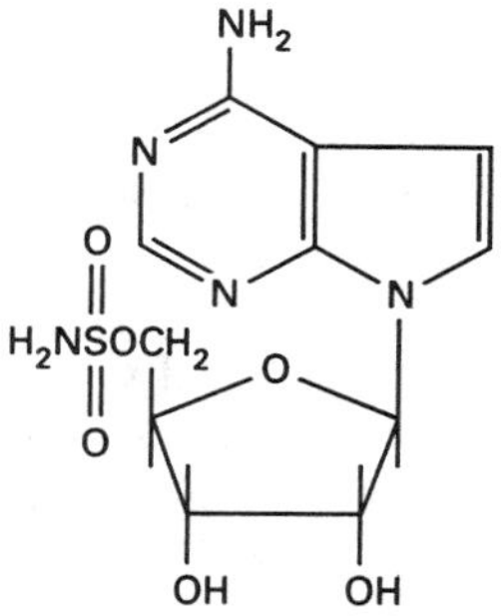

SF 2494

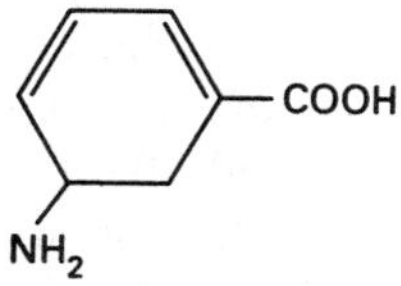

Gabaculine

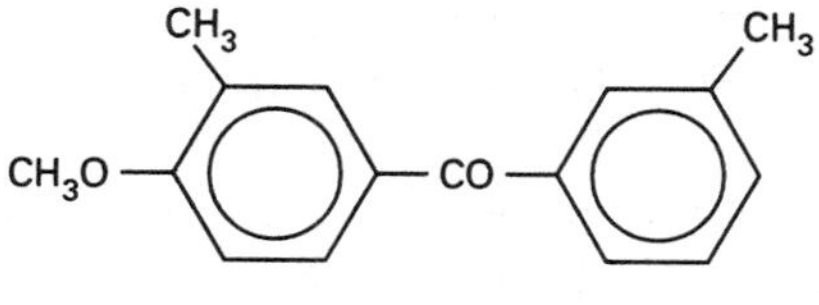

Methoxyphenone

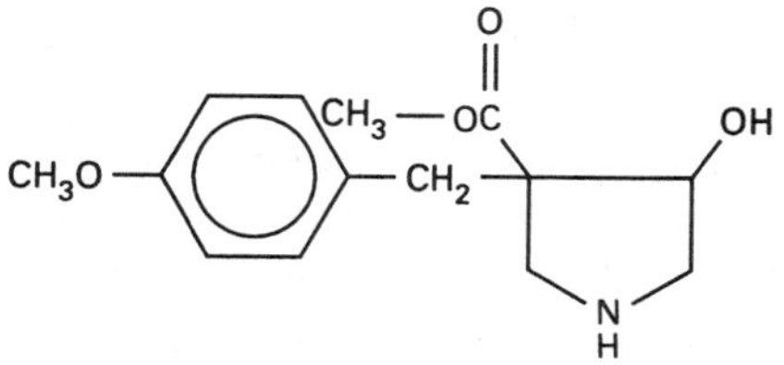

Anisomycin

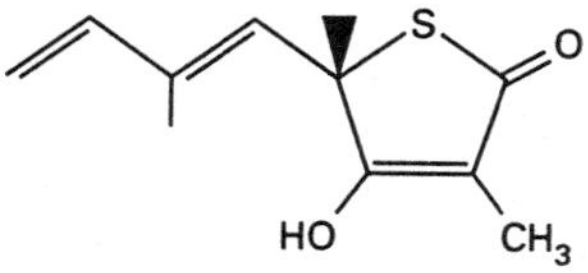

Thiolactomycin

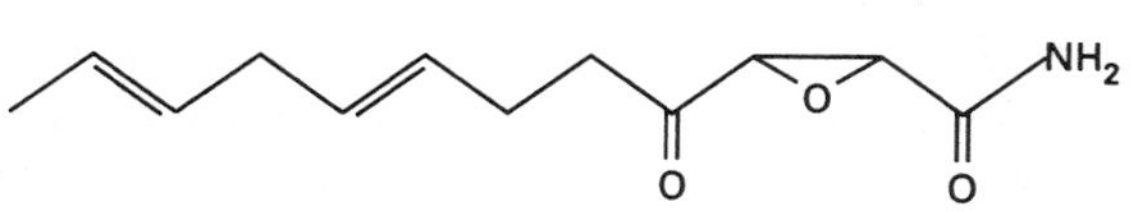

Cerulenin

Figure 18.2 Structures of several nonpathogenic-microbe-derived phytotoxins.

cide with activity equal to that of bialaphos [108]. Without the 5′-*O*-sulfamoyl moeity, tubercin has no herbicidal activity. The mechanism of action is unknown; however, the structural similarity to nucleosides suggests that it might interfere with nucleic acid metabolism.

A streptomycin-like product of *Streptomcyces,* SF-701 (Figure 18.2), is herbicidally active against barnyard millet (*Panicum crus-galli*), but has little activity against rice [109]. It does not affect photosynthesis, but is a potent inhibitor of starch synthesis. It has been hypothesized that the effect is a secondary effect of inhibited protein synthesis, because of the structural similarity to streptothricin antibiotics that are known protein synthesis inhibitors.

Gabaculine (Figure 18.2), a benzoic acid derivative of *Streptomcyces toyocaenis,* is a potent inhibitor of synthesis of all porphyrins, including chlorophyll [110]. It inhibits glutamine-1-semialdehyde conversion to ALA [111], as well as other pyridoxyl phosphate-requiring reactions.

Hadacidin (Figure 18.2), a metabolite of *Penicillium* sp., is a plant growth inhibitor at relatively high levels [112]. It inhibits adenylsuccinate synthetase with a K_i of 1 μM, thus inhibiting AMP synthesis [113]. It also inhibits starch synthesis [114].

The myxobacterium *Stigmatella aurantiaca* produces two types of structurally related compounds that inhibit PS II, stigmatellin and the aurachins [115]. These compounds behave like phenol or diuron-type PS II inhibitors. However, stigmatellin and two aurachins also inhibit electron flow through the cytochrome b_6/f complex.

The cyanobacterium *Scytonema hofmanni* produces a halogenated compound, cyanobacterin (Figure 18.2), that inhibits growth of other algae [116]. Growth of duckweed (*Lemna gibba*) is completely inhibited by 2.3 μM cyanobacterin [117]. In limited studies of the effects on terrestrial plants, it was more phytotoxic to peas and corn than to four weed species [117]. It is a PS II inhibitor with a level of activity comparable to that of DCMU [117]. However, lack of cross-resistance of triazine and DCMU-resistant mutants to cyanobacterins, and of cyanobacterin-resistant plants to DCMU, indicates that this compound affects a PS II site that is different from that of the triazine/urea site [118, 119]. The site appears to be between Q_A and Q_B [119]. Attempts to improve the activity of cyanobacterin in the Hill reaction through modification of the molecule have not been successful [120]; no attempt has been made to determine the selectivity of analogs of cyanobacterin.

Thiolactomycin (Figure 18.2), an antibiotic from *Cephalosporium caerulens* and cerulenin (Figure 18.2), from *Nocardia* sp., both inhibit fatty acid biosynthesis in chloroplasts of higher plants; however, their efficacy is much lower than that of synthetic graminicides such as the cyclohexanediones or aryl-propanoic acids [121]. Rather than inhibiting acetyl-CoA carboxylase, as do these synthetic herbicides, thiolactomycin inhibits acetyl-CoA transacylase [122, 123] and cerulenin inhibits 3-oxoacyl-acyl carrier protein synthase [124].

18.3.3 Toxins from Plant Pathogens

Plant pathogens are a source of many potent phytotoxins [125, 126]. The phytotoxins apparently function to disable the plant's defenses and in many cases to kill plant tissues before they are utilized by the pathogen as a food source. They have been implicated as "virulence factors" for many plant pathogens. There are many cases in which a single pathogenic species produces several phytotoxins with completely different chemical structures. These toxins can be categorized as either host-specific or non-host-specific.

Host-specific toxins are those that are produced by plant pathogens and are phytotoxic only to the plant host of the producing organism. To date, the structures of around 20 of these compounds have been elucidated. Of those identified, all but one are from pathogens that only infect cultivated crops. Their level of selectivity is often remarkable, sometimes being relegated to only one variety of a species. Comparatively little effort has been made to discover host-specific toxins from weed pathogens. Industry is not enthusiastic about this approach because these compounds are too selective and those identified have generally been chemically complex.

Regardless, government and university laboratories have recently begun to search for host-selective toxins from pathogens that infest weeds which cause a high level of economic damage [16]. The only host-selective phytotoxin from a weed-infesting fungal pathogen to be discovered thus far is maculosin, a diketopiperazine (cyclo-(L-Pro-L-Tyr-) (Figure 18.3) [127, 128]. This relatively simple compound has a very high level of specificity for spotted knapweed (*Centaurea maculosa*). Maculosin is synergized by tenuazonic acid, a product of *Alternaria alternata* [127]. The mechanism of phytotoxic action of maculosin is unknown.

Most of the phytotoxins produced by plant pathogens have a host range beyond that of the producing organism. Some of these compounds have highly desirable selectivity patterns. For example, the cyclic tetrapeptide tentoxin (Figure 18.3), produced by *A. alternata,* causes severe chlorosis in all of the major weeds in soybeans and corn without affecting these crops [2, 129]. The mechanism of action of tentoxin is unique. Although it is a potent inhibitor of the ATPase activity of CF_1 of the chloroplast [130], thereby inhibiting photophosphorylation [131], this effect cannot wholly account for the chlorosis that it causes. This is obvious when a mutant of *Oenothera* lacking functional CF_1 is treated with tentoxin. The seedlings of this mutant are light green until storage reserves are depleted, indicating that, initially, greening is not entirely dependent on photophosphorylation [132]. Treatment of the seeds with tentoxin during germination results in completely chlorotic (yellow) seedlings. Another dramatic effect of the toxin is the complete interruption of transport from the cytoplasm to the plastid and posttranslational processing of polyphenol oxidase in sensitive species [133, 134]. A specific plastid envelope ATPase is also inhibited by tentoxin [129, 135]. Thus, polyphenol oxidase movement from the cytoplasm to the plastid may be dependent

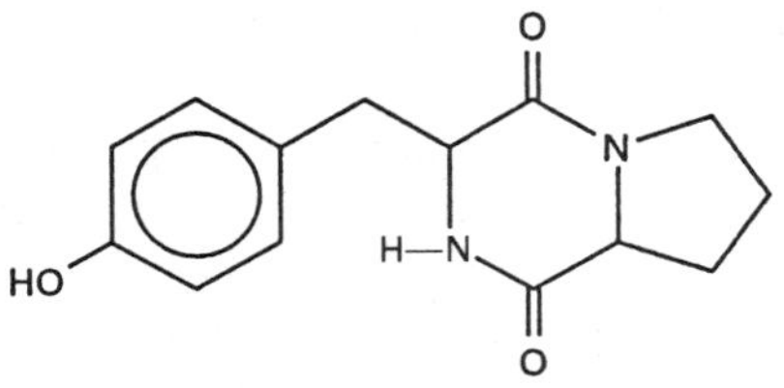

Maculosin

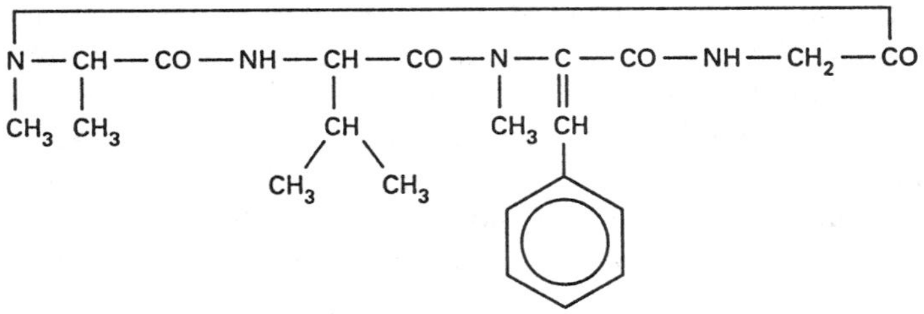

Tentoxin

Figure 18.3 Structures of a host-specific phytotoxin (maculosin) and a non-host-specific phytotoxin (tentoxin).

on this ATPase. Tentoxin does not affect import of most other nuclear-coded proteins into the plastid; however, several differences, both absences and presences, exist between protein profiles of etioplasts from untreated and tentoxin-treated plants [129, 135]. Lack of polyphenol oxidase may result indirectly in loss of chlorophyll accumulation in light, since this enzyme has been speculated to play a role in mediation of oxygen metabolism in the plastid [136].

Tentoxin is only one of many nonselective plant pathogen-derived phytotoxins that have been described [126], and some of these have been patented as herbicides [18]. The modes of action of some of these are listed in Table 18.3.

18.3.4 Discovery and Development of Microbial Phytotoxins

Discovery. The choice of a source of microbial compounds to be exploited is the first of a myriad of decisions that must be made in industrial exploitation of this resource. Thousands of microbial secondary products with biological activity have been chemically characterized. There is public documentation of the evaluation of relatively few of these compounds for herbicidal activity, and less than 100 of these compounds have been patented as herbicides. Known compounds represent perhaps the most readily accessible group of compounds for exploitation. Despite the fact that the most successful microbial toxin herbicide, phosphinothricin, was discovered by exploitation of a known compound, there is little evidence that this strategy is being pursued by industry. Nevertheless, a survey of the modes of action of some of these compounds (Table 18.3) provides a great deal of information about herbicide target sites that are not currently shared by commercial herbicides (compare with Table 15.1). This comparison of the

TABLE 18.3 KNOWN SITES OF ACTION OF MICROBIAL TOXINS (COMPARE WITH TABLE 15.1).

Affected site (molecular site)	Microbial toxin*
Amino acid metabolism	
Methionine synthesis (β-cystathionase)	Rhizobitoxine [137]
Arginine synthesis (Ornithine carbamoyl transferase)	Phaseolotoxin [138]
Glutamine synthesis (Glutamine synthetase)	Phosphinothricin
	Oxetin
	Tabtoxin
	Phosalacine
All transaminases	Gabaculine
Glutamate synthesis	Gostatin
Plastid functions	
Electron transport (D-1 protein)	Cyanobacterin
	Aurachins
	Stigmatellin
Energy transfer (CF_1 ATPase)	Tentoxin
Chlorophyll sythesis (ALA synthesis)	Gabaculine
Nuclear-coded protein uptake (Envelope ATPase)	Tentoxin
Plastid nucleic acid synthesis	Tagetitoxin (Chapter 12)
Other sites	
Lipid synthesis (Acetyl-CoA transacylase)	Thiolactomycin
(3-oxoacyl-ACP synthase)	Cerulenin
Membrane ATPase activity	Fusicoccin [139]
K^+/H^+ antiporter?	Syringomycin [140]
RNA Polymerase?	Sirodesmin PL [141]
Pyrimidine synthesis (Aspartate carbamyltransferase)	AAL toxin [142]

* Followed by reference if not covered elsewhere in this chapter.

known mechanisms of action of microbial phytotoxins and commercial herbicides suggests that synthetic screening programs have had very limited success in exploiting all viable sites of herbicide action.

Plant pathogens are an excellent reservoir of phytotoxins. However, industry is wary of this source because of the difficulty in screening large numbers of plant pathogens (they are relatively difficult to collect, isolate, and culture), and the compounds produced often are extremely complex. A compound with more than one chiral center, as many phytotoxins from plant pathogens are, would generally not be considered as a potential pesticide that can be synthesized economically by industrial processes. Conversely, microbial systems (immobilized cell or enzyme reactors) are now being considered for the synthesis of agricultural chemicals with chiral centers [143]. Host-specific toxins of weed pathogens are of little interest because of their extreme selectivity. Actinomycetes, and in particular the genus *Streptomyces,* produce large numbers of secondary metabolites and are easy to culture and, as a result, represent the bulk of the industrial screening effort for new agricultural chemicals. The same species of a soil microorganism will often produce many different antibiotics, depending on its geographic origin and the method used to culture it. Thus, soil microorganisms from extreme and exotic environments are obtained in an effort to increase the likelihood of discovering novel metabolites.

After isolation, the microorganism is usually cultured in a variety of ways and in a variety of media in order to increase the chances that a culture regime conducive to secondary metabolite production is chosen. The culture broths are then bioassayed for herbicidal activity. The bioassay is generally limited by the amount of material available. Thus, microbioassays of various types are often utilized. These include assays of root growth of small-seeded plants, phytotoxicity to *Lemna* sp. or other small aquatic plants, or small-scale potted plant studies. Although microbioassays involving excised plant tissues or isolated plant cells require less material, microbioassays involving whole plants are best for identifying herbicidal potential. If the results from bioassays are sufficiently promising, another cycle of culturing and bioassay is run to determine the stability of production of the phytotoxic principle. Promising activity can be the result of large amounts of one weak phytotoxin, small amounts of one potent phytotoxin, or combinations of effects of several phytotoxins with different selectivities, sites of action, and efficacies.

If a culture broth of a microorganism is deemed promising, it is then fractionated to isolate the active compound(s). A major problem is the rediscovery of common phytotoxins such as cycloheximide. Many of these can be identified at this point by their analytical and biological activity profiles. After isolation, further bioassay must be conducted and correlated with the activity of the crude preparation. Bioassays that require very small amounts of material are helpful at this point because of the minute amounts of material available from chromatographic fractions. Results from whole-plant bioassays of the culture medium can be used to pick the appropriate microbiossay. After isolation of the active compound, improvement of fermentation yields of the compound can be obtained more rapidly because of elimination of the need for bioassay. At this point, the compound is usually structurally identified.

Development. If the compound is novel, several possibilities for exploitation exist. If the complexity of the compound precludes industrial synthesis, patenting of the natural product and production by fermentation is a possibility. If this approach is taken, as has been the case for bialaphos [144], development of the best fermentation production procedures and genetic improvement of the producing microorganism must coincide with greenhouse and field studies.

Another approach is to determine if a more easily synthesized analog will have herbicidal activity. In this case, patenting the natural compound might not be wise, particularly if the phytotoxic properties of this or related compounds have been reported previously in the scientific literature. However, in most structure-activity relationship (SAR) studies of complex microbial phytotoxins, any substantial simplification of the molecule results in great or complete loss of activity. Evolution has apparently optimized the biological effect of many natural phytotoxins. Considerable investment can be made in discovery and SAR study before one learns that a simple derivative with herbicidal activity cannot be found.

If the natural compound is relatively simple, the natural compound might be patented, but produced by industrial synthesis rather than by fermentation, as with glufosinate. However, even in this case, the company that produces glufosinate is carrying out research to produce glufosinate by enzyme reactor technology [145]. SAR studies of microbial phytotoxins might reveal analogs with better herbicidal properties than the natural compound (e.g., methoxyphenone).

There are several real and potential disadvantages of microbial toxins as herbicides. As mentioned earlier, the structures of many of these compounds are so complicated that they cannot be synthesized economically. There is no guarantee that natural compounds will be any more environmentally safe than synthesized compounds. In fact, some of the most potent mammalian toxins known are microbial products. Many natural products have an environmental half life that may be too short for an effective herbicide. Therefore, many microbial toxins would have to be chemically modified to decrease their degradation rate. Also, resistance may develop more rapidly than for synthetic compounds. Since microbial toxins have been present in the environment for eons, genes for resistance to many of them probably exist (e.g., [146]) but, because of the lack of sufficient selection pressure, not throughout entire populations. This could change rapidly with widespread spraying of large areas with levels of the toxins that would never be encountered in nature.

18.4 THE FUTURE

Plants and microbes contain a virtually untapped reservoir of potential herbicides that can be used directly or as templates for synthetic herbicides. Numerous factors have increased the interest of the herbicide industry and the herbicide market in this source of natural products as herbicides. These include diminishing returns with traditional herbicide discovery methods, increased environmental and toxicological concerns with synthetic herbicides, and the high level of reliance of modern agriculture on herbicides. Despite the relatively small amount of previous effort in the development of natural compounds as herbicides, they have made a large impact in the area of insecticides. Thus far, relatively minor successes can be found as herbicides. The number of options that must be considered in discovery and development of a natural product as a herbicide is larger than for a synthetic herbicide. Furthermore, the molecular complexity and limited environmental stability of many natural phytotoxins, compared to synthetic herbicides, are discouraging. Structural modifications may alleviate these problems in some cases. However, advances in chemistry and biotechnology are increasing the speed and ease with which we can discover and develop naturally occurring compounds as herbicides.

REFERENCES

1. Duke, S. O. 1986. "Naturally occurring chemical compounds as herbicides." *Rev. Weed Sci.*, 2, 15–44.
2. Duke, S. O., and J. Lydon. 1987. "Herbicides from natural compounds." *Weed Technol.*, 1, 122–28.
3. Lydon, J., and S. O. Duke. 1989. "Potential of plants for pesticide use." In *Herbs, Spices and Medicinal Plants—Recent Advances in Botany, Horticulture, and Pharmacology, Vol. 3.*, L. Craker and J. E. Simon, eds. Phoenix, AZ: Oryx Press, pp. 1–41.
4. Putnam, A. R. 1988. "Allelochemicals from plants as herbicides." *Weed Technol.*, 2, 510–18.
5. McLaren, J. S. 1986. "Biologically active substances from higher plants: status and future potential." *Pestic. Sci.*, 17, 559–78.
6. Grainage, M., and S. Ahmed. 1988. *Handbook of Plants with Pest-Control Properties.* New York: Wiley-Interscience.
7. Russell, G. G. 1986. "Phytochemical resources for crop protection." *New Zealand J. Technol.*, 2, 127–34.
8. Duke, S. O. 1986. "Microbial phytotoxins as herbicides—a perspective." In *The Science of Allelopathy*, A. R. Putnam, and C. S. Tang, eds. New York: Wiley.
9. Cutler, H. G. 1984. "Biologically active natural products from fungi: templates for tomorrow's pesticides." *Amer. Chem. Soc. Symp. Ser.*, 257, 153–70.
10. Cutler, H. G. 1988. "Perspectives on discovery of microbial phytotoxins with herbicidal activity." *Weed Technol.*, 2, 525–32.
11. Sekizawa, Y., and T. Takematsu. 1983. "How to discover new antibiotics for herbicidal use." In *Pesticide Chemistry, Human Welfare and the Environment. Vol. 2. Natural Products*, N. Takahashi, H. Yoshioka, T. Misato, and S. Matsunaka, eds. Oxford: Pergamon Press, pp. 261–68.

12. Fischer, H.-P., and D. Bellus. 1983. "Phytotoxicants from microorganisms and related compounds." *Pestic. Sci.,* 14, 334–46.
13. Poole, N. J., and E. J. T. Chrystal. 1985. "Microbial phytotoxins." Brit. Crop Protect. Conf.—Weeds, 591–600.
14. Misato, T., and I. Yamaguchi. 1984. "Pesticides of microbial origin." *Outlook Agric.,* 13, 136–39.
15. Cutler, H. G. 1986. "Isolating, characterizing, and screening mycotoxins for herbicidal activity." In *The Science of Allelopathy,* A. Putnam and C.-S. Tang, eds. New York: Wiley, pp. 147–70.
16. Kenfield, D., G. Bunkers, G. A. Strobel, and F. Sugawara. 1988. "Potential new herbicides—phytotoxins from plant pathogens." *Weed Technol.,* 2, 519–524.
17. Ayer, S. W., B. G. Isaac, D. M. Krupa, K. E. Crosby, L. J. Letendre, and R. J. Stonard. 1989. "Herbicidal compounds from micro-organisms." *Pestic. Sci.,* 27, 221–23.
18. Poole, N. J., and E. J. T. Chrystal. 1985. "Microbial phytotoxins." Brit. Crop Protect. Conf.—Weeds, 591–600.
19. Rice, E. L. 1984. *Allelopathy,* 2nd ed. New York: Academic.
20. Yang, R. Z., and C. S. Tang. 1988. "Plants used for pest control in China: A literature review." *Econ. Bot.,* 42, 376–406.
21. Rice, E. L. 1983. *Pest Control with Nature's Chemicals.* Norman, OK: University of Oklahoma Press.
22. Einhellig, F. A. 1987. "Interactions among allelochemicals and other stress factors of the plant environment." Amer. Chem. Soc. Symp. Ser. 330, 343–57.
23. Lodhi, M. A. K., R. Bilal, and K. A. Malik. 1987. "Allelopathy in agroecosystems: Wheat phytotoxicity and its possible roles on crop rotation." *J. Chem. Ecol.,* 13, 1881–91.
24. Davidson, J. 1915. "A comparative study of the effect of cumarin and vanillin on wheat grown in soil, sand and water cultures." *J. Amer. Soc. Agron.,* 7, 221–38.
25. Krogmeier, M. J., and J. M. Bremmer. 1989. "Effects of phenolic acids on seed germination and seedling growth in soil." *Biol. Fertil. Soils,* 8, 115–22.
26. B. R. Dalton, U. Blum, and S. B. Weed. 1989. "Differential sorption of exogenously applied ferulic, *p*-coumaric, *p*-hydroxybenzoic, and vanillic acids in soil." *Soil Sci. Amer. J.,* 53, 757–62.
27. T. S. C. Wang, S. W. Li, and Y. L. Ferng. 1978. "Catalytic polymerization of phenolic compounds by clay mineral." *Soil Sci.,* 126, 15–21.
28. Vaughan, D., G. P. Starling, and B. G. Ord. 1983. "Amelioration of the phytotoxicity of phenolic acids by some soil microbes." *Soil. Biol. Biochem.,* 15, 613–14.
29. Schmidt, S. K. 1988. "Degradation of juglone by soil bacteria." *J. Chem. Ecol.,* 14, 1561–71.
30. Shettel, N. L., and N. E. Balke. 1983. "Plant growth response to several allelopathic chemicals." *Weed Sci.,* 31, 293–98.
31. Kaminsky, R., and W. H. Muller. 1978. "Recommendation against use of alkaline soil extractions in study of allelopathy." *Plant Soil,* 49, 641–45.
32. Duke, S. O., R. D. Williams, and A. H. Markhart. 1983. "Interactions of moisture

stress and three phenolic compounds on lettuce seed germination." *Ann. Bot.,* 52, 923–26.

33. Blum, U., T. M. Gerig, and S. B. Weed. 1989. "Effects of mixtures of phenolic acids on leaf area expansion of cucumber seedlings grown in different pH Portsmouth A_1 soil materials." *J. Chem. Ecol.,* 15, 2413–23.
34. Muzafarov, E. N., and E. K. Zolotareva. 1989. "Uncoupling effect of hydroxycinnamic acid derivatives on pea chloroplasts." *Biochem. Physiol. Pflanzen,* 184, 363–69.
35. Tissut, M., D. Chevallier, and R. Douce. 1980. "Effet de différents polyphénols sur les mitochondries et les chloroplastes isolès." *Phytochemistry,* 19, 495–500.
36. Demos, E. K., M. Woolwine, R. H. Wilson, and C. McMillan. 1975. "The effects of ten phenolic compounds on hypocotyl growth and mitochondrial metabolism of mung bean." *Amer. J. Bot.,* 62, 97–102.
37. Glass, A. D. M. 1974. "Influence of phenolic acids upon ion uptake. III. Inhibition of potassium ion absorption." *J. Exp. Bot.,* 25, 1104–13.
38. Yakushkina, N. I., and V. T. Starikova. 1977. "Effects of coumarin and gibberellin on certain aspects of the energy metabolism of corn seedlings." *Fiziol. Rast.,* 24, 1211–16.
39. Burgess, J., and P. J. Linstead. 1977. "Coumarin inhibition of microfibril formation at the surface of cultured protoplasts." *Planta,* 133, 267–73.
40. Schilling, D. G., and F. Yoshikawa. 1987. "A rapid seedling bioassay for the study of allelopathy." Amer. Chem. Soc. Symp. Ser., 330, 334–42.
41. Koeppe, D. E. 1972. "Some reactions of isolated corn mitochondria influenced by juglone." *Physiol. Plant.,* 27, 89–94.
42. Niemeyer, H. M. 1988. "Hydroxyamic acids (4-hydroxy-1,4-benzoxazin-3-ones), defence chemicals in the Gramineae." *Phytochemistry,* 27, 3349–58.
43. Barnes, J. P., and A. R. Putnam. 1987. "Role of benzoxazinones in allelopathy by rye (*Secale cereale* L.)." *J. Chem. Ecol.,* 13, 889–907.
44. Putnam, A. R. 1988. "Allelochemicals: blessings or curse for agriculture?" *Comments Agric. Food Chem.,* 1, 183–206.
45. Queirolo, C. B., C. S. Andreo, R. H. Vallejos, H. M. Niemeyer, and L. J. Corcuera. 1981. "Effects of hydroxamic acids isolated from Gramineae on adenosine-5′-triphosphate synthesis in chloroplasts." *Plant Physiol.,* 68, 941–43.
46. Nair, M. G., C. J. Whitenack, and A. R. Putnam. 1990. "2,2′-oxo-1,1′-azobenzene. A microbially transformed allelochemical from 2,3-benzoxazolinone." *J. Chem. Ecol.,* 16, 353–64.
47. Ravanel, P., Tissut, M., and R. Douce. 1986. "Platanetin: a potent natural uncoupler and inhibitor of the exogenous NADH dehydrogenase in intact plant mitochondria." *Plant Physiol.,* 80, 500–04.
48. Cline, E. I., S. A. Adesanya, S. K. Ogundana, and M. F. Roberts. 1989. "Induction of PAL activity and dihydrostilbene phytoalexins in *Dioscorea alata* and their plant growth inhibitory properties." *Phytochemistry,* 28, 2621–25.
49. Gorham, J. 1978. "Effect of lunularic acid and analogues on liverwort growth and IAA oxidation." *Phytochemistry,* 17, 99–105.
50. Yoshida, S., T. Asami, T. Kawano, K. Yoneyama, W. D. Crow, D. M. Paton, and N. Takahashi. 1988. "Photosynthetic inhibitors in *Eucalyptus gradis.*" *Phytochemistry,* 27, 1943–46.

51. Muller, W. H., and C. H. Muller. 1964. "Volatile growth inhibitors produced by *Salvia* species." *Bull. Torrey Bot. Club.,* 91, 327–30.
52. Duke, S. O., R. N. Paul, and S. M. Lee. 1988. "Terpenoids from the genus *Artemisia* as potential pesticides." Amer. Chem. Soc. Symp. Ser., 380, 318–34.
53. Lorber, P., and W. H. Muller. 1980. "Volatile growth inhibitors produced in *Salvia leucophylla*—effects on metabolic activity in mitochondrial suspensions." *Comp. Physiol. Ecol.,* 5, 68–75.
54. Grayson, B. T., K. S. Williams, P. A. Freehauf, R. R. Pease, W. T. Ziesel, R. L. Sereno, and R. E. Reinsfelder. 1987. "The physical and chemical properties of the herbicide cinmethylin (SD 95481)." *Pestic. Sci.,* 21, 143–53.
55. El-Deek, M. H., and F. D. Hess. 1986. "Inhibited mitotic entry is the cause of growth inhibition by cinmethylin." *Weed Sci.,* 34, 684–88.
56. Sherman, M. E., L. Thompson, and R. E. Wilkinson. 1983. "Sicklepod (*Cassia obtusifolia*) management in soybeans (*Glycine max*)." *Weed Sci.,* 31, 622–27.
57. Duke, S. O. 1991. "Plant terpenoids as pesticides." In *Handbook of Natural Toxins. Vol. 6. Plant and Fungal Toxins,* 2nd ed., R. F. Keeler and A. T. Tu, eds. New York: Marcel Dekker, pp. 269–96.
58. Duke, S. O., K. C. Vaughn, E. M. Croom, Jr., and H. N. Elsohly. 1987. "Artemisinin, a constituent of annual wormwood (*Artemisia annua*), is a selective phytotoxin." *Weed Sci.,* 35, 499–505.
59. DiTomaso, J. M., and S. O. Duke. 1991. "Is polyamine biosynthesis a possible site of action of cinmethylin or artemisinin?" *Pestic. Biochem. Physiol.,* 39, 158–67.
60. Whaun, J., N. Brown, W. Milhous, C. Lambros, J. Scovill, A. Lin, and D. Klayman. 1985. "Qinghaosu, a potent antimalarial, perturbs polyamine metabolism in human malaria cultures." In *Polyamines: Basic and Clinical Aspects,* K. Imahori, F. Suzuki, O. Suzuki, and U. Bachrach, eds. Utrecht, The Netherlands: VNU Science Press, pp. 301–10.
61. Fischer, N. H., and L. Quijano. 1985. "Allelopathic agents from common weeds. *Amaranthus palmari, Ambrosia artemisiifolia,* and related weeds." Amer. Chem. Soc. Symp. Ser., 268, 133–47.
62. Miller, G. W., J. M. Sasse, C. J. Lovelace, and K. S. Rowan. 1984. "Effects of podolactone-type inhibitors and abscisic acid on chlorophyll synthesis in barley leaves." *Plant Cell Physiol.,* 25, 635–42.
63. Vaughn, K. C. and M. A. Vaughan. 1988. "Mitotic disrupters from higher plants." Amer. Chem. Soc. Symp. Ser., 380, 273–93.
64. Vaughan, M. A., and K. C. Vaughn. 1988. "Mitotic disrupters from plants and their potential uses as herbicides." *Weed Technol.,* 2, 533–39.
65. Rizvi, S. J. H., D. Mukerji, and S. N. Mathur. 1981. "Selective phytotoxicity of 1,3,7-trimethyl xanthine between *Phaseolus mungo* and some weeds." *Agric. Biol. Chem.,* 45, 1255–56.
66. Rizvi, S. J. H., V. Rizvi, D. Mukerjee, and S. N. Mathur. 1987. "1,3,7-Trimethylxanthine, an allelochemical from seeds of *Coffea arabica:* some aspects of its mode of action as a natural herbicide." *Plant Soil,* 98, 81–91.
67. Liso, R., L. De Gara, F. Mommasi, and O. Arrigoni. 1985. "Ascorbic acid requirement for increased peroxidase activity during potato tuber slice aging." *FEBS Lett.,* 187, 141–45.

68. Petroski, R. J., D. L. Dornbos, Jr., and R. G. Powell. 1990. "Germination and growth inhibition of annual ryegrass (*Lolium multiflorum* L.) and alfalfa (*Medicago sativa* L.) by loline alkaloids and synthetic *N*-acylloline derivatives." *J. Agric. Food Chem.*, 38, 1716–18.

69. Dodge, A. D., and J. P. Knox. 1986. "Photosensitizers from plants." *Pestic. Sci.*, 17, 579–86.

70. Towers, G. H. N., and J. T. Arnason. 1988. "Photodynamic herbicides." *Weed Technol.*, 2, 545–49.

71. Rebeiz, C. A., A. Montazer-Zouhoor, H. J. Hopen, and S.-M. Wu. 1984. "Photodynamic herbicides: 1: Concept and phenomenology." *Enzyme Microb. Technol.*, 6, 390–401.

72. Rebeiz, C. A., A. Montazer-Zouhoor, J. M. Mayasich, B. C. Tripathy, S.-M. Wu, and C. C. Rebeiz. 1988. "Photodynamic herbicides and chlorophyll biosynthesis modulators." Amer. Chem. Soc. Symp. Ser., 339, 295–328.

73. Rebeiz, C. A., A. Montazer-Zouhoor, J. M. Mayasich, B. C. Tripathy, S.-M. Wu, and C. C. Rebeiz. 1988. "Phytodynamic herbicides, recent developments and molecular basis of selectivity." *CRC Crit. Rev. Plant Sci.*, 6, 385–436.

74. Rebeiz, C. A., J. A. Juvik, and C. C. Rebeiz. 1988. "Porphyric insecticides 1. Concept and phenomenology." *Pestic Biochem. Physiol.*, 30, 11–27.

75. Duggan, J., and M. Gassman. 1974. "Induction of porphyrin synthesis in etiolated bean leaves by chelators of iron." *Plant Physiol.*, 53, 206–15.

76. Huang, L., and P. A. Castelfranco. 1990. "Regulation of 5-aminolevulinic acid (ALA) synthesis in developing chloroplasts. III. Evidence for functional heterogeneity of the ALA pool." *Plant Physiol.*, 92, 172–78.

77. Masuda, T., H. Kouji, and S. Matsunaka. 1990. "Diphenyl ether herbicide-decreased heme contents stimulate 5-aminolevulinic acid synthesis." *Pestic. Biochem. Physiol.*, 36, 106–14.

78. Finckh, B. F., and K. J. Kunert. 1985. "Vitamins C and E: an antioxidative system against herbicide-induced lipid peroxidation in higher plants." *J. Agric. Chem.*, 33, 574–77.

79. Yaronskaya, E. B., N. V. Shalygo, and N. G. Averina. 1989. "Selectivity of action of a photodynamic herbicide containing 5-aminolevulinic acid and 2,2′-dipyridyl." *Vesti Akad. Navuk BSSR, Ser. Biyal. Navuk*, 1989(4), 38–40.

80. Averina, N. G., N. V. Shalygo, E. B. Yaronskaya, and V. V. Rassadina. 1989. "Porphyrin accumulation in monocots and dicots in the presence of 5-aminolevulinic acid and 2,2′ dipyridyl." *Vesti Akad. Navuk BSSR, Ser. Biyal. Navuk*, 1989(1), 100–02.

81. Lydon, J., and S. O. Duke. 1988. "Porphyrin synthesis is required for photobleaching activity of the *p*-substituted diphenyl ether herbicides." *Pestic. Biochem. Physiol.*, 31, 74–83.

82. Engvild, K. C. 1986. "Chlorine-containing natural compounds in higher plants." *Phytochemistry*, 25, 781–91.

83. Scheuer, P. J., and J. Darius, eds. 1978–81. *Marine Natural Products. Vols. I–IV.* New York: Academic.

84. Kuniyoshi, M., K. Yamada, and T. Higa. 1985. "A biologically active diphenyl ether from the green alga *Cladophora fascicularis*." *Experientia*, 41, 523–24.

85. Salva, J., and D. J. Faulkner. 1990. "A new brominated diphenyl ether from a Philippine *Dysidea* species." *J. Nat. Prod.,* 53, 757–60.
86. Wium-Anderson, U. Anthoni, C. Christopherson, and G. Houen. 1982. "Allelopathic effects on phytoplankton by substances isolated from aquatic macrophytes (Charales)." *Oikos,* 39, 187–90.
87. Hagin, R. D. 1989. "Isolation and identification of 5-hydroxyindole-3-acetic acid and 5-hydroxytryptophan, major allelopathic aglycons in quackgrass (*Agropyron repens* L. Beauv.)." *J. Agric. Food Chem.,* 37, 1143–49.
88. Miyakado, M. 1986. "The search for new insecticidal and fungicidal compounds from plants." *J. Pestic. Sci.,* 11, 484–92.
89. Kreis, W., and E. Reinhard. 1989. "The production of secondary metabolites by plant cells cultivated by bioreactors." *Planta Medica,* 55, 409–92.
90. Signs, M. W., and H. E. Flores. 1990. "The biosynthetic potential of plant roots." *BioEssays,* 12, 1–7.
91. Templeton, G. E., R. J. Smith, Jr., and D. O. TeBeest. 1986. "Progress and potential of weed control with mycoherbicides." *Rev. Weed Sci.,* 2, 1–14.
92. Charudattan, R. 1990. "Pathogens with potential for weed control." Amer. Chem. Soc. Symp. Ser., 439, 132–54.
93. Templeton, G. E., and D. K. Heiny. 1990. "Mycoherbicides." In *New Directions in Biological Control: Alternatives for Suppressing Agricultural Pests and Diseases,* R. Baker and P. Dunn, eds. New York: Alan D. Liss, Inc., pp. 279–86.
94. Daigle, D. J., W. J. Connick, Jr., P. C. Quimby, Jr., J. Evans, B. Trask-Morrell, and F. E. Fulghum. 1990. "Invert emulsions: carrier and water source for the mycoherbicide, *Alternaria cassiae*." *Weed Technol.,* 4, 327–31.
95. Heisey, R. M., J. DeFrank, and A. R. Putnam. 1985. "A survey of soil microorganisms for herbicidal activity." Amer. Chem. Soc. Symp. Ser., 268, 337–49.
96. Omura, S., M. Murata, N. Imamura, Y. Iwai, H. Tanaka, A. Furusaki, and T. Matsumoto. 1984. "Oxetin, a new antimetabolite from an actinomycete. Fermentation, isolation, structure and biological activity." *J. Antibiot.,* 37, 1324–32.
97. Willms, L. 1989. "Glufosinate, a new amino acid with unexpected properties." *Pestic. Sci.,* 27, 219–21.
98. Tachibana, K. 1987. "Herbicidal characteristics of bialaphos." In *Pesticide Science and Biotechnology,* R. Greenhalgh and T. R. Roberts, eds. Oxford: Blackwell Sci. Publi., pp.145–48.
99. Tebbe, C. C., and H. H. Reber. 1988. "Utilization of the herbicide phosophinothricin as a nitrogen source by soil bacteria." *Appl. Microbiol. Biotechnol.,* 29, 103–05.
100. Nishino, T., and S. Murao. 1983. "Isolation and some properties of an aspartate aminotransferase inhibitor, gostatin." *Agric. Biol. Chem.,* 47, 1961–66.
101. Yamada, O., Y. Kaise, F. Futatsuya, S. Ishida, K. Ito, H. Yamamoto, and K. Munakata. 1974. "Studies on plant growth-regulating activities of anisomycin and toyocamycin." *Agric. Biol. Chem.,* 36, 2013–15.
102. Yamada, O., A. Kurozumi, F. Futatsuya, K. Ito, S. Ishida, and K. Munakata. 1979. "Studies on chlorosis-inducing activities and plant growth inhibition of benzophenone derivatives." *Agric. Biol. Chem.,* 43, 1467–71.
103. Heisey, R. M., and A. R. Putnam. 1986. "Herbicidal effects of geldanamycin and nigericin antibiotics from *Streptomyces hygroscopicus*." *J. Nat. Prod.,* 49, 859–65.

104. Heisey, R. M., and A. R. Putnam. 1990. "Herbicidal activity of the antibiotics geldanamycin and nigericin." *J. Plant Growth Regul.*, 9, 19–25.

105. Shavit, N., and A. San Pietro. 1967. "K^+-dependent uncoupling of photophosphorylation by nigericin." *Biochem. Biophys. Res. Commun.*, 28, 277–83.

106. Sze, H. 1980. "Nigerin-stimulated ATPase activity in microsomal vesicles of tobacco callus." *Proc. Natl. Acad. Sci. USA*, 77, 5904–08.

107. Fushimi, S., S. Nishikawa, N. Mito, M. Ikemoto, M. Sasaki, and H. Seto. 1989. "Studies on a new herbicidal antibiotic, homoalanosine." *J. Antibiot.*, 42, 1370–78.

108. Iwata, M., Sasaki, T., H. Iwamatsu, S. Miyadoh, K. Tachibana, K. Matsumoto, T. Shomura, M. Sezaki, and T. Watanabe. 1987. "A new antibiotic, SF 2494 (5′-O-sulfamoyltubercidin) produced by *Streptomyces mirabilis*." *Sci. Rep. Meija Seika Kaisha*, 26, 17–22.

109. Kida, T., T. Ishikawa, and H. Shibai. 1985. "Isolation of two streptomycin-like antibiotics, Nos. 6241-A and B, as inhibitors of *de novo* starch synthesis and their herbicidal activity." *Agric. Biol. Chem.*, 49, 1839–44.

110. Hill, C. M., S. A. Pearson, A. J. Smith, and L. J. Rogers. 1985. "Inhibition of chlorophyll synthesis in *Hordeum vulgare* by 3-amino 2,3,-diydrobenzoic acid (gabaculin)." *Biosci. Rep.*, 5, 775–81.

111. Kannangara, C. G., and A. Schoube. 1985. "Biosynthesis of Δ-aminolevulinate in greening barley leaves. VII. Glutamate 1-semialdehyde accumulation in gabaculine-treated leaves." *Carlsberg Res. Commun.*, 50, 179–91.

112. Gray, R. A., G. W. Gauger, E. L. Dulaney, E. A. Kaczka, and H. B. Woodruff. 1964. "Hadacidin, a new plant growth inhibitor produced by fermentation." *Plant Physiol.*, 39, 204–07.

113. Hatch, M. D. 1967. "Inhibition of plant adenylosuccinate synthetase by hadacidin and the mode of action of hadacidin and structurally related compounds on plant growth." *Phytochemistry*, 6, 115–19.

114. Kida, T., and H. Shibai. 1985. "Inhibition by hadacidin, duazomycin A, and other amino acid derivatives of *de novo* starch synthesis." *Agric. Biol. Chem.*, 49, 3231–37.

115. Oettmeier, W., R. Dostatni, C. Majewski, G. Höfle, T. Fecker, B. Kunze, and H. Reichenbach. 1990. "The aurachins, naturally occurring inhibitors of photosynthetic electron flow through photosystem II and the cytochrome b_6/f-complex." *Z. Naturforsch.*, 45c, 322–28.

116. Mason, C. P., K. R., Edwards, R. E. Carlson, J. Pignatello, F. K. Gleason, and J. M. Wood. 1982. "Isolation of chlorine-containing antibiotic from the freshwater cyanobacterium, *Scytonema hofmani*." *Science*, 215, 400–02.

117. Gleason, F. K., and D. E. Case. 1986. "Activity of the natural algicide, cyanobacterin, on angiosperms." *Plant Physiol.*, 80, 834–37.

118. Gleason, F. K., D. E. Case, K. D. Sipprell, and T. S. Magnuson. 1986. "Effect of the natural algicide, cyanobacterin, on a herbicide-resistant mutant of *Anacystis nidulans* R2." *Plant Sci.*, 46, 5–10.

119. Mallipudi, L. R., and F. K. Gleason. 1989. "Characterization of a mutant of *Anacystis nidulans* R2 resistant to the natural herbicide, cyanobacterin." *Plant Sci.*, 60, 149–54.

120. Carlson, J. L., T. A. Leaf, and F. K. Gleason. 1987. "Synthesis and activity of analogs of the natural herbicide cyanobacterin." Amer. Chem. Soc. Symp. Ser., 355, 141–50.

121. Feld, A., K. Kobek, and H. K. Lichtenthaler. 1989. "Inhibition of *de novo* fatty-acid biosynthesis in isolated chloroplasts by different antibiotics and herbicides." *Z. Naturforsch.*, 44c, 976–78.

122. Kato, M., T. Ehara, A. Kawaguchi, and M. Yamada. 1987. "Effect of thiolactomycin on lipid synthesis in higher plants." *Plant Cell. Physiol.*, 28, 857–65.

123. Nishida, I., A. Kawaguchi, and M. Yamada. 1986. "Effect of thiolactomycin on the individual enzymes of the fatty acid synthase system in *Echerischia coli*." *J. Biochem.*, 99, 1447–54.

124. Packter, N. M., and P. K. Stumpf. 1975. "Fat metabolism in higher plants. The effect of cerulenin on the synthesis of medium- and long-chain acids in leaf tissue." *Arch. Biochem. Biophys.*, 167, 655–67.

125. Strobel, G., F. Sugawara, and J. Clardy. 1987. "Phytotoxins from plant pathogens of weedy plants." Amer. Chem. Soc. Symp. Ser., 330, 516–23.

126. Harborne, J. B. 1983. "Toxins of plant-fungal interactions." In *Handbook of Natural Toxins, Vol. 1. Plant and Fungal Toxins,* R. F. Keeler and A.T. Tu, eds. New York: Marcel Dekker, pp. 743–82.

127. Stierle, A. C., J. H. Cardellina II, and G. A. Strobel. 1989. "Phytotoxins from *Alternaria alternata,* a pathogen of spotted knapweed." *J. Nat. Prod.,* 52, 42–47.

128. Stierle, A. C., J. H. Cardellina II, and G. A. Strobel. 1988. "Maculosin, a host-specific phytotoxin for spotted knapweed from *Alternaria alternata*." *Proc. Natl. Acad. Sci. USA*, 85, 8008–11.

129. Lax, A. R., H. S. Shepherd, and J. V. Edwards. 1988. "Tentoxin, a chlorosis-inducing toxin from *Alternaria* as a potential herbicide." *Weed Technol.,* 2, 540–44.

130. Steele, J. A., T. F. Uchytil, R. B. Durbin, P. Bhatnagar, and D. H. Rich. 1976. "Chloroplast coupling factor 1: A species specific receptor for tentoxin." *Biochemistry,* 73, 2245–48.

131. Arntzen, C. J. 1972. "Inhibition of photophosphorylation by tentoxin, a cyclic tetrapeptide." *Biochim. Biophys. Acta,* 283, 539–42.

132. Lax, A. R., and K. C. Vaughn. 1986. "Lack of correlation between effects of tentoxin on chloroplast coupling factor and chloroplast ultrastructure." *Physiol. Plant.,* 66, 384–91.

133. Vaughn, K. C., and S. O. Duke. 1981. "Tentoxin induced loss of plastidic polyphenol oxidase." *Physiol. Plant.,* 53, 421–28.

134. Vaughn, K. C., and S. O. Duke. 1984. "Tentoxin stops processing of polyphenol oxidase into an active protein." *Physiol. Plant.,* 60, 257–62.

135. Lax, A. R., and H. S. Shepherd. 1988. "Tentoxin: a cyclic tetrapeptide having potential herbicidal usage." Amer. Chem. Soc. Symp. Ser., 380, 24–33.

136. Vaughn, K. C., A. R. Lax, and S. O. Duke. 1988. "Polyphenol oxidase: the chloroplast oxidase with no established function." *Physiol. Plant.,* 72, 659–65.

137. Giovanelli, J., L. Owens, and S. Mudd. 1971. "Mechanism of action of β-cystathionase by rhizobitoxine." *Biochim. Biophys. Acta,* 227, 671–84.

138. Turner, J. G., and R. E. Mitchell. 1985. "Association between symptom development and inhibition of ornithine carbamoyltransferase in bean leaves treated with phaseolotoxin." *Plant Physiol.,* 79, 468–73.

139. Rasi-Caldogno, F., and M. C. Pugliarello. 1985. "Fusicoccin stimulates the H^+-ATPase of plasmalemma in isolated membrane vesicles from radish." *Biochem. Biophys. Res. Commun.*, 133, 280–85.

140. Reidl, H. H., and J. Y. Takemoto. 1987. "Mechanism of action of bacterial syringomycin. Simultaneous measurement of early responses in yeast and maize." *Biochim. Biophys. Acta,* 898, 59–69.

141. Rouxel, T., Y. Chupeau, R. Fritz, A. Kollmann, and J.-F. Bousquet. 1988. "Biological effects of sirodesmin PL, a phytotoxin produced by *Leptosphaeria maculans.*" *Plant Sci.*, 57, 45–53.

142. Gilchrist, D. G. 1983. "Molecular mode of action." In *Toxins and Plant Pathogenesis,* J. M. Daly and B. J. Deverall, eds. New York: Academic, pp. 81–136.

143. Calton, G. J. 1987. "Use of microorganisms and enzymes in the synthesis and production of optically active agricultural chemicals." Amer. Chem. Soc. Symp. Ser., 334, 181–89.

144. Takebe, H., S. Imai, H. Ogawa, A. Satoh, and H. Tanaka. 1989. "Breeding of bialaphos producing strains from a biochemical engineering viewpoint." *J. Ferment. Bioengin.*, 67, 226–32.

145. Bartsch, K., R. Dichmann, P. Schmitt, E. Uhlmann, and A. Schulz. 1990. "Stereospecific production of the herbicide phosphinothricin (glufosinate) by transamination: cloning, characterization, and overexpression of the gene encoding a phosphinothricin-specific transaminase from *Escherichia coli.*" *Appl. Environ. Microbiol.*, 56, 7–12.

146. Durbin, R. D., and P. J. Langston-Unkefer. 1988. "The mechanisms for self-protection against bacterial phytotoxins." *Annu. Rev. Phytopath.*, 26, 313–29.

Index